CERNAS – Current Evolution and Research Novelty in Agricultural Sustainability

CERNAS – Current Evolution and Research Novelty in Agricultural Sustainability

Editors

Raquel P. F. Guiné
António Dinis Ferreira
António Moitinho Rodrigues

MDPI

Basel • Beijing • Wuhan • Barcelona • Belgrade • Novi Sad • Cluj • Manchester

Editors

Raquel P. F. Guiné
CERNAS and Department of Food Industry, Polytechnic Institute of Viseu
Viseu
Portugal

António Dinis Ferreira
CERNAS—Centro de Estudos de Recursos Naturais, Ambiente e Sociedade
Polytechnic Institute of Coimbra
Coimbra
Portugal

António Moitinho Rodrigues
CERNAS Research Centre
Polytechnic Institute of Castelo Branco
Castelo Branco
Portugal

Editorial Office
MDPI
St. Alban-Anlage 66
4052 Basel, Switzerland

This is a reprint of articles from the Special Issue published online in the open access journal *Agronomy* (ISSN 2073-4395) (available at: https://www.mdpi.com/journal/agronomy/special_issues/evolution_agricultural_sustainability).

For citation purposes, cite each article independently as indicated on the article page online and as indicated below:

Lastname, A.A.; Lastname, B.B. Article Title. *Journal Name* **Year**, *Volume Number*, Page Range.

ISBN 978-3-0365-9602-0 (Hbk)
ISBN 978-3-0365-9603-7 (PDF)
doi.org/10.3390/books978-3-0365-9603-7

Cover image courtesy of Raquel P. F. Guiné

Contents

Preface

Climate changes pose overwhelming impacts on primary production and, consequently, on agricultural and animal farming. Additionally, at present, agriculture still depends strongly on fossil fuels both for energy and production factors ,such as synthetized inorganic fertilizers and harmful chemicals such as pesticides. The need to feed the growing world population poses many challenges. The need to reduce environmental impacts to a minimum, maintain healthy ecosystems, and improve soil microbiota are central to ensuring a promising future for coming generations. Livestock production under cover crop systems helps to alleviate compaction so that oxygen and water can sufficiently flow in the soil, add organic matter, and help hold soil in place, reducing crusting and protecting against erosion. The use of organic plant production practices allied to the control of substances used in agriculture also decisively contributes to alleviating the pressure on ecosystems. Some of the goals of this new decade are to use enhanced sustainable production methodologies to improve the input/output ratios of primary production, reduce environmental impacts, and rely on new innovative technologies.

This reprint addresses original studies and reviews focused on the current evolution and research novelty in agricultural sustainability. New developments are discussed on issues related to quality of soil, natural fertilizers, or the sustainable use of land and water. Also, crop protection techniques are pivotal for sustainable food production under the challenges of the Sustainable Development Goals of the United Nations, allied to innovative weed control methodologies as a way to reduce the utilization of pesticides. The role of precision and smart agriculture is becoming more pertinent as communication technologies improve at a rapid rate. Waste management, reuse of agro-industrial residues, extension of shelf life, and use of new technologies are ways to reduce food waste, all contributing to higher sustainability in food supply chains, leading to a more rational use of natural resources. The unquestionable role of bees as pollinators and contributors to biodiversity is adjacent to characterizing beekeeping activities, which in turn contributes, together with the valorization of endemic varieties of plant foods, to the development of local communities. Finally, the short circuits and local food markets have a decisive role in the preservation and enhancement of rural economies.

Raquel P. F. Guiné, António Dinis Ferreira, and António Moitinho Rodrigues

Editors

agronomy

MDPI

Editorial

CERNAS—Current Evolution and Research Novelty in Agricultural Sustainability

Raquel P. F. Guiné

CERNAS Research Centre, Polytechnic University of Viseu, 3504-510 Viseu, Portugal; raquelguine@esav.ipv.pt

Climate changes have overwhelming impacts on primary production and, consequently, on agricultural and animal farming. Additionally, at present, agriculture still depends strongly on fossil fuels, both for energy and production factors such as synthetized inorganic fertilizers and harmful chemicals like pesticides.

The need to feed the growing world population poses many challenges. Additionally, it is mandatory to reduce environmental impacts to a minimum, maintain healthy ecosystems, and improving soil microbiota as a way of ensuring a promising future for the coming generations. Food security is a major concern of modern times, and different disciplines are devoted to the effective combat to this eminent threat. The work by Necula et al. [1] focuses on the commercialization of food products originating from agricultural farming or mountain activities, highlighting that new consumers, especially after the COVID-19 pandemic and its food systems' impact, tend to attribute a higher value to local and organic food products.

Current changes in the use of land for agricultural purposes in many regions of the globe has an impact on carbon storage, and therefore an improved understanding of this reality is beneficial to comprehend and quantify carbon emissions. Kong et al. [2] suggest that the development of agricultural fields with a high carbon density or, alternatively, the conversion of lands with low carbon density are essential to improve carbon sequestration in the future as a result of cropland transformation.

Livestock production under cover crop systems helps to alleviate compaction so that oxygen and water can sufficiently flow in the soil, adding organic matter, and helping hold soil in place, reducing crusting and protecting against erosion.

The use of organic plant production practices allied to the control of substances used in agriculture also decisively contributes to alleviating the pressure on ecosystems.

The use of animal manure has been a traditional practice to improve the nitrogen content of agricultural soils. However, new developments in this area have been reported lately, namely the use of technologically advanced treatments, like plasma technology, which can improve microbial quality without harming the soil-dwelling organisms [3].

Water is a more and more valuable resource, becoming scarce in many regions of the globe, so that its application in agriculture sometimes has to be highly regulated. In the Mediterranean, this problem is assuming particular relevance, so Valenzuela-Mahecha et al. [4] alert us to the need to establish measures of both a technical as well as a financial nature to enable farmers who deal with irrigation problems cope with this reality. They presented a new index-based drought insurance scheme that will help farmers to deal with the negative economic impacts of the more and more frequent water scarcity events.

Some of the goals of this new decade are to use enhanced sustainable productive methodologies to improve the input/output ratios of primary production, to reduce environmental impacts, and to rely on new innovative technologies.

The need to improve plant production ratios, allied to the reduction in agricultural losses and crop problems, is pivotal to increase the food production of vegetable origin. The work by Fadiji et al. [5] brings some insights into the symbiotic integration of endophytic

Citation: Guiné, R.P.F. CERNAS—Current Evolution and Research Novelty in Agricultural Sustainability. *Agronomy* **2023**, *13*, 2829. https://doi.org/10.3390/agronomy13112829

Received: 6 November 2023
Accepted: 10 November 2023
Published: 16 November 2023

viruses into maize crops, and how diversified agricultural practices can impact their abundance. Not only is the improvement of crop production ratios pivotal, but the conservation of food and agricultural products along the food chain is also essential to diminish the current huge amounts of food not consumed and disposed along the food chain. The use of appropriate packaging or a controlled atmosphere can greatly extend shelf-life and improve food quality [6] while reducing organic waste. The review by Fadiji et al. [7] focuses on the role of nanotechnology in helping cope with food insecurity and minimizing post-harvest losses that have both economic as well as environmental impacts.

The work by Ferrão et al. [8] provides new knowledge about the characterization of hazelnut fruits grown in Portugal, some of them coming from native varieties, which usually are well adapted to the local edaphoclimatic conditions, thus presenting good production ratios while requiring low cultural interventions.

The minimization of agricultural or industrial waste is another of the main goals of sustainability, as established by circular economy principles. The use of agrifood waste in crop fields with the aims to, on the one hand control, weeds and, on the other hand, complement soil nutrients is discussed by Lorenzo et al. [9] as a sustainable option to reduce agricultural residues. The management of weeds through alternative ecological ways is essential in reducing the harmful intensive usage of synthetic chemical weed-controlling substances. As shown by Monteiro and Santos [10], precision weed management assumes a role in more sustainable weed control. Also, concepts of precision agriculture, smart agriculture, and innovative technologies such as nanotechnology, artificial intelligence, genetic modification, or others bring endless possibilities for the intervening actors in agricultural domains to improve crop production, efficiently manage natural resources, and control the ecologic footprint of products and activities [11].

The reutilization of waste, such as that coming from food industrial processing plants, can have important environmental relevance while also providing sources for valuable compounds of possible utilization in diverse industries in the food, pharmaceutical, or cosmetic fields. This is the case in the extraction of polyphenolic compounds from cherry seeds [12]. Also, the use of agricultural or industrial waste of a lignocellulosic nature, such as cherry seeds, can have positive environmental impacts and provide new sources of materials at the same time to obtain more natural polyurethane foams while reducing the use of petroleum-derived ones [13].

Extreme marine environments, like saline areas in coastal zones, can provide new opportunities for the agricultural farming of species especially adapted to these high salt concentrations, like halophyte plants [14]. Due to the effects of global warming, the high salinity of some agricultural environments also brings challenges, as discussed in the review by Shultana et al. [15].

The pivotal role of bees as pollinators is vastly recognized, and many threats are presently having harmful effects on bee colonies, such as the varroa mite (*Varroa destructor*) or the Asian wasp (*Vespa velutina*), which influence the life of bees and threaten entire colonies. The characterization of beekeepers and beekeeping activities helps understand how the beekeepers are coping with the present challenges as a way to also provide more helpful support to this essential activity from ecological as well as agronomic points of view. The work by Guiné et al. [16] presents the characterization of beekeeping in different European countries, highlighting the differences encountered.

Acknowledgments: This work was supported by the FCT—Foundation for Science and Technology, I.P. Furthermore, we would like to thank the CERNAS Research Center (Ref. UIDB/00681/2020) and the Polytechnic University of Viseu for their support.

Conflicts of Interest: The author declares no conflict of interest.

References

1. Necula, D.; Ungureanu-Iuga, M.; Ognean, L. The Agri-Food and Mountain Products Market: Insights beyond the COVID-19 Pandemic. *Agronomy* **2023**, *13*, 2739. [CrossRef]
2. Kong, J.; Chen, L. The Changes in Cropland Pattern Enhanced Carbon Storage in Northwest China. *Agronomy* **2023**, *13*, 2736. [CrossRef]
3. Mousavi, H.; Cottis, T.; Pommeresche, R.; Dörsch, P.; Solberg, S.Ø. Plasma-Treated Nitrogen-Enriched Manure Does Not Impose Adverse Effects on Soil Fauna Feeding Activity or Springtails and Earthworms Abundance. *Agronomy* **2022**, *12*, 2314. [CrossRef]
4. Valenzuela-Mahecha, M.A.; Pulido-Velazquez, M.; Macian-Sorribes, H. Hydrological Drought-Indexed Insurance for Irrigated Agriculture in a Highly Regulated System. *Agronomy* **2022**, *12*, 2170. [CrossRef]
5. Fadiji, A.E.; Galeemelwe, O.; Babalola, O.O. Unravelling the Endophytic Virome Inhabiting Maize Plant. *Agronomy* **2022**, *12*, 1867. [CrossRef]
6. Pinto de Andrade, L.; Veloso, A.; Espírito Santo, C.; Dinis Gaspar, P.; Silva, P.D.; Resende, M.; Beato, H.; Baptista, C.; Pintado, C.M.; Paulo, L.; et al. Effect of Controlled Atmospheres and Environmental Conditions on the Physicochemical and Sensory Characteristics of Sweet Cherry Cultivar Satin. *Agronomy* **2022**, *12*, 188. [CrossRef]
7. Fadiji, A.E.; Mthiyane, D.M.N.; Onwudiwe, D.C.; Babalola, O.O. Harnessing the Known and Unknown Impact of Nanotechnology on Enhancing Food Security and Reducing Postharvest Losses: Constraints and Future Prospects. *Agronomy* **2022**, *12*, 1657. [CrossRef]
8. Ferrão, A.C.; Guiné, R.P.F.; Ramalhosa, E.; Lopes, A.; Rodrigues, C.; Martins, H.; Gonçalves, R.; Correia, P.M.R. Chemical and Physical Properties of Some Hazelnut Varieties Grown in Portugal. *Agronomy* **2021**, *11*, 1476. [CrossRef]
9. Lorenzo, P.; Guilherme, R.; Barbosa, S.; Ferreira, A.J.D.; Galhano, C. Agri-Food Waste as a Method for Weed Control and Soil Amendment in Crops. *Agronomy* **2022**, *12*, 1184. [CrossRef]
10. Monteiro, A.; Santos, S. Sustainable Approach to Weed Management: The Role of Precision Weed Management. *Agronomy* **2022**, *12*, 118. [CrossRef]
11. Martinho, V.J.P.D.; Guiné, R. de P.F. Integrated-Smart Agriculture: Contexts and Assumptions for a Broader Concept. *Agronomy* **2021**, *11*, 1568. [CrossRef]
12. Dulyanska, Y.; Cruz-Lopes, L.P.; Esteves, B.; Ferreira, J.V.; Domingos, I.; Lima, M.J.; Correia, P.M.R.; Ferreira, M.; Fragata, A.; Barroca, M.J.; et al. Extraction of Phenolic Compounds from Cherry Seeds: A Preliminary Study. *Agronomy* **2022**, *12*, 1227. [CrossRef]
13. Cruz-Lopes, L.; Dulyanska, Y.; Domingos, I.; Ferreira, J.; Fragata, A.; Guiné, R.; Esteves, B. Influence of Pre-Hydrolysis on the Chemical Composition of Prunus Avium Cherry Seeds. *Agronomy* **2022**, *12*, 280. [CrossRef]
14. Nelson, D.R.; Chaiboonchoe, A.; Hazzouri, K.M.; Khraiwesh, B.; Alzahmi, A.; Jaiswal, A.; Friis, G.; Burt, J.A.; Amiri, K.M.A.; Salehi-Ashtiani, K. Tissue-Specific Transcriptomes Outline Halophyte Adaptive Strategies in the Gray Mangrove (*Avicennia Marina*). *Agronomy* **2022**, *12*, 2030. [CrossRef]
15. Shultana, R.; Zuan, A.T.K.; Naher, U.A.; Islam, A.K.M.M.; Rana, M.M.; Rashid, M.H.; Irin, I.J.; Islam, S.S.; Rim, A.A.; Hasan, A.K. The PGPR Mechanisms of Salt Stress Adaptation and Plant Growth Promotion. *Agronomy* **2022**, *12*, 2266. [CrossRef]
16. Guiné, R.P.F.; Mesquita, S.; Oliveira, J.; Coelho, C.; Costa, D.T.; Correia, P.; Correia, H.E.; Dahle, B.; Oddie, M.; Raimets, R.; et al. Characterization of Beekeepers and Their Activities in Seven European Countries. *Agronomy* **2021**, *11*, 2398. [CrossRef]

agronomy

MDPI

Review

The Agri-Food and Mountain Products Market: Insights beyond the COVID-19 Pandemic

Doru Necula [1,2], Mădălina Ungureanu-Iuga [1,3,*] and Laurenț Ognean [2]

1 Mountain Economy Center (CE-MONT), "Costin C. Kirițescu" National Institute of Economic Researches (INCE), Romanian Academy, 49th, Petreni Street, 725700 Vatra Dornei, Romania; doru.necula@usamvcluj.ro
2 Department of Physiology, University of Agricultural Sciences and Veterinary Medicine of Cluj-Napoca, 3–5th, Mănăștur Street, 400037 Cluj-Napoca, Romania; lognean@usamvcluj.ro
3 Integrated Center for Research, Development and Innovation in Advanced Materials, Nanotechnologies, and Distributed Systems for Fabrication and Control (MANSiD), "Ştefan cel Mare" University of Suceava, 13th University Street, 720229 Suceava, Romania
* Correspondence: madalina.iuga@ce-mont.ro

Abstract: Food security is one of the main concerns in the context of a global crisis such as the COVID-19 pandemic. The reduction in people's mobility determined changes in consumers' behavior and underlined the need for the re-organization of the food supply chains. This paper aims to summarize the effects of the COVID-19 pandemic on the global, Romanian and mountain food markets, as well as to discuss the mountain agriculture potential and the food democracy model. The trend in the post-pandemic era is heading toward the digitalization of agriculture and food distribution, with great attention on product sustainability. People are more and more aware of healthy food and the environmental impact of this sector. Many studies revealed the need for specific policies to counteract the effects of the pandemic on food quality and security and on the economic welfare of people. In the post-pandemic period in mountain areas, there is a need for the valorization of food products that originate from here since they have great health and financial potential. Supporting mountain agriculture could ensure the production of high-value products, which are generally preferred by consumers. The COVID-19 pandemic contributed to the re-orientation of consumers towards local and organic foods. Future research regarding the efficiency of the programs and policies implemented in some mountain areas after the pandemic is necessary.

Keywords: COVID-19 pandemic; traditional foods; mountain products; food policies; food system resilience

Citation: Necula, D.; Ungureanu-Iuga, M.; Ognean, L. The Agri-Food and Mountain Products Market: Insights beyond the COVID-19 Pandemic. *Agronomy* **2023**, *13*, 2739. https://doi.org/10.3390/agronomy13112739

Academic Editors: António Dinis Ferreira, Raquel P. F. Guiné and António Moitinho Rodrigues

Received: 28 September 2023
Revised: 25 October 2023
Accepted: 25 October 2023
Published: 30 October 2023

1. Introduction

Humanity has survived throughout its evolutionary history, from the Paleolithic period or the Stone Age to the Neolithic or the agricultural age, continuing on to the industrial age and culminating at the beginning of the XXI century. At this time, there was a revolution of cybernetic and informational systems, causing famines, and on a global scale, anthropocentric economic crises that mankind produced [1]. This style of living has certainly been harmful, both for the environment and for the human species [2]. Among the problems that occurred, the rise of the sea and ocean levels since 1990 by 10–20 cm [3] and the melting of glaciers, which registered values of about 28 billion tons of ice melted between 1994 and 2017, must be mentioned [4]. Also, soil fertility has decreased greatly in recent decades due to the advanced modernization of agricultural activities, and more than 35% of arable land has been degraded due to uncontrolled anthropogenic activities [5].

In this context, what are the implications for humanity? An answer can be given: since the emergence of humanity until today, man has learned to adapt. Going through an extreme event can be traumatic, but at the same time, it creates opportunities to design new valuable mechanisms for society at present. It can be said that the COVID-19 pandemic has introduced new adaptations, new changes, or better said, the beginning of an era of change [6], creating the opportunity to evaluate systems, traditions, and values. The pandemic started and caused restrictive measures to be taken by governments, such as social distancing, isolation, and restrictions on space circulation, highlighting, from the beginning, the vulnerability of the food system and its slowdown. In the context of food security, this fact has caused fear all over the world. During the COVID-19 pandemic, hunger levels in the world increased in just one year from 8.4% to around 10% [7]. The largest undernourished populations are in Africa and Asia; at the end of 2020, they represented around 700 million more severely affected people compared to those in 2019 [8]. This number has increased by approximately 100 million on these two continents. The COVID-19 pandemic has had many effects on various aspects of life and disturbed all sectors, including agriculture and food [9]. The agri-food systems changed continuously under the pressures of the COVID-19 pandemic and climate change, and the actual concern is focused on viable solutions for the sustainable development of this sector [10,11].

There is great potential for the food industry, especially if we take into account the rural and especially mountainous landscape. Small family farms represent a huge potential for modeling and transforming local food systems. The mountainous areas are characterized by highly heterogeneous farms. Applying differentiated policies specific to these areas would allow for an increase in this type of farming and ensure the raised potential of traditional biocultural food.

There are some studies presenting the predilection of consumers for online shopping [12,13] and for local organic food [14]. In addition, the changes in the food market during and after the COVID-19 pandemic have been evaluated, and some directions were evidenced: transparency and the tradability demand from consumers, the sustainable development of the agri-food system, the support of traditional and authentic foods, and digitalization for enhancing communication between actors [15]. However, it is important to evaluate all these aspects in the mountain context, which presents many more particularities compared to those of lowland areas and cities. Other studies reported the increased interest of consumers in organic foods [14] obtained in less-polluted areas and without agrochemicals and additives. Mountain products usually meet these criteria since they are made on a smaller scale via extensive systems. Thus, it is important to outline the advantages of these products that may contribute to food security and the local economy. Based on the existing literature, some hypotheses can be noted:

1. The COVID-19 pandemic changed consumers' behavior related to agri-food mountain products.
2. Mountain products have an important potential for the development of the communities and the satisfaction of consumers' demand for organic foods.

To our knowledge, the impact of the COVID-19 pandemic on mountain farmers, the product market, and consumers has not yet been assessed, but there is the premise that farmers, in general, must produce regardless of the context. They are the first ones who have to produce food since this is their basic activity, besides the proper functioning of the food chain being ensured. There is a gap in the literature regarding the synthesis of the mountain agri-food products market in the post-pandemic context. Furthermore, there is no review paper presenting consumers' behavior related to mountain products during the COVID-19 pandemic and post-pandemic eras. Such a synthesis would be helpful for institutions to design and adopt food policies specific to mountain areas. Furthermore, this information could help mountain farmers and producers to adapt their marketing strategies and food quality to the consumers' demand. Thus, the aim of this review was to summarize the main characteristics of the global and Romanian post-pandemic food market, as well as to underline the potential of mountain products and mountain agriculture in this context. The paper comprises six sections apart from the introduction, methods, and conclusion: the first section presents the food market in the post-pandemic period at global level and in Romania; the second section is related to the evolution of the mountain products market in the post-pandemic era; the next section presents the importance of mountain products and a case study supporting their potential; the last section presents the food democracy as a model of food governance in the post-pandemic era.

2. Materials and Methods

Scientific publications were considered from the existing literature indexed in databases such as Web of Science, Scopus, and Google Academic. After reading the abstracts, papers that did not align with the aim of this research were eliminated. A total of 93 articles and reports were included in this review after a thorough assessment of the abstracts and the provided information. Only papers in English were considered and the complete papers not available were excluded. The keywords used for the search included: *agri-food system, *COVID-19 pandemic, *post-pandemic, *food system resilience, *sustainable development, *mountain products, *food policy, *mountain agriculture, *food democracy.

A bibliometric analysis was conducted by using VOSviewer software (version 1.6.19, Leiden University). A map based on textual data (Figure 1) was generated using data from reference manager files with RIS extensions, specifically drawing from titles and abstract sections. The full counting method was selected, with a minimum requirement of a term occurring at least three times.

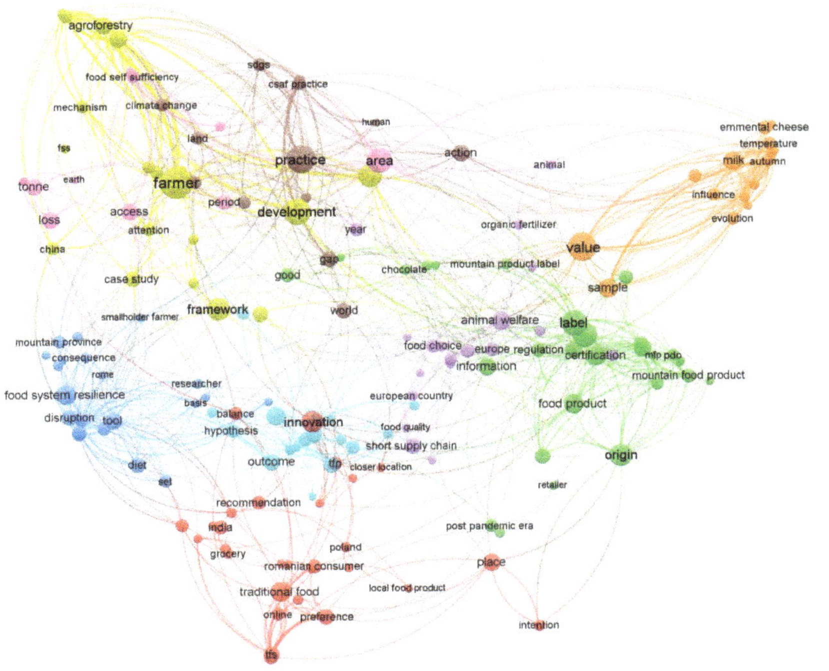

Figure 1. Key elements network map.

The keywords related to the COVID-19 post-pandemic situation of the food system and mountain products were grouped into 11 clusters. The largest cluster, represented with a red color, contains the most important elements related to the topic, with 30 items. It is followed by the dark green cluster, comprising 29 items. The third cluster comprised 21 items, in dark blue, followed by the fourth cluster with 20 elements in light green, the fifth cluster with 19 items in purple, the sixth cluster with 18 items in turquoise, the seventh cluster with 16 items in orange, the eighth cluster with 14 items in brown, and the last one with 12 elements in pink.

The second map generated highlights the most significant authors who contributed to the topic (Figure 2).

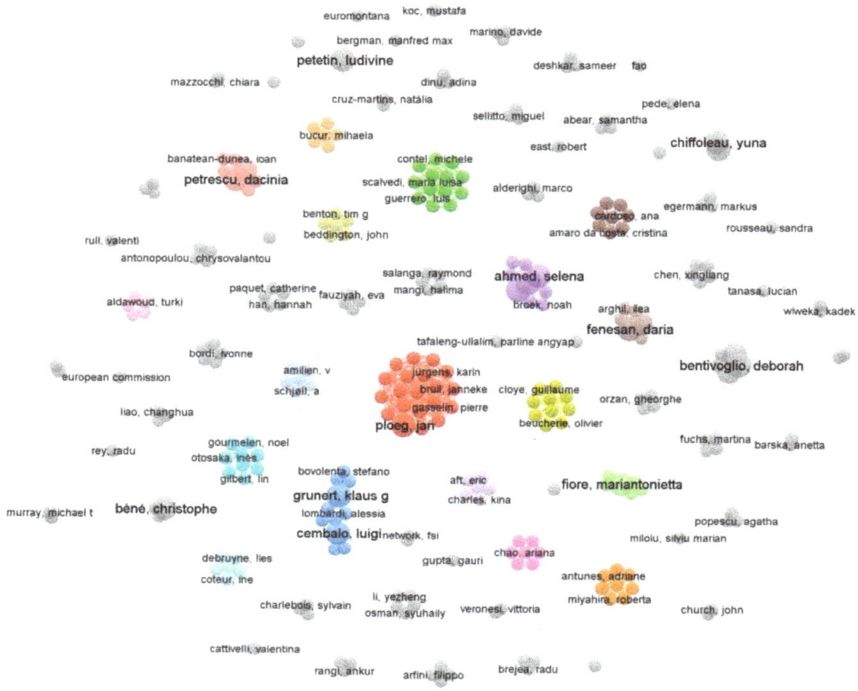

Figure 2. Map of the authors concerned about the COVID-19 post-pandemic food market.

The number of papers related to the food market in the post-pandemic era included in this study is depicted in Figure 3, and the papers are grouped based on their publication year. The highest number of papers was published after 2019.

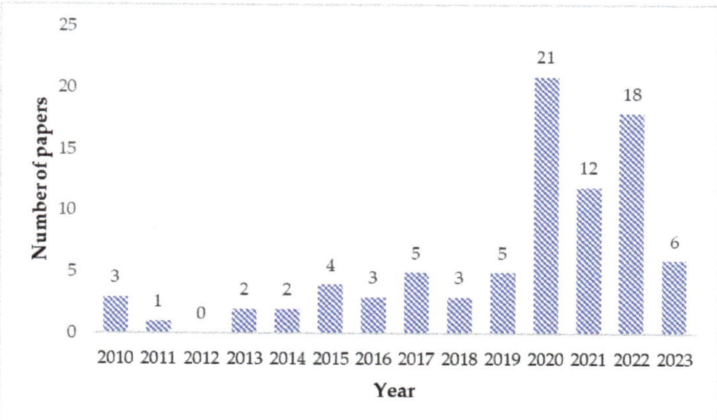

Figure 3. Number of papers related to the topic considered in this review and published between 2010 and 2023.

3. Post-Pandemic Food Market

3.1. The Global Post-Pandemic Food Market

The achievement of the second Sustainable Development Goal (SDG) which aims to eliminate hunger by 2030, encountered complications due to the COVID-19 pandemic. The food system requires immediate transformation to become sustainable. In this context, the UN Food Systems Summit (FSS) aimed to "launch bold new actions, solutions and strategies to deliver progress on all 17 SDGs, each of which relies on healthier, more sustainable and more equitable food systems" [16]. It is crucial to support mountain farming beyond industrialized agriculture because people in these areas possess essential knowledge on sustainable livestock management [17]. One-third of the global food production is managed by smallholder farmers, yet they face challenging working conditions that do not allow them to earn enough income [17]. Hubeau et al. [18] summarized the pathways for food system transformation, along with the necessary strategies and actions:

- Promoting innovations and chain-wide partnerships: establishing sustainability definitions for particular food chains, supporting financial innovations, and creating novel distribution and business models [18];
- Supporting food system efficiency and resilience: establishing sustainability standards, product differentiation, developing innovative food products, integrating modern technology, implementing risk management systems, and diversifying markets [18];
- Closing mineral cycles and valorizing by-products: enhancing the relationship between agriculture and the food industry, efficiently using energy heat sources, and valorizing waste [18];
- Promoting renewable resources: increasing the use of renewable sources and reduction the use of depletable ones [18];
- Increasing transparency and promoting equitable relationships within the agri-food system: supporting demand-driven production, cooperation with authorities to design/implement/supervise sustainable added value, and assessing the code of conduct in the agri-food system [18];
- Supporting co-creation related to sustainable practices: developing community practices, linking best practices to innovation and research, and creating communications platforms for agri-food actors and authorities [18];
- Valorizing food by raising community involvement: disseminating sustainable practices, communicating with consumers, enhancing working conditions within the agri-food system, and encouraging co-creation within the organization [18].

The priorities addressed by the summits held in 2021 included the following: "the development of inter-governmental and global institutional mechanisms to provide credible and authoritative consensuses on scientific evidence to support decisive and effective policies; the improvement of research efficiency and linkages across various scientific fields such as climate, natural resources, food, health, and nutrition, to support multi-sectoral policies; the implementation of robust synthesis and assessment processes to strengthen the legitimacy of scientific advice through transparency that incorporates the perspectives of low- and middle-income countries" [19]. There is a pressing need to translate the research findings into policies within the agri-food system and to develop national solutions that can be adapted to the global context by intensifying international cooperation [19]. The aspects for cooperation (namely target-setting, enhancing the promotion and use of science to govern practice), as well as the principles for engagement between public and private entities, should be thoroughly investigated, defined, and evaluated from a financial perspective [19].

After more than two years since the global health crisis began, with its repercussions affecting all areas, the food market, a fundamental component of our current society, has undergone significant changes and is unlikely to return to its initial state. In the pre-pandemic context, the food market was gradually experimenting with digitization. The market for food products relied heavily on in-person shopping, as consumers preferred to see, touch, taste, and smell fresh products and personally select their favorite items. While traditional trade of these products remains the most preferred choice, the online market share has been steadily growing in recent years. However, during the pandemic and the post-pandemic period, this shift accelerated significantly [20]. One of the reasons for this change in behavior during the pandemic was the raised popularity of the internet and the proliferation of smart devices, which facilitated the slow but sure transition to digital commerce for food products. Furthermore, the modification of lifestyle and consumption patterns, coupled with the time-saving convenience, contributed to the growth of online food shopping [20]. As a result, the population became accustomed to this shift and formed a habit of continuing to engage in online trading for traditional food products, even in the post-pandemic period.

Online shopping offers great flexibility in meeting various needs while eliminating the queues, the traffic jams, and increased costs. During the pandemic, a period characterized by uncertainty, many people felt safer avoiding trips to the local grocery store and physical contact with others. Stay-at-home recommendations and restrictions significantly boosted online demand [12]. Consumers have two options for buying food online: "the business-to-consumer (B2C) model or the online-to-offline (O2O) models" [21]. In the B2C model, which is a traditional online shopping model, people make purchases on various web pages and receive their parcel within a few days (usually 3–10), while the O2O is a newer approach that combines online shopping with local businesses, where people buy the desired food online and eat it offline [22]. During the COVID-19 pandemic, the food delivery system experienced significant growth, with consumers utilizing third-party O2O platforms and/or mobile applications to find restaurants and access a wide variety of food products [21]. Li et al. [21] identified various factors influencing consumer behavior during the pandemic, including technical and practical aspects, system related characteristics, emotional and subjective factors, individual characteristics, products or service quality, risk management, social influences, and food properties. Nielsen et al. [23] conducted a study on consumer behavior related to food values, purchases, and eating habits during the COVID-19 pandemic and concluded that the dieticians should consider the mental and emotional status of individuals, as well as the period of lockdown, when providing dietary guidance. Furthermore, the authors recommend supporting local food products to promote healthy eating habits, sustainable development, and enhanced food systems resilience in the post-COVID-19 era [23]. Consumer choices regarding food were found to be influenced by health, social, and psychological factors, with an increased preference for organic food, self-cooking tendency, health, and food quality and safety being reported [24]. Liao et al. [25] revealed that consumer demand for traceability information led to raised government implications in pandemic control efforts, subsidies, higher demand, improved traceability, enhanced human welfare, and increased consumer satisfaction.

The global food security chains were significantly impacted by the COVID-19 pandemic, resulting in substantial disruptions that sent shockwaves throughout the entire supply chain, from manufacturing to the commercialization stage [26]. In their paper on "Food security and disruptions of the global food supply chains during COVID-19", Alabi & Ngwenyama [26] proposed solutions to increase the resilience of global food security chains: they recommended decentralization of the system, the use of commerce platforms, adoption of cloud-based technology, achieving end-to-end supply chain visibility, and the application of Industry 4.0 principles. Priyadarshini & Abhilash [27] put forward suggestions to enhance resilience in the post-COVID-19 era in India: improving digitalization and internet connectivity for local retail and shops in both cities and villages, providing functional foods and immune supplements to the economically disadvantaged population through government programs already implemented, and marketing of "planetary healthy nutrition" to control food insecurity and improve nutrition security, guaranteeing long-term sustainability in the food industry. In the post-pandemic period, innovations in the food industry should be considered. Some of these innovations relate to the Industry 4.0 instruments, such as the Internet of Things, internet and communication technologies, and blockchain technology, while others relate to novel ingredients and technologies like lab-produced meat, plant-based meat substitutes, and the utilization of a wide range of by-products [28]. Additionally, the use of supplements to bolster the immune system to support the recovery of COVID-19 patients, the digitalization and integration of artificial intelligence in the food production, and education efforts aimed at emerging technologies and accelerating initiatives are key directions for enhancing food system sustainability in the post-pandemic era [28]. Serrano et al. [13] concluded that food delivery services and the take-out system were vital for restaurants during the pandemic. The authors recommended retaining technological solutions found for online and in-person food industry such as: improving e-commerce food platforms and delivery services, facilitating contactless cards payments, digitalizing services (such as online reservations, digital menus, QR code use), implementing food and beverages traceability, and using air purifiers [13]. Furthermore, the stringent food quality control practices that were adopted during the pandemic to reduce virus spread and enhance consumer trust should be kept for an extended time [13]. Guiné et al. [14] studied consumer behavior in Portugal and Turkey and found that people prefer to buy organic vegetables, fruits, dairy products, and wild fish from captures mainly due to the absence of chemical pesticides and fertilizers, the smaller environmental impact, the positive effects on health and farmer welfare, the convenience of proximity to home, awareness of the sustainability of organic food, and its perceived higher value.

Food system resilience is defined as "something more akin to flexibility, the ability to respond to disruption in a way that leaves the functioning of the entire food supply chain system unaffected" [29]. Various strategies proposed by the research community to increase food system resilience in the post-pandemic period are presented in Table 1. The assessment of food chain resilience can involve the consideration of multiple indicators, including household food insecurity access measures, the degree of household dietary diversity, z-score, the presence of mycotoxins after harvest, post-harvest mass losses, losses of nutritive compounds in food, the presence of agro-chemicals in agri-food products, price volatility indices, food by-products [30], etc. The adoption of agroecology in mountain areas and the preservation of agrobiodiversity can contribute to bolstering the resilience of the agri-food system [31].

Table 1. Strategies to increase food system resilience in the post-pandemic era.

Region	Focus	Proposed Strategy	References
Worldwide	Impact of the COVID-19 pandemic on food security and resilience of local food systems	- acquiring capacities like assets, savings, etc. - access to insurance; - improved access to information for stakeholders - better cooperation between community members; - higher inclusion level and greater expectancies and self-efficacy of the community actors; - evaluation of the potential ripple effects when a part of the food system is affected.	[30]
India	Suggestions to increase agri-food system resilience after the COVID-19 pandemic	- marketing of healthy diet habits; - valorization of underutilized wild crops; - development of decentralized heating, ventilation, air conditioning systems based on regenerable energy for food storage; - encouraging proper crop diversification with price assurance; - digitalization of the agri-food system; - promotion of nutraceuticals, functional foods, and herbs consumption; - encouraging volunteering and local food production.	[27]

Table 1. *Cont.*

Region	Focus	Proposed Strategy	References
North America	Impact of the COVID-19 pandemic on food security and global food supply chains	- digitization of global food supply chains; - a good balance between the existing governmental policies for COVID-19 pandemic effects diminishing and creation of a long-term base for food supply chains resilience; - development of online shopping platforms; - end-to-end supply chain transparency; - Industry 4.0 instruments application; - use of cloud technology to increase interoperability and data management; - system decentralization to reduce transport and storage costs and to diminish environmental impact.	[26]
Europe	Sustainable food supply chains contribution to agri-food system changes in the actual context	- combination of short with long food supply chains; - increase community's self-sufficiencies.	[32]
Latin America	Food policy after the COVID-19 pandemic in areas of indigenous people	- increase community access to public goods such as novel technologies, irrigation systems, roads, transport, etc.; - accessibility of financing programs, and productive resources; - development of local biodiversity; - food quality control; - development of food marketing instruments; - development of consumer's feed-back instruments.	[2]
Italy	Roman Solidarity Purchasing Groups' contribution to food system resilience during the COVID-19 pandemic	- implementation of Solidarity Purchasing Groups to increase products handling flexibility and local producers' remuneration.	[33]
China and United States	Resilience of household food system in the COVID-19 pandemic context	- increase agri-food system sustainability; - reducing food losses and waste.	[34]
China	Evaluation of food system resilience during the COVID-19 pandemic	- promotion of traditional and ecological products; - support of wild food environment; - implementation of ecological agriculture by supporting landscape diversity; - production and consumption of local agri-food products.	[35]

Throughout history, physical shopping has been regarded as a pleasant experience with its inherent advantages. However, when we analyze the cost–benefit ratio, online commerce maintains its advantages. Buying groceries online, especially in our increasingly busy and fast-paced world, remains the easiest and quickest alternative for obtaining groceries. It is highly unlikely that people will abandon this habit once they have become accustomed to these conveniences, and over time these practices will become deeply ingrained habits [36]. The COVID-19 pandemic and post-COVID-19 period continue to reshape the food market permanently. There are several directions outlined for the post-pandemic era, as illustrated in Figure 4: a growing demand for transparency, encompassing an end-to-end perspective to satisfy discerning consumers and assess critical aspects in the food industry; the promotion of sustainable development and purpose-driven consumption that impacts the environment, society, and human health; emphasis on authenticity by promoting traditional and high-quality authentic products; increased digitalization to optimize the interaction between consumers and the seller [15].

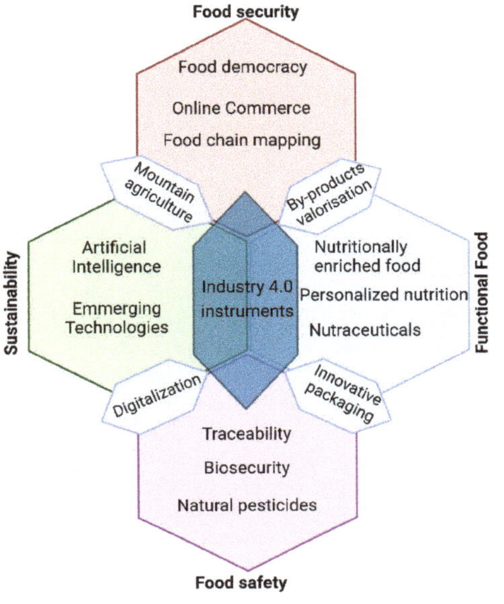

Figure 4. Directions outlined after the COVID-19 pandemic in the food sector.

However, farmers alone may lack the capacity to address the challenges of sustainable agri-food system development, thus it is necessary to create effective public–private partnerships across the agri-food value chain [37]. Hege et al. [38] asserted that urgent measures must be taken by authorities and community members to promote collaborations among stakeholders in the food chain, to provide financial support for healthy foods, encourage policy flexibility in nutrition programs, and develop community-based models involving various stakeholders.

3.2. The Romanian Post-Pandemic Food Market

Today, there is a growing awareness among the population about the importance of a healthier lifestyle, which is increasingly reflected in the interest in a more natural and healthier diet [39]. Historically, the food market has been negatively affected during pandemics and epidemics, often resulting in disastrous consequences for food consumption [40]. The COVID-19 pandemic had particularly adverse effects, especially for financially disadvantaged individuals, impacting food security. The food system comprises farmers, processors, traders, distributors, and consumers, with these key actors interacting across various stages of the food chain, including production, storage, processing, distribution, and transport [30,41,42].

Short supply chains are viewed as innovative and in a continuous process of reinvention [32]. Since the communist period, short supply chains have served as a survival solution for urban populations. During that era, people sourced their supplies from relatives, friends, or small farmers in the countryside, benefiting both sides [43]. Today, this system is well-established and can be defined as a "system of production, processing and marketing aimed at ecologically sustainable means and methods that govern economic, social, environmental and health benefits for local communities" [44]. The current COVID-19 health crisis through which we have all lived did not have as detrimental an impact on the food market. On the contrary, it proved to be functional due to the presence of these short supply systems within local communities.

The crisis situation fostered trusting relationships between local producers and end consumers, which will likely contribute in the future to food safety and security. Moreover, food produced and supplied by small local producers and marketed through these short chains is perceived as a healthier alternative [44,45]. Existing literature in this field reports and analyzes the success factors of short supply chains with all types of products, particularly dairy products [46]. Most authors have identified the primary critical success factor in the short food supply chain as the traditional specificity of local brands, the natural and ecological aspects, direct and reciprocal relations between producers and consumers, safety and traceability, specific local craftmanship, culinary and cultural heritage, cooperation, and consumer health assurance [47]. In a study by Burlea-Schiopoiu et al. [48], the impact of food delivery applications on Romanian consumers' behavior was examined, and the authors recommended the implementation of consumer loyalty strategies, underlying the great visibility of such applications reflecting the consumers' empathy and loyalty driven by their convenience and money-saving characteristics.

Social distancing rules imposed during the COVID-19 pandemic have boosted online shopping around the world, driving the digitalization trend. In 2020, online products purchases rose by 30% compared to 2019, while retail sales decreased by 17.9% [49]. This trend has been most notable in the food supply chain [50]. Furthermore, digital technologies have been successfully employed by farmers to satisfy the changing preferences of consumers or even restaurants [49]. Romania also experienced a major increase in online deliveries during the pandemic and the post-pandemic period, rising by over 30% [51,52]. Around 44% of Romanians made online purchases during the pandemic, with Romania ranking a lower position in the European Union compared to the Netherlands, Denmark, and Sweden with respective figures of 94%, 92%, and 89% [53].

Several authors investigated the importance of traditional products in the context of the COVID-19 pandemic and consumer preferences for purchasing them. Key factors influencing consumer behavior during the COVID-19 pandemic with respect to traditional products were: the effect on health, the ease of online purchasing, the taste, the effects on the environment, contribution on the local or national economy [50]. The production and consumption of traditional foods are often linked to geographical identity [54]. For example, in the Dorna area, the Emmental type cheese is known as "Schweitzer de Dorna" [55]. Authenticity and origin are typically the hallmark qualities of traditional products, which simultaneously contribute to their cultural and economic renown, thus supporting local agricultural economies [56]. Traditional products are regarded as a model of identity, culture, and heritage passed down through generations, making significant contributions to the sustainable development of rural mountain areas [57]. The Romanian village associated with food traditions is closely connected to archaic production methods, featuring specific local ingredients and recipes that contribute not only to the local economy but also to the environmental preservation [58]. Consequently, the pandemic period spurred the search for these products directly from the producers or farmers.

There is a clear need to implement government programs to support mountain farmers. However, it is worth nothing that the role of farmers in decision-making entities within the traditional food supply chain in mountainous areas is relatively weak. Despite being the primary providers of raw product, they often rely on other actors in the supply chain for selling their products.

4. Mountain Food Market in the Post-Pandemic Period

Worldwide, food consumers are experiencing an increasing awareness of the food they eat and the water they drink, along with their health effects [59]. This heightened awareness has led to a shift in the modern lifestyle, where consumers are no longer inclined to favor processed foods laden with numerous additives. Instead, there is a growing interest in mountain food products due to their natural characteristics and high nutritional quality [60,61]. Certified mountain products hold significant market potential (Figure 5) and have gained substantial attention from consumers in recent years. These products offer a more complex sensory and nutritional profile, delivering benefits not typically found in other products, with their quality being largely influenced by the environment, climate, and processing conditions [62].

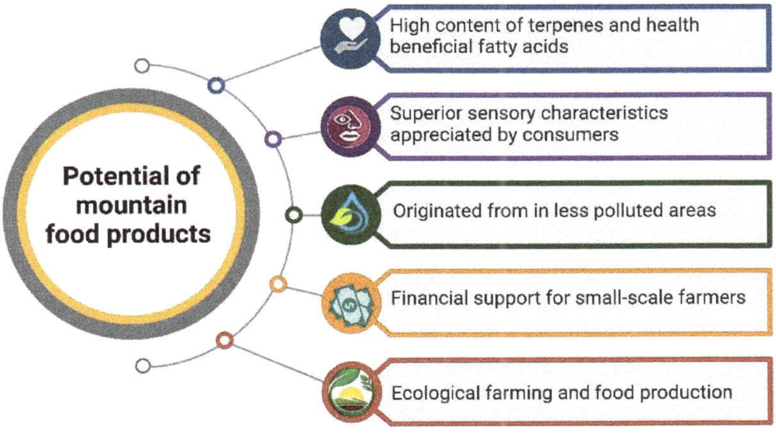

Figure 5. Potential of mountain food products for sustainable development of food system.

Sustainable Development Goal (SDG) 2.4 underscores the importance of ensuring "sustainable food production and resilient agricultural practices by 2030" [63]. The pandemic has boosted the vulnerabilities of mountain communities, whose main occupation is agriculture. Kumar et al. [63] investigated the vulnerability of small farmers in mountain areas in India, considering the pandemic and other challenging circumstances. Their findings highlight the pressing need for policies aimed at enhancing ecosystem resilience in mountainous regions. They also recommend harnessing artificial intelligence to improve efficiency and assist mountain farmers in addressing various challenges in different contexts (Figure 6).

Govinda Gowda [64] conducted a study on the effects of COVID-19 pandemic in Bangalore, India, and concluded that people were significantly affected by the economic consequences of the pandemic, rather than medical issues. This highlights the necessity for policies aimed at mitigating the economic and financial difficulties in the post-COVID-19 period. A case study conducted in the Mount Bromo region, Indonesia, reported that the small-scale farmers developed a mechanism to raise their welfare and secure their family's economic situation during the COVID-19 pandemic through a rational choice: people owning small land diversified their workplaces by doing off-farm and non-farm activities, others focused on livestock raising, while some individuals even provided support services for tourism [65]. In the mountain province of the Philippines, people faced income lowering, limited farm inputs, deterioration, and oversupply of vegetables during the pandemic [66]. To help farmers overcome these difficulties, the authors proposed solutions such as community trainings, quality control of the local food, trainings in food processing and preservation to reduce food waste and to ensure food safety, and the establishment of a local food storage facility or a food bank for preserved food [66].

In a study on consumers' behavior in of the Sibiu region of Romania, it was found that the main factors influencing people's buying decisions were related to food properties, origin, freshness, sensory profile, traceability, and reliability; young people were particularly awareness of health, and food composition [67]. During the COVID-19 pandemic, young people limited their interaction with others and experienced a certain level of stress and anxiety which led them to order food online [67]. The same authors asserted that short food supply chains can serve as examples of best practices for changing the food system to meet sustainability requirements. Research conducted by Covaci et al. [68] revealed that the COVID-19 pandemic stimulated the honey and apiculture products market, with the Manuka and mountain honey being preferred as people recognized the beneficial effects of honey consumption in the COVID-19 context. Cattivelli [69] presented a decision support system that calculates the food self-sufficiency cities in South Tyrol with urban gardens.

The study areas were selected based on the impact of changes in the food industry in the context of food insecurity and COVID-19 mobility. The results indicated that this tool can be used for food planning, determination of the 'foodprint' and land suitability, and achieving people's food self-sufficiency. One of its most important advantages is the framework created that encompasses all stages of the food chain and territorial characteristics, including the morphological and climatic properties of the area of implementation [69]. Restructuring the food industry in the mountainous area studied is necessary, along with the application of social innovation strategies in cooperation with the community members.

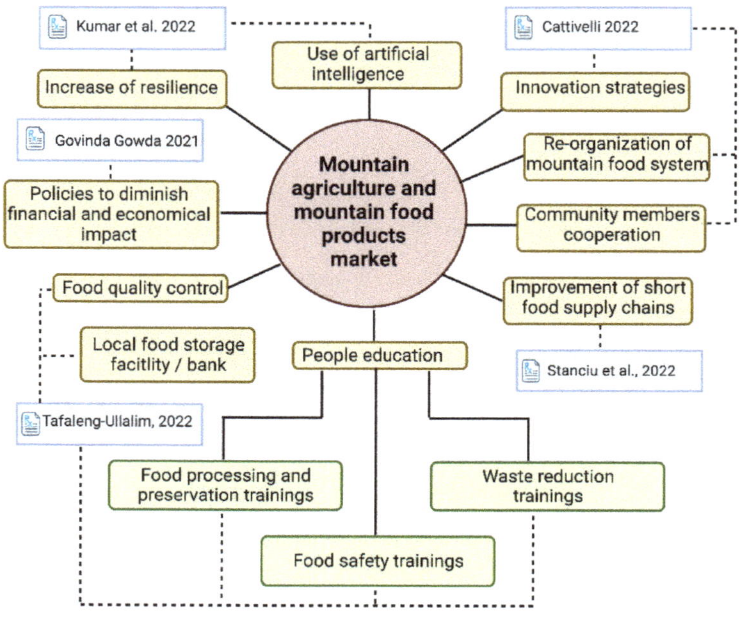

Figure 6. Strategies for the mountain agriculture and food market in the post-pandemic period [63–69].

5. Supporting Mountain Agriculture to Grow the Agri-Food Market with Healthy Products

The effective consumption of typical mountain products significantly contributes to the sustainability of local economies in mountain areas.. Consumers' preference for products with sustainable production characteristics consistently centers around key features: a mountain product label with "mountain product" certification, ecological certification, and specific information on animal welfare [70]. In the European Union, some products are certified as Protected Designation of Origin (PDO), or Protected Geographical Indication (PGI), the Guaranteed Traditional Specialty label (STG) or the optional term "Food from the mountains" [71]. These quality labels enable producers to preserve the integrity of traditional food and avert falsification, while also allowing them to convey products' quality attributes to consumers [72].

Incorporating territorial and traditional values, local food products contribute to the sustainable evolution of local economies, especially in most the developed areas within mountainous regions [73]. Unfortunately, in recent years, the environment, rural economy, and cultural heritage have been adversely affected by depopulation in mountains, leading to the destruction of pastures and meadows in these areas [70].

Consumers have developed an idealized perception that includes mountain green spaces, clean air, crystal clear waters, traditions and cultural identity, and traditional product processing, which leads them to choose traditional, healthy mountain products [74].

European mountain product certifications aim to promote the production boost in the economies of disadvantaged areas, especially in those mountain regions [75]. However, these certifications do not always fully align with the options available to producers in mountain areas; the high cost associated with PDO or IGP labels deters many farmers from adopting them [76]. That is why the European Union (EU) introduced the optional label "mountain product" which was regulated by the European Commission in 2012. The goal was to facilitate and promote the development of agricultural systems and mountain economies, as well as the sustainable development of the entire food chain [77]. Farmers who trade their products with the "mountain product" label have influenced and strengthened consumer behavior regarding ecological mountain products [78,79]. In the EU, more than a third of production with a geographical indication originates in mountain agricultural regions, with 50–75% being cheeses with the designation of protected origin DOP [76]. These cheeses are typical, inherent to the mountain territory, traditionally and historically produced by grazing animals in diverse natural mountain meadows and hayfields [70].

Mountain agriculture operates on a smaller scale compared to conventional agriculture, making it better suited to penetrate niche markets due to its more limited economic capacity [80]. The most suitable approach for farmers and local producers in mountain areas is to utilize the "mountain product" label, as it is the most recommended tool for entering the food market, offering the opportunity to launch these products and justify an adequate pricing [80]. Bentivoglio et al. [81] contend in their work that the mountain product label can support the local agricultural economy, enhance mountain territories, and protect the origin of mountain products, biodiversity, and the environment. The Euromontana association also asserts that mountain products contribute to traceability, biodiversity conservation, and environmental quality, while contributing to the connection of mountain products with environmental protection. Traditional mountain agriculture is perceived by consumers as a system that ensures animal welfare and holds significant interest [82,83].

Case Study of the Bioalpin Cooperative in Tyrol

The potential of mountain food products is exemplified in the case study of the Bioalpin cooperative from Tyrol. The cooperative was founded in 2002 as a regional platform for organic food products sourced from small farms and it sells a complete range of mountain products by using its own brand, Bio vom Berg, which translates to "organic from the mountain" [84]. About 60% of the Bioalpin turnover is generated through the regional supermarket MPreis which primary stocks products from Tyrol and adjacent regions, giving preference to Bioalpin products over other organic brands. The Bioalpin cooperative offers a large range of food products like fruits and vegetables, eggs, cereals, meat products, honey, and herbs, with an emphasis on milk and dairy products [85]. The sales volume of Bioalpin increased from EUR 672 000 in 2003 to EUR 14 million in 2022 [84,86], making a substantial contribution to the local economy. The cooperative members are small dairy producers, producer groups, and individual farmers, with more than six hundred small-scale farms associated to Bioalpin. In addition to income coming from the cooperative, farmers also earn money from off-farm employment, public payments, and complementary direct sales [85]. The cooperative ensures access to a large retail chain and supports the functioning of artisanal factories. Thereby, Bioalpin not only gives financial benefits to small-scale farmers but also contributes to the development of a network of local processing and trading units that help the local economy [84].

The Bioalpin cooperative manages both the horizontal and vertical levels of the supply chain; the horizontal one ensures enough quantities of goods and the use of collective packaging, while vertical coordination implies price negotiations, established volumes, and retailer implication in the product range and innovation [84]. As an example, for dairy producers, horizontal coordination has the advantage of creating a collective identity and decreasing competition among members by specializing each dairy in a small number of product types carrying on the same collective label. This allows them to become more professional and to increase their specialized know-how. In vertical coordination, the cooperative empowers members and users of the Bioalpin brand to negotiate the prices with MPreis as a function of the cost calculations [85].

The Bioalpin cooperative increased its network by applying a "multiplicative growth" base on long-term relationships, resulting in costs diminishing. Furthermore, the purpose of Bioalpin is to enhance conditions for all its participants, not to get the maximum profit [84]. By extending and increasing its turnover, the cooperative has enabled the development of the central hub in a professional way. This fact allows small farmers to focus on their own management and raises product differentiation [84]. This example of good practices in Tyrol could serve as a model that can be adapted in many other mountain areas in the world. It is important to document all the particularities of each region and develop support programs for traditional product manufacturing and trade. This case study underscores the significant potential of mountain products in contributing to regional economies and overall well-being.

6. Food Democracy—A Model of Food Governance in the Post-Pandemic Period

Given that the food system faced significant challenges during the pandemic, food security has been limited, partly due to long supply chains with little flexibility. A new system oriented towards food democracy offers opportunities for both consumer producers to participate in building food systems that are as sustainable as possible and support alternatives for how food is produced and consumed [87]. The term "food democracy" was created in 1999 by Lang [88]. Petetin [87] noted that food democracy creates the base for creating an alternative and transformative food system, to stimulate consumers to seek and choose sustainable food systems that reflect common values. The ability of individuals to make choices about where and how they purchase food reflects the degree of control that consumers can exert over food systems [89]. Due to the restrictions introduced during the COVID-19 pandemic, small family farms experienced significant growth, with demand surging for various products, especially among rural households [87]. Today, agricultural producers and small stores have demonstrated a great degree of resilience and adaptability to manage the increased demand. This alternative approach to food consumption, in which consumers have the opportunity to choose what they put on their plates, transforms them from "consumers" into "active citizens" and food democrats [90].

The food safety policy of many countries aim to design future global food security and safety policies and strategies based on the best and most nutritious foods, while discouraging unhealthy options through relentless promotion in the mass media [91]. Buying food directly from small local farmers reflects much healthier and more nutritious food consumption, with far-reaching implications for public health over the long term [87]. Engaging in activities such as gardening by growing different varieties of vegetables, raising poultry, rabbits, or other animals enhances food security and provides alternatives for individuals with varying incomes [92].

Food democracy offers the potential to restructure the food supply chain centered on dairy and regional products, fostering stronger networks between commercial and local farmers. Once the COVID-19 health crisis subsides, there will be a need to create new strategies to improve democratic agri-food systems. Governments should provide greater financial support to small family farms, as they play a crucial role in food security, revitalization of the cultural landscape, rural tourism, and recreational activities, which are job-generators activities. Small family farms also maintain a harmonious relationship with the environment, fostering greater biodiversity with more habitats, which enhances the rural landscape [93]. The COVID-19 pandemic promoted the transformation of food supply chains, making them more sustainable, resilient, and democratic, while the post-pandemic period created a powerful framework for a food democracy, with a focus on locally and regionally sourced food and the promotion of natural and healthy products' consumption [87].

7. Conclusions and Further Research Directions

People's established habits have been significantly affected by the COVID-19 pandemic. The emerge of the new coronavirus has instilled fear and prompted a shift from a normal lifestyle to a more protective one. This change in behavior has influenced consumer preferences, particularly towards online shopping, which offers various conveniences such as cashless transactions, home delivery, and access to a wide range of products that are no longer readily available. The current COVID-19 crisis should prompt us to explore the challenges and opportunities in order to make informed decisions about the future of agriculture and food systems. Many consumers have embraced online ordering and direct purchasing from manufacturers. In the future, small-scale producers will play a particularly important role in the production of healthy food, especially in mountain regions. These areas often rely on individual households for animal husbandry, and short supply chains for mountain products are emerging as a promising solution. These chains ensure a market where consumers are increasingly conscious of the quality of life, food security, safety, and overall food health. It is essential to develop food policies that support mountain agriculture, involving public authorities, producers, and community members. Private-public partnerships and financial aid for farmers would be helpful in supporting the shift toward sustainable agri-food systems. Further research should focus on evaluating the effectiveness of policies implemented in the post-pandemic period with the goal of enhancing food security and community resilience. Furthermore, mountain areas require closer examination as awareness grows regarding the availability of healthy and sustainable products in these regions.

Author Contributions: Conceptualization, D.N. and M.U.-I.; methodology, M.U.-I., D.N., and L.O.; software, M.U.-I.; validation, M.U.-I., L.O. and D.N.; formal analysis, M.U.-I. and D.N.; investigation, D.N. and M.U.-I.; resources, M.U.-I., L.O. and D.N.; data curation, M.U.-I.; writing—original draft preparation, M.U.-I. and D.N.; writing—review and editing, M.U.-I., L.O. and D.N.; visualization, M.U.-I., L.O. and D.N.; supervision, M.U.-I., L.O. and D.N.; project administration, M.U.-I. and D.N.; funding acquisition, L.O. and D.N. All authors have read and agreed to the published version of the manuscript.

Funding: The University of Agricultural Sciences and Veterinary Medicine of Cluj-Napoca supported a part of the APC.

Data Availability Statement: Not applicable.

Acknowledgments: Figures 4–6 were created with BioRender.com: https://www.biorender.com/ (accessed on 15 March 2023).

Conflicts of Interest: The authors declare no conflict of interest.

References

1. Rull, V. *Quaternary Ecology, Evolution, and Biogeography*; Academic Press: London, UK, 2020; ISBN 0128204737.
2. Lugo-Morin, D.R. Innovate or Perish: Food Policy Design in an Indigenous Context in a Post-Pandemic and Climate Adaptation Era. *J. Open Innov. Technol. Mark. Complex.* **2022**, *8*, 34. [CrossRef]
3. Church, J.A.; White, N.J. Sea-Level Rise from the Late 19th to the Early 21st Century. *Surv. Geophys.* **2011**, *32*, 585–602. [CrossRef]
4. Slater, T.; Lawrence, I.R.; Otosaka, I.N.; Shepherd, A.; Gourmelen, N.; Jakob, L.; Tepes, P.; Gilbert, L.; Nienow, P. Review article: Earth's ice imbalance. *Cryosphere* **2021**, *15*, 233–246. [CrossRef]
5. Gupta, G.S. Land Degradation and Challenges of Food Security. *Rev. Eur. Stud.* **2019**, *11*, 63. [CrossRef]
6. Bergman, M.M. The World after COVID. *World* **2020**, *1*, 45–48. [CrossRef]
7. FAO. *Tracking Progress on Food and Agriculture-Related SDG Indicators 2021: A Report on the Indicators under FAO Custodianship*; FAO: Rome, Italy, 2021. [CrossRef]
8. FAO. *Global Report on Food Crises: Joint Analysis for Better Decision*; FAO: Rome, Italy, 2021. Available online: https://www.wfp.org/publications/2020-global-report-food-crises (accessed on 27 September 2023).
9. Sanudin; Widiyanto, A.; Fauziyah, E.; Sundawati, L. Agroforestry farmers' resilience in social forestry and private Forest programs during the COVID-19 pandemic in Indonesia. *For. Sci. Technol.* **2023**, *19*, 197–209.
10. Dou, Y.; Li, Y.; Li, M.; Chen, X.; Zhao, X. The Role of Agroforestry in Poverty Alleviation: A Case Study from Nujiang Prefecture, Southwestern China. *Sustainability* **2023**, *15*, 12090. [CrossRef]

11. Zafeiriou, E.; Chatzissavvidis, C.; Antonopoulou, C.; Arabatzis, G. Sweet chestnut and agricultural development: A farmers' perspective for Northern Greece. *Int. J. Agric. Sustain.* **2022**, *20*, 199–215. [CrossRef]
12. Dannenberg, P.; Fuchs, M.; Riedler, T.; Wiedemann, C. Digital Transition by COVID-19 Pandemic? *The German Food Online Retail. Tijdschr. Voor Econ. En Soc. Geogr.* **2020**, *111*, 543–560. [CrossRef]
13. De Souza, T.S.P.; Miyahira, R.F.; Matheus, J.R.V.; de Brito Nogueira, T.B.; Maragoni-Santos, C.; Barros, F.F.C.; Antunes, A.E.C.; Fai, A.E.C. Food services in times of uncertainty: Remodeling operations, changing trends, and looking into perspectives after the COVID-19 pandemic. *Trends Food Sci. Technol.* **2022**, *120*, 301–307. [CrossRef]
14. Guiné, R.P.F.; Florença, S.G.; Costa, D.T.V.A.; Çelik, S.; Ferreira, M.; Cardoso, A.P.; Çetin, S.; Costa, C.A. Comparative Study about the Consumption of Organic Food Products on Samples of Portuguese and Turkish Consumers under the COVID-19 Pandemic Context. *Agronomy* **2022**, *12*, 1385. [CrossRef]
15. Veronesi, V.; Schiavello, M. Post-Pandemic Trends in the Food & Beverage Industry. In *The Evolution of Made in Italy*; Palgrave Macmillan, C., Ed.; Springer International Publishing: Cham, Switzerland, 2023; pp. 17–25. ISBN 978-3-031-15667-0.
16. UN FSS. United Nations Food Systems Summit 2021. 2021. Available online: https://sc-fss2021.org/community/relevant-actors/ (accessed on 23 October 2023).
17. Tanzer, M.; Gläsel, A.; Egermann, M. Elucidating the capabilities of international mechanisms to foster procedural just system change—The case of the 2021 UN Food System Summit. *Environ. Innov. Soc. Transit.* **2022**, *45*, 72–82. [CrossRef]
18. Hubeau, M.; Marchand, F.; Coteur, I.; Mondelaers, K.; Debruyne, L.; Van Huylenbroeck, G. A new agri-food systems sustainability approach to identify shared transformation pathways towards sustainability. *Ecol. Econ.* **2017**, *131*, 52–63. [CrossRef]
19. Webb, P.; Benton, T.G.; Beddington, J.; Flynn, D.; Kelly, N.M.; Thomas, S.M. The urgency of food system transformation is now irrefutable. *Nat. Food* **2020**, *1*, 584–585. [CrossRef] [PubMed]
20. Alaimo, L.S.; Fiore, M.; Galati, A. How the COVID-19 pandemic is changing online food shopping human behaviour in Italy. *Sustainability* **2020**, *12*, 9594. [CrossRef]
21. Li, Y.; Yao, P.; Osman, S.; Zainudin, N.; Sabri, M.F. A Thematic Review on Using Food Delivery Services during the Pandemic: Insights for the Post-COVID-19 Era. *Int. J. Environ. Res. Public Health* **2022**, *19*, 15267. [CrossRef]
22. Wang, O.; Somogyi, S.; Charlebois, S. Food choice in the e-commerce era: A comparison between business-to-consumer (B2C), online-to-offline (O2O) and new retail. *Br. Food J.* **2020**, *122*, 1215–1237. [CrossRef]
23. Nielsen, D.E.; Karamanoglu, I.; Yang Han, H.; Labonté, K.; Paquet, C. Food Values, Food Purchasing, and Eating-Related Outcomes Among a Sample of Quebec Adults During the COVID-19 Pandemic. *Can. J. Diet. Pract. Res.* **2022**, *84*, 1–8. [CrossRef]
24. Wachyuni, S.S.; Wiweka, K. The Changes in Food Consumption Behavior: A Rapid Observational Study of COVID-19 Pandemic. *Int. J. Manag. Innov. Entrep. Res.* **2020**, *6*, 77–87. [CrossRef]
25. Liao, C.; Lu, Q.; Shui, Y. Governmental Anti-Pandemic and Subsidy Strategies for Blockchain-Enabled Food Supply Chains in the Post-Pandemic Era. *Sustainability* **2022**, *14*, 9497. [CrossRef]
26. Alabi, M.O.; Ngwenyama, O. Food security and disruptions of the global food supply chains during COVID-19: Building smarter food supply chains for post COVID-19 era. *Br. Food J.* **2023**, *125*, 167–185. [CrossRef]
27. Priyadarshini, P.; Abhilash, P.C. Agri-food systems in India: Concerns and policy recommendations for building resilience in post COVID-19 pandemic times. *Glob. Food Sec.* **2020**, *29*, 100537. [CrossRef] [PubMed]
28. Galanakis, C.M.; Rizou, M.; Aldawoud, T.M.S.; Ucak, I.; Rowan, N.J. Innovations and technology disruptions in the food sector within the COVID-19 pandemic and post-lockdown era. *Trends Food Sci. Technol.* **2021**, *110*, 193–200. [CrossRef]
29. Chenarides, L.; Manfredo, M.; Richards, T.J. COVID-19 and Food Supply Chains. *Appl. Econ. Perspect. Policy* **2021**, *43*, 270–279. [CrossRef]
30. Béné, C. Resilience of local food systems and links to food security—A review of some important concepts in the context of COVID-19 and other shocks. *Food Secur.* **2020**, *12*, 805–822. [CrossRef] [PubMed]
31. Romeo, R.; Manuelli, S.; Abear, S. The International Year of Sustainable Mountain Development 2022: An opportunity to promote action for mountains. *Front. Sustain. Food Syst.* **2022**, *6*, 933080. [CrossRef]
32. Chiffoleau, Y.; Dourian, T. Sustainable food supply chains: Is shortening the answer? A literature review for a research and innovation agenda. *Sustainability* **2020**, *12*, 9831. [CrossRef]
33. Tarra, S.; Mazzocchi, G.; Marino, D. Food System Resilience during COVID-19 Pandemic: The Case of Roman Solidarity Purchasing Groups. *Agriculture* **2021**, *11*, 156. [CrossRef]
34. Dou, Z.; Stefanovski, D.; Galligan, D.; Lindem, M.; Rozin, P.; Chen, T.; Chao, A.M. Household Food Dynamics and Food System Resilience Amid the COVID-19 Pandemic: A Cross-National Comparison of China and the United States. *Front. Sustain. Food Syst.* **2021**, *4*, 577153. [CrossRef]
35. Ahmed, S.; Downs, S.M.; Yang, C.; Chunlin, L.; Ten Broek, N.; Ghosh-Jerath, S. Rapid tool based on a food environment typology framework for evaluating effects of the COVID-19 pandemic on food system resilience. *Food Secur.* **2020**, *12*, 773–778. [CrossRef]
36. East, R. Online Grocery Sales after the Pandemic. *Int. J. Mark. Res.* **2022**, *64*, 13–18. [CrossRef]
37. Ntawuruhunga, D.; Ngowi, E.E.; Mangi, H.O.; Salanga, R.J.; Shikuku, K.M. Climate-smart agroforestry systems and practices: A systematic review of what works, what doesn't work, and why. *For. Policy Econ.* **2023**, *150*, 102937. [CrossRef]
38. Hege, A.; McCormick, N.; Robinson, P.; Charles, K.; Jones, J.; Aft, E. Perspectives from the front line: The post-pandemic emergency food system in North Carolina. *J. Agric. Food Syst. Community Dev.* **2021**, *10*, 241–245. [CrossRef]

39. Voinea, L.; Popescu, D.V.; Bucur, M.; Negrea, T.M.; Dina, R.; Enache, C. Reshaping the traditional pattern of food consumption in Romania through the integration of sustainable diet principles. A qualitative study. *Sustainability* **2020**, *12*, 5826. [CrossRef]
40. Bucur, B. Bogdan Murgescu, Romania and Europe: The accumulation of economic gaps (1500–2010). *Sociol. Românească* **2010**, *8*, 165–168.
41. Sukhwani, V.; Deshkar, S.; Shaw, R. Covid-19 lockdown, food systems and urban–rural partnership: Case of Nagpur, India. *Int. J. Environ. Res. Public Health* **2020**, *17*, 5710. [CrossRef]
42. Devereux, S.; Béné, C.; Hoddinott, J. Conceptualising COVID-19's impacts on household food security. *Food Secur.* **2020**, *12*, 769–772. [CrossRef]
43. Sabates-Wheeler, R. Safety in small numbers: Local strategies for survival and growth in Romania and the Kyrgyz Republic. *J. Dev.'Stud.* **2007**, *43*, 1423–1447. [CrossRef]
44. Barska, A.; Wojciechowska-Solis, J. E-consumers and local food products: A perspective for developing online shopping for local goods in Poland. *Sustainability* **2020**, *12*, 4958. [CrossRef]
45. Tanasă, L. Benefits of short food supply chains for the development of rural tourism in Romania as emergent country during crisis. *Agric. Econ. Rural Dev.* **2014**, *11*, 181–193.
46. Sellitto, M.A.; Vial, L.A.M.; Viegas, C.V. Critical success factors in Short Food Supply Chains: Case studies with milk and dairy producers from Italy and Brazil. *J. Clean. Prod.* **2018**, *170*, 1361–1368. [CrossRef]
47. Migliore, G.; Schifani, G.; Cembalo, L. Opening the black box of food quality in the short supply chain: Effects of conventions of quality on consumer choice. *Food Qual. Prefer.* **2015**, *39*, 141–146. [CrossRef]
48. Burlea-Schiopoiu, A.; Puiu, S.; Dinu, A. The impact of food delivery applications on Romanian consumers ' behaviour during the COVID-19 pandemic. *Socioecon. Plann. Sci.* **2020**, *82*, 101220. [CrossRef] [PubMed]
49. OECD. E-Commerce in the Time of COVID-19, in OECD Policy Responses to Coronavirus (COVID-19). Available online: https://www.oecd.org/coronavirus/policy-responses/e-commerce-in-the-time-of-covid-19-3a2b78e8/ (accessed on 27 September 2023).
50. Petrescu-Mag, R.M.; Vermeir, I.; Petrescu, D.C.; Crista, F.L.; Banatean-Dunea, I. Traditional foods at the click of a button: The preference for the online purchase of Romanian traditional foods during the COVID-19 pandemic. *Sustainability* **2020**, *12*, 9956. [CrossRef]
51. Butu, A.; Brumă, I.S.; Tanasă, L.; Rodino, S.; Vasiliu, C.D.; Doboș, S.; Butu, M. The impact of COVID-19 crisis upon the consumer buying behavior of fresh vegetables directly from local producers. Case study: The quarantined area of Suceava County, Romania. *Int. J. Environ. Res. Public Health* **2020**, *17*, 5485. [CrossRef] [PubMed]
52. Morosan-Danila, L.; Bordeianu, O.-M. The Need For Change And Shaping The Post-COVID Business Environment in Romania. In Proceedings of the 16th Economic International Conference New Challenges and Opportunities for the Economy 4.0, Suceava, Romania, 7–8 May 2020; Volume 13, pp. 387–397. [CrossRef]
53. Eurostat. Largest Share of e-Shoppers in the North of the EU. Available online: https://ec.europa.eu/eurostat/de/web/products-eurostat-news/-/ddn-20180117-1/ (accessed on 27 September 2023).
54. Bojnec, Š.; Petrescu, D.C.; Petrescu-Mag, R.M.; Rădulescu, C.V. Locally produced organic food: Consumer preferences. *Amfiteatru Econ.* **2019**, *21*, 161. [CrossRef]
55. Necula, D.; Tamas-Krumpe, O.; Fenesan, D.; Todoran, D.; Ognean, L. Aspects regarding the production and the hygiene- sanitary control of the Dorna Swiss Cheese. *Sci. Pap. Ser. D Anim. Sci.* **2022**, *65*, 517–524.
56. Serrano-Cruz, M.R.; Espinoza-Ortega, A.; Sepúlveda, W.S.; Vizcarra-Bordi, I.; Thomé-Ortiz, H. Factors associated with the consumption of traditional foods in central Mexico. *Br. Food J.* **2018**, *120*, 2695–2709. [CrossRef]
57. Guerrero, L.; Guàrdia, M.D.; Xicola, J.; Verbeke, W.; Vanhonacker, F.; Zakowska-Biemans, S.; Sajdakowska, M.; Sulmont-Rossé, C.; Issanchou, S.; Contel, M.; et al. Consumer-driven definition of traditional food products and innovation in traditional foods. A qualitative cross-cultural study. *Appetite* **2009**, *52*, 345–354. [CrossRef]
58. Purcărea, T.V.; Orzan, G.; Orzan, M.; Stoica, I. Romanian Consumer Behavior Regarding Traditional Foods: Contributing to the Rebuilding of a Healthier Food Culture. *J. Food Prod. Mark.* **2013**, *19*, 119–137. [CrossRef]
59. Murray, M.T.; Pizzorno, J.; Pizzorno, L. *The Encyclopedia of Healing Foods*; Atria Books: New York, NY, USA, 2005; ISBN 9780743480529.
60. Alderighi, M.; Bianchi, C.; Lorenzini, E. The impact of local food specialities on the decision to (re)visit a tourist destination: Market-expanding or business-stealing? *Tour. Manag.* **2016**, *57*, 323–333. [CrossRef]
61. Rey, R. 'Mountain Product', of High- biological Quality. Meadows' Poliflora, Organic Fertilizer and a Sustainable Mountain Economy. *Procedia Econ. Financ.* **2014**, *8*, 622–631. [CrossRef]
62. Necula, D.; Feneșan, D.; Tamas-Krumpe, O.; Buta, A.; Todoran, C.; Ognean, L. The evaluation of specific climatic factors in the mountainous region of Dorna depression on raw cow milk fat and protein content. *Sci. Pap. Vet. Med.* **2021**, *54*, 65–70.
63. Kumar, P.P.; Rangi, A.; Roy, A. Assessment of Sustainable Food Systems in Mountain Regions of India Through Climate Smart Agriculture During Covid-19 Pandemic: The Case Study of AI for Rural. *Why It Matters* **2022**.
64. Govinda Gowda, H. Impact of Pandemic—COVID-19 in Agrarian Part of the Country—A Study across Villages Adopted by Mount Carmel College, Autonomous, Bangalore. *Int. J. Eng. Manag. Res.* **2021**, *11*, 22–26. [CrossRef]
65. Susanti, A. Farmer survival mechanism during the pandemic: A case study of the tengger tribal community, east Java. *J. Sosiol. Reflektif* **2022**, *17*, 261–278.

66. Tafaleng-Ullalim, P.A. Food Security in Times of COVID-19 Pandemic: Experiences of Mountain Province Through the Agriculture Lens. *J. Posit. Sch. Psychol.* **2022**, *6*, 129–139.
67. Stanciu, M.; Popescu, A.; Rasvan, R.I. Short Food Supply Chains and Young People'S Attitude Towards Healthy Eating. *Sci. Pap. Manag. Econ. Eng. Agric. Rural Dev.* **2022**, *22*, 625–635.
68. Covaci, B.; Brejea, R.; Covaci, M. Sweeteners World Trade and Behaviour in the Pandemic. Evidence from Honey Remedies Nexus Mountain Apis Mellifera Product. *Sugar Tech* **2023**, *25*, 938–949. [CrossRef]
69. Cattivelli, V. The contribution of urban garden cultivation to food self-sufficiency in areas at risk of food desertification during the COVID-19 pandemic. *Land Use Policy* **2022**, *120*, 106215. [CrossRef]
70. Mazzocchi, C.; Sali, G. Supporting mountain agriculture through "mountain product" label: A choice experiment approach. *Environ. Dev. Sustain.* **2022**, *24*, 701–723. [CrossRef]
71. Necula, D.; Ilea, A.; Coman, I.; Tamas-Krumpe, O.; Feneşan, D.; Ognean, L. Characteristics of the Compostion and Bioactive Properties of Mountain Milk Used for Emmental Cheese Making-Review. *Sci. Pap. Ser. D Anim. Sci. Int. Sess. Sci. Commun. Fac. Anim. Sci.* **2021**, *64*, 191–198.
72. Glogoveţan, A.I.; Dabija, D.C.; Fiore, M.; Pocol, C.B. Consumer Perception and Understanding of European Union Quality Schemes: A Systematic Literature Review. *Sustainability* **2022**, *14*, 1667. [CrossRef]
73. Tregear, A.; Arfini, F.; Belletti, G.; Marescotti, A. The impact of territorial product qualification processes on the rural development potential of small-scale food productions. In Proceedings of the XI World Congress of Rural Sociology, Trondheim, Norway, 25–30 July 2004.
74. Schjøll, A.; Amilien, V.; Arne Tufte, P.; Revoredo-Giha, C.; Leat, P.; Kupiec, B.; Lamprinopoulou, C. Promotion of mountain food: An explorative a study about consumers' and retailers' perception in six European countries. In Proceedings of the 9th European IFSA Symposium, Vienna, Austria, 4–7 July 2010; pp. 1558–1567.
75. Euromontana. Implementation of the EU Optional Quality Term "Mountain Product": Where Do We Stand in the Different Member States? 2020. Available online: https://www.euromontana.org/wp-content/uploads/2020/06/2020-05-26-Implementation-of-the-OQT_EN.pdf (accessed on 27 September 2023).
76. Martins, N.; Ferreira, I.C.F.R. Mountain food products: A broad spectrum of market potential to be exploited. *Trends Food Sci. Technol.* **2017**, *67*, 12–18. [CrossRef]
77. McMorran, R.; Santini, F.; Guri, F.; Gomez-y-Paloma, S.; Price, M.; Beucherie, O.; Monticelli, C.; Rouby, A.; Vitrolles, D.; Cloye, G. A mountain food label for Europe? *Rev. Géographie Alp.* **2015**, *103–104*, 1–22. [CrossRef]
78. Bazzani, C.; Caputo, V.; Nayga, R.M.; Canavari, M. Revisiting consumers' valuation for local versus organic food using a non-hypothetical choice experiment: Does personality matter? *Food Qual. Prefer.* **2017**, *62*, 144–154. [CrossRef]
79. Rousseau, S. The role of organic and fair trade labels when choosing chocolate. *Food Qual. Prefer.* **2015**, *44*, 92–100. [CrossRef]
80. Finco, A.; Bentivoglio, D.; Bucci, G. A label for mountain products? Let's turn it over to producers and retailers. *Qual.-Access Success* **2017**, *18*, 198–205.
81. Bentivoglio, D.; Savini, S.; Finco, A.; Bucci, G.; Boselli, E. Quality and origin of mountain food products: The new European label as a strategy for sustainable development. *J. Mt. Sci.* **2019**, *16*, 428–440. [CrossRef]
82. Cembalo, L.; Caracciolo, F.; Lombardi, A.; Del Giudice, T.; Grunert, K.G.; Cicia, G. Determinants of Individual Attitudes Toward Animal Welfare-Friendly Food Products. *J. Agric. Environ. Ethics* **2016**, *29*, 237–254. [CrossRef]
83. Zuliani, A.; Esbjerg, L.; Grunert, K.G.; Bovolenta, S. Animal welfare and mountain products from traditional dairy farms: How do consumers perceive complexity? *Animals* **2018**, *8*, 207. [CrossRef] [PubMed]
84. Van der Ploeg, J.D.; Barjolle, D.; Bruil, J.; Brunori, G.; Costa Madureira, L.M.; Dessein, J.; Drąg, Z.; Fink-Kessler, A.; Gasselin, P.; Gonzalez de Molina, M.; et al. The economic potential of agroecology: Empirical evidence from Europe. *J. Rural Stud.* **2019**, *71*, 46–61. [CrossRef]
85. Schermer, M. From 'Additive' to 'Multiplicative' Patterns of Growth. *Int. J. Sociol. Agric. Food* **2017**, *24*, 57–76.
86. Issuu Ranking Top 500. 2023. Available online: https://issuu.com/zielgruppenverlag.at/docs/top.tirol_ranking_top500_2310 (accessed on 24 October 2023).
87. Petetin, L. The COVID-19 crisis: An opportunity to integrate food democracy into post-pandemic food systems. *Eur. J. Risk Regul.* **2020**, *11*, 326–336. [CrossRef]
88. Koc, M. *For Hunger-Proof Cities: Sustainable Urban Food Systems*; IDRC: Ottawa, ON, Canada, 1999; ISBN 0889368821.
89. Booth, S.; Coveney, J. *Food Democracy: From Consumer to Food Citizen*; Springer: Berlin/Heidelberg, Germany, 2015; ISBN 9812874232.
90. Petetin, L. *Food Democracy in Food Systems*; Springer: Berlin, Germany, 2016.
91. Hawkes, C. Healthy Diets for Human Resilience in the Age of COVID-19. 2020. Available online: https://nutritionconnect.org/news-events/healthy-diets-human-resilience-age-covid-19/ (accessed on 27 September 2023).

92. Pungas, L. Food self-provisioning as an answer to the metabolic rift: The case of 'Dacha Resilience' in Estonia. *J. Rural Stud.* **2019**, *68*, 75–86. [CrossRef]
93. Van der Ploeg, J.D. Ten qualities of family farming. *Farming Matters* **2013**, *29*, 8–11.

agronomy

MDPI

Article

The Changes in Cropland Pattern Enhanced Carbon Storage in Northwest China

Junqia Kong [1] and Longfei Chen [2,*]

1 College of Landscape Architecture, Zhejiang A&F University, Hangzhou 311300, China; kongjunqia@lzb.ac.cn
2 Linze Inland River Basin Research Station, Chinese Ecosystem Research Network, Key Laboratory of Eco-Hydrology of Inland River Basin, Northwest Institute of Eco-Environment and Resources, Chinese Academy of Sciences, Lanzhou 730000, China
* Correspondence: chenlongfei_mail@163.com

Abstract: Northwest China has experienced dramatic changes in agricultural land area in recent years. The effects of these changes on carbon storage are unknown, and this ambiguity hinders land development policies related to carbon emissions. In this study, we evaluated the effects of annual cropland changes (expansion and abandonment) during 2000 to 2020 on carbon storage in Northwest China by using land use data, carbon density data, and statistical yearbooks using the Intergovernmental Panel on Climate Change (IPCC) method. The results indicated that the area of cropland increased by 1.47×10^6 ha from 2000 to 2020, in that the area of cropland expansion and abandonment are 3.58×10^6 and -2.11×10^6 ha, respectively. Cropland expansion was mainly from other land and grassland, and the conversion of cropland to grassland made up the largest proportion of cropland abandonment, followed by built-up land. The cropland changes resulted in a total carbon sequestration of 4.05 Tg (0.20 Tg C year^{-1}), including a 17.66 Tg decrease and 21.71 Tg increase in carbon storage due to, respectively, cropland expansion and cropland abandonment, in which the conversion of forest to cropland (−8.60 Tg) and cropland to forest (11.16 Tg) were the main causes of the increase and decrease in carbon storage. Specifically, regional carbon storage due to cropland changes exhibited an increasing variation characteristic during 2000 to 2007, a gradually decreasing variation characteristics during 2007 to 2014, and fluctuated stabilization since then (during 2014 to 2020). In addition, the highest carbon emission was found in Xinjiang (−3.68 Tg), followed by Ningxia (−0.21 Tg) province, while Shanxi (3.44 Tg), Gansu (3.17 Tg) and Qinhai (1.33 Tg) had carbon accumulation. Overall, cropland changes acted as a carbon sink in Northwest China from 2000 to 2020. We suggest that the development of high-carbon-density lands or the conversion of low-carbon-density lands are critical to increasing future carbon sequestration due to cropland change.

Keywords: abandonment; area; carbon sequestration; expansion

Citation: Kong, J.; Chen, L. The Changes in Cropland Pattern Enhanced Carbon Storage in Northwest China. *Agronomy* **2023**, *13*, 2736. https://doi.org/10.3390/agronomy13112736

Academic Editors: António Dinis Ferreira, Raquel P. F. Guiné and António Moitinho Rodrigues

Received: 8 October 2023
Revised: 25 October 2023
Accepted: 25 October 2023
Published: 30 October 2023

1. Introduction

Land-use and cover change (LUCC), as the direct driver of human activities disturbing natural ecosystems [1], is one of the critical sources of greenhouse gas emissions to the atmosphere, with direct or potential impacts on global climate [2]. It is estimated that LUCC lead to 145 ± 16 Pg C of carbon emissions between 1850 and 2020 and accounts for approximately 10% of all anthropogenic carbon emissions [3]. The change of agricultural land area is the most widespread form of LUCC [4]. It is reported that global cropland expansion accelerated over the past two decades, with a near doubling of the annual expansion rate [5]. Furthermore, cropland abandonment has become a universal phenomenon of the economic development in the world [6]. It can clearly be seen the cropland patterns have changed substantially, and these changes consequently affected the carbon balance in terrestrial ecosystems [7,8]. For example, it is reported that greenhouse gas emissions due to cropland expansion accounted for about 25% of global greenhouse gas emissions [9].

Over the last 20 years in China, large areas of cropland have been lost due to urban expansion because of rapid industrialization and urbanization that began in 1980s [10,11]. Further, agricultural expansion claimed additional areas, including those in arid regions, due to the increasing demand for food [12]. The effects of this later expansion on carbon emissions cannot be ignored. It estimated that the total carbon emissions from cropland expansion in China ranged from 2.94 to 5.61 $\times$ 10^3 Tg during the past 300 years [13]. However, China's carbon emissions have risen sharply with rapid industrialization over the past 30 years, making it the world's largest emitter of CO_2 [14,15]. At present, China is facing global pressure to reduce carbon emissions, and it pledged to strive for the reversal of increasing carbon emissions by approximately the year 2030. Therefore, it is of critical importance that the impact of cropland changes over the past 20 years or so on carbon stocks in terrestrial ecosystems is clearly determined, as it will serve as the baseline for the future optimization of the agricultural land-use structure that will be used to relieve the present pressure of carbon emissions.

Uncertainty characterizes current estimates of impacts of cropland changes on carbon budgets for distinct types of conversions alike, due to the diversity of associated environmental and anthropogenic factors [16,17]. For example, conversion from forest and grassland to cropland often leads to a major loss of carbon because carbon density decreases [18,19]. Nevertheless, there are also rare cases in which converting natural land to cropland can actually increase carbon storage if the productivity of the ensuing cropland is substantially greater than that of the prior natural land cover. For example, the conversion from sparsely-covered grassland (e.g., desert grassland) to cropland may lead to an increase in carbon pool [20]. However, the current literatures to date tends to concentrate exclusively on the effects of a single type of cropland expansion or abandonment on carbon budgets. Additionally, research to assess the effects of LUCC on carbon budgets at varying scales has resulted in the development of some commonly used methods [3,21]. At regional or global spatial scales, empirical statistical models (e.g., bookkeeping), remote sensing models (e.g., CASA) or process-based ecosystem models (e.g., TEM and LPJ) are usually used for evaluating the effects of LUCC on carbon budgets [22–25]. However, the accuracy of the representation of both the temporal evolution and spatial heterogeneity of carbon storage with modeling approaches is limited by the availability of land use data, and most of the estimates of carbon storage are based on long time intervals rather than annual intervals [22]. Hence, more precise and annual information on LUCC is needed to analyze the annual temporal evolution of carbon budgets.

Northwest China, characterized by an arid and semi-arid climate, is known for its long history of irrigation dependence [26]. Over the past 20 years, the use of agricultural water and soil resource reached unprecedented levels in Northwest China to accommodate population growth and the continuous expansion of urbanization [27,28]. For instance, the process of oasis development was greatly promoted, resulting in the expansion of artificial oasis areas (cropland landscape with large-scale desert background) from 2.1 $\times$ 10^5 to 10.4 $\times$ 10^5 ha [29]. However, carbon budget estimates of land cover change in arid and semi-arid areas at regional scales are still under-represented in these efforts. Therefore, in this study, LUCC data from 2000, 2010 and 2020 were selected due to the cropland transition phenomenon having been very apparent in Northwest China over the last 20 years or so. Land use data and annual statistical yearbook data were combined to quantify temporal and spatial dynamics in cropland expansion and abandonment throughout Northwest China. Then, we calculated annual carbon storage induced by cropland expansion and abandonment from 2000 to 2020 by matching different vegetation carbon density and soil carbon density of soil types to the annual area of cropland conversion. The aim of the present research was to analyze annual temporal evolution and spatial variability in carbon storage induced by cropland expansion and abandonment in Northwest China, and provide an example to calculate land-use change data year by year when there are no more time-frequency land use data available. We hypothesized that the cropland biomass carbon after cropland expansion was zero since crops are harvested each year as their carbon is quickly returned to the atmosphere via oxidation (burning or decomposition), which does not represent a permanent C stock [8], and the changes in cropland pattern would therefore result in a carbon sequestration. The results of this study can provide a reference for rational land-use management based on assessment of annual regional carbon storage in areas that may have previously been overlooked, which is conducive to stable and sustainable development in arid and semi-arid regions.

2. Materials and Methods

The overall approach used in this study consisted of two parts. The first part includes land use data and provincial statistical yearbooks for calculating annual area of cropland change, and mapping spatial distribution. The second part involves vegetation and soil carbon density data combined with the area of annual change in cropland for calculating carbon storage. The structure of this study is shown in Figure 1.

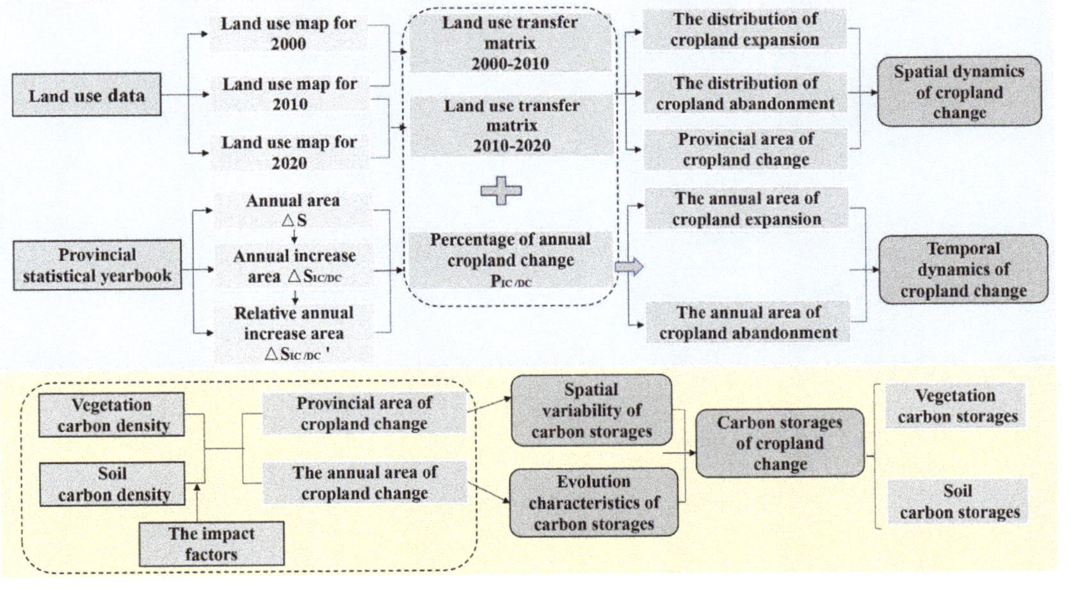

Figure 1. Research framework for calculations of carbon storage induced by cropland changes.

2.1. Study Area

This study was performed in Northwest China (73°15′ E–111°15′ E, 31°32′ N–49°10′ N), located in the innermost part of the Eurasian continent [30]. Administrative divisions in this area include Xinjiang, Qinghai, Gansu, Ningxia and Shanxi provinces, accounting for 32.2% of the total land area of China, and an area of approximately 3.10 million km^2 (Figure 2). In Northwest China, there are four climate types from south to north: subtropical monsoon climate, temperate monsoon climate, temperate continental arid climate and plateau mountain climate. Most of the area exhibits a typical continental climate, with very low mean annual precipitation (below 250 mm), mean annual temperature from −2 to 19 °C [31], and annual evapotranspiration ranging from 225 to 285 mm [32]. The main land cover types in this region consist of temperate evergreen forest (i.e., *Picea* spp., *Abies sibirica*, etc.), temperate deciduous forest (i.e., *Populus*, *Betula*, etc.), temperate shrub land (i.e., *Haloxylon*, etc.), grassland (i.e., alpine meadow and desert steppe), cropland (i.e., oasis, etc.), built-up lands, bare land and basins [33]. The natural landscape changes from east to west from forest and typical grassland, to a grassland-desert and desert; vegetation cover shows a gradually decreasing trend. Most of the rivers are inland rivers, including the

Tarim River, Irtysh River, Ili River, Manas River, Heihe, etc., of which the Tarim River is the largest inland river in China. The development of agriculture is mainly for irrigated agriculture with the famous Hetao Plain, Ningxia Plain, and Hexi Corridor, etc.

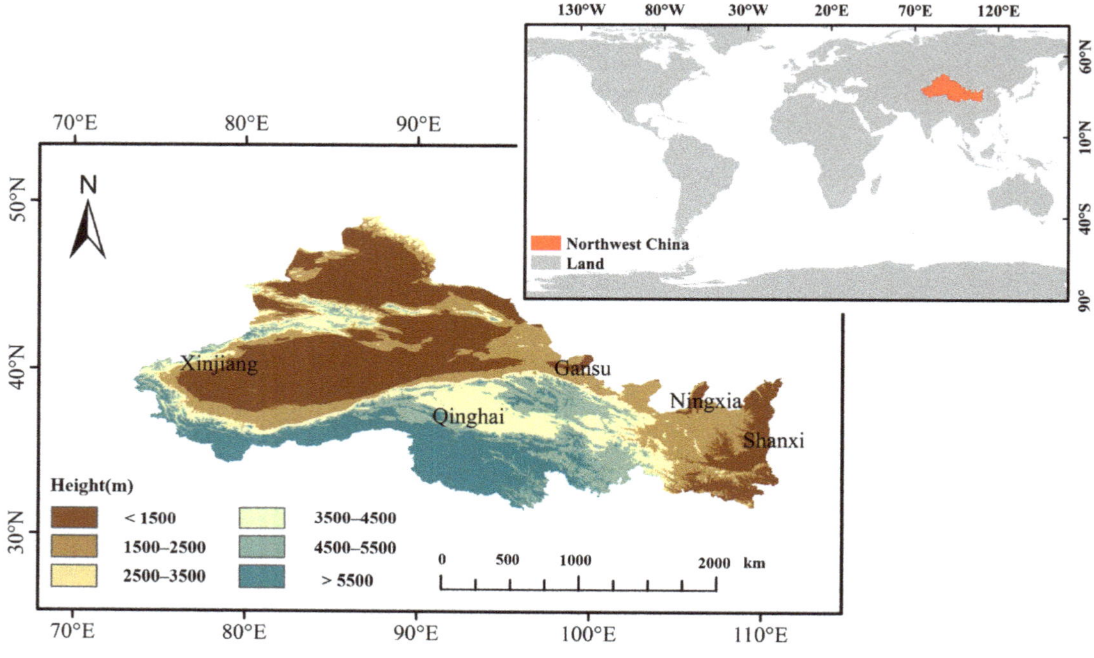

Figure 2. Location of Northwest China and its administrative divisions.

2.2. Data Sets and Pre-Processing

2.2.1. Land-Use Category and Area

We used a land-use dataset for 2000, 2010 and 2020, derived from the Land Cover (LC) project of the European Space Agency (ESA) Climate Change Initiative (CCI) Climate Research Data Package (CRDP) (https://www.esa-landcover-cci.org/ (accessed on 1 January 2022)), with a spatial resolution of 300 m. The Land Cover Classification System (LCCS) proposed by the Food and Agriculture Organization (FAO) of the United Nations was adopted. The land-use dataset was used as follows: (1) The change is detected between CCI land cover classes (the original 22 land-use types) grouped into the six IPCC land categories according to Land Cover CCI product user guide: cropland, forest, grassland, water, built-up land, and other land (see Table S1). (2) Two land-cover transition matrixes were derived from three periods (2000, 2010 and 2020) of land-use images (Figure S1) for the purpose of calculating the area of cropland expansion and abandonment during 2000 to 2010 and 2010 to 2020. (3) The spatial analysis tool "overlay" in Arcgis 10.2 was used to visualize and analyze the distribution of cropland expansion and abandonment for the years 2000 to 2010 and 2010 to 2020. (4) Accuracy assessment was quantitatively described by sub-pixel fractional error matrices and Kappa coefficient. In this study, 600 sample points were randomly selected in the study area and superimposed with the images in Google Earth Pro to verify the properties of land-use types and establish accuracy assessment in Northwest China. For 2000, 2010 and 2020, the accuracy of the main types of cropland in Northwest China is 83.2%, 85.1% and 84.6%, respectively, and the overall accuracy is 70.2%, 71.1% and 70.9%, respectively, while the Kappa coefficient is 0.753, 0.769 and 0.786, respectively, which meet the requirements of this study.

2.2.2. Annual Area of Cropland Change

Land use data acquired from remote sensing images can reflect the spatial and temporal patterns of LUCC, but not the annual area of LUCC. Therefore, in this study, the annual provincial statistical yearbooks (http://data.cnki.net/ (accessed on 1 January 2022); see Table S2) were combined with remote sensing data with high spatial resolution to obtain the annual cropland reclamation and transfer area in five northwestern provinces from 2000 to 2020.

The annual cropland expansion and abandonment area (annual cropland expansion area was defined as a positive value; annual abandonment area was defined as negative values) in the study period was calculated using the following equations:

$$\Delta S = (S_1, S_2, S_3 \ldots, S_n) \tag{1}$$

where ΔS is annual cropland area matrixes, S_i is annual cropland area (ha), and $i = 1 \ldots n$.

$$\Delta S_{IC} = (\Delta S_1, \Delta S_2, \Delta S_3 \ldots, \Delta S_{n-1}) \tag{2}$$

where ΔS_{IC} is the annual increase in cropland area due to expansion, $\Delta S_{IC_i} = \Delta S_{i+1} - \Delta S_i$ (ha), and $i = 1 \ldots n - 1$. Considering that ΔS_{IC} may be <0, and annual increase in cropland area was defined as positive values, it was necessary to smooth and revise the annual increase in cropland:

$$\Delta S_{IC}' = (\Delta S_1 + |\Delta S_{IC\min}|, \Delta S_2 + |\Delta S_{IC\min}|, \Delta S_3 + |\Delta S_{IC\min}| \ldots, \Delta S_{n-1} + |\Delta S_{IC\min}|) \tag{3}$$

$$P_{IC} = (\Delta S_{IC_1}'/sum(\Delta S_{IC}'), \Delta S_{IC_2}'/sum(\Delta S_{IC}') \ldots, \Delta S_{IC_{n-1}}'/sum(\Delta S_{IC}')) \tag{4}$$

where P_{IC} (%) is the coefficient of annual increase in cropland area due to expansion. The calculation method of P_{DC} (%) (the coefficient of annual decrease in cropland area due to abandonment) is similar to P_{IC} (%), except that the Formula (2) is changed to the following:

$$\Delta S_{DC} = (\Delta S_1, \Delta S_2, \Delta S_3 \ldots, \Delta S_{n-1}) \tag{5}$$

where ΔS_{DC} is the annual decrease in cropland area due to abandonment, and $\Delta S_{DC_i} = \Delta S_i - \Delta S_{i+1}$ (ha). The subsequent calculation method is the same as in Formulas (3) and (4).

The annual area of cropland expansion or abandonment (ha) was obtained by multiplying P_{IC} (%) or P_{DC} (%) by cropland expansion or abandonment area (ha) from two land cover transition matrixes for the years 2000 to 2010 and 2010 to 2020. The annual area of different types of cropland conversion was obtained by multiplying P_{IC} (%) or P_{DC} (%) by different types of cropland conversion area (ha) from two land cover transition matrixes for the years 2000 to 2010 and 2010 to 2020.

2.3. Calculation of Changes in Carbon Storage Induced by Cropland Change

Carbon storage induced by cropland change was calculated with the following equation [34]:

$$\Delta C = \Delta VC + \Delta SOC \tag{6}$$

where ΔC (Tg) represents the change in carbon storage caused by cropland change; ΔVC (Tg) represents change in biomass carbon storage; ΔSOC (Tg) represents the change in soil organic carbon (SOC).

2.3.1. Calculation of Change in Biomass Carbon Storage

We obtained vegetation carbon density information for each land-use type (Table 1) from published literature [35]. Changes in biomass carbon storage caused by cropland change was calculated with the following formula [34], and belowground biomass was not included in the calculation:

$$\Delta VC = \sum_{1}^{i} [(VD_{Afteri} - VD_{Beforei}) \times \Delta A_{To-otheri}] \quad (7)$$

where VD_{Afteri} and $VD_{Beforei}$ (t C ha^{-1}) represent carbon density in vegetation for land use i after and before the conversion, and $\Delta A_{To-otheri}$ (ha) represents the area of land use i converted to another land type.

Table 1. Vegetation carbon density in each land cover type.

Land-Use Type	Forest	Grassland	Cropland	Water	Built-Up Land	Other Land
vegetation carbon density (t C ha^{-1})	79.22	3.46	5.70	0	0	0.55

2.3.2. Calculation of Changes in Soil Carbon Storage

For analytical purposes all land in a given stratum should have common biophysical conditions (e.g., soil type). We obtained the soil carbon density of each soil type from the 1:1,000,000 soil-type map of the China Second National Soil Survey (Table S3). Based on soil carbon density and the impact factors for soil carbon change [36] (Table 2), we applied the Tier 1 method from IPCC (2006) to calculate soil carbon storage caused by cropland change using the following formula [34]:

$$\Delta SOC = \sum_{i,s} (SD_{i,s} \times F_{impacti,s} \times \Delta A_{to-otheri,s}) \quad (8)$$

where $SD_{i,s}$ represents soil carbon density for land-use type i with soil type s; $\Delta A_{To-otheri,s}$ represents the transformed area of land-use type i with soil type s; and $F_{impacti,s}$ represents the impact factors of SOC change during cropland change (Table 2) [36].

Table 2. The impact factors of SOC change in cropland change.

Items	Forest	Grassland	Other Land	Cropland
Forest	-	-		−27%
Grassland	-	-		−20%
Other land				80%
Cropland	90%	100%	−20%	-

3. Results

3.1. Spatio-Temporal Dynamics of Cropland Change

3.1.1. Temporal Dynamics of Cropland Change

Northwest China has experienced continuous cropland changes during the period from 2000 to 2020 (Table 3). Overall, the total area of cropland in Northwest China increased by 1.47×10^6 ha (4.2%) between 2000 and 2020, in that the area of cropland expansion and abandonment increased by 3.58×10^6 and 2.11×10^6 ha, respectively, from 2000 to 2020. Separately, the net increase in cropland area from 2010 to 2020 was 1.01×10^5 ha more than in the previous period. However, the area of cropland expansion and abandonment decreased by 29.33% and 50.79%, respectively, from 2010 to 2020. Cropland expansion was mainly from other land and grassland, contributing, respectively 56.68 and 39.88% of the reclaimed cropland. Meanwhile, the conversion of cropland to grassland made up the largest proportion of cropland abandonment, followed by built-up land, then other land. The three LUCC, respectively accounted for 67.44, 14.86 and 10.58% of the area of cropland abandonment. Except for the increase in the conversion of built-up land to cropland (slightly) and cropland to built-up land, the conversion of other land uses showed a decreasing trend from 2010 to 2020 over the previous period.

Table 3. Changes in cropland area and type in Northwest China for different time periods ($\times 10^2$ ha).

Type	2000–2005	Proportion (%)	2005–2020	Proportion (%)	2000–2020	Proportion (%)
Forest to cropland	656.15	3.13	408.85	2.76	1065.00	2.97
Grassland to cropland	8527.23	40.63	5757.72	38.82	14,284.95	39.88
Water to cropland	64.20	0.31	39.12	0.26	103.32	0.29
Built-up land to cropland	12.51	0.06	50.35	0.34	62.86	0.18
Other land to cropland	11,725.91	55.87	8575.27	57.82	20,301.18	56.68
Cropland expansion	20,986.00	100.00	14,831.31	100.00	35,817.31	100.00
Cropland to forest	1125.01	7.97	291.65	4.20	1416.66	6.73
Cropland to grassland	10,144.49	71.88	4056.76	58.42	14,201.25	67.44
Cropland to water	47.83	0.34	33.78	0.49	81.61	0.39
Cropland to built-up land	1064.67	7.54	2064.12	29.72	3128.79	14.86
Cropland to other land	1730.52	12.26	498.38	7.18	2228.90	10.58
Cropland abandonment	14,112.53	100.00	6944.70	100.00	21,057.23	100.00
Net increase	6873.48		7886.61		14,760.08	

The trend in cropland expansion was contrary to the trend in cropland abandonment (Figure 3). The area of cropland expansion exhibited a downward trend from 2000 to 2007 (-1.68×10^5 ha), then a gradual increase from 2007 to 2014, with a peak in 2014 (6.71×10^5 ha). Since then, the area of cropland expansion has been on a downward trend. However, the area of cropland abandonment gradually increased from 2000 to 2007 (2.12×10^5 ha), then gradually decreased after 2007, and reached an all-time low in 2014 (1.32×10^4 ha). Since then, the area of cropland abandonment fluctuated. The trend in cropland expansion was consistent with the trend in other land, grassland and forest conversion, while the trend in cropland abandonment was consistent with the trend of cropland to grassland from 2000 to 2020.

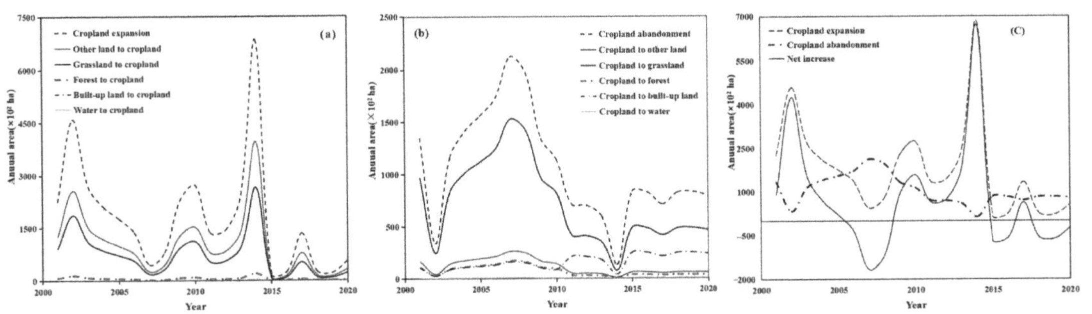

Figure 3. Temporal dynamics of cropland expansion and abandonment. (**a**) annual area of cropland expansion; (**b**) annual area of cropland abandonment; (**c**) annual area of net increase.

3.1.2. Spatial Variability in Cropland Change

Cropland change across Northwest China exhibited spatial variability (Figure 4). During the period of 2000–2020, the cropland area in Xinjiang, Ningxia and Qinghai provinces increased by 2.10×10^6, 6.93×10^4 and 1.50×10^4 ha, respectively, while that in Shanxi and Gansu provinces decreased by 3.20×10^5 and 4.09×10^5 ha, respectively. Specifically, Qinghai province exhibited an increase (6.22×10^4 ha) in the early period, and then a decrease (-4.72×10^4 ha) in the later period, while trends in Shanxi and Gansu provinces continued to show a steadily decreasing trend, and Xinjiang and Ningxia provinces continued to show an increasing trend between 2010 to 2020.

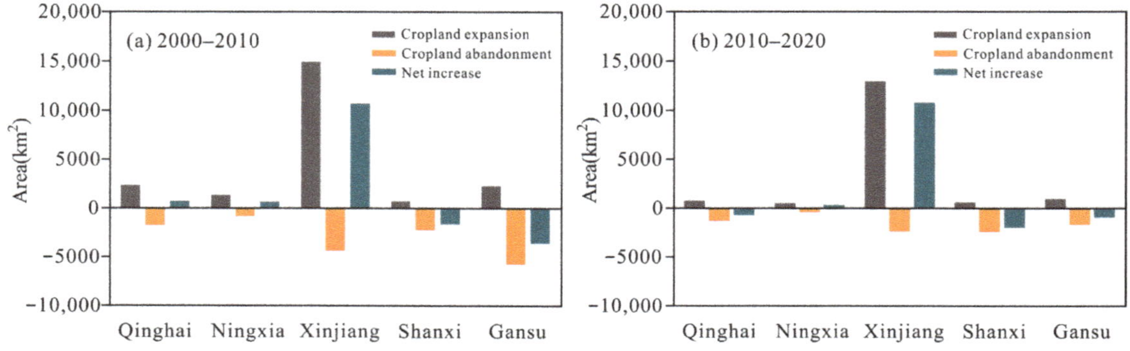

Figure 4. Provincial area of cropland changes from 2000 to 2010 (**a**), 2011 to 2020 (**b**).

The spatial distribution of cropland change was similar for both periods (Figure 5). The most dramatic changes took place in Xinjiang province, especially in its northern part. Cropland change in Qinghai province, which is mainly covered by grassland, occurred mainly in areas of that cover type. Additionally, cropland changes were notable in the northern part of Ningxia and Shanxi provinces in both periods. Cropland change in Gansu province occurred mainly in the southeastern part.

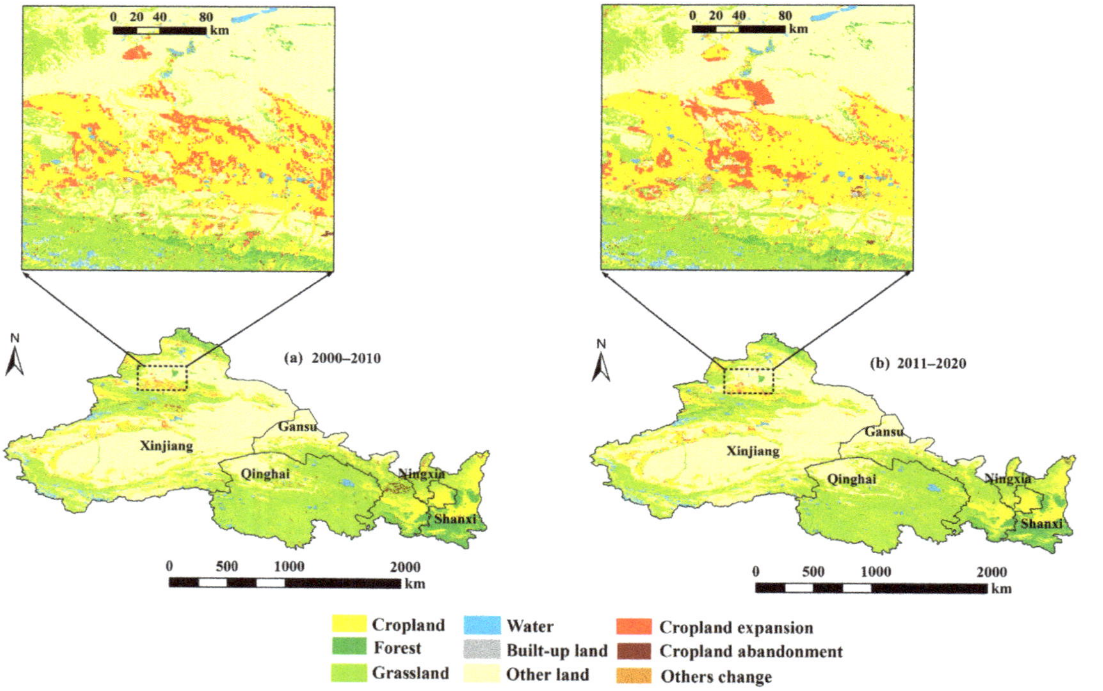

Figure 5. Spatial distribution of cropland change from 2000 to 2010 (**a**), 2011 to 2020 (**b**). The red fields represent cropland expansion and abandonment, the last fields represent other types of land-use change. Others represent the lands remaining unchanged.

3.2. *Effects of Cropland Change on Carbon Storage*

3.2.1. Changes in Carbon Storage over Time

Calculated changes in carbon storage suggested that during 2000 to 2020, cropland changes led to about 4.05 Tg (2.46 Tg in 2000–2010 and 5.49 Tg in 2010–2020) of total carbon accumulation, including a 2.28 and 1.76 Tg increase due to, respectively, soil and vegetation carbon storage, corresponding to an increase in storage rates of approximately 0.20 Tg yr^{-1}. Meanwhile, carbon storage due to cropland expansion decreased by 17.66 Tg from 2000 to 2020, which corresponds to a decrease of 3.16 Tg in soil and 14.50 Tg in vegetation. Moreover, carbon storage induced by cropland abandonment increased by 21.71 Tg from 2000 to 2020, in which carbon storage in soil and vegetation increased by 5.45 and 16.26 Tg, respectively.

Overall, variation characteristics in carbon storage caused by cropland change over time in Northwest China was affected by carbon storage caused by cropland abandonment and expansion over time (Figure 6). Overall, the carbon storage caused by cropland changes showed an increasing variation characteristic from 2000 to 2007, reached a maximum in 2007 (2.22 Tg), gradually decreased after 2007, reached a minimum in 2014 (−3.09 Tg), and fluctuated since then. Specifically, carbon storage caused by cropland expansion and abandonment over time both exhibited an increasing variation characteristic from 2000 to 2007 (−0.23 and 2.46 Tg), a gradual decrease from 2007 to 2014, a minimum in 2014 (−3.19 and 0.10 Tg), and a gradual increase with fluctuations since then (2014–2020).

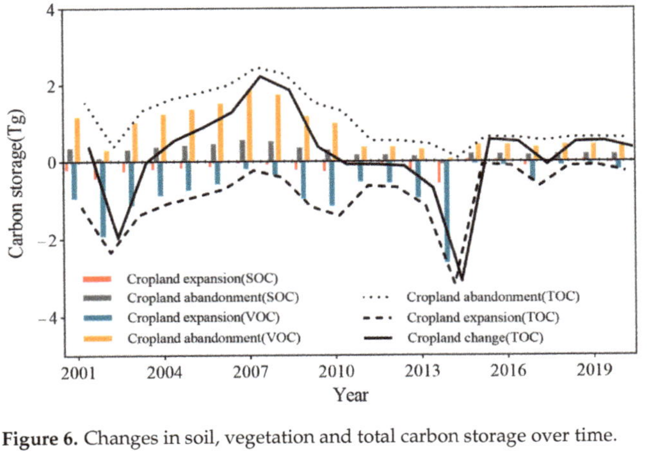

Figure 6. Changes in soil, vegetation and total carbon storage over time.

3.2.2. Carbon Storage in Different Types of Cropland Conversion

The effects on carbon storage differed across different types of cropland expansion in Northwest China (Figure 7a). The decrease in carbon storage was mainly from forest, grassland and other land expansion. Specifically, the conversion of forest to cropland had the greatest impact, a 8.60 Tg decrease in carbon storage, accounting for 48.73% of the decrease in carbon storage due to cropland expansion. Carbon storage induced by the conversion of grassland and other land to cropland decreased by 6.40 and 2.65 Tg, accounting for 36.25% and 14.90% of the decrease in carbon storage due to cropland expansion.

Effects on carbon storage differed across different types of cropland abandonment in Northwest China (Figure 7b). The increase in carbon storage was mainly from the conversion of cropland to forest and grassland; the conversion of cropland to built-up land and other land led to a decrease in carbon storage. Specifically, the conversion of cropland to forest had the greatest impact, with an 11.16 Tg increase in carbon storage, accounting for 50.48% of the increase in carbon storage caused by cropland abandonment. Carbon storage changed upon the abandonment of cropland to grassland, other land, or built-up land by 10.74; −0.76; and −0.24 Tg, respectively, accounting for 62.33; 3.2; and 1.01% of the increase in carbon storage caused by cropland abandonment.

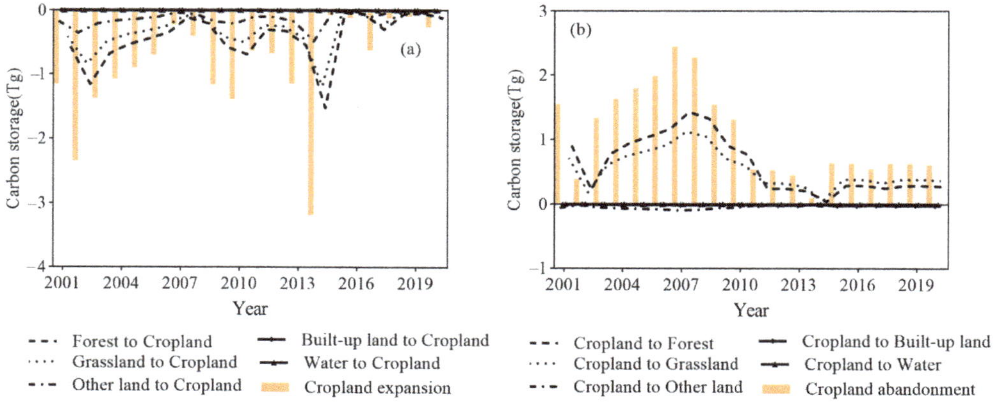

Figure 7. Changes in carbon storage due to different cropland conversion types. (**a**) changes in carbon storage in cropland expansion; (**b**) changes in carbon storage in cropland abandonment.

3.2.3. Spatial Variability in Carbon Storage

The changes in area and carbon storage caused by cropland changes were relatively consistent in spatial characteristics for both periods of study (Figure 8). In general, the highest carbon emission was found in Xinjiang (−3.68 Tg), followed by Ningxia (−0.21 Tg) province, while Shanxi (3.44 Tg), Gansu (3.17 Tg) and Qinhai (1.33 Tg) had carbon accumulation. Specifically, the largest decrease in carbon storage in Xinjiang province was −0.69 Tg from 2000 to 2010, and −2.99 Tg from 2010 to 2020. The smallest amount of carbon emissions occurred in Ningxia province with a decrease of 0.13 Tg from 2000 to 2010, and 0.08 Tg from 2010 to 2020. Additionally, vegetation carbon storage in Shanxi and Xinjiang provinces increased in the first period (3.06 and 0.01 Tg) and decreased in the second period (−0.09 and −2.28 Tg).

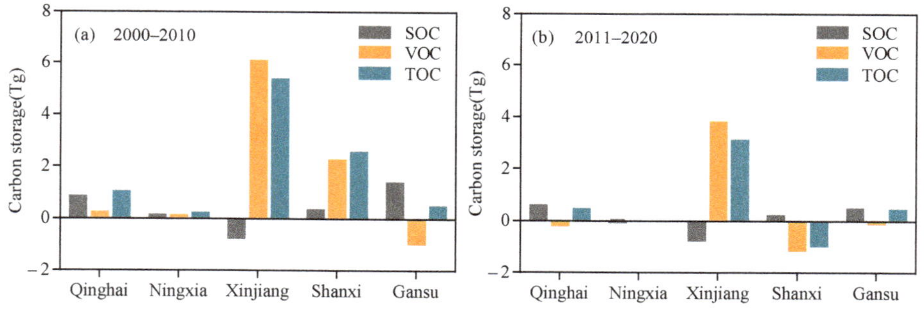

Figure 8. Carbon storage due to cropland change in each province from 2000 to 2010 (**a**), and 2011 to 2020 (**b**).

4. Discussion

4.1. Characteristics of Cropland Change from 2000 to 2020

Over the last several decades, the area and distribution of cropland in Northwest China has changed because of social and economic transitions (e.g., urbanization, population growth, etc.) [25,26]. Our results indicated that the area of cropland in Northwest China increased by 1.48 × 10^6 ha during the period of 2000–2020, with a mean increase of 7.4×10^4 ha per year. However, expansion and abandonment resulted in fluctuations in cropland area in different periods. During the first period from 2000 to 2007, cropland was mostly abandoned. With rapid urbanization and the expiration of new land contract periods, cropland occupation became more prominent, leading to land encroachment by urbanization. Meanwhile, farmers abandon land-use rights to other farmers or economic organizations, also contributing to a decrease in cropland area [26,37]. During the second period, from 2007 to 2014, cropland was reclaimed at a large-scale. The Chinese government successively established the strict boundaries for cropland (120 million ha), and increased the protection of cropland. At the same time, advancements in technologies (e.g., sprinkler and drip irrigation) improved the utilization efficiency of water and soil resources [38], enabling a gradual increase in cropland area from 2007 to 2014. Finally, with the improvement of cropland protection policy and ecological engineering programs (promoting the conversion of cropland to forest and to grassland) [39], the change in cropland area stabilized when cropland expansion and abandonment reached an equilibrium from 2014 to 2020. Considering the role of socio-economic factors in different stages of cropland change together with a numerical evaluation can help determine the change characteristics of annual cropland area. In summary, agricultural land development in Northwest China was likely to increase and be affected by socio-economic factors during 2000 to 2020.

Even though the area of cropland land has steadily increased over time due to the balance of cropland expansion and abandonment, the occupation of cropland by built-up land still showed an increasing trend in 2010–2020 over the previous period in our study. As we know, cropland abandonment driven by rapid urbanization is an irreversible trend across the globe. At the global scale, Huang et al. (2020) showed that the global urban expansion occupied a total of 159,170 km^2 of cropland, in which China witnessed the largest cropland losses from urban expansion, accounting for 15.5‰ of the total cropland area from 1992 to 2016 (about 0.65‰/a; our study was 0.43‰/a) [12]. At the national scale, Liu et al. (2019) also found that croplands were the primary contributor to urban expansion in China since the 1970s [40], and Ju et al. (2018) found that 42,822 km^2 of cropland was converted to built-up land in China, accounting for 43.8% of total cropland area loss during 1987 to 2010 (about 1.9%/a; our study was 0.74%/a) [11]. The above illustrates that compared to the expansion of cities on a national scale in China, urban expansion occupies only a small portion of cropland area in Northwest China in the present. However, the increasing occupation of cropland by urban expansion may increase the vulnerability of food security in Northwest China due to the importance of cropland resource for food production [41]. In the future, Northwest China should balance urban expansion with cropland protection by severely restricting the occupation of cropland.

4.2. Effects of Cropland Change on Carbon Storage

In arid and semiarid regions, cropland change (including expansion and abandonment) has been shown to have an important effect on carbon storage in both the biosphere and the pedosphere [42]. Our results demonstrated that cropland change from 2000 to 2020 resulted in a cumulative carbon sequestration of 4.05 Tg or 0.20 Tg yr^{-1}, of which 2.28 Tg and 1.76 Tg were in soil and vegetation, respectively. This is consistent with other studies that have demonstrated that Northwest China is a sink for carbon as a result of cropland change [27].

Carbon storage caused by cropland expansion decreased by 17.66 Tg (0.88 Tg yr^{-1}/ 0.24×10^{-6} Tg ha^{-1} yr^{-1}) from 2000 to 2020. Similar results also have been found in many studies regarding the considerable losses of carbon storage caused by cropland expansion [13]. Globally, cropland is predicted to expand by 21% during 2010–2050 [43]. Cropland expansion is expected to lead to biomass and soil carbon emission of 13.7% and 4.6% during 2010–2050, respectively [7]. Of course, the loss of total carbon storage caused by cropland expansion is noticeable at national scales. For example, in China, cropland expansion was shown to result in annual carbon emissions of approximately 5.04×10^{-5} Tg ha^{-1} yr^{-1} [44]. In the United States, cropland expansion resulted in total carbon emissions of 1.38×10^{-5} TgC ha^{-1} yr^{-1} [8]. At regional scales, carbon emissions caused by cropland expansion in Hubei China were 3.31×10^{-6} TgC ha^{-1} yr^{-1} [45]. These indicated that carbon emissions per unit area caused by cropland expansion in Northwest China are lower than those in previous studies in different scales. In arid and semi-arid area, water resources are the main natural factors restricting the development of agriculture [28]. Crop irrigation presumably requires ground- or surface-water pumping, which entails additional fossil carbon emissions, and these emissions should be attributed to cropland given their dependence [46]. Also, the impact of other agricultural management, such as tillage and fertilization, on carbon emissions is not calculated in our study. Future improvements could reduce the uncertainty associated with this part of the carbon emission and facilitate more accurate assessments.

Carbon emission caused by cropland expansion in Northwest China was due mainly to the conversion of forest into cropland. The conversion of forest with high carbon density to cropland has also resulted in a major decline in carbon storage of about 8.60 Tg, accounting for 48.73% of the decrease in carbon storage due to cropland expansion. The loss of forest due to cropland expansion is noticeable not only in Northwest China but worldwide. It is reported that deforestation largely driven by cropland expansion have been the second largest source of anthropogenic greenhouse gas emissions globally [47]. Thus, if measures are taken to control deforestation in specified regions, the rate of carbon loss could be reduced. Meanwhile, carbon storage induced by the conversion of grassland to cropland decreased by 6.40 Tg, accounting for 36.25% of the decrease in carbon storage due to cropland expansion. Many studies have also shown that grassland conversion to cropland can cause sizable carbon emissions although it has often received less attention than deforestation [8,20]. This estimate may reflect the transition trend of relatively carbon-rich grassland or sensitive grassland. Although the carbon intensity is lower than that of forest conversion to cropland, the carbon loss caused by extensive grassland conversion to cropland in Northwest China cannot be ignored. In addition, desert-grassland, sandy and other lands occupy a large fraction of the area of Northwest China; with their low carbon density, these land types were reclaimed into cropland, changing the carbon cycle of desert ecosystem [48,49]. Thus, the carbon loss caused by large areas of other lands conversion in Northwest China also cannot be ignored. Of course, the new Chinese Environmental Protection Law proposed by the Chinese government has also emphasized the conservation of forests, grasslands and other natural ecosystems [50]. Over all, we think that further land-use policies to effectively support reasonable and restricted cropland expansion (such as deforestation, grassland and other land reclamation) will be able to potentially relief pressure on carbon emission caused by cropland expansion.

Meanwhile, it is worth noting that 3.13×10^5 ha of cropland was occupied by built-up area, with a loss of approximately 0.24 Tg (0.012 Tg yr^{-1}) of carbon storage. A preliminary estimate suggested that the occupied cropland was the source of carbon storage loss during the process of built-up land expansion [51]. However, the indirect emission effects of urbanization (such as waste products, population migration and land degradation) should attract our attention more than the variability in carbon storage caused by the conversion of built-up land into cropland [52]. These indirect emissions are likely to increase the uncertainty of carbon emissions. In terms of this issue, the New Urbanization policy proposed by the Chinese government highlighted the reduction in natural disturbance and the promotion of the reasonable development of land use [53]. Therefore, measures should be taken to balance urban expansion with the cropland protection policy in Northwest China, as the rate of carbon emission could be slowed in the future.

In addition, only 1.41×10^5 ha of cropland was converted to forest, becoming the main source of carbon storage (11.16 Tg) in cropland change. These changes can be attributed to the implementation of a series of ecological engineering programs by the Chinese government in Northwest China, including the Natural Forest Conservation Program and Grain for Green Program. These programs promoted the conversion of cropland to forest and to grassland, and significantly affected carbon storage [37,54]. This indicates that the implementation of ecological engineering programs in Northwest China can promote the development of land cover types with high carbon density (forest and grassland) increasing regional carbon storage [55]. Undoubtedly, ecological engineering can play an important role in efforts to address environmental crises, improve human well-being, and achieve sustainability, and these effects will gradually expand as investment in ecological engineering continues to grow [37]. Above all, we propose that the capacity for carbon sequestration in our study area will benefit from the optimization of land-use structure with land-use policies (i.e., ecological engineering programs). This can be accomplished especially by increasing the area of ecologically valuable land with high capacity for carbon storage such as forest, limiting or decreasing deforestation, and restricting the abandonment of cropland to low-carbon-density land use, such as built-up land. Thus, we should also focus on the trade-off relationship between cropland protection and ecological construction.

4.3. Strengths and Limitations of This Study

This study used a novel approach to investigate carbon storage changes caused by cropland change (cropland expansion and abandonment) in Northwest China by combining land use data, carbon density data, and statistical yearbooks between 2000 and 2020. This is the first attempt to concentrate on the effect of both cropland expansion and cropland abandonment on the carbon storage in Northwest China. Compared to previous studies, this study mainly focused on annual the evaluation of carbon storage caused by annual cropland expansion or abandonment, a few arbitrary time points of land use data (2000, 2010 and 2020) combined with a series of provincial statistical yearbooks to obtain the annual cropland expansion and abandonment area in five northwestern provinces from 2000 to 2020, which provide an example to calculate land-use change data year by year when there are no more time-frequency land use data available. In addition, the results of this study can provide a reference for rational land-use management based on the assessment of annual regional carbon storage in areas that may have previously been overlooked, which is conducive to stable and sustainable development in arid and semi-arid regions.

Several uncertainties in this study will be the focus in future research. First, remote sensing data with spatial detail and statistical data with temporal frequency were used to obtain the annual area of cropland expansion and abandonment; this may not accurately reflect the annual area in cropland change. Thus, obtaining high-resolution land-use data with shorter time intervals may be a more effective method. Second, this study only involved aboveground biomass, and carbon emissions brought by agricultural management, such as tillage, fertilization and irrigation, on carbon emissions are not calculated. Belowground biomass and carbon emissions brought by agricultural management should be considered in the evaluation in such regional studies. Third, the carbon density data are set at a fixed level in this study, and those adopted in this study were from published studies and not from a sampling method varies from region to region. Furthermore, the time interval of 10 years used in this study may be insufficient to detect changes and stabilization in soil carbon density due to carbon density varying from time to time [56]. For future research, it is better to consider the spatio-temporal heterogeneity of carbon density if the data are accessible [42].

More importantly, in arid and semi-arid areas, water resources are the main natural factors restricting the utilization of land resources and the development of agriculture and forestry [29]. Although the conversion of cropland to forest and grassland are the most important sources of carbon sinks in cropland changes, the potential impact of an increase in vegetation cover on water demand needs to be considered. Thus, in the future, there is a need to consider how to make cropland changes result in reasonable carbon sinks within the confines of limiting water resources. Cropland changes over the long term can be simulated based on water resource limitation scenarios, which can provide valuable information for decision making processes involved in cropland change in Northwest China.

5. Conclusions

These results indicate that cropland changes acted as a carbon sink (4.05 Tg) in Northwest China, despite the dominance of cropland expansion area, carbon sequestration from the conversion of cropland to forest (11.16 Tg, affected by ecological engineering programs) contributed to the most increase. Thus, it is essential to promote the development of the area of ecologically valuable land with a high capacity for carbon storage, such as forests, or limit the conversion of low-carbon-density lands to cropland, such as by deforestation. More importantly, this is the first attempt to evaluate inter-annul carbon storage change due to both cropland expansion and cropland abandonment by combining land use data, carbon density data, and statistical yearbooks with IPCC method in Northwest China that area may have previously been overlooked.

Supplementary Materials: The following supporting information can be downloaded at: https://www.mdpi.com/article/10.3390/agronomy13112736/s1, Figure S1: Land cover maps: (a) 2000, (b) 2010, (c) 2020; Table S1: Land-use categories; Table S2: Annual cropland area in five northwestern provinces; Table S3: SOC density of different soil types (arranged alphabetically).

Author Contributions: J.K. performed the laboratory experiments, analyzed the data, coordinated the study, participated in conceptual design and prepared the manuscript. L.C. performed most of the work for conceptual design and prepared the manuscript. All authors have read and agreed to the published version of the manuscript.

Funding: This study was funded by the Second Tibetan Plateau Scientific Expedition and Research (STEP) program (Grant No. 2019QZKK0303), the National Natural Science Foundation of China (Grant No. 32301561) and the National Office for Philosophy and Social Sciences (Grant No. 21BGL058).

Data Availability Statement: Data are included in the manuscript or will be available upon request.

Conflicts of Interest: The authors declare no conflict of interest.

References

1. Lembi, R.C.; Cronemberger, C.; Picharillo, C.; Koffler, S.; Sena, P.H.A.; Felappi, F.C.; Moraes, A.R.; Arshad, A.; Santos, J.P.; Mansur, A.V. Urban expansion in the Atlantic Forest: Applying the Nature Futures Framework to develop a conceptual model and future scenarios. *Biota Neotrop.* **2020**, *20*, e20190904. [CrossRef]
2. Friedlingstein, P.; O'sullivan, M.; Jones, M.W.; Andrew, R.M.; Zaehle, S. Global carbon budget 2020. *Earth Syst. Sci. Data* **2020**, *12*, 3269–3340.
3. Houghton, R.A.; Nassikas, A.A. Global and regional fluxes of carbon from land use and land cover change 1850–2015. *Glob. Biogeochem. Cycles* **2017**, *31*, 456–472. [CrossRef]
4. IPBES. *Global Assessment Report on Biodiversity and Ecosystem Services of the Intergovernmental Science-Policy Platform on Biodiversity and Ecosystem Services*; Brondizio, E.S., Settele, J., Díaz, S., Ngo, H.T., Eds.; IPBES Secretariat: Bonn, Germany, 2019; p. 28.
5. Potapov, P.; Turubanova, S.; Hansen, M.C.; Tyukavina, A.; Zalles, V.; Khan, A.; Song, X.P.; Pickens, A.; Shen, Q.; Cortez, J. Global maps of cropland extent and change show accelerated cropland expansion in the twenty-first century. *Nat. Food.* **2022**, *3*, 19–28. [CrossRef]
6. Huang, Y.; Li, F.; Xie, H. A scientometrics review on farmland abandonment research. *Land* **2020**, *9*, 263. [CrossRef]
7. Molotoks, A.; Stehfest, E.; Doelman, J.; Albanito, F.; Fitton, N.; Dawson, T.P.; Smith, P. Global projections of future cropland expansion to 2050 and direct impacts on biodiversity and carbon storage. *Glob. Chang. Biol.* **2018**, *24*, 5895–5908. [CrossRef] [PubMed]
8. Spawn, S.A.; Lark, T.J.; Gibbs, H.K. Carbon emissions from cropland expansion in the United States. *Environ. Res. Let.* **2019**, *14*, 045009. [CrossRef]
9. Johnson, B.G.; Zuleta, G.A. Land-use land-cover change and ecosystem loss in the Espinal ecoregion, Argentina. *Agric. Ecosyst. Environ.* **2013**, *181*, 31–40. [CrossRef]
10. Liu, F.; Zhang, Z.; Zhao, X.; Wang, X.; Zuo, L.; Wen, Q.; Ling, J.; Bin, S. Chinese cropland losses due to urban expansion in the past four decades. *Sci. Total Environ.* **2019**, *650*, 847–857. [CrossRef]
11. Ju, H.; Zhang, Z.; Zhao, X.; Wang, X.; Wu, W.; Yi, L.; Wen, Q.; Liu, F.; Xu, J.; Hu, S. The changing patterns of cropland conversion to built-up land in China from 1987 to 2010. *J. Geogr. Sci.* **2018**, *28*, 1595–1610. [CrossRef]
12. Huang, Q.; Liu, Z.; He, C.; Gou, S.; Bai, Y.; Wang, Y.; Shen, M. The occupation of cropland by global urban expansion from 1992 to 2016 and its implications. *Environ. Res. Lett.* **2020**, *15*, 084037. [CrossRef]

13. Yang, X.; Jin, X.; Xiang, X.; Fan, Y.; Liu, J.; Shan, W.; Zhou, Y. Carbon emissions induced by farmland expansion in China during the past 300 years. *Sci. China Earth Sci.* **2019**, *62*, 423–437.
14. He, Q.; Zeng, C.; Xie, P.; Liu, Y.; Zhang, M. An assessment of forest biomass carbon storage and ecological compensation based on surface area: A case study of Hubei Province, China. *Ecol. Indic.* **2018**, *90*, 392–400.
15. Shi, L.; Sun, J.; Lin, J.; Zhao, Y. Factor decomposition of carbon emissions in Chinese megacities. *J. Environ. Sci.* **2019**, *75*, 209–215. [CrossRef] [PubMed]
16. Liu, S.; Li, F.; Zhang, Q.; Li, Z.; Tian, C.; Qiao, Y.; Qiao, Y.; Du, K.; Cheng, H.; Chen, G.; et al. Soil Quality Improvement with Increasing Reclamation Years in the Yellow River Delta. *Agronomy* **2023**, *13*, 2576. [CrossRef]
17. Wang, Y.; Yang, J.; Duan, C. Research on the Spatial-Temporal Patterns of Carbon Effects and Carbon-Emission Reduction Strategies for Farmland in China. *Sustainability* **2023**, *15*, 10314.
18. Li, Z.; Tang, Q.; Wan, X.; Chen, B.; Sun, C.; Xin, X. Grassland carbon change in Northern China due to contemporary and future land use and land cover change. *Agronomy* **2023**, *13*, 2180. [CrossRef]
19. Qiu, L.; Wei, X.; Zhang, X.; Cheng, J.; Long, T. Soil organic carbon losses due to land use change in a semiarid grassland. *Plant Soil.* **2012**, *355*, 299–309.
20. Li, X.; Li, Y.; Li, F.; Ma, Q.; Zhang, Q.; Yin, P. Changes in soil organic carbon, nutrients and aggregation after conversion of native desert soil into irrigated arable land. *Soil Tillage Res.* **2009**, *104*, 263–269.
21. Fu, C.; Yu, G.; Fang, H.; Wang, Q. Effects of Land use and cover change on terrestrial carbon balance of China. *Prog. Geogr.* **2012**, *31*, 88–96. (In Chinese)
22. Oelbermann, M.; Voroney, R.P. An evaluation of the century model to predict soil organic carbon: Examples from Costa Rica and Canada. *Agroforest Syst.* **2011**, *82*, 37–50. [CrossRef]
23. Qin, Z.; Zhuang, Q.; Zhu, X. Carbon and nitrogen dynamics in bioenergy ecosystems: 1. Model development, validation and sensitivity analysis. *GCB Bioenergy* **2014**, *6*, 740–755. [CrossRef]
24. Sitch, S.B.; Prentice, I.C.; Arneth, A.; Bondeau, A.; Cramer, W.; Kaplan, J.O.; Levis, S.; Lucht, W.; Sykes, M.T.; Thonicke, K.; et al. Evaluation of ecosystem dynamics, plant geography and terrestrial carbon cycling in the LPJ dynamic global vegetation model. *Glob. Chang. Biol.* **2010**, *9*, 161–185. [CrossRef]
25. Zhao, Y.; Wang, M.; Hu, S.; Zhang, X.; Ouyang, Z.; Zhang, G.; Huang, B.; Zhao, S.; Wu, J.; Xie, D.; et al. Economics-and policy-driven organic carbon input enhancement dominates soil organic carbon accumulation in Chinese croplands. *Proc. Natl. Acad. Sci. USA* **2018**, *115*, 4045–4050. [CrossRef]
26. Zhang, Y.; Li, C.; Wang, T.; Cai, C.; Bao, Y. County-level patterns of cropland and their relationships with socio-economic factors in northwestern China. *Agric. Ecosyst. Environ.* **2015**, *203*, 11–18. [CrossRef]
27. Zhang, M.; Huang, X.J.; Chuai, X.W.; Yang, H.; Lai, L.; Tan, J.Z. Impact of land use type conversion on carbon storage in terrestrial ecosystems of China: A spatial-temporal perspective. *Sci. Rep.* **2015**, *5*, 10233. [CrossRef]
28. Wang, Y.J.; Qin, D.H. Influence of climate change and human activity on water resources in arid region of Northwest China: An overview. *Adv. Clim. Chang. Res.* **2017**, *8*, 268–278. [CrossRef]
29. Wang, T. Review and prospect of research on oasification and desertification in arid regions. *J. Desert. Res.* **2014**, *29*, 1–9. (In Chinese)
30. Pan, X.; Zhang, L.; Huang, C. Future climate projection in northwest china with regcm4.6. *Earth Space Sci.* **2020**, *7*, e2019EA000819.
31. Yang, P.; Xia, J.; Zhang, Y.; Hong, S. Temporal and spatial variations of precipitation in Northwest China during 1960–2013. *Atmos. Res.* **2017**, *183*, 283–295. [CrossRef]
32. Deng, X.; Liu, Y.; Liu, Z. Temporal-spatial dynamic change characteristics of evapotranspiration in arid region of Northwest China. *Acta Ecol. Sin.* **2017**, *37*, 2994–3008. (In Chinese)
33. Chen, Y.; Luo, G.; Maisupova, B.; Chen, X.; Mukanov, B.M.; Wu, M.; Mambetov, B.T.; Huang, J.; Li, C. Carbon budget from forest land use and management in Central Asia during 1961–2010. *Agric. For. Meteorol.* **2016**, *221*, 131–141. [CrossRef]
34. IPCC. Good Practice Guidance for Land Use, Land-Use Change and Forestry. In *IPCC Guidelines for National Greenhouse Gas Inventories (Prepared by the National Greenhouse Gas Inventories Programme)*; IPCC: Geneva, Switzerland, 2006.
35. Fang, J.; Guo, Z.; Piao, S.; Chen, A. Terrestrial vegetation carbon sinks in China, 1981–2000. *Sci. China Ser. D* **2007**, *50*, 1341–1350. [CrossRef]
36. IPCC. *Good Practice Guidance for Land Use, Land Use Change and Forestry*; Institute for Global Environmental Strategies: Hayama, Japan, 2003; p. 617.
37. Ye, X.; Chuai, X. Carbon sinks/sources' spatiotemporal evolution in China and its response to built-up land expansion. *J. Environ. Manag.* **2022**, *321*, 115863. [CrossRef]
38. Wang, X.F.; Hao, L.; Qin, H.B.; Su, L.Y.; Liu, Z. Textual quantitative analysis of cultivated land ecological management and protection policies in China from the perspective of policy tools. *China Land Sci.* **2018**, *32*, 15–23.
39. Lichtenberg, E.; Ding, C. Chapter 5: Assessing Farmland Protection Policy in China. *Land Use Policy* **2008**, *25*, 59–68. [CrossRef]
40. Liu, X.; Wang, S.; Wu, P. Impacts of urban expansion on terrestrial carbon storage in China. *Environ. Sci. Technol.* **2019**, *53*, 6834–6844. [CrossRef]
41. EIU. Global Food Security Index 2018: Building Resilience in the Face of Rising Food-Security Risks. Economist Intelligence Unit, Ed. 2018. Available online: https://nonews.co/wp-content/uploads/2018/10/GFSI2018.pdf (accessed on 1 May 2022).

42. Ke, S.; Wu, Y.; Cui, H.; Lu, X. The mechanisms and spatial-temporal effects of farmland spatial transition on agricultural carbon emission: Based on 2018 counties in China. *Environ. Sci. Pollut. Res.* **2023**, *30*, 107716–107732. [CrossRef]
43. Mogollón, J.M.; Beusen, A.H.; Van Grinsven, H.J.M.; Westhoek, H.; Bouwman, A.F. Future agricultural phosphorus demand according to the shared socioeconomic pathways. *Glob. Environ. Chang.* **2018**, *50*, 149–163. [CrossRef]
44. Yang, F.; He, F.; Li, S.; Li, M.; Wu, P. A new estimation of carbon emissions from land use and land cover change in China over the past 300 years. *Sci. Total Environ.* **2023**, *863*, 160963. [CrossRef]
45. Tang, L.; Ke, X.; Zhou, T.; Zheng, W.; Wang, L. Impacts of cropland expansion on carbon storage: A case study in Hubei, China. *J. Crop. Improv.* **2020**, *265*, 110515. [CrossRef]
46. Dubey, A.; Lal, R. Carbon footprint and sustainability of agricultural production systems in Punjab, India, and Ohio, USA. *J. Crop. Improv.* **2014**, *23*, 332–350. [CrossRef]
47. Li, Y.; Brando, P.; Morton, D.; Lawrence, D.; Yang, H.; Randerson, J. Deforestation-induced climate change reduces carbon storage in remaining tropical forests. *Nat. Commun.* **2022**, *13*, 1964. [CrossRef]
48. Su, Y.Z.; Yang, R.; Liu, W.J.; Wang, X.F. Evolution of soil structure and fertility after conversion of native sandy desert soil to irrigated cropland in arid region, China. *Soil. Sci.* **2010**, *175*, 246–254. [CrossRef]
49. Yang, H.; Huang, X. Enforcement key to China's environment. *Science* **2015**, *347*, 834–835. [CrossRef] [PubMed]
50. Molotoks, A.; Henry, R.; Stehfest, E.; Doelman, J.; Smith, P. Comparing the impact of future cropland expansion on global biodiversity and carbon storage across models and scenarios. *Philos. Trans. R. Soc. B* **2020**, *375*, 20190189. [CrossRef] [PubMed]
51. Xie, E.; Zhang, X.; Lu, F.; Peng, Y.; Zhao, Y. Spatiotemporal changes in cropland soil organic carbon in a rapidly urbanizing area of southeastern china from 1980 to 2015. *Land Degrad. Dev.* **2022**, *9*, 33. [CrossRef]
52. Bai, X.; Shi, P.; Liu, Y. Society: Realizing China's urban dream. *Nature* **2014**, *509*, 158. [CrossRef] [PubMed]
53. Wang, Z.; Xu, L.; Shi, Y.; Ma, Q.; Wu, Y.; Lu, Z.; Mao, L.; Pang, E.; Zhang, Q. Impact of Land Use Change on Vegetation Carbon Storage during Rapid Urbanization: A Case Study of Hangzhou, China. *Chin. Geogr. Sci.* **2021**, *31*, 209–222. [CrossRef]
54. Yuan, J.; Ouyang, Z.; Zheng, H.; Su, Y. Ecosystem carbon storage following different approaches to grassland restoration in south-eastern Horqin Sandy Land, northern China. *Glob. Ecol. Conserv.* **2021**, *26*, e01438. [CrossRef]
55. Lei, L.; Li, Y.; Zhou, Z.; Li, N.; Zhao, C.; Li, Q. Cropland abandonment alleviates soil carbon emissions in the North China Plain. *Environ. Monit. Assess.* **2023**, *34*, 1859–1868. [CrossRef] [PubMed]
56. Jiang, W.; Deng, Y.; Tang, Z.; Lei, X.; Chen, Z. Modelling the potential impacts of urban ecosystem changes on carbon storage under different scenarios by linking the CLUE-S and the InVEST models. *Ecol. Model.* **2017**, *345*, 30–40. [CrossRef]

agronomy

MDPI

Article

Plasma-Treated Nitrogen-Enriched Manure Does Not Impose Adverse Effects on Soil Fauna Feeding Activity or Springtails and Earthworms Abundance

Hesam Mousavi [1,*], Thomas Cottis [1], Reidun Pommeresche [2], Peter Dörsch [3] and Svein Øivind Solberg [1]

1 Faculty of Applied Ecology, Agricultural Sciences, and Biotechnology, Inland Norway University of Applied Sciences, 2418 Elverum, Norway
2 Norwegian Centre for Organic Agriculture, 6630 Tingvoll, Norway
3 Faculty of Environmental Sciences and Natural Resource Management, Norwegian University of Life Sciences, 1432 Aas, Norway
* Correspondence: hemousavi65@gmail.com

Abstract: Plasma treatment of animal manure is a new technology, enriching the manure with plant-available nitrogen. Therefore, the product is termed nitrogen-enriched organic fertilizer (NEO). The producer (N2 Applied) claims that NEO can be a sustainable alternative to conventional fertilizers used in agriculture. However, the effect of this product on soil-dwelling organisms is unknown. This study investigates and compares the effects of NEO on changes in soil fauna feeding activity, the abundance of springtails, and the abundance and weight of earthworms to mineral fertilizer, organic fertilizer (cattle slurry), and no fertilizer in pot and field experiments with sandy clay loam soil. Early effect evaluation (week 7) indicated influences on soil fauna feeding activity; among treatments, higher amounts of fertilizers went along with lower feeding activity, regardless of fertilizer type. However, the initial fertilizer application stimulation was transient and stabilized with time after fertilization towards mid-term (week 14) and late effect evaluations (week 21). Accordingly, differences between feeding activities were less than five percent at late effect evaluation. Similarly, none of the fertilizers used imposed adverse effects on the abundance of springtails and the abundance and weight of earthworms; these parameters were almost identical among all fertilizing treatments. After two years of application in field trials and in a pot experiment, NEO and the other used fertilizers seem not to harm the selected soil-dwelling organisms.

Keywords: sustainable agriculture; nitrogen; fertilizing; organic farming; soil fauna; NEO

Citation: Mousavi, H.; Cottis, T.; Pommeresche, R.; Dörsch, P.; Solberg, S.Ø. Plasma-Treated Nitrogen-Enriched Manure Does Not Impose Adverse Effects on Soil Fauna Feeding Activity or Springtails and Earthworms Abundance. *Agronomy* **2022**, *12*, 2314. https://doi.org/10.3390/agronomy12102314

Academic Editors: Raquel P. F. Guiné, António Dinis Ferreira and António Moitinho Rodrigues

Received: 27 August 2022
Accepted: 21 September 2022
Published: 26 September 2022

1. Introduction

The current trends of population growth and resource scarcity underline the importance of using new sustainable technologies in agriculture. Agri-food systems worldwide depend severely on mineral fertilizers [1], and current plant production systems are intensively fertilized with nitrogen (N) [2,3]. According to the FAO, the global nitrogen input into agriculture is eight times higher today than in the 1960s [4].

The favorable effect of N fertilization on plant productivity is well recognized [5–7]. However, N intensification has many severe trade-offs [3,8]. Although increased food production is crucial for sustaining an increasing human population, preserving soil fertility is also critical. While increasing food production per area is commonly highlighted, the effects of excessive fertilizing on soil organisms and their functions are often neglected [9,10]. Overuse of fertilizers can lead to air, soil, and water pollution, as well as adverse effects on biodiversity and the climate [11]. Besides, soil nutrients are manipulated through fertilization, and changes in functional soil groups are stimulated by favoring some groups over others [12]. Soil fauna and soil microorganisms contribute to various ecosystem services such as plant health, disease protection, pathogenicity, and nutrient turnover [13].

Considering the European Green Deal and the "Farm to Fork strategy," the European Commission aims for at least a 50% reduction in nutrient leaching by 2030. Accordingly, a 20% reduction in fertilizer use is anticipated [11]. Moreover, identifying fertilization regimes with the least possible adverse impact on soil organisms is fundamental as it enhances sustainability in food production. Hence, there is a necessity for the purposeful use of fertilizers, e.g., fertilizing with mineral and organic fertilizers, fertilizer efficiency enhancement [14], and developing novel high-tech fertilizers.

Nitrogen Enriched Organic fertilizer (NEO) has been introduced as a novel fertilizer with potentially advantageous properties [15]. Atmospheric nitrogen is fixed as nitrogen oxides (NOx) by a plasma process using green electricity and added to organic fertilizers (e.g., manure, slurry, digestate) using N2 Applied's (Asker, Norway) patented unit. Once the NOx reacts with water, it forms nitrous acid (HNO_2) and nitric acid (HNO_3), which lowers the pH of the slurries and stabilizes it. The units are small enough to allow farmers to produce their NEO locally, resulting in self-sufficiency and enhanced agricultural sustainability when substituting conventional organic and mineral fertilizers, nonetheless because of lowered ammonia (NH_3) and methane (CH_4) emissions [15–17]. NEO is highly fluid, holding less than 10% solid particles due to filtration during the production process. In the present study, we investigate and compare the effects of NEO on soil living organisms with those of other fertilizers.

Hypothetically, mineral fertilizers can enhance soil biological activity by increasing plant productivity and residue return [10,18]. Studies on fertilizer effects on the abundance and weight of earthworms found that a combination of mineral and organic fertilizers was even more significant than mineral fertilizer alone [19–22]. Another study indicated a positive effect of mineral fertilizer on springtails and mite abundance, despite the reduction in species richness [23]. On the contrary, N fertilizers, mainly ammonium N, can potentially contribute to diminishing soil biological activity by acidifying soil and inducing changes in soil functional communities [10,18,24,25]. Besides, repressing certain soil enzymes involved in nutrient cycles, e.g., the amidase involved in the N cycle, is likely due to the repeated application of mineral fertilizers [24].

Similarly, perhaps due to reduced plant species richness, reduced soil microbial weight was reported in perennial grassland under high N fertilizing rates. However, results from annual croplands do not support this conclusion [9,26–28], despite results being highly dependent on fertilization rates [18,21]. Regardless, the functional activity of soil organisms is a complex trait controlled by a multitude of environmental and management factors and recurrent mineral N fertilization [25]. It was shown that repeated application of organic fertilizers stimulates soil microbial and faunal growth and activity [18,19,29,30]. Indeed, organic amendments provide carbon for soil living communities and improve productivity and residue return. In addition, organic fertilizers enhance soil microbial and faunal communities more than chemical fertilizers. Albeit, this positive effect is expressed more when combining organic and mineral fertilizers [19,20,29]. However, these effects vary between annual and perennial production systems [29].

The susceptibility of soil fauna and invertebrates to elevated nitrogen levels differs. For example, soil fauna feeding activity under fertilization has been reported to be reduced in the short-term [9,24] and the long term [31]. Another study found that the abundance of springtails decreased following cattle slurry application; however, it recovered, but not entirely to initial numbers, later during the same growing season [32]. Furthermore, there are positive reports about fertilizers enhancing soil faunal structure, diversity, and feeding activity [26,27,31], specifically for springtails in the topsoil layer [33]. The mentioned controversies highlight the importance of investigating fertilizer effects on soil biota, especially when dealing with novel fertilizers such as NEO.

The current study aimed to (1) identify if NEO has any detrimental effects on soil fauna compared to conventional fertilizers and (2) develop a method for evaluating the immediate effect of fertilizers on earthworm abundance and weight. A preliminary study showed that NEO did not negatively affect soil fauna feeding activity more than other fertilizers [34]. In the present study, we expand our research using a different type of NEO and compare effects on the abundance of earthworms and springtails as "bioindicators of soil quality" [35–37]. Fertilization treatments included mineral fertilizer, NEO, untreated cattle slurry, and a combination of organic and mineral fertilizers.

2. Materials and Methods

In this study, we conducted three sets of experiments. First, a growing chamber experiment to identify and compare the fertilizer effects on soil fauna feeding activity; second, a field experiment to identify and compare fertilizer effects on the abundance of springtails. Third, an outdoor experiment to identify and compare the immediate effect of fertilization on the abundance and weight of earthworms.

2.1. Soil Fauna Feeding Activity

2.1.1. Experimental Design

We conducted pot experiments in a growing chamber at Inland Norway University of Applied Sciences. Perforated pots (13 × 18 cm^2 with 2.5 L of field soil were used. Treatments were distributed randomly among the four replicates after the pots were fertilized at loading time.

The trial consisted of five fertilization regimes distributed in seven fertilization treatments (Table 1); no fertilizer; mineral fertilizer (Yara Mila 18-3-15) [38]; NEO type D (N2 Applied) [15]; organic fertilizer (untreated cattle slurry); and organic fertilizer + mineral fertilizer (Yara Liva 16-0-0) [39]. NEO and untreated slurry were applied in liquid form, while Yara Mila and Yara Liva were pelleted. Treatments were (1) no fertilizer, (2) mineral fertilizer 73 kg N ha^{-1}, (3) mineral fertilizer 175 kg N ha^{-1}, (4) NEO type D 73 kg N ha^{-1}, (5) NEO type D 175 kg N ha^{-1}, (6) organic fertilizer 73 kg N ha^{-1}, (7) organic fertilizer + MF 175 kg N ha^{-1}.

Table 1. Fertilizing treatments and application rates used in the growing chamber trial.

	Fertilizing Treatment	Organic Fertilizer (Tons ha^{-1})	Kg N in Yara Mila18-3-15 (kg ha^{-1})	Kg N in Organic Fertilizer (kg ha^{-1})	Kg N in Yara Liva 16-0-0 (kg ha^{-1})	Total kg N (kg ha^{-1})
1	No fertilizer	-	-	-	-	0
2	Mineral fertilizer 73 kg N ha^{-1}		73			73
3	Mineral fertilizer 175 kg N ha^{-1}		175			175
4	NEO type D 73 kg N ha^{-1}	22		73		73
5	NEO type D 175 kg N ha^{-1}	50		175		175
6	Organic fertilizer 73 kg N ha^{-1}	55		73		73
7	Organic fertilizer + mineral fertilizer 175 kg N ha^{-1}	55		73	102	175

NEO type D had a pH of 5.22 and contained 1746 mg L^{-1} NH_4^+ –N, 1131 mg L^{-1} NO_2^- –N, and 1562 mg L^{-1} NO_3^- –N, totaling 4439 mg L^{-1} N. The untreated slurry had a pH of 7.32, containing 1804 mg L^{-1} NH_4^+ –N and 149 mg L^{-1} NO_2^- –N, totaling 1953 mg L^{-1} N. Therefore, we targeted a slurry amount of 55 tons ha^{-1}; nonetheless, during production, the N2 applied apparatus excludes all dry materials bigger than 5 mm; as a result, NEO's quantity decreases by 10% to 50 tons ha^{-1}. Therefore, each ton of untreated slurry contained 1.95 kg of plant-available N, while each ton of NEO contained 4.44 kg of plant-available N.

The soil was acquired from the adjacent experimental farm and analyzed at Eurofins soil lab (https://www.eurofins.no/agro-testing/ (accessed on 20 July 2022)), indicating a sandy clay loam texture, more than 10% clay, and soil organic matter of 4.5%. The soil pH was 7.4, which is relatively high, with a normal phosphorus status (P-AL = 11 mg/100 g), and a low potassium status (K-AL = 5 mg/100 g). Moreover, we estimated the soil's field capacity at 33.6% VWC, with a total pore volume of 41.4%.

In order to simulate field conditions, we planted seeds in the pots. Pots were prepared following the protocol that we developed before [34]. First, a soaked paper tissue was laid at the lowermost of the perforated pots to prevent soil outpour. Then, an initial 0.6 L (5 cm) soil load into the pots. Next, a soil load of 0.8 L (6 cm) was mixed with the fertilizer. Fertilizers were dosed following the advised field application rates (measured in tons per hectare), accounting for the soil surface in each pot (169 cm^2) (Table 1). Afterward, three rows of Italian ryegrass seeds (*Lolium multiflorum Lam.*), variety "Barpluto" (NAK Nederland/Ref. DE148-214011) per pot were then sown over the top of 0.9 L (6 cm) of additional soil. Finally, 0.2 L (1 cm) of soil was added to form the surface soil.

In the growing chamber, we used Lumatek ATS300W 80 × 80 cm LED light pads (https://lumatek-lighting.com/ (accessed on 21 July 2022)) that delivered a complete visible light spectrum (380–780 nm wavelength) recommended for plantation under controlled conditions [40]. Three adjacent LED pads were positioned 35 cm above the plants and were uplifted alongside plant growth. The light and dark intervals were adjusted according to Nordic summer days with 16 h light and 8 h darkness. Additionally, 16 pots per LED pad were confirmed to receive equal light using a digital light intensity meter. Throughout the experiment, the growth chamber had a temperature of 16 °C.

Five hundred milliliters of water, or 55% of the field capacity for our dry soil, was used to irrigate the pots at first since the soil was not entirely dry at pot preparation. After that, pots were irrigated with 200 mL of water thrice a week for the first four weeks. However, as plants progressed in the developmental stages, irrigation frequency was increased the weeks before harvest upon visual inspection [41].

The pots were positioned for the first two weeks adhering to each other. However, from week three, there was a five cm distance between the pots to avoid plants competing for light and space. Thinning was performed following germination; 24 vigorous plants per pot were kept (3 × 8 rows). Moreover, a few germinated weeds were removed by hand.

2.1.2. Evaluating Feeding Activity

Bait-lamina strips (Terra Protecta GmbH, Berlin, Germany) [42] were used to evaluate soil fauna feeding activity. This method is considered efficient, rapid, and reproducible with high statistical applicability [43,44]. This method evaluates the functional activity of soil fauna, feeding activity as one of the critical factors in soil nutrient cycling [44–46]. The technique has helped researchers screen various soil management practices and has given valuable information on the feeding activity of soil fauna. [44]. In this method, 16 1.5 mm diameter holes are located 5 mm apart on perforated PVC strips (1 mm × 6 mm × 120 mm). The holes are filled with bait substrate. The bait substrate comprises 5% activated carbon, 25% wheat bran, and 70% cellulose powder [47]. After a certain period of exposure to soil, the degree to which substrate is used up in the holes reveals the feeding activity of soil fauna, whereas soil microorganisms (e.g., bacteria, nematodes, fungi) have a negligible effect [42,48–50].

We conducted preliminary tests to determine proper intervals for the bait-lamina sampling in a pot experiment. In these tests, we noticed that after four weeks of soil fauna feeding activity, the percentage of bait consumption in the strips varied from 3–29%. Moreover, after extending the period to eight weeks, we had several strips with all holes empty, showing 100% feeding activity. Therefore, we determined seven weeks as an appropriate test period.

In this experiment, we planted three Bait-lamina strips diametrically in each pot (replicate) when watering for the first time to assess and compare the early effect of fertilization on soil fauna feeding activities [47]. Seven weeks later, the first set of strips was taken out, and the second set was inserted in the same order as the first set to evaluate the mid-term effect. Seven weeks later (week 14 after plantation/fertilizing), the second set was taken out, and the third set was inserted to evaluate the late effect of fertilizing soil fauna feeding activity. At last, this set was taken out seven weeks later (week 21 after plantation/fertilizing). Plant growth and watering were sustained until the experiment's termination to simulate the conditions seen in an actual field. The strips were visually examined for the removal of the bait substrate on each sampling [47]. Three categories—empty (1), partially empty (0.5), or filled (0)—were assessed and used to describe the disappearance of the bait substrate [9]. With a maximum of 100% feeding activity (all 16 empty holes), each empty hole (score 1) was equivalent to 6.25% feeding activity.

2.2. The Abundance of Springtails (Collembola)

Springtails live in all soil layers, depending on soil moisture, and have diverse life forms in different soil strata and nutrition types [51]. They graze on fungi, algae, and bacteria or feed on plant detritus or other organic substances. [52]. They are great soil bio-indicators, especially in shallow soils. As a pray for other arthropods, they play a central role in the food chain [35].

2.2.1. Experimental Design

We conducted springtail sampling in two different field trials that had been fertilized with NEO and other fertilizers one year before the first sampling. The first trial was a cereal field located at Blæstad experimental farm at Inland Norway University of Applied Sciences (60°49′11.7″ N 11°10′48.4″ E). The second trial was on a grass field located at Stjørdal, Trøndelag (63°20′33.4″ N 10°17′56.9″ E). The experimental design in both trials was a traditional randomized complete block design with four replicates. Both trials consisted of four fertilization regimes; mineral fertilizer (Yara Mila 18-3-15) [38]; NEO type B (N2 Applied) [15]; organic fertilizer (untreated cattle slurry); and no fertilizer. Both fields were fertilized for two consecutive years. The grain field was fertilized once a year: before sowing: (22 April 2020 and 27 April 2021). The grass field was fertilized twice a year: in early spring (27 April 2020 and 4 May 2021) and after the first harvest (24 June 2020 and 15 June 2021).

Fertilizer doses in the grain field were (1) mineral fertilizer 666.6 kg ha^{-1}, (2) NEO 37.6 tons ha^{-1}, (3) organic fertilizer 41 tons ha^{-1}, and (4) no fertilizer. Then again, doses in the grass field were (1) mineral fertilizer 650 kg ha^{-1} in spring + 500 kg ha^{-1} after the first harvest, (2) NEO 37.5 tons ha^{-1} + 28 tons ha^{-1} after the first harvest, (3) organic fertilizer 41 tons ha^{-1} + 30.5 tons ha^{-1} after the first harvest, and (4) no fertilizer. NEO type B had a pH of 5.35 and contained 1480 mg L^{-1} NH_4^+ –N, 777 mg L^{-1} NO_2^- –N, and 1250 mg L^{-1} NO_3^- –N, totaling 3507 mg L^{-1} N. The cattle slurry used in this experiment had a pH of 7.32, and it contained 1804 mg L^{-1} NH_4^+ –N and 149 mg L^{-1} NO_2^- –N, totaling 1953 mg L^{-1} N.

NEO and untreated slurry were applied in liquid form while mineral fertilizer was pelleted. For a homogenous liquid fertilizer, all the barrels were stirred well prior to bottling/spreading to dissolve the sediments. Next, the fertilizers were dispersed manually using containers and rapidly harrowed with the soil using a tractor before sowing in the grain field and spreading on the grass field surface. The grain field was seeded with barley 'Rødhette' (180 kg ha^{-1}) in 2020 and spring wheat 'Mirakel' (220 kg ha^{-1}) in 2021. The grass field was a mixture of timothy, meadow fescue, and red clover seeded in 2019. In the grain field, herbicides were applied once in June with Ariane S (Corteva Agriscience, Puerto Rico) and once at the end of the growing season with Roundup (Bayer, Germany). No irrigation was applied. The 2021 season at Blæstad was decent regarding cereal growth, with a relatively cool May and a little over average precipitation: 78 mm in May and 62 mm in June. Moreover, in Trøndelag, the season was good, with precipitation around normal.

The soil in the field trial at Blæstad was identical to the soil we used in the growing chamber experiment (see Section 2.2.1). The soil in the grass field trial at Trøndelag was classified as clay loam. The organic matter content was 5.1%, pH was 6,2, and plant available phosphorus and potassium were normal (P-AL = 8, K-AL = 7), but the potassium reserve was high ($KHNO_3$ = 140).

2.2.2. Evaluating the Abundance of Springtails

The soils from experimental plots in the grain and grass fields were sampled twice in 2021 for springtail abundance. The first sampling occurred on 15 June 2021, some weeks after fertilization. The second sampling was on 20 October 2021, after harvesting. The temperature on the first sampling day in June was 20 °C and 6 °C after some rainy days in October when the soil was sampled again.

Three diametric samples were collected on each replicate's corners and in the center of the field plots. First, soil sampling was conducted by hammering down a corer (5 cm high, 5 cm diameter = 98.17 cm^3 volume) in the surface soil and collecting the sample into a zipper bag using a spade [35]. The samples were transferred directly to the lab, and the three samples from each replicate were placed upside down on slightly modified Berlese funnels [35,53,54]. In this method, moisture, heat, and light gradient drive the soil organisms to move away from the heat source (60 W lamp) [55], passing a mesh screen and falling into the vessel, ending in the collection tube filled with 91% ethanol. This way, animals can be preserved for further investigation for a long time. The samples remained in extraction units for a week on each sampling occasion. Following completion of extraction, the abundance of springtails in samples was counted and registered using a light microscope. Springtail abundances were scaled up to 1 m^2 and −5 cm depth, estimating the number of soil cores fitting into 1 m^2 (169.8).

2.3. *The Fate of Earthworms (Lumbricidae)*

Earthworms are medium to large oligochaetes that are substrate feeders and play an essential role in decomposition in the soil [35]. Due to several essential functions, their activity increases soil fertility. However, their distribution highly depends on moisture, soil type, pH, and vegetation. Earthworms are categorized into three ecological groups: litter dwellers, horizontal burrowers, and deep burrowers [56]. Because of their size, a significant fraction of the biomass in loamy meadows is composed of earthworms; however, they are scarce in shallow or acid soils [57,58]. In this study, we developed a protocol for evaluating the immediate effect of fertilizers on earthworms. The earthworms used in the experiments were a mixture of juveniles and adults from the most common Norwegian earthworms: geophagous (soil eating) field worms (*Aporrectodea caliginosa*) and pink worm (*Aporrectodea rosea*). In addition, other common species in Norwegian arable soil include dew worm *Lumbricus terrestris*, *L. rubellus*, and a few individuals of the less common *Allolobophora chlorotica* [30]. The earthworms used in our trials were found in an organic vegetable garden adjacent to the experimental field.

2.3.1. Experimental Design

The experiment was repeated twice in June 2021 and June 2022, with three replicates. The study location was at Blæstad experimental farm, Innlandet (60°49′11.7″ N 11°10′48.4″ E); the soil analysis was identical to the growing chamber experiment (Section 2.1.1), and the fertilizer treatments were (1) no fertilizer, (2) mineral fertilizer (Yara Mila) 666 kg ha^{-1}, (3) NEO type B (2021), and type D (2022) 3.4 tons ha^{-1}, and (4) organic fertilizer (untreated slurry) 3.7 tons ha^{-1}. Over, NEO and mineral fertilizer contained almost equal N per hectare in both experiments. The duration of the experiments was eight days.

2.3.2. Changes in Abundance and Weight of Earthworms

The developed protocol is as follows:

1. Holes with 30 cm diameter and 20 cm are dug out in the field. The soil from the holes was visually inspected to exclude present earthworms.
2. Earthworm-proof but water-permeable textile (tested before the experiment) is inserted into the hole.
3. Earthworms (Lumbricidae) used in the experiment were excavated from the same experimental farm two days before and stored in a pile of soil pending the experiment. On the day of starting the experiment, the earthworms were detached from the soil pile, sorted, weighed, and an equal number of worms (11 in the 2021 trial and 13 in the 2022 trial) making up a similar total weight were deposited in separate containers and marked. The worms were not rinsed before weighing. The worms were handled cautiously and remained detached from the soil for the shortest possible period. After counting and weighing, earthworms were transferred to other containers with soil.
4. Next, 10 cm of soil was filled back into the holes, and the earthworms were placed over the top.
5. The next 5 cm of soil was carefully mixed with the fertilizer and filled back into the hole.
6. As a supplementary food source for the worms, 100 g of grass was spread on this layer.
7. Two cm loose soil scattered over.
8. Finally, the last 3 cm of loose soil was scattered on the top.
9. Outer edges of the textile were fetched together and closed over. At last, a heavy substance (a stone) was placed over the top to inhibit wind opening or bird feeding. Finally, the experimental units were covered with white plastic tarpaulin on days of intense sun to avoid excessive temperature caused by the sun and the black textile.
10. At the end of the experiment (8 days), the soil bags were lifted out of the soil and dispersed over a flat surface. The living earthworms were carefully handpicked, counted, and weighed in less than five minutes to avoid desiccation.

2.4. Data Handling and Statistical Analyzes

We registered and sorted the data from each respective experiment. Using Minitab 20 statistical software (2021 Minitab, LLC (State College, PA, USA)), the differences in soil fauna feeding activity, the abundance of springtails, and variations in weight and abundance of earthworms were examined. The differences among fertilizing treatments were assessed using a one-way ANOVA and Welch's test. Games–Howell pairwise comparison was further utilized to compare, categorize, and plot data at a 95% confidence interval for the means. The error bars in the graphs are calculated by using individual standard deviations.

3. Results

3.1. Soil Fauna Feeding Activity

Using a growth chamber experiment where all variables except fertilization were held constant, we examined and evaluated the impact of different fertilization treatments on the feeding activity of soil fauna. We investigated the early effects (seven weeks), mid-term effects (14 weeks), and late effects (21 weeks).

There was a significant early fertilization effect on the feeding activity (p = 0.001). However, this effect was not associated with the type of fertilizer but with the amount of fertilizer applied (Figure 1A, Table S1). After seven weeks, mineral fertilizer 73 kg N ha^{-1} (78.13%), organic fertilizer 73 kg N ha^{-1} (74.74%), NEO type D 73 kg N ha^{-1} (73.70%), and organic + mineral fertilizer 175 kg N ha^{-1} (61.98%) exhibited increased soil fauna feeding activity relative to no fertilizer (54.17%). Both NEO type D 175 kg N ha^{-1} (49.22%) and mineral fertilizer 175 kg N ha^{-1} (46.35%) tended to have lower soil fauna feeding activity than soil without fertilizer. Additionally, the lead was insignificant even though the mixture of organic and mineral fertilizers was the only high N content treatment that improved soil fauna feeding activity above that of no fertilizer (Figure 1A, Table S1).

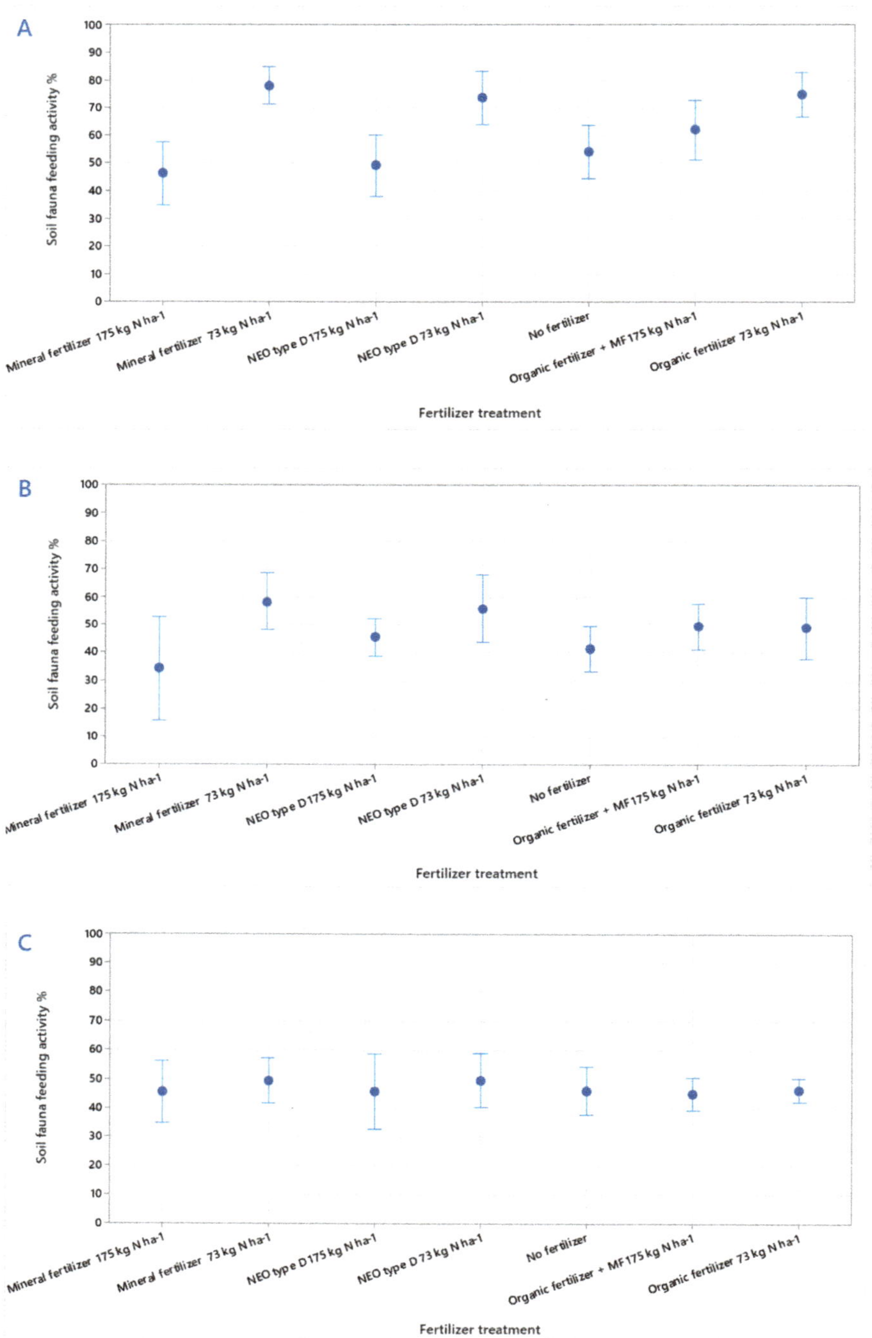

Figure 1. Effects of different fertilization treatments on soil fauna feeding activity (%) (**A**) seven weeks after fertilizing, (**B**) 14 weeks after fertilizing (7–14 weeks), and (**C**) 21 weeks after fertilizing (14–21 weeks). Error bars are individual standard deviations at a 95% confidence interval.

Regarding the mid-term effect, 14 weeks after fertilizing, initial (week 7) differences in soil faunal feeding activity converged and became more even between treatments. However, the reduction was more evident among those treatments with higher feeding activity during the initial weeks (Figures 1B and 2). Hence, the average feeding activity was 62.61% among all treatments seven weeks after fertilizing, which dropped significantly ($p = 0.001$) to 47.75% at the mid-term evaluation (Figure S1, Table S1).

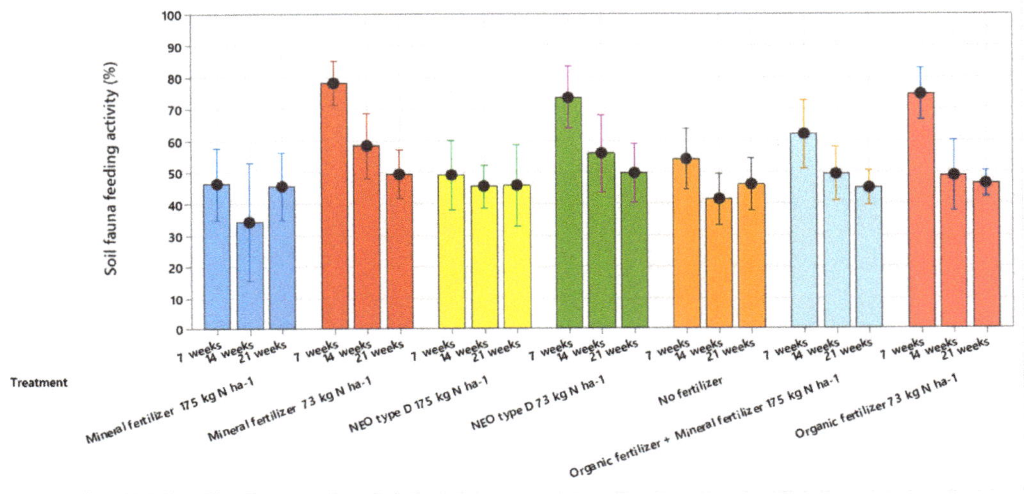

Figure 2. Early (0–7 weeks), mid-term (7–14 weeks), and late effects (14–21 weeks) on soil fauna feeding activity (%) between different fertilization treatments. Error bars are individual standard deviations at a 95% confidence interval.

Furthermore, at mid-term evaluation, feeding activities were not significantly different ($p = 0.08$) among fertilization treatments. Instead, mineral fertilizer 73 kg N ha^{-1} (58.44%), NEO type D 73 kg N ha^{-1} (55.99%), organic + mineral fertilizer 175 kg N ha^{-1} (49.48%), organic fertilizer 73 kg N ha^{-1} (48.96%), and NEO type D 175 kg N ha^{-1} (45.57%) had higher soil faunal feeding activity than no fertilizer (41.41%). By comparison, mineral fertilizer 175 kg N ha^{-1} (34.38%) had the lowest feeding activity. Thus, only the mineral fertilizer with a high N content reduced the ability of soil fauna to feed; however, as mentioned earlier, this was not statistically significant (Figure 1B, Table S1).

The late fertilization effect on feeding activity resembled the mid-term effect, i.e., despite a slight insignificant average reduction from mid-term to late effect among all fertilizing treatments (47.75% to 46.88%), soil fauna feeding activity appeared to stabilize seven weeks after fertilization without showing any significant effects (Figures 2 and S1).

The lower amounts of fertilizer, regardless of fertilizer type, supported higher feeding activity, NEO type D 73 kg N ha^{-1} (49.74%), mineral fertilizer 73 kg N ha^{-1} (49.48%), organic fertilizer 73 kg N ha^{-1} (46.35%) showed higher feeding activity than no fertilizer (46.09%). On the other hand, higher fertilizer amounts, NEO type D 175 kg N ha^{-1} (45.83%), mineral fertilizer 175 kg N ha^{-1} (45.57%), and organic + mineral fertilizer 175 kg N ha^{-1} (45.05%) had lower feeding activity. However, the difference between the highest and lowest feeding activities was a maximum of 4.2% and insignificant (Figure 1C, Table S1).

Lastly, low amounts of NEO type D, mineral fertilizer, organic fertilizer, and to some extent, the combination of organic and mineral fertilizer seemed to stimulate soil fauna feeding activity in the initial weeks after fertilization. However, this early effect gradually disappeared, whereas other treatments, including no fertilizer, had more or less constant soil faunal feeding activity throughout the experiment (Figure 2).

3.2. The Abundance of Springtails (Collembola)

We investigated and compared the effects of different fertilization treatments on the abundance of springtails at two field locations; one under cereal and another under grass cultivation. Both fields were fertilized for two consecutive years. Moreover, two samplings were performed, once just before fertilization in early summer and another during fall.

During summer, the abundance of springtails in the cereal field was slightly higher for organic fertilizer than NEO and mineral fertilizer; 509.4, 467, and 297.1 per m^2, respectively (Figure 3A, Table S1). However, the difference was insignificant (p = 0.25). The same pattern was observed during the fall. The number of springtails was slightly higher for organic fertilizer and no fertilizer than NEO and mineral fertilizer; 213.3, 213.3, 169.8, and 169.8 per m^2, respectively (Figure 3B, Table S1). Likewise, the difference between fertilization treatments in the fall sampling was insignificant (p = 0.669).

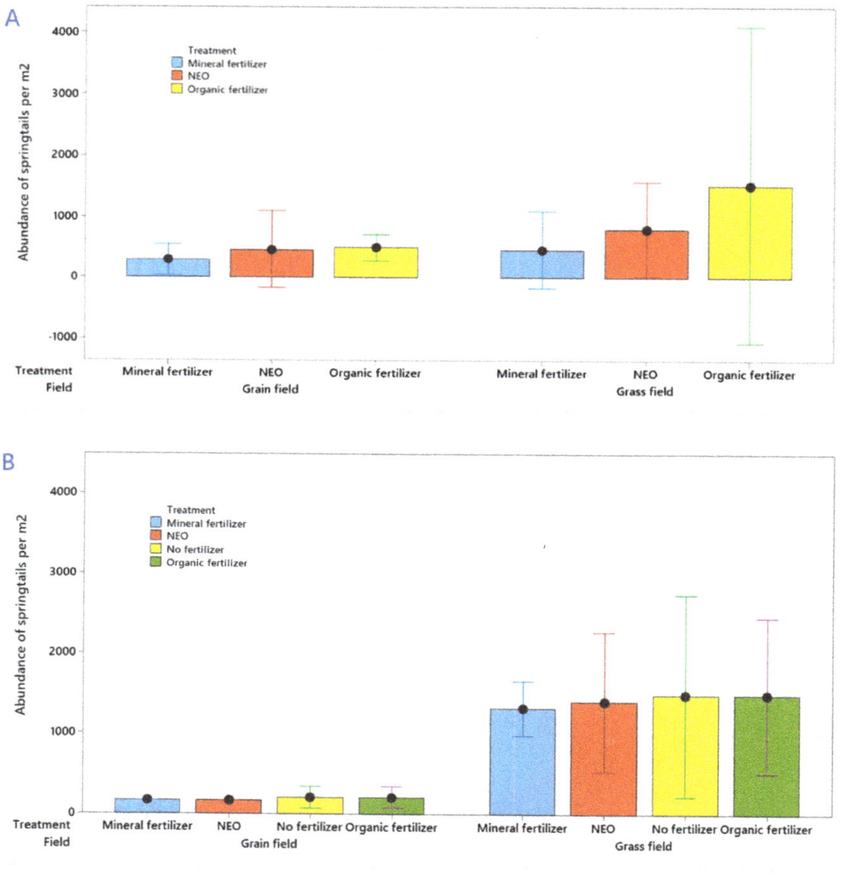

Figure 3. The effect of different fertilization treatments on the abundance of springtails per m^2 in (**A**) summer and (**B**) fall samplings. Error bars are individual standard deviations at a 95% confidence interval.

Similarly, no fertilization effects were found regarding springtail abundance in the grass field during summer and fall (p = 0.404, p = 0.943). However, during summer, higher abundance was observed for organic fertilizer than NEO and mineral fertilizer; 1528, 807, and 467 per m^2, respectively (Figure 3A, Table S1). In the fall, a slightly higher abundance was observed for organic fertilizer and no fertilizer than NEO and mineral fertilizer; 1486, 1486, 1401, and 1316, respectively (Figure 3B, Table S1).

Generally speaking, the springtail abundance was higher in the grass field during fall than in summer and almost identical in the grain field during both seasons. Nevertheless, none of the fertilizer treatments affect springtail abundance regardless of field and season.

3.3. The Fate of Earthworms (Lumbricidae)

In the 2021 trial, the abundance of earthworms for all treatments increased after eight days, except for the treatment with no fertilizer. However, the difference was insignificant among fertilizing treatments ($p = 0.38$). Organic fertilizer showed an average increase of 4.33 worms, followed by mineral fertilizer, 4, and NEO, 2.33, while no fertilizer had one fewer living earthworm than the beginning (Figure 4A, Table S1). Moreover, like for abundance, a similar pattern was observed for the average weight change. Among all captured earthworms, mineral fertilizer had an increment of 4.07 g after eight days, followed by organic fertilizer 2.87 g, NEO 1.93 g; however, the no fertilizer control had 1.67 g fewer earthworms (Figure 4A, Table S1). Nevertheless, the difference in weight change among fertilizing treatments was insignificant ($p = 0.34$).

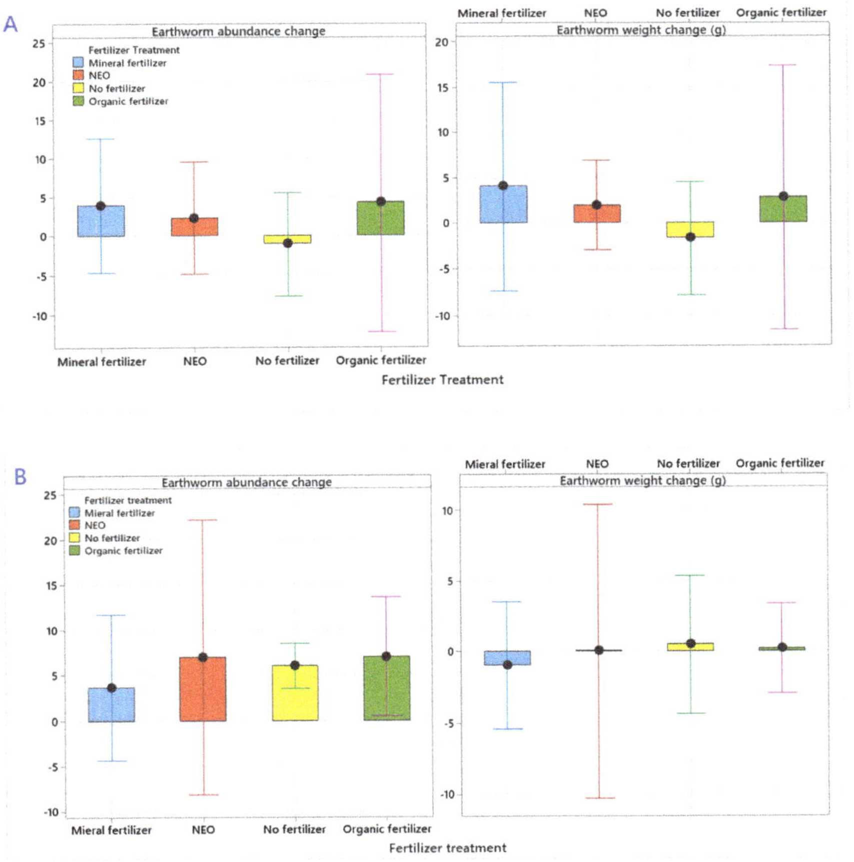

Figure 4. The immediate effect of different fertilization treatments on the abundance (left) and weight (right) of earthworms in (**A**) June 2021 and (**B**) June 2022. Error bars are individual standard deviations at a 95% confidence interval.

In the 2022 trial, however, the outcomes were slightly different. Although the abundance was increased for all fertilization treatments, the weights of living worms were almost identical to the beginning. The average abundance increment for organic fertilizer and NEO was 7, no fertilizer 6, and mineral fertilizer 3.67 earthworms (Figure 4B, Table S1). However, the average weight of all earthworms increased by 0.5 g for no fertilizer, 0.22 g for organic fertilizer, and 0.05 g for NEO. In comparison, mineral fertilizer reduced the total weight by 0.09 g (Figure 4B, Table S1). Nevertheless, changes in abundance or weight were insignificant between fertilizing treatments ($p = 0.69$ and $p = 0.83$, respectively).

Thus, the results indicated no adverse effects of fertilizing on the abundance and weight of earthworms, regardless of the fertilizer type used in both experiments.

4. Discussion

It is known that N fertilization affects the taxonomic composition of soil faunal communities, their population dynamics, and their feeding activity. However, it is not well understood if soil-dwelling organisms adapt to these external factors [59], especially when the external factor is a newly developed fertilizer (NEO). In this study, we investigated and compared the effects of different fertilization regimes on soil-dwelling organisms. We screened changes in soil faunal feeding activity under controlled conditions, the abundance of springtails, and the immediate effect of different fertilizers on the abundance and weight of earthworms under field conditions. The goal was to detect if NEO, a novel fertilizer with potentially toxic contents of nitrite, has any detrimental effects on soil-dwelling organisms compared to conventional fertilizers.

4.1. Soil Fauna Feeding Activity

The soil fauna feeding activity was evaluated at three intervals in pot experiments under controlled conditions; 0–7 weeks (early effect), 7–14 weeks (mid-term effect), and 14–21 weeks (late effect). Low (73 kg N ha^{-1}) and high (175 kg N ha^{-1}) rates of mineral fertilizer, NEO, organic fertilizer (untreated slurry), and a mixture of organic and mineral fertilizer were used as the fertilizing treatments.

Early effect analysis showed that low doses of fertilizer stimulated feeding activity irrespective of fertilizer type, while high amounts of fertilizer resulted in slightly less feeding activity. The only exception was the combination of organic and mineral fertilizer, which tended to have a higher feeding activity than no fertilizer. In line with our results, a grassland study showed that a high amount of organic fertilizer reduced soil fauna feeding activity within days after fertilizing [59]. Except for this, we could not detect any beneficial or detrimental early effect of fertilizers on soil fauna feeding activity. It may be argued that microbial biomass is promoted within the first weeks after high N fertilization resulting in alternate food sources for soil mesofauna, and they may have shifted away from the bait substrate [59], which explains lower feeding activity under higher fertilization.

The mid-term evaluation showed almost the same pattern as the evaluation of the early effect. The only difference was that at the higher N application rates, given organic and mineral fertilizer combination showed a slightly higher feeding activity than the organic fertilizer alone. Additionally, compared to no fertilizer, a high concentration of NEO showed a slightly increased soil faunal feeding activity. However, variations across fertilization treatments were smaller at the mid-term assessment than at the early effect evaluation, demonstrating that the initial stimulation gradually faded with time after fertilizing.

Finally, at the late effect evaluation, the initial stimulation by low amounts of fertilizer disappeared. Like during the early effect evaluation, higher amounts of fertilizer had lower feeding activities, and lower amounts of fertilizer had higher feeding activity irrespective of fertilizer type; whereas the difference among treatments was much smaller than in short- and mid-term evaluations, with less than five percent difference between the highest and lowest feeding activities.

Although similar to an earlier study [34], higher amounts of fertilizer, regardless of fertilizer type, initially showed a somewhat negative effect on soil faunal feeding activity, this detrimental effect progressively stabilized with time after fertilization. Furthermore, after some weeks of fluctuations in soil faunal feeding activity, a similar stabilizing effect has been reported in an oil palm plantation fertilized with different amounts of mineral N fertilizer [26]. The rationale for these transient effects might be the soil's buffering capacity and other soil chemical responses that gradually diminish fertilization's perturbation effect. Moreover, the soil fauna may be functionally redundant, conveying resilience to transient perturbations. Thus, we can summarize that neither NEO nor conventional fertilizers used in our experiment adversely affected the soil faunal feeding activity.

4.2. The Abundance of Springtails (Collembola)

In the grain field during summer, some weeks after fertilization, the numbers of springtails were almost identical among the plots fertilized with organic fertilizer and NEO. There were slightly fewer springtails than the latter two in the plots fertilized with mineral fertilizer. However, in the grass field, the plots fertilized with organic fertilizer supported an almost double and quadruple number of springtails compared to the plots fertilized with NEO and mineral fertilizer, respectively. Correspondingly, in the cereal field during fall, there were slightly more springtails in the plots fertilized with organic fertilizer and no fertilizer than in the plots fertilized with NEO and mineral fertilizer. The same pattern was observed in the grass field during the fall.

NEO had no adverse effects on the number of springtails after fertilization or after harvest; the number of springtails was generally lower during the summer than during the fall. It has been indicated that the abundance of springtails decreases after cattle slurry application [32]. However, this does not have been the case in our study. There might be two reasons for the apparent lack of response to organic fertilization. Springtails are moisture-dependent organisms [37]; therefore, it is a valid argument that sampling on a warm sunny Scandinavian day with limited moisture in the surface soil forced a major part of the springtail community to move deeper in the soil to avoid desiccation [60]. Moreover, more decaying plant matter as food for springtails may be available deeper in the soil after harvest [61].

Nonetheless, in line with our findings, another study showed almost no fertilization effect on the abundance of springtails [62], while another study indicated that fertilization increases the abundance of springtails [63]. Our study showed no adverse effect of NEO or other fertilizers on the abundance of springtails.

4.3. The Fate of Earthworms (Lumbricidae)

We used our developed method to investigate and compare the immediate effects of NEO and other fertilizers on earthworms. The experiment was repeated twice, once during summer 2021 and then in summer 2022. We targeted both the changes in the abundance and the average total weight of earthworms. The treatments were mineral fertilizer, NEO, organic fertilizer, and no fertilizer.

The results from the first experiment indicated no negative effect of fertilizer treatments on the abundance or weight of earthworms after eight days. Only the treatment with no fertilizer showed a slight reduction in the abundance and weight of earthworms. Moreover, roughly the same results were drawn from the second-year experiment. The only difference was that no fertilizer treatment did not lead to any reduction in the abundance or weight of earthworms.

In the case of NEO and organic fertilizer, a promoting effect of adding organic matter to the soil on the earthworm population was expected [18,30]. However, the concern was that excessive liquid slurry in a single dose might adversely affect earthworms [20]. Nonetheless, this did not occur with the amounts applied in our experiments. Moreover, mineral fertilizers might benefit earthworms through direct or indirect effects [19,20,22,64]. However, ammonia-based fertilizer potentially could have adverse effects on the earthworm population in the long run by lowering soil pH [18,65]; this was not the case in our experiments.

The concern might arise from the increasing number of earthworms in our experimental plots after eight days. Earthworms might have escaped their confinement even though we tested this before the study. Another possibility might be that tiny juveniles were contained in the soil before starting the experiment, which grew larger and became discoverable after eight days. Moreover, the most unlikely scenario might be that there were juveniles hatched from the cocoons within the experimental period. Notwithstanding, it is reasonable to argue that these error sources should have been identical for all experimental plots. Thus, it is logical to conclude that fertilization with NEO or any other fertilizer did not inhibit earthworms in the soil but supported an increase in number and activity.

5. Conclusions

NEO, the novel, plasma-treated nitrogen-enriched organic fertilizer, did not adversely affect soil faunal feeding activity, the abundance of springtails, and the abundance and weight of earthworms, as observed in pot and field trials. Moreover, fertilization with organic and mineral fertilizers was not seemed to harm the selected soil-dwelling organisms. Hence, NEO does not adversely affect the selected soil-dwelling organisms compared to conventional fertilization regimes commonly used in plant production today.

Supplementary Materials: The following supporting information can be downloaded at: https://www.mdpi.com/article/10.3390/agronomy12102314/s1; Table S1: Effects of different fertilizing treatments, including mineral fertilizer, NEO, organic fertilizer (untreated cattle slurry), organic fertilizer + mineral fertilizer (MF), and no fertilizer on soil fauna feeding activity (%) in the early effect, mid-term, and late effect evaluations, springtail abundance in summer and fall samplings at crop and grass fields, and the abundance and weight change (g) of earthworms, respectively. The Games–Howell pairwise comparison method at a 95% confidence interval is used to compare the differences between means. Means that do not share a letter are significantly different.; Figure S1: Effects of all fertilizing treatments on soil fauna feeding activity (%) at seven weeks, 14 weeks, and 21 weeks after fertilizing. Individual standard deviations at a 95% confidence interval are used in the graphs.

Author Contributions: Conceptualization, H.M., T.C. and S.Ø.S.; methodology, T.C., H.M., S.Ø.S. and R.P.; software, H.M.; validation, H.M., T.C., S.Ø.S. and R.P.; formal analysis, H.M.; investigation, H.M. and R.P.; resources, T.C. and S.Ø.S.; data curation, H.M.; writing—original draft preparation, H.M.; writing—review and editing, H.M., S.Ø.S., P.D., T.C. and R.P.; visualization, H.M.; supervision, S.Ø.S. and T.C.; project administration, T.C.; funding acquisition, T.C. and S.Ø.S. All authors have read and agreed to the published version of the manuscript.

Funding: This research was funded by the Research Council of Norway, grant number 309640 Plasmabehandlet husdyrgjødsel—gjødselvirkning, miljøpåvirkning og klimagassutslipp, and the APC was funded by Inland Norway University of Applied Sciences.

Institutional Review Board Statement: Not Applicable.

Informed Consent Statement: Not Applicable.

Data Availability Statement: Data is accessible at https://osf.io/n4dyc/?view_only=25142c8ddd854758b8808d14fc792a76 (accessed on 3 August 2022).

Acknowledgments: The authors would like to acknowledge the administrative efforts of Morten Tofastrud and Elisabeth Røe throughout the project. Besides, we would like to thank the funders of the project, the Research Council of Norway and N2 Applied, along with those who provided us equipment and infrastructure required for our study: Norwegian Agricultural Extension service (NLR), Inland Norway University of Applied Sciences (INN), and Norwegian Centre for Organic Agriculture (NORSØK).

Conflicts of Interest: The authors declare no conflict of interest. The funders had no role in the design of the study; in the collection, analyses, or interpretation of data; in the writing of the manuscript, or in the decision to publish the result.

References

1. Mancus, P. Nitrogen Fertilizer Dependency and Its Contradictions: A Theoretical Exploration of Social-Ecological Metabolism*. *Rural Sociol.* **2007**, *72*, 269–288. [CrossRef]
2. Matson, P.A.; Parton, W.J.; Power, A.G.; Swift, M.J. Agricultural intensification and ecosystem properties. *Science* **1997**, *277*, 504–509. [CrossRef] [PubMed]
3. Lu, C.; Tian, H. Global nitrogen and phosphorus fertilizer use for agriculture production in the past half century: Shifted hot spots and nutrient imbalance. *Earth Syst. Sci. Data* **2017**, *9*, 181–192. [CrossRef]
4. FAO. Fertilizers Indicators. Available online: https://www.fao.org/faostat/en/#data/EF (accessed on 18 July 2022).
5. Harris, S.L.; Thom, E.R.; Clark, D.A. Effect of high rates of nitrogen fertiliser on perennial ryegrass growth and morphology in grazed dairy pasture in northern New Zealand. *New Zealand J. Agric. Res.* **1996**, *39*, 159–169. [CrossRef]
6. Cinar, S.; Özkurt, M.; Cetin, R. Effects of nitrogen fertilization rates on forage yield and quality of annual ryegrass (*Lolium multiflorum* L.) in central black sea climatic zone in turkey. *Appl. Ecol. Environ. Res.* **2020**, *18*, 417–432. [CrossRef]
7. Abraha, A.B.; Truter, W.F.; Annandale, J.G.; Fessehazion, M.K. Forage yield and quality response of annual ryegrass (*Lolium multiflorum*) to different water and nitrogen levels. *Afr. J. Range Forage Sci.* **2015**, *32*, 125–131. [CrossRef]
8. Savci, S. Investigation of Effect of Chemical Fertilizers on Environment. *APCBEE Procedia* **2012**, *1*, 287–292. [CrossRef]
9. Siebert, J.; Sunnemann, M.; Auge, H.; Berger, S.; Cesarz, S.; Ciobanu, M.; Guerrero-Ramirez, N.R.; Eisenhauer, N. The effects of drought and nutrient addition on soil organisms vary across taxonomic groups, but are constant across seasons. *Sci. Rep.* **2019**, *9*, 639. [CrossRef]
10. Bünemann, E.K.; Schwenke, G.D.; Van Zwieten, L. Impact of agricultural inputs on soil organisms—A review. *Soil Res.* **2006**, *44*. [CrossRef]
11. European Commission. Sustainable Use of Nutrients: The Common Agricultural Policy Supports Farmers in the Safe and Efficient Use of Nutrients. Available online: https://ec.europa.eu/info/food-farming-fisheries/sustainability/environmental-sustainability/low-input-farming/nutrients_en (accessed on 18 July 2022).
12. Bell, C.W.; Asao, S.; Calderon, F.; Wolk, B.; Wallenstein, M.D. Plant nitrogen uptake drives rhizosphere bacterial community assembly during plant growth. *Soil Biol. Biochem.* **2015**, *85*, 170–182. [CrossRef]
13. Yuan, J.; Zhao, M.; Li, R.; Huang, Q.; Rensing, C.; Shen, Q. Lipopeptides produced by B. amyloliquefaciens NJN-6 altered the soil fungal community and non-ribosomal peptides genes harboring microbial community. *Appl. Soil Ecol.* **2017**, *117-118*, 96–105. [CrossRef]
14. Körschens, M.; Albert, E.; Armbruster, M.; Barkusky, D.; Baumecker, M.; Behle-Schalk, L.; Bischoff, R. Effect of mineral and organic fertilization on crop yield, nitrogen uptake, carbon and nitrogen balances, as well as soil organic carbon content and dynamics: Results from 20 European long-term field experiments of the twenty-first century. *Arch. Agron. Soil Sci.* **2013**, *59*, 1017–1040. [CrossRef]
15. N2Applied. Nitrogen Enriched Organic Fertiliser. Available online: https://n2applied.com/ (accessed on 19 July 2022).
16. Graves, D.B.; Bakken, L.B.; Jensen, M.B.; Ingels, R. Plasma Activated Organic Fertilizer. *Plasma Chem. Plasma Processing* **2019**, *39*, 1–19. [CrossRef]
17. Ingels, R.; Graves, D.B. Improving the Efficiency of Organic Fertilizer and Nitrogen Use via Air Plasma and Distributed Renewable Energy. *Plasma Med.* **2015**, *5*, 257–270. [CrossRef]
18. Edwards, C.A.; Bohlen, P.J. *Biology and Ecology of Earthworms*; Springer Science & Business Media: Berlin/Heidelberg, Germany, 1996; Volume 3.
19. Tiwari, S.C. Effects of organic manure and NPK fertilization on earthworm activity in an Oxisol. *Biol. Fertil. Soils* **1993**, *16*, 293–295. [CrossRef]
20. Edwards, C.A.; Lofty, J.R. Nitrogenous fertilizers and earthworm populations in agricultural soils. *Soil Biol. Biochem.* **1982**, *14*, 515–521. [CrossRef]
21. Edwards, C.A.; Lofty, J.R. *Biology of Earthworms*; Chapman and Hall; Wiley: London, UK; New York, NY, USA, 1977.
22. Hendrix, P.F.; Mueller, B.R.; Bruce, R.R.; Langdale, G.W.; Parmelee, R.W. Abundance and distribution of earthworms in relation to landscape factors on the Georgia Piedmont, U.S.A. *Soil Biol. Biochem.* **1992**, *24*, 1357–1361. [CrossRef]
23. Guðleifsson, B. Impact of long term use of fertilizer on surface invertebrates in experimental plots in a permanent hayfield in Northern-Iceland. *Agric. Soc. Icel.* **2002**, *15*, 37–49.
24. McLaughlin, A.; Mineau, P. The impact of agricultural practices on biodiversity. *Agric. Ecosyst. Environ.* **1995**, *55*, 201–212. [CrossRef]
25. Geisseler, D.; Scow, K.M. Long-term effects of mineral fertilizers on soil microorganisms–A review. *Soil Biol. Biochem.* **2014**, *75*, 54–63. [CrossRef]
26. Wahyuningsih, R.; Marchand, L.; Pujianto, S.; Caliman, J. Impact of inorganic fertilizer to soil biological activity in an oil palm plantation. *IOP Conf. Ser. Earth Environ. Sci.* **2019**, *336*, 012017. [CrossRef]
27. Graenitz, J.; Bauer, R. The effect of fertilization and crop rotation on biological activity in a 90 year long-term experiment. *Bodenkultur* **2000**, *51*, 99–105.
28. Geisseler, D.; Lazicki, P.A.; Scow, K.M. Mineral nitrogen input decreases microbial biomass in soils under grasslands but not annual crops. *Appl. Soil Ecol.* **2016**, *106*, 1–10. [CrossRef]
29. Wei, M.; Hu, G.; Wang, H.; Bai, E.; Lou, Y.; Zhang, A.; Zhuge, Y. 35 years of manure and chemical fertilizer application alters soil microbial community composition in a Fluvo-aquic soil in Northern China. *Eur. J. Soil Biol.* **2017**, *82*, 27–34. [CrossRef]

30. Pommeresche, R.; Løes, A.-K. Relations between Agronomic Practice and Earthworms in Norwegian Arable Soils. *Dyn. Soil Dyn. Plant Glob. Sci. Books* **2009**, *3*, 129–142.
31. Pelosi, C.; Boros, G.; van Oort, F.; Schmidt, O. Soil Oligochaeta communities after 9 decades of continuous fertilization in a bare fallow experiment. *Soil Org.* **2020**, *92*, 129–141. [CrossRef]
32. Pommeresche, R.; Løes, A.-K.; Torp, T. Effects of animal manure application on springtails (Collembola) in perennial ley. *Appl. Soil Ecol.* **2017**, *110*, 137–145. [CrossRef]
33. Silva, D.M.d.; Jacques, R.J.S.; Silva, D.A.A.d.; Santana, N.A.; Vogelmann, E.; Eckhardt, D.P.; Antoniolli, Z.I. Effects of pig slurry application on the diversity and activity of soil biota in pasture areas. *Ciênc. Rural.* **2016**, *46*, 1756–1763. [CrossRef]
34. Mousavi, H.; Cottis, T.; Hoff, G.; Solberg, S.Ø. Nitrogen Enriched Organic Fertilizer (NEO) and Its Effect on Ryegrass Yield and Soil Fauna Feeding Activity under Controlled Conditions. *Sustainability* **2022**, *14*, 2005. [CrossRef]
35. Schinner, F.; Öhlinger, R.; Kandeler, E.; Margesin, R. *Methods in Soil Biolology*, 1st ed.; Springer: Berlin/Heidelberg, Germany, 1996. [CrossRef]
36. Fründ, H.-C.; Graefe, U.; Tischer, S. Earthworms as Bioindicators of Soil Quality. In *Biology of Earthworms*; Karaca, A., Ed.; Springer: Berlin/Heidelberg, Germany, 2011; pp. 261–278.
37. Machado, J.; Oliveira Filho, L.C.; Santos, J.; Paulino, A.; Baretta, D. Morphological diversity of springtails (Hexapoda: Collembola) as soil quality bioindicators in land use systems. *Biota Neotrop.* **2019**, *19*, e20180618. [CrossRef]
38. Yara. YaraMila™. Available online: https://www.yara.com/crop-nutrition/products-and-solutions/global-fertilizer-brands/yaramila/ (accessed on 20 July 2022).
39. Yara. YaraLiva™. Available online: https://www.yara.com/crop-nutrition/products-and-solutions/global-fertilizer-brands/yaraliva/ (accessed on 20 July 2022).
40. Lumatek. ATS300W Specs Sheet. Available online: https://lumatek-lighting.com/lumatek-ats-300w/ (accessed on 21 July 2022).
41. Yiotis, C.; McElwain, J.C.; Osborne, B.A. Enhancing the productivity of ryegrass at elevated CO2 is dependent on tillering and leaf area development rather than leaf-level photosynthesis. *J. Exp. Bot.* **2020**, *72*, 1962–1977. [CrossRef] [PubMed]
42. TerraProtecta. The Bait-Lamina Test. Available online: http://www.terra-protecta.de/en/bait_strips.html (accessed on 24 July 2022).
43. Törne, E.v. Assessing feeding activities of soil-living animals. Bait-lamina-tests. *Pedobiologia* **1990**, *34*, 89–101.
44. Hamel, C.; Schellenberg, M.P.; Hanson, K.; Wang, H. Evaluation of the "bait-lamina test" to assess soil microfauna feeding activity in mixed grassland. *Appl. Soil Ecol.* **2007**, *36*, 199–204. [CrossRef]
45. Kratz, W. Bait-lamina test, general aspect, applications and perspectives. *Environ. Sci. Pollut. Res.* **1998**, *5*, 3.
46. Jänsch, S.; Scheffczyk, A.; Römbke, J. The bait-lamina earthworm test: A possible addition to the chronic earthworm toxicity test? *Euro-Mediterr. J. Environ. Integr.* **2017**, *2*, 5. [CrossRef]
47. ISO18311(E). Soil Quality–Method for Testing Effects of Soil Contaminants on The Feeding Activity Of Soil Dwelling Organisms in The Field–Bait-Lamina Test" for Approval for The Next Stage (FDIS) until 2015-04-17 (see Recommendation Paris-2). 2016. Available online: https://www.iso.org/standard/62102.html (accessed on 21 July 2022).
48. Gardi, C.; Montanarella, L.; Arrouays, D.; Bispo, A.; Lemanceau, P.; Jolivet, C.; Mulder, C.; Ranjard, L.; Römbke, J.; Rutgers, M.; et al. Soil biodiversity monitoring in Europe: Ongoing activities and challenges. *Eur. J. Soil Sci.* **2009**, *60*, 807–819. [CrossRef]
49. Simpson, J.E.; Slade, E.; Riutta, T.; Taylor, M.E. Factors Affecting Soil Fauna Feeding Activity in a Fragmented Lowland Temperate Deciduous Woodland. *PLoS ONE* **2012**, *7*, e29616. [CrossRef]
50. Rożen, A.; Sobczyk, Ł.; Liszka, K.; Weiner, J. Soil faunal activity as measured by the bait-lamina test in monocultures of 14 tree species in the Siemianice common-garden experiment, Poland. *Appl. Soil Ecol.* **2010**, *45*, 160–167. [CrossRef]
51. Verhoef, H.A.; Brussaard, L. Decomposition and nitrogen mineralization in natural and agroecosystems: The contribution of soil animals. *Biogeochemistry* **1990**, *11*, 175. [CrossRef]
52. Hopkin, S.P. *Biology of the Springtails (Insecta-Collembola)*; Oxford University Press: Oxford, UK, 1997; p. 330.
53. Tullgren. Ein sehr einfacher Ausleseapparat für terricole Tierformen. *Z. Angew. Entomol.* **1918**, *4*, 149–150. [CrossRef]
54. Berlese, A. *Apparecchio per Raccogliere Presto ed in Gran Numero Piccoli Arthropodi*; Kessinger Publishing, LLC: Whitefish, MT, USA, 1905; Volume 2, pp. 85–89.
55. O'Connor, F.B. The extraction of Enchytraeidae from soil. *Prog. Soil Zool.* **1962**, 279–285.
56. Bouché, M.B. Strategies lombriciennes. *Ecol. Bull.* **1977**, *25*, 122–132.
57. Graff, O. Unsere Regenwürmer (Lexikon für Freunde der Bodenbiologie). *Z. Pflanz. Bodenkd.* **1984**, *147*, 530. [CrossRef]
58. Sims, R.W.; Gerard, B.M.; London, L.S.; Association, E.B.W.S. *Earthworms: Keys and Notes for the Identification and Study of the Species; Linnean Society of London and the Estuarine and Brackish-Water Sciences Association*; Linnean Society of London and the Estuarine and Brackish-Water Sciences Association, Brill Archive: London, UK.
59. Birkhofer, K.; Baulechner, D.; Diekötter, T.; Zaitsev, A.; Wolters, V. Fertilization Rapidly Alters the Feeding Activity of Grassland Soil Mesofauna Independent of Management History. *Front. Ecol. Evol.* **2022**, *10*, 864470. [CrossRef]
60. Butterfield, J. Changes in decomposition rates and Collembola densities during the forestry cycle in conifer plantations. *J. Appl. Ecol.* **1999**, *36*, 92–100. [CrossRef]
61. Christiansen, K.A.; Bellinger, P.; Janssens, F. Chapter 55 - Collembola: (Springtails, Snow Fleas). In *Encyclopedia of Insects*, 2nd ed.; Resh, V.H., Cardé, R.T., Eds.; Academic Press: San Diego, CA, USA, 2009; pp. 206–210.
62. Gergócs, V.; Flórián, N.; Tóth, Z.; Szili-Kovács, T.; Mucsi, M.; Dombos, M. Crop species and year affect soil-dwelling Collembola and Acari more strongly than fertilisation regime in an arable field. *Appl. Soil Ecol.* **2022**, *173*, 104390. [CrossRef]

63. Wang, S.; Chen, H.Y.H.; Tan, Y.; Fan, H.; Ruan, H. Fertilizer regime impacts on abundance and diversity of soil fauna across a poplar plantation chronosequence in coastal Eastern China. *Sci. Rep.* **2016**, *6*, 20816. [CrossRef]
64. Lalthanzara, H.; Ramanujam, S.N. Effect of fertilizer (NPK) on earthworm population in the Agroforestry system of Mizoram, India. *Sci. Vis.* **2010**, *10*, 159–167.
65. Ernst, D. *The Farmer's Earthworm Handbook: Managing Your Underground Money-Makers*; Lessiter Publications, Inc.: Brookfield, WI, USA, 1995.

agronomy

Article

Hydrological Drought-Indexed Insurance for Irrigated Agriculture in a Highly Regulated System

Miguel Angel Valenzuela-Mahecha [1,2,*], Manuel Pulido-Velazquez [1] and Hector Macian-Sorribes [1]

[1] Research Institute of Water and Environmental Engineering (IIAMA), Universitat Politècnica de València (UPV), 46022 Valencia, Spain
[2] Civil and Agricultural Engineering Department, Universidad Nacional de Colombia (UNAL), Bogotá D.C. 111321, Colombia
* Correspondence: mavalenzuelam@unal.edu.co

Abstract: Water scarcity is an increasingly recurring problem for irrigated agriculture in Mediterranean regions. It is, therefore, necessary to establish technical and financial measures to enable irrigators to deal with this problem. This study presents a new index-based drought insurance scheme in an irrigation district in the Jucar river basin in Spain, a highly regulated water system. Three insurance scheme options were evaluated and, the values of the fair risk premiums, the maximum compensation, and the deductible franchise were established. These insurance schemes were designed in agreement with the preexisting drought system operating rules to reduce moral hazard and adverse selection. Risk-reducing and effective evaluation methods were used to determine the insurance coverage's viability for irrigators: standard deviation gross margin, minimum gross margin, and RMSL. The proposed insurances were also evaluated using synthetic hydrological time series generated with a stochastic ARMA model through a basin-wide water resource simulation model developed in the DSS Shell AQUATOOL. Financial indicators, such as the basis risk and claim ratio were applied to analyze the economic feasibility for insurance companies. The results show that a suitable and efficient option is an early-bird contract combined with a trigger of emergency or alert state in a multi-year contract. This type of specialized insurance helps to fill the existing gap in traditional insurance schemes for irrigated crops and offered additional coverage to farmers under drought and water scarcity conditions.

Keywords: hydrological drought; index insurance; irrigation water management; decision support system; Jucar river basin

Citation: Valenzuela-Mahecha, M.A.; Pulido-Velazquez, M.; Macian-Sorribes, H. Hydrological Drought-Indexed Insurance for Irrigated Agriculture in a Highly Regulated System. *Agronomy* **2022**, *12*, 2170. https://doi.org/10.3390/agronomy12092170

Academic Editors: António Dinis Ferreira, Raquel P. F. Guiné and António Moitinho Rodrigues

Received: 19 July 2022
Accepted: 1 September 2022
Published: 13 September 2022

1. Introduction

Irrigated agriculture plays a key role in food production, and therefore, in the food and nutrition security of the world's population [1,2]. With the impact of climate change, the already low profitability of most rainfed crops, as well as their vulnerability to climatic events (drought, heatwaves), will continue to increase the pressure to irrigate more land [3]. Extreme weather events, major biodiversity loss and ecosystem collapse, food crises, water crises, and, the failure of climate change mitigation and adaptation are currently the main global threats [4]. These risks affect the agricultural sector directly since it is not only the productive sector with the highest use and demand for water resources [5] but is also the most exposed to droughts and water scarcity. This makes it necessary to implement production schemes that allow farmers to produce more food while using the minimum amount of water possible. It is also crucial to create risk mitigation strategies that contemplate the technical and economic implications that their implementation would entail.

The agricultural production sector is affected by a variety of drought types: a meteorological drought takes place when there is a continuous shortage of rainfall; agricultural drought is associated with the deficit of moisture in the root zone of a crop in a certain place and time and a hydrological drought entails the decrease in the availability of surface and groundwater in a management system during a given period (compared to the average values) [6–8]. Irrigated agriculture mostly faces the risk of hydrological drought, since its water supply depends directly on the water available in regulatory reservoirs [9].

The sustainable management of water resource systems requires both environmental and financial sustainability. Therefore, economic factors must be considered in the rational decision-making about the use of water [10]. For instance, the European Union's Water Framework Directive (WFD) already contemplates these factors in its proposal to achieve sustainability and manage water scarcity [11–13]. At the watershed level, economic instruments, such as subsidies, water prices, water banks, and water markets have been studied [11,14–17]. Agricultural insurance has been successfully implemented as a financial instrument of agro-climatic risk management. It aims to be a system of protection for agricultural production by transferring different risks, such as drought events, hailstorms, winds, rains, and frost, to the insurance company [18,19].

In recent years, indexed or parametric insurance plans have been increasingly applied worldwide. The compensation scheme of these plans is based on the behavior of a certain index or variable linked to the risk to be covered by the insurance company, that is, payment will be made when values above or below certain pre-established thresholds of said index or variable considered are reported [18,20,21]. However, this insurance model has only been standardized for rainfed crops, and its application to irrigated crops is still being discussed. The relatively high overall costs of actuarial data capture, risk classification, moral hazard monitoring, and claim validation have hindered the development of conventional agricultural insurance markets, which is why indexed insurance is proposed as a low-cost alternative to conventional insurance products [22–26]. The difficulty of applying indexed insurance schemes lies in designing indexes or triggers that correlate as close as possible with the occurrence of claims. The chosen index must have a high correlation with potential losses and meet quality standards, such as being transparent, verifiable, easily measurable, and timely, and officially reported [24,27–32].

There are studies related to indexed insurance in irrigation districts using various index types, such as the river flow accumulations index [33], the Standardized Precipitation Index (SPI) [34], and drought indexes established in the river basin management plan [35,36], the volume of water stored in reservoirs [31,37], or a combination of rainfall and the water storage available for irrigation [38].

The Jucar river basin, the main watershed in the Valencian region, is a highly regulated Mediterranean basin with a large share of water use for irrigation (around 80%) and recurrent long and severe droughts. The economic losses caused by meteorological disasters in the Valencian region have increased by 95 million Euros in the last ten years, reaching values of almost 380 million in 2018. During this period the crops of the region were affected mostly by heavy rains and hail (50%). The losses caused by drought events were around 15%. This has led farmers to adopt conventional insurance. The total insured production has oscillated from 2.5 to 3.0 million tons every year during the last 15 years [39]. There is already a tradition in the development of institutions and multi-actor partnerships for drought management [40] in the basin, with a well-established and innovative set of drought indicators [41,42] that trigger a set of water management measures according to the river basin drought plan.

This paper proposes an alternative insurance scheme that allows farmers to face the economic impacts due to water scarcity events in irrigation districts, especially in those located in highly regulated basins, such as that of the Jucar river. This proposal is based on the implementation of indexed insurance. For the insurance design, the Jucar river basin Scarcity State index was used in three insurance scheme options: (1) variable premium and/or variable franchise based on the forecast of water availability for the insured irrigation campaign, (2) multiannual insurance contract, and (3) advance contract with a constant premium. The viability of the insurance scheme, both for the insurance companies and the beneficiaries, is assessed using financial indicators.

2. Materials and Methods

The implementation framework for the hydrological index insurance is shown in Figure 1. It consists of two main processes: the insurance design and its subsequent evaluation. It begins with general information about the basin, system state variables (such as reservoir storage and inflows), and Precipitation, which are combined and weighted to estimate the drought state index. At the same time, the irrigation district is characterized by the historical data on the crops (sown area, production, yields, production costs, and revenue), and the historical water delivered allows for an economic assessment of the drought impact.

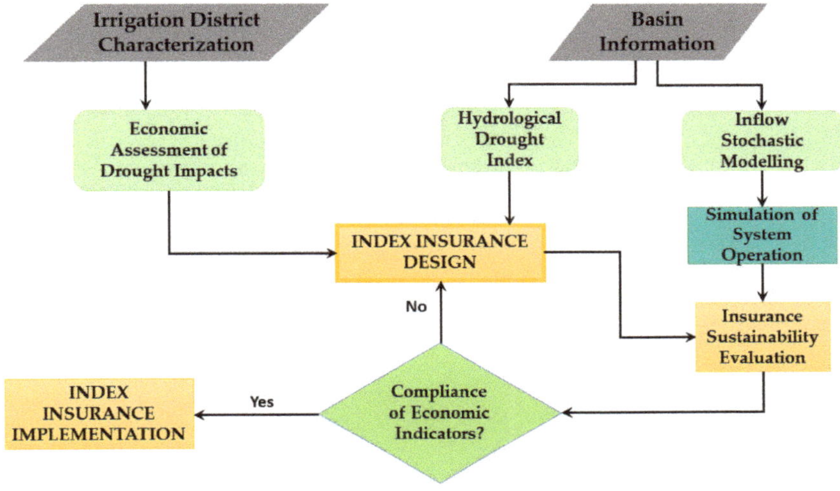

Figure 1. Framework for the design of the insurance.

For the insurance sustainability evaluation, a basin-level operation system simulation is performed. It begins with the development of an autoregressive stochastic time series model to forecast the inflows of the system and their interaction with changes in the storage in the main reservoirs. This simulation also includes demand units (agricultural, urban, industrial and ecological flows) as well as all the normativity and regulations that exist within the management of the water resource in the basin. A technical and financial sustainability assessment of the proposed insurance is made using economic indicators. If this design option meets the defined economic indicators, it is possible to validate the proposed design scheme.

The estimation of a pure (actuarially fair) hydrological drought insurance premium is usually done in two steps: (i) quantifying the risk associated with the irrigation water allocated to farmers, and (ii) quantifying the impact of water allocation on the farmers' income or revenue [36]. Afterward, premium rates, deductible franchises, and insurance contract periods of different insurance schemes are compared to design the best-fitting alternative.

To implement the insurance scheme proposed in this paper, the following principles must be taken into account: (1) the Drought Index must be included in the River Basin Management Plan; (2) the operational regulations and supply reduction rates defined in the drought plans should not change during the term of the insurance contract; (3) the administrators of the irrigation districts should be the ones that hold the insurance policy instead of each farmer; [9,31]; (4) only irrigation districts with a single source of water (surface) are fit for this insurance scheme; (5) although other factors—such as crop variety, phenological phase, the chemical and physical soil qualities, nutrition, and high temperatures—determine crop yields in some measure, water shortage should be considered the primary cause of diminishing crop yields.

2.1. Economic Assessment of Drought Impacts

This analysis is based on the study carried out by [31], where they tackle some of the main problems that arise from the design of indexed insurance, such as base risk, moral hazard, and adverse selection [43–47].

Initially, is necessary to determine the relationship between the selected hydrological drought index DSI and the historical water deliveries in the irrigation district (w_d) during the period to be analyzed. Afterward, the water shortage (w_{st}) in the irrigation district in m^3/ha is measured by subtracting the water delivered (w_d) in a given year (t) from the guaranteed water delivered (GWA). GWA is the average water delivered to the irrigation district in the period of analysis, which depends on water rights, water availability and the system operation. By comparing the volume of delivered water with the volume of stored water and the consolidated demand, the years in which there is a water shortage can be identified. Once the periods of water scarcity have been identified, the economic impact generated by droughts in the irrigation districts can be established.

To estimate the compensation that a farmer would receive from this insurance scheme, the Net Value of Agricultural Production ($NVAP_i$) had to be calculated. To do this, the area sown (s_{ic}), yield (y_c), and prices (p_c) for each crop (c)were considered [48]. There exist C crops, indexed by c = 1, . . . , C. Afterward, the irrigation costs in each crop scenario (VCI) were subtracted from this value:

$$VAP_i = \sum_{c=1}^{C} s_{ic} * y_c * p_c \tag{1}$$

$$VCI_i = \sum_{c=1}^{C} s_{ic} * VCI_{ic} \tag{2}$$

$$NVAP_i = VAP_i - VCI_i \tag{3}$$

The $NVAP_i$ was calculated for two hypothetical scenarios based on the drought State Index DSI in the River Basin Agency: (a) a normal state, where DSI > 0.5, and (b) an emergency state, where DSI < 0.15. To this end, the value of the water (wv_i) in years in which a state of emergency was declared was compared with these values in a normal scenario I where the total water needed for irrigation can be allocated:

$$wv_i = wv_i(wd_i) = \frac{NVAP_I - NVAP_i}{\mathrm{Wst}_i} \tag{4}$$

$$wv_t = wv_t(wd_t) \tag{5}$$

where w_{di} and $NVAP_i$ are water delivered and the net value of agricultural production in scenario i, respectively, and $NVAP_I$ is the net value of agricultural production in a normal scenario with fully guaranteed water allocation.

2.2. Design of the Insurance Scheme

The compensation that a farmer would receive (€/ha) in a given year (t) results from multiplying w_{st} by a unit compensation equal to the value of water (w_v) in €/m^3, where the deductible franchise (γ) is the minimum amount of loss that can be incurred before insurance coverage applies [31,49].

$$ind_t = \begin{cases} 0, \ if \ wd_t \geq (1-\gamma) * GWA \\ ws_t * wv_t \ \ if \ wd_t < (1-\gamma) * GWA \end{cases} \tag{6}$$

That is, compensation is triggered by the DSI:

$$(1-\gamma) * GWA = f(DSI = Trigger) \tag{7}$$

The liability or maximum compensable value of the insurance scheme in €/ha is determined from the expression:

$$Liability = GWA * wv_t \tag{8}$$

Finally, the insurance premium is calculated based on the expected compensation, where t represents the year in which compensation would need to be paid and T is the total number of years of the insurance analysis.

$$Premium = E(Ind_t) = \frac{1}{T} * \sum_{t=1}^{t=T} Ind_t \tag{9}$$

This methodology is implemented in the three hydrological drought insurance options described in Table 1.

Table 1. Insurance scheme options for designing.

Option	Characteristics
Option 1: Variable premium and/or variable franchise based on the forecast of water availability for the secured irrigation season.	Farmers may purchase insurance based on the conditions of scarcity presented before the start of the irrigation season (1 April). The value of the premium would depend on two franchises (state of alert and emergency scarcity) [31].
Option 2: Pre-season index contract. Multi-year insurance contract	Farmers may purchase insurance that uses certain indexes to adapt the value of its premium to the real risk at the time of the purchase. This can either be a one-year or a multi-year policy. The period of the policy would be set in October, i.e., the beginning of the harvest season. That is, the value of the premium is estimated with the DSI measured in October t and the compensation is calculated with the DSI in 1 April t + 1 [50].
Option 3: Early Contract with a Constant Premium (Early Bird)	Farmers may purchase insurance at a constant premium. However, they would have to buy it early, before the drought can be predicted [51].

2.3. Selection of the Most Appropriate Insurance Plan

In agriculture index-based insurance, the high risk-reducing effectiveness of the contract was most frequently assessed from a minimized variance or downside risk in the income, with and without insurance [32]. Financial indicators are used to determine the insurance scheme's viability, both for the insurance companies and the beneficiaries. Risk-reducing and effective evaluation methods were used to determine the insurance coverage's viability for irrigators: standard deviation gross margin, minimum gross margin, and mean root square loss RMSL (Table 2).

In the analysis, it is important to consider additional loadings on the fair premiums of 10% and 37% which may represent acquisition effort, administrative expense, risk-bearing (i.e., reinsurance costs), and profit allocation [37].

Table 2. Performance Indicators for assessing the insurance schemes.

Indicator	Description
% Basis Risk	Expected difference between the compensation received by the farmer with the insurance (ind_t) and the current losses calculated based on the historical water deliveries. The base risk can be broken down into base loss (probable farmer losses due to index insurance) and base gain (probable farmer gains due to index insurance) [31].
RMSL (Gross margin)	It is a simple function of the semivariance (i.e., losses) with respect to the trend of the gross margin without Insurance [31,52,53].
Minimum Gross Margin	Minimum gross margin in productive systems, with and without insurance (revenue—direct costs), in historical time series [31,54].
Claim ratio	The most commonly used indicator for assessing the performance of an insurance or reinsurance undertaking is the claim rate [18]. This index makes it possible to determine whether the price fixed for a given insurance scheme is correct; that is, whether it actually allows for the settlement of claims arising in a given period. Claim Ratio = claim incurred/premium collected

Simulation of the Insurance Design through a Water Resource System Model

The feasibility of the designed insurance schemes, as well as the potential benefits for both farmers and insurers, is determined through a retrospective analysis that calculates the value of the claims in a hypothetical scenario where coverage had been in operation during previous years [55]. For this, financial indicators, such as the basis risk and claim ratio let insurance companies analyze their feasibility (Table 2).

To evaluate these insurance schemes, synthetic inflow time series are generated with a stochastic model and included in a simulation model that includes drought and water management procedures of the water resource system. This DSS is used to analyze the impacts of droughts, including management rules based on drought indicators [56–58].

If the claim ratio were less than 1, it would indicate that the premium collected would be higher than the compensation paid, and therefore, the insurer would make a profit. Were it to be 1, this would imply a balance between the compensation paid and the premium. Were it to be higher than 1, the insurer would incur some losses in as much as the compensation paid would be greater than the premium collected during the evaluated period.

2.4. *Case Study*

2.4.1. Description

The Real Acequia de Escalona (RAE), an irrigation canal of 23 km located in the lower Jucar river basin Figure 2, brings water from a dam located downstream of the Tous reservoir and ends in the municipality of Villanueva de Castellón (Source: https://sequiaescalona.org/quienes-somos) (accessed on 15 September 2021). Four irrigation districts (Comunidad de Regantes, CR), about 2700 ha in total, benefit from the water provided by the RAE: CR Real Acequia de Escalona (66.07%), CR Sumacárcer (12.69%), CR del Valle de Cárcer y Sellent (12.58%) and CR La Defensa de los Derechos de Riego de las Tierras del Valle de Cárcer (8.66%). We chose two citrus crops (orange and tangerine) to analyze in this study since they are the most representative in the area (approximately 90% of the total crops). The irrigation districts in the RAE have particular qualities whose analysis may be very useful for insurance design. For instance, these communities have the right to administer and distribute the water allocated to them, as well as a legal status that grants them the economic management of the water in that area.

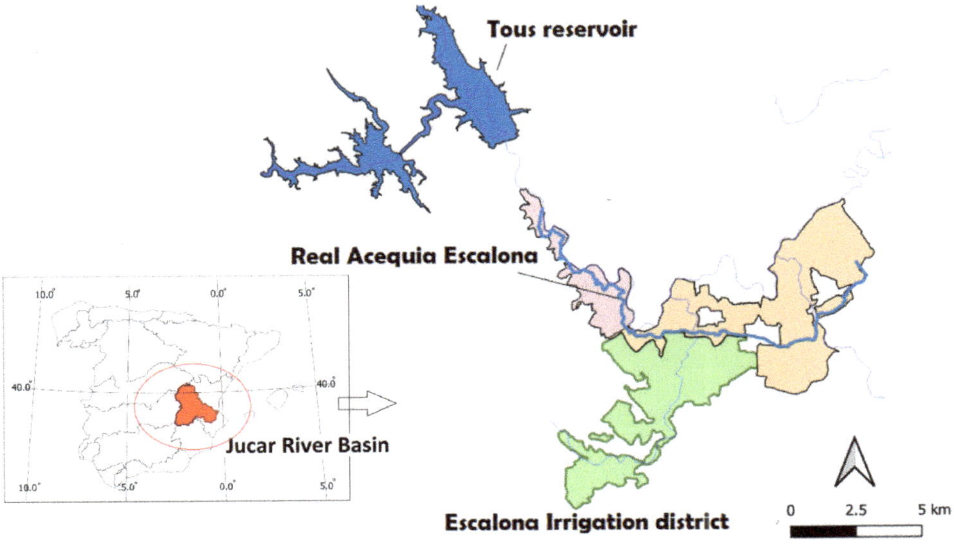

Figure 2. Area of influence Real Acequia de Escalona RAE.

2.4.2. Drought State Index

The current drought indicator system in Spain is determined by an operational index, the Drought State Index (DSI). This index reflects the amount of available water for the end-users in each month, concerning the amount of available water for that month historically. The index consists of a combination of some selected control variables distributed throughout the river basin, including storage in surface reservoirs; piezometric levels; river discharges; reservoir inflows, and precipitations in those areas where they are significant in relation to water resources availability [59,60]. This is done by relating ten measurements of representative variables of the river basin (precipitation (one variable), piezometric levels (three variables), flows (four variables), and storage (two variables). Previously, to give a dimensionless numerical value that allows them to be compared on a single scale, all variables should be stationarized in order to filter out the seasonal component and avoid the influence of the annual meteorological cycle in the calculation of the indicator in a given month. The DSI is then calculated as [59]:

$$if\ V_i \geq V_{med},\quad DSI = \frac{1}{2}\left[1 + \frac{V_i - V_{med}}{V_{p95} - V_{med}}\right] \tag{10}$$

$$if\ V_i < V_{med},\quad DSI = \frac{V_i - V_{p5}}{2(V_{med} - V_{p5})} \tag{11}$$

In this equation, V represents the value of the seasonal variable in each month being considered (i); Vmed represents the mean value of the whole series, which in this case goes from 1980 to 2012; and Vp_5 and Vp_{95} represent the 5th and 95th percentile of that series. The DSI may range from 0 to 1, which allows the scarcity situation to be classified into four levels: Normal (DSI > 0.5), Prealert (0.5 > DSI > 0.3), Alert (0.3 > DSI > 0.15) and Emergency (DSI < 0.15). When the DSI falls into the Prealert or Alert status, this means that there is a moderate to severe water shortage, while an Emergency status means that there is a serious shortage.

Afterward, the impact of each one of the ten variables is weighted to combine them properly into the same equation. The river basin agency determines the weight of each variable from the relative volume of demand to be supplied by the water resources represented by each indicator, which was then adjusted and validated by comparison to previous drought situations [59]. This is done to ensure that, based on the historical data logged or supplied by simulation models, the alert and emergency scenarios detected by the system represent the historical scarcity circumstances gathered in the basin as closely as feasible. This index allows us to determine when and how often water scarcity occurs in the river basin, as well as the impact it has on a community of irrigators since operating rules or restrictions on the use of water are introduced based on it.

Figure 3 presents the DSI used in this study for insurance analysis (1980–2019). According to this index, the following periods were identified in the Jucar river basin: there was a drought from 1982/1983 to 1985/1986, followed by a wet period between 1987/1988 and 1990/1991. Then came the drought of 1991/1992–1994/1995, followed by some rainy years (they were, however, not so rainy as the previous wet period). This, in turn, prevented the years 1997/1998–2000/2001 from being particularly dry. Afterward came the drought from 2004/2005 to 2007/2008, which, despite not taking place during the years with the least rainfall recorded in the Júcar region during the analyzed period, was the most severe [59]. Finally, another drought period began in early 2018.

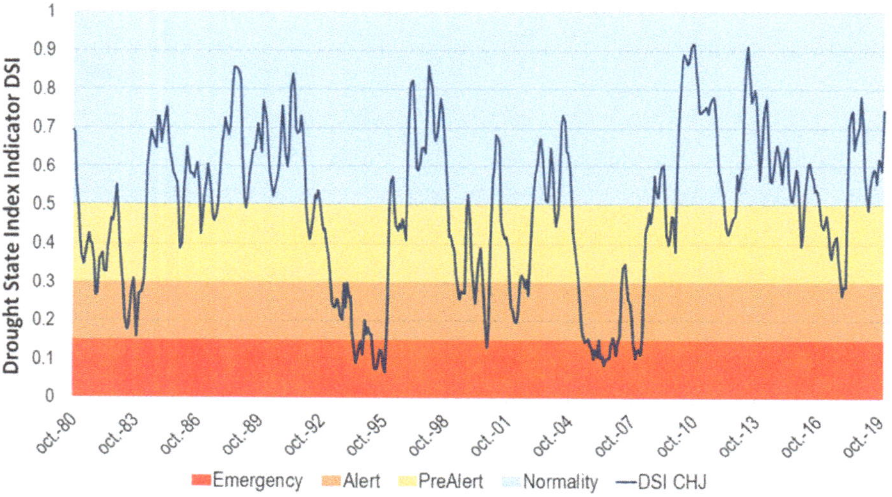

Figure 3. Drought State Index in the Jucar river basin DSI_{CHJ} (1980–2019). Source: modified from [58].

2.4.3. Water Scarcity in the RAE

According to the Júcar River Basin Management Plan, the RAE has a gross demand of 25.36 Hm^3 of water per year. However, an analysis of the historical records (2006–2019), shows that a lower volume of water has been delivered in the irrigation district in 9 of those 14 years. This may be due to periods of water scarcity, deliberate decisions from the planners of the Jucar River Basin Agency (CHJ Confederacion Hidrografica del Jucar), or the advancement in irrigation techniques, which nowadays require a lower volume of water to irrigate wider areas.

Given the above, to identify actual periods of water scarcity, we determined the storage in the three main reservoirs that regulate the water supply in the Jucar river basin on 1 April and compared it to the actual volume of water delivered to the RAE (Figure 4). The drawn red line represents the minimum amount of water that ought to be delivered during a drought period (16.5 Hm^3). There is a direct relationship between the amount of water stored and the amount of water that was delivered for irrigation. According to this analysis, between 2005 and 2019, there were three years in which less water than the minimum volume was delivered, which means that a hydrological drought took place during those years (2005–2008). Moreover, even though there was a relatively high initial storage in the reservoirs during 2008/2009, 2009/2010, and 2018/2019, the actual amount of water delivered during those years was very close to the minimum, which generated economic impacts in the irrigation districts.

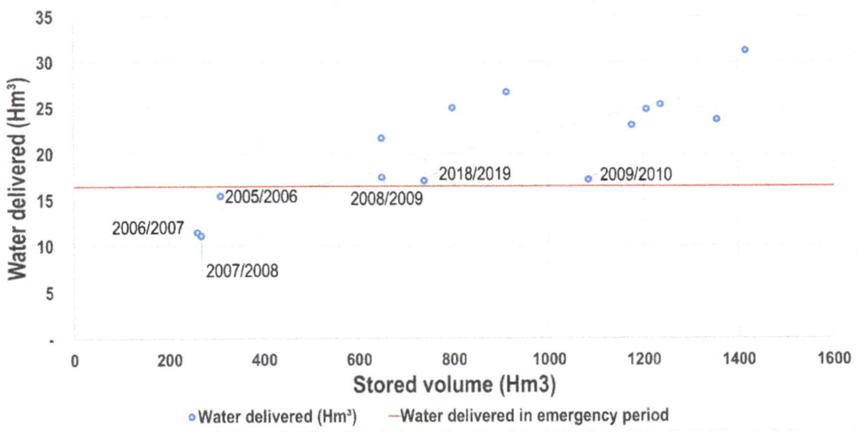

Figure 4. Water stored on 1 April vs water delivered to the RAE.

3. Results

3.1. Correlation between Drought Index DSI and Water Deliveries

The model used to estimate the irrigation water deliveries in comparison with the DSI measured on 1 April, i.e., before the irrigation season, in the RAE (Equation (12)) is obtained by considering the historical water deliveries to the irrigation districts (wd_t). Figure 5 shows the high correlation between these variables obtained in the model.

$$Wdt = 28.77 * DSI_{CHJ}{}^{0.4273}\ R^2 = 0.75 \quad (12)$$

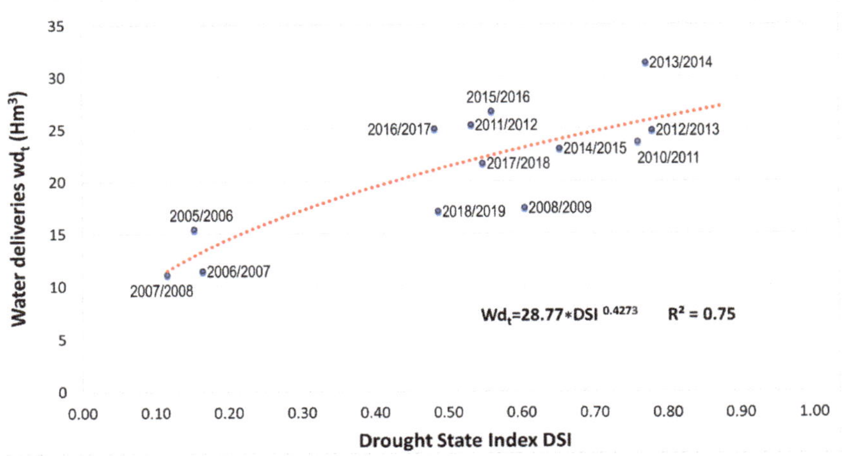

Figure 5. Water deliveries vs. Drought State Index.

3.2. Measuring the Economic Impact of Hydrological Droughts

To determine the impact of a drought, the management measures established in the river basin drought management plan must be taken into account, since they dictate that the volume of water delivered may be reduced up to 35% when a drought brings about a state of emergency [59].

The second step is to analyze the impact of the reduced water availability on crop yields, for which we consider the work carried out by [15], who simulated the change in citrus-fruit yield according to different changes in water availability in the Jucar Basin. These changes were the consequence of deficit irrigation and allow for the determination of the relationship between yield and net irrigation (In) (m^3/ha).

$$Yield\left(\frac{Kg}{ha}\right) = -0.00000003In^3 + 0.0003In^2 - 0.0868In + 21495 \quad (13)$$

To assess the cost of scarcity in €/m^3 of water, historical records of production, sown area, yields, sales prices, and production costs from 2006 to 2019 in the Valencian Community were considered. Since there are several sources of information, after a selection process, the reports of the Valencian agricultural sector were chosen as the most fitting (These reports can be accessed at: http://www.agroambient.gva.es/es/informes-sector-agrario-anos-anteriores) (accessed on 15 October 2020).

A regular net demand of 4107.33 m^3/ha-year has been established in the Júcar Hydrological Plan for the RAE. However, during a state of emergency, the allocated water is reduced by 35%. By applying the López citrus water productivity model (2017), insurers may find that crop yields would be reduced by approximately 1290 kg/ha during these periods, i.e., 5.35% of the production. With this in mind, a compensation value can be established. In this case, it was set to 0.09 €/m^3, which, considering that 7789 m^3/ha of water had been historically guaranteed in the region, results in a maximum compensable value of 701 €/ha.

3.3. Hydrological Index Insurance Design

- Option 1: Variable premium and/or variable franchise based on the forecast of water availability for the insurance irrigation season.

In this proposal, the value of the premium is calculated based on the correlation of the expected yield, the historical values of the DSI, and the expected amount of water delivered (Equation (12)) in the event of an alert or emergency scenario (triggered by a 0.3 and a 0.15 DSI, respectively) additionally, an intermediate scenario (DSI = 0.2) was used. The above allows for the establishment of three different deductible franchise values (according to the DSI), which are presented in Table 3. There is a significant difference in the premium values.

Table 3. Premium rates for a Hydrological Drought Insurance Option 1.

DSI_{CHJ} Trigger **Deductible Franchise**	**0.5** **γ = 0%**	**0.3** **γ = 19%**	**0.2** **γ = 31%**	**0.15** **γ = 37%**	**0.3** **γ = 0%**	**0.2** **γ = 0%**	**0.15** **γ = 0%**
Premium rate (€/ha)	293.00	170.00	171.00	158.00	127.00	60.00	30.00
% liability	41.80	24.25	24.39	22.54	18.19	8.56	4.28

- Option 2: Pre-season index contract. Multi-year insurance contract

Three states of scarcity were selected: Normal, Alert, and Emergency. As depicted in Table 4, the value of the premiums differs significantly depending on the pre-season shortage scenario and the selected index. Considering the emergency state as an example, the premium value for a 1-year contract ranges between 292 and 947 €/ha/year, which corresponds to 41.65% and 135.09% of the maximum compensable value. Similarly, Table 4 shows that another possibility to reduce the premium rate in the years when a state of emergency is reached is to contract a multiannual policy, depending on the scenario or state of scarcity taking place at the time of purchase of the policy. With a 3-year contract, premiums could range between 268 and 871 €/ha (38.23% and 124.25% of the maximum compensable value). The latter being such a high value, it is not viable for implementation.

Table 4. Premium rates for a Hydrological Drought Insurance Option 2, using a deductible franchise of 37% for 1, 2, and 3 years of the policy contract.

Pre-Season Index Contracts DSI_{CHJ}	Premium 1 Year-Contract		Premium 2 Year-Contract		Premium 3 Year-Contract	
	€/ha/year	% Liability	€/ha/year	% Liability	€/ha/year	% Liability
Normality	292.00	41.65	243.00	34.66	268.00	38.23
Pre alert—Alert	542.00	77.32	451.00	77.32	498.00	71.04
Emergency	947.00	135.09	788.00	112.41	871.00	124.25

- Option 3: Early Contract

The correlation between past and future water volumes is no longer significant (that is, the system loses its memory) when data from 32 months before the start of the irrigation season are considered. In those cases, the degree of significance of the correlation would be lower than 90%. For a significance of 99%, the water-allocation data considered for predicting storage in the present cannot be more than 16 months. Therefore, to guarantee a reasonable degree of significance in the correlation, these types of multiannual contracts should be renewed every 2 years.

3.4. Assessment of Insurance Schemes Proposed in the RAE Irrigation District

To establish the effectiveness of the proposed insurance scheme, Table 5 presents a comparison of different options, for a period of analysis between 2006 and 2018, across the following parameters: standard deviation gross margin, the RMSL, and the minimum gross margin with and without the insurance contract.

- Basis Risk

Ideally, the base loss and base gain should be similar, so the insurance system does not favor either the farmer or the insurance company [31]. The basis risk of the insurance scheme is calculated by breaking it down into basis loss and basis gain. This is done, in turn, by comparing insurance system indemnities, based on drought indexes, with the potential compensation (calculated from the records of delivered water between 2006 and 2019). In the current insurance proposal, the basis gain is, in most cases, higher than the basis loss. This entails that the insured farmer would receive more than the expected compensation, which does not favor the insurance company. However, there is one case where the insurance company would benefit greatly. When a DSI of 0.15 is reached and the γ is 0%, the value of basis loss will be negative 9.13%. This means that despite suffering economic losses under these circumstances the farmer that took the insurance would not receive any compensation from the insurance company (Table 5).

- Standard Deviation Gross Margin (€/ha), Minimum Gross Margin (€/ha), RMSL (€/ha)

The lower the RMSL value for comparing two structures, the greater the structural similarity of the structures. The Mean Root Square Loss (RMSL) is appropriate because minimizing semivariance instead of complete variance is relevant as farmers are primarily interested in managing their losses downwards [49,50]. Therefore, if the RMSL reduces with insurance, the contract is efficient [50]. Considering the three criteria expressed above, the insurance condition with a value of the DSI index = 0.2 and no deductible franchise γ = 0% would be selected. Although it does not meet all the conditions, another option that approaches is the value of the index DSI = 0.3 and no deductible franchise γ = 0%.

Table 5. Risk-reducing effectiveness evaluation indicators applied in RAE.

Insurance Scheme	Additional Loadings	Standard Deviation (€/ha)	Minimum Gross Margin (€/ha)	RMSL (€/ha)	Basis Risk (Loss) (%)	Basis Risk (Gain) (%)
No insurance		567.9	1193.1	84.1		
γ = 37% DSI = 0.15	0% 10% 37%	551.5 551.5 551.5	1156.6 1140.8 1098.1	96.1 103.2 122.7	3.85	12.55
γ = 19% DSI = 0.3	0% 10% 37%	550.8 550.8 550.8	1144.6 1127.6 1081.7	101.5 109.2 130.3	3.85	12.84
γ = 0% DSI = 0.5	0% 10% 37%	597.3 597.3 597.3	1021.6 992.3 913.2	158.3 172.4 211.9	2.71	25.96
γ = 0% DSI = 0.3	0% 10% 37%	550.8 550.8 550.8	1187.1 1174.3 1139.9	82.6 88.2 103.6	3.85	12.84
γ = 0% DSI = 0.15	0% 10% 37%	626.6 626.6 626.6	1163.1 1160.1 1152.0	99.2 100.7 104.9	9.13	3.85
γ = 0% DSI = 0.2	0% 10% 37%	550.8 550.8 550.8	1254.6 1248.6 1232.4	54.9 57.2 63.5	3.85	12.84
γ = 31% DSI = 0.2	0% 10% 37%	550.8 550.8 550.8	1143.6 1126.5 1080.3	102.0 109.7 130.9	3.85	12.84

3.5. Simulation of Insurance Design Based on the Implementation of a Management Model

The MASHWIN and SIMGES modules of the AQUATOOL software were implemented. The demands included in the model were both agricultural and urban. Figure 6 presents the scheme of the simulation. MASHWIN included a monthly stochastic analysis model intended for the study of the water inflows to a hydraulic system of different sources during a certain period. MASHWIN combines a multivariate ARMA auto-regressive and moving-mean model with the monthly spatial disaggregation analysis provided by the condensed Lane model [60].

Then, by generating 100 synthetic series of water inflows, with SIMGES the volumes of different reservoirs, as well as their water inputs, outputs, and evaporation rates were obtained. Finally, these data were used as input to calculate the DSI and simulate an insurance scenario.

When the simulation was run with a DSI trigger of 0.2, the average value of the claim ratio was 0.74 and the claim rate ranged between 0.6 and 1.4 in 82% of the cases (Figure 7). On the other hand, when the simulation was run with a SI of 0.15 and a γ = 37%, the average value of the claim ratio was 1.1. The claim rate peaked at 1.6 under these conditions, and no claims were made in 52% of the cases.

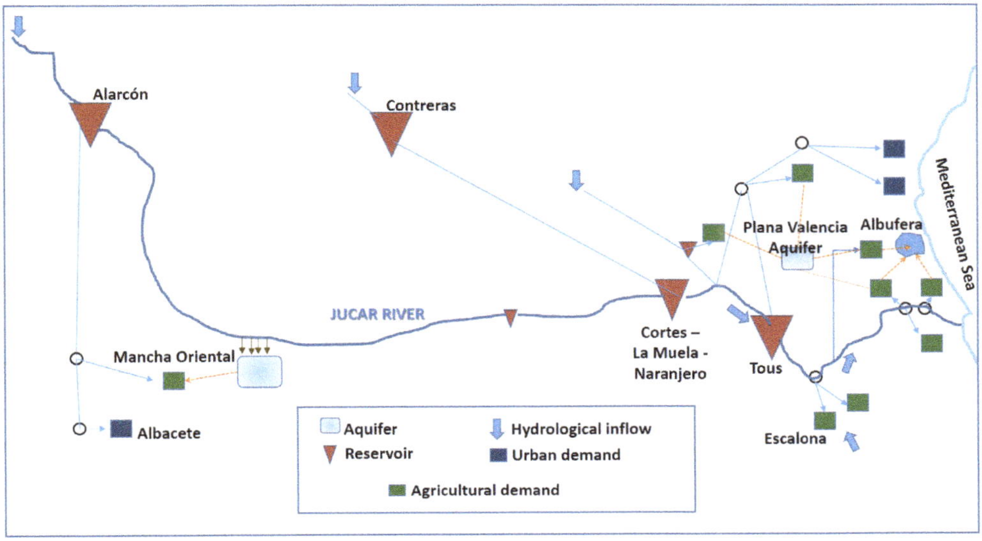

Figure 6. Simplified schematic of the Júcar river system. Source: [56,61].

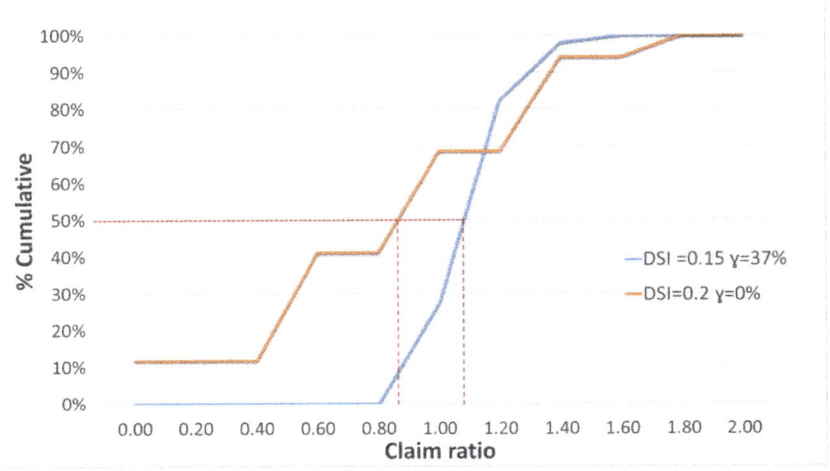

Figure 7. % Claim Ratio accumulated for two values of Drought State Index and two percentages of deductible Franchise.

4. Discussion

Traditional agricultural insurances are contracts between the farmer and the insurer, indemnifying either the cost of production, planting, or installations or a combination of them. However, drought or water scarcity impacts on irrigated crops are not included. In this study, drought insurance covers the value of the economic impact generated by reducing water deliveries to the irrigation district as the insurable capital. This type of specialized insurance attempted to offer additional coverage to farmers under drought and water scarcity conditions [50], filling the existing gap in traditional insurance schemes for irrigated crops.

A key novelty of the approach is the use of a water resource management model, which allows the assessment of the insurance considering the systemwide features and the variation of resources and demands. This improvement is important for evaluating indexed hydrological drought-insurance schemes since the exposure to drought events of each particular agricultural demand will depend on the water resource system operating rules and the variability in resources and demands over time and space. Unlike the case of rainfed agriculture, water deliveries to the irrigated demands will be driven by water availability in the system. Even in a situation of meteorological drought, there might be enough water storage to cope with it without restrictions, and on the contrary, water storage could be heavily depleted even in the absence of a meteorological drought. This is a challenge for the development of hydrological insurance.

The evaluation of the proposed insurance scheme was carried out using stochastic modeling, which allows for the generation of a wide range of inflow scenarios and drought status index values. The suitability of the suggested methodology requires a strong correlation between the drought status index and water deliveries to irrigation. A weak correlation between the mentioned variables would negatively affect the basis risk, premium values, and claim indemnity results and, as a consequence, the insurance applicability.

By including an official drought state index established by the River Basin Agency, together with proper design schemes, trigger values, and deductible franchises, the incidence of moral hazard problems and adverse selection in irrigated-drought insurance is reduced. With this approach, both individual and combined options for indexed drought insurance were considered. A suitable option is an early-bird contract combined with a trigger of emergency or alert state in a multi-year contract. On the other hand, in the analysis of the insurance scheme based on a "pre-index" (an index based on the values at the beginning of the hydrological year), the premiums get high for the farmers. For instance, if an alert or emergency scenario were declared in October before the irrigation season (April–September), the whole irrigation season would most likely be affected by water scarcity. Insurance companies would not see it profitable to offer insurance in such cases, which is why the present study suggests offering multi-year contracts with at least three years of validity.

5. Conclusions

The proposed framework assesses hydrological drought insurance schemes for irrigated agriculture considering water scarcity at the basin level through an economic assessment of drought impacts. This approach includes financial indicators to evaluate the viability of the insurance scheme for the insurance companies and the beneficiaries, using multiple stochastic realizations of inflow time series and the simulation of the water system's operation.

The comparison between purchasing or not purchasing insurance was assessed by integrating financial indicators from the perspective of the farmer and insurance companies. The selection of the design alternative is based on minimizing the semi-variance (RMSL) and gross margins standard deviation and maximizing the minimum gross margin. In the basis risk analysis in the current insurance proposal, the basis risk gain is in most cases higher than the loss. This entails that the insured farmer would receive more than the expected compensation, which does not favor the insurance company. The basis risk and claim ratio analysis should have a balance between the benefits for the insurance companies and for the farmers.

Given the frequent and severe drought events that affect water availability in the Jucar river basin, the implementation of this type of insurance is not meant to replace the agricultural insurance currently provided, but rather to serve as a complementary insurance alternative for irrigated agriculture. Its premium values are reasonable for farmers compared to the benefits it can provide, and also with multi-year contract options it can become a viable business alternative for insurance companies.

The estimation of the economic impacts of water scarcity on agriculture is challenging due to the variability of driving factors, such as prices, quality, and demand for products. As future research, this study can be complemented with an estimation of the impact of drought analysis on revenue and production costs. For instance, water scarcity may decrease product quality, lowering its price. The level of adoption of the proposed insurance must also be studied from a social-economic perspective, considering its viability and aspects, such as willingness to pay and the inclusion of possible subsidies from the basin agencies or government. Additionally, future studies can also define drought-related indexes with a higher correlation with water deliveries for irrigation and the corresponding economic losses.

Author Contributions: Conceptualization, M.A.V.-M. and M.P.-V.; methodology, M.A.V.-M., M.P.-V. and H.M.-S.; validation, M.A.V.-M., M.P.-V. and H.M.-S.; formal analysis, M.A.V.-M., M.P.-V. and H.M.-S.; investigation, M.A.V.-M., M.P.-V. and H.M.-S.; writing—original draft preparation, M.A.V.-M.; writing—review and editing, M.A.V.-M., M.P.-V. and H.M.-S.; supervision, M.P.-V. and H.M.-S.; funding acquisition, M.A.V.-M. All authors have read and agreed to the published version of the manuscript.

Funding: This research and the APC were funded by Universidad Nacional de Colombia and the doctoral student training program of Ministry of Science, Technology, and Innovation of Colombia (MinCiencias). It has also received funding from the European Union's Horizon 2020 research and innovation programme under the GoNEXUS project (grant agreement No 101003722) and the ADAPTAMED project (RTI2018-101483-B-I00) funded by the Ministerio de Economia y Competitividad (MINECO) of Spain including EU FEDER funds.

Data Availability Statement: The data presented in this study are available on request from the corresponding author.

Conflicts of Interest: The authors declare no conflict of interest.

References

1. FAO. Agua. Available online: https://www.fao.org/water/es/ (accessed on 1 February 2022).
2. The World Bank. Water in Agriculture. Available online: https://www.bancomundial.org/es/topic/water-in-agriculture#1 (accessed on 1 February 2022).
3. Berbel, J.; Espinosa Tasón, J. La gestión del regadío ante la escasez del agua: El caso de España. In *Inundaciones y Sequías. Análisis Multidisciplinar Para Mitigar el Impacto de los Fenómenos Climáticos Extremos*; Universitat d'Alacant: Joaquín Melgarejo, Ma Inmaculada López-Ortiz, Patricia Fernández-Aracil, 2021; pp. 411–416. Available online: http://hdl.handle.net/10317/10392 (accessed on 1 February 2022).
4. Future Earth. Our Future on Earth. 2020. Available online: www.futureearth.org/publications/our-future-on-earth (accessed on 1 September 2022).
5. Alamanos, A.; Mylopoulos, N.; Loukas, A.; Gaitanaros, D. An Integrated Multicriteria Analysis Tool for Evaluating Water Resource Management Strategies. *Water* **2018**, *10*, 1795. [CrossRef]
6. MITECO. Tipos de Sequía. In Observatorio Nacional de la Sequía. Ministerio Para la Transición Ecológica y el Reto Demográfico; Madrid, Spain. 2018. Available online: https://www.miteco.gob.es/es/agua/temas/observatorio-nacional-de-la-sequia/ (accessed on 15 December 2021).
7. Van Loon, A.F. Hydrological drought explained. *WIREs Water* **2015**, *2*, 359–392. [CrossRef]
8. Tallaksen, L.M.; Van Lanen, H.A. *Hydrological drought. Processes and estimation methods for streamflow and groundwater*; Tallaksen, L., Van Lanen, H.A.J., Eds.; Elsevier: Amsterdam, The Netherlands, 2004.
9. Gómez-Limón, J.A.; Guerrero-Baena, M.D. Diseño de un seguro indexado para la cobertura del riesgo de sequía hidrológica en la agricultura de regadío. *Agua Territ.* **2019**, *3*, 79–92. [CrossRef]
10. Pulido-Velazquez, M.; Cabrera Marcet, E.; Garrido Colmenero, A. Economía del agua y gestión de recursos hídricos. *Ing. Agua* **2014**, *18*, 95. [CrossRef]
11. Expósito, A.; Beier, F.; Berbel, J. Hydro-Economic Modelling for Water-Policy Assessment Under Climate Change at a River Basin Scale: A Review. *Water* **2020**, *12*, 1559. [CrossRef]
12. Comisión Europea. *Informe Sobre la Revisión de la Política Europea de Lucha Contra la Escasez de Agua y la Sequía*; Comisión Europea: Bruselas, Belgium, 2012; Available online: https://eur-lex.europa.eu/legal-content/ES/TXT/PDF/?uri=CELEX:52012DC0672&from=ES (accessed on 15 January 2020).
13. Smith, R. Directive 2008/94/EC of the European Parliament and of the Council of 22 October 2008. In *Core EU Legislation*; Macmillan Education UK: London, UK, 2015; pp. 423–426. [CrossRef]

14. Harou, J.J.; Pulido-Velazquez, M.; Rosenberg, D.E.; Medellín-Azuara, J.; Lund, J.R.; Howitt, R.E. Hydro-economic models: Concepts, design, applications, and future prospects. *J. Hydrol.* **2009**, *375*, 627–643. [CrossRef]
15. López Nicolás, A.F.L. Métodos y Herramientas Para Diseño y Evaluación de Instrumentos Económicos para la Gestión a Escala de Cuenca. Ph.D. Thesis, Universitat Politècnica de València, Valencia, Spain, 2017. Available online: https://riunet.upv.es/handle/10251/91143 (accessed on 10 October 2019).
16. Mechler, R.; Pulido-Velazquez, M.; Koehler, M.; Jenkins, K.; Surminski, S.; Williges, K.; Botzen, W.; Cremades, R.; Dransfeld, B.; Hudson, P.; et al. Economic instruments. In *Novel Multi-Sector Partnerships in Disaster Risk Management*; Aerts, J., Mysiak, J., Eds.; Enhance: Brussel, Belgium, 2017; pp. 70–97. [CrossRef]
17. Rey, D.; Pérez-Blanco, C.D.; Escriva-Bou, A.; Girard, C.; Veldkamp, T.I.E. Role of economic instruments in water allocation reform: Lessons from Europe. *Int. J. Water Resour. Dev.* **2019**, *35*, 206–239. [CrossRef]
18. FAO. *Seguros Agricolas Para la Agricultura Familiar en America Latina y el Caribe-Lineamientos Para su Desarrollo e Implementación; Santiago, Chile.* 2018. Available online: http://www.fao.org/publication/es (accessed on 15 June 2019).
19. Nnadi, F.N.; Chikaire, J.; Echetama, J.A.; Ihenacho, R.A.; Umunnakwe, P.C. Agricultural insurance: A strategic tool for climate change adaptation in the agricultural sector. *Net J. Agric. Sci.* **2013**, *1*, 1–9.
20. Cole, S.A.; Xiong, W. Agricultural Insurance and Economic Development. *Annu. Rev. Econ.* **2017**, *9*, 235–262. [CrossRef]
21. Hess, U.; Hazell, P.; Kuhn, S. Innovations and Emerging Trends in Agricultural Insurance: How can we transfer natural risks out of rural livelihoods to empower and protect people? Conference paper. Draft for Discussion Purposes. GPFI SME Finance SubGroup. In Proceedings of the G-20 Round Table on Innovations in Agricultural Finance, Antalya, Turkey, 9 September 2015.
22. Barnett, B.J.; Barrett, C.B.; Skees, J.R. Poverty Traps and Index-Based Risk Transfer Products. *World Dev.* **2008**, *36*, 1766–1785. [CrossRef]
23. Alderman, H.; Haque, T. Insurance Against Covariate Shocks: The Role of Index-Based Insurance in Social Protection in Low-Income Countries of Africa. In *World Bank Working Papers*; The World Bank: Columbia, DC, USA, 2007. [CrossRef]
24. Hazell, P.; Skees, J. Insuring against Bad Weather Recent Thinking. January 2005. Available online: https://biotech.law.lsu.edu/blog/INSURING_AGAINST_BAD_WEATHER_RECENT_THINKING.pdf (accessed on 10 October 2021).
25. Mahul, O.; Stutley, C.J. *Government Support to Agricultural Insurance: Challenges and Options for Developing Countries*; The World Bank: Washington, DC, USA, 2010. [CrossRef]
26. Jensen, N.; Barrett, C. Agricultural Index Insurance for Development. *Appl. Econ. Perspect. Policy* **2017**, *39*, 199–219. [CrossRef]
27. World Bank. 2005. Available online: https://openknowledge.worldbank.org/handle/10986/8797 (accessed on 1 September 2022).
28. Roth, J.; McCord, M. *Agricultural Microinsurance. Global Practices and Prospects*, 1st ed.; Berold, R., Ed.; The MicroInsurance Centre, LLC: Appleton, WI, USA, 2008; Available online: https://www.findevgateway.org/sites/default/files/publications/files/mfg-en-paper-agricultural-microinsurance-global-practices-and-prospects-2008.pdf (accessed on 1 September 2022).
29. Luna, A.; El Seguro Agrícola en Colombia. Reflexiones Comparadas Sobre su Desarrollo a Partir de Experiencias Internacionales. Fasecolda. 2013, 26. Available online: https://fdocuments.co/document/el-seguro-agrcola-en-colombia.html?page=1 (accessed on 25 July 2022).
30. Boucher, S.; Moya, A. Recomendaciones para el Diseño e Implementación de un Seguro Indexado para Pequeños Productores Cafeteros. In *Misión Estudios Competitividad Caficultura en Colombia*; Universidad del Rosario: Bogota, Colombia, 2014; Available online: https://www.urosario.edu.co/Mision-Cafetera/Archivos/Gestion-de-Riesgo-Steve-Boucher-Andres-Moya.pdf (accessed on 12 January 2020).
31. Maestro, T.; Bielza, M.; Garrido, A. Hydrological drought index insurance for irrigation districts in Spain. *Span J. Agric. Res.* **2016**, *14*, e0105. [CrossRef]
32. Abdi, M.J.; Raffar, N.; Zulkafli, Z.; Nurulhuda, K.; Rehan, B.M.; Muharam, F.M.; Khosim, N.A.; Tangang, F. Index-based insurance and hydroclimatic risk management in agriculture: A systematic review of index selection and yield-index modelling methods. *Int. J. Disaster Risk Reduct.* **2022**, *67*, 102653. [CrossRef]
33. Leiva, A.J.; Skees, J.R. Using Irrigation Insurance to Improve Water Usage of the Rio Mayo Irrigation System in Northwestern Mexico. *World Dev.* **2008**, *36*, 2663–2678. [CrossRef]
34. Quiroga, S.; Garrote, L.; Fernandez-Haddad, Z.; Iglesias, A. Valuing drought information for irrigation farmers: Potential development of a hydrological risk insurance in Spain. *Span J. Agric. Res.* **2011**, *9*, 1059. [CrossRef]
35. Pérez Blanco, C.D.; Gómez, C.M. Insuring water: A practical risk management option in water-scarce and drought-prone regions? *Water Policy* **2014**, *16*, 244–263. [CrossRef]
36. Ruiz, J.; Bielza, M.; Garrido, A.; Iglesias, A. Dealing with drought in irrigated agriculture through insurance schemes: An application to an irrigation district in Southern Spain. *Span J. Agric. Res.* **2015**, *13*, e0106. [CrossRef]
37. Gómez-Limón, J.A. Hydrological drought insurance for irrigated agriculture in southern Spain. *Agric. Water Manag.* **2020**, *240*, 106271. [CrossRef]
38. Arandara, R.; Gunasekera, S.; Mookerjee, A. *Index Insurance: A Viable Solution for Irrigated Farming?* World Bank: Washington, DC, USA, 2019.
39. Generalitat Valenciana. *Informe del Sector Agrario Valenciano 2020*; Informes del Sector Agrario Valenciano; Generalitat Valenciana. Conselleria de Agricultura, Desarrollo Rural, Emergencia Climática y Transición Ecológica: Valencia, Spain, 2021; Available online: https://agroambient.gva.es/es/informes-del-sector-agrario-valenciano (accessed on 2 February 2022).

40. Carmona, M.; Máñez Costa, M.; Andreu, J.; Pulido-Velazquez, M.; Haro-Monteagudo, D.; Lopez-Nicolas, A.; Cremades, R. Assessing the effectiveness of Multi-Sector Partnerships to manage droughts: The case of the Jucar river basin: Assessing Governance Performance. *Earth's Future* **2017**, *5*, 750–770. [CrossRef]
41. Ortega-Gómez, T.; Pérez-Martín, M.A.; Estrela, T. Improvement of the drought indicators system in the Júcar River Basin, Spain. *Sci. Total Environ.* **2018**, *610–611*, 276–290. [CrossRef]
42. Zaniolo, M.; Giuliani, M.; Castelletti, A.F.; Pulido-Velazquez, M. Automatic design of basin-specific drought indexes for highly regulated water systems. *Hydrol. Earth Syst. Sci.* **2018**, *22*, 2409–2424. [CrossRef]
43. Barnett, B. Agricultural Index Insurance Products: Strengths and Limitations; No 32971, Agricultural Outlook Forum 2004; United States Department of Agriculture: 2004. Available online: https://EconPapers.repec.org/RePEc:ags:usaofo:32971 (accessed on 12 January 2020).
44. Zhang, J.; Tan, K.S.; Weng, C. Index Insurance Design. *ASTIN Bull.* **2019**, *49*, 491–523. [CrossRef]
45. Carpenter, R. *Index Insurance. Status and Regulatory Challenges*; International Association of Insurance Supervisors (IAIS): Basel, Germany, 2018; Available online: https://a2ii.org/en/home (accessed on 16 February 2022).
46. Suchato, P.; Mieno, T.; Schoengold, K.; Foster, T. The potential for moral hazard behavior in irrigation decisions under crop insurance. *Agric. Econ.* **2022**, *53*, 257–273. [CrossRef]
47. Carter, M.; Elabed, G.; Serfilippi, E. Behavioral economic insights on index insurance design. *Agric. Financ. Rev.* **2015**, *75*, 8–18. [CrossRef]
48. Lopez-Nicolas, A.; Pulido-Velazquez, M.; Macian-Sorribes, H. Economic risk assessment of drought impacts on irrigated agriculture. *J. Hydrol.* **2017**, *550*, 580–589. [CrossRef]
49. International Risk Management Institute, Inc. (IRMI). Glossary of Insurance & Risk Management Terms. Available online: http://www.irmi.com/online/insurance-glossary/default.aspx (accessed on 20 January 2019).
50. Maestro Villarroya, T. Hydrological Drought Index Insurance for Irrigated Agriculture. Ph.D. Thesis, Universidad Politécnica de Madrid, Madrid, Spain, 2016. [CrossRef]
51. Maestro, T.; Barnett, B.J.; Coble, K.H.; Garrido, A.; Bielza, M. Drought Index Insurance for the Central Valley Project in California. *Appl. Econ. Perspect. Policy* **2016**, *38*, 521–545. [CrossRef]
52. Vedenov, D.; Barnett, B.J. Efficiency of Weather Derivatives as Primary Crop Insurance Instruments. *J. Agric. Resour. Econ.* **2004**, *29*, 387–403.
53. Adeyinka, A.A.; Krishnamurt, C.; Maraseni, T.; Chantarat, S. Viability of Weather Index Insurance in Managing Drought Risk in Australia. In Proceedings of the Actuaries Summit 2013, Sydney, Australia, 20–21 May 2013; Available online: http://eprints.usq.edu.au/id/eprint/23781 (accessed on 16 February 2022).
54. Kellner, U.; Musshoff, O. Precipitation or water capacity indices? An analysis of the benefits of alternative underlyings for index insurance. *Agric. Syst.* **2011**, *104*, 645–653. [CrossRef]
55. World Bank Group. *Seguro Agropecuario Catastrófico en Colombia. Anexo "As If"*; World Bank Group: Bogotá, DC, USA, 2018; p. 12.
56. Andreu, J.; Ferrer-Polo, J.; Pérez, M.A.; Solera, A.; Paredes-Arquiola, J. Drought Planning and Management in the Júcar River Basin, Spain. In *Drought in Arid and Semi-Arid Regions*; Schwabe, K., Albiac, J., Connor, J.D., Hassan, R.M., Meza González, L., Eds.; Springer: Dordrecht, The Netherlands, 2013; pp. 237–249. [CrossRef]
57. Rubio-Martin, A.; Pulido-Velazquez, M.; Macian-Sorribes, H.; Garcia-Prats, A. System Dynamics Modeling for Supporting Drought-Oriented Management of the Jucar River System, Spain. *Water* **2020**, *12*, 1407. [CrossRef]
58. Sechi, G.M.; Sulis, A. Drought mitigation using operative indicators in complex water systems. *Phys. Chem. Earth Parts A/B/C* **2010**, *35*, 195–203. [CrossRef]
59. CHJ. Plan Especial De Sequía. Demarcación Hidrográfica Del Júcar. Confederación Hidrográfica del Júcar; Valencia, Spain. 2018. Available online: https://www.chj.es/es-es/medioambiente/gestionsequia/Paginas/PlanEspecialdeAlertayEventualSequia.aspx (accessed on 10 December 2018).
60. Alvarez, J.A.; Solera, A.S.; Romá, J.C.; Polo, J.F. *SIMGES Simulation Model of Water Resource Management including Conjunctive Use. Version 3.02 User Manual*; Valencia Polytechnic University: Valencia, Spain, 2007; p. 106.
61. Confederación Hidrográfica del Júcar. *Plan Hidrográfico de la Demarcación Hidrográfica del JÚCAR. Ciclo de Planificación Hidrológica 2015–2021. Anejo VI. Sistemas de Explotación y Balances*; Confederación Hidrográfica del Júcar: Valencia, Spain, 2015; p. 582. Available online: https://www.chj.es/Descargas/ProyectosOPH/Consulta%20publica/PHC-2015-2021/PHJ1521_Anejo06_SE-Balances_151126.pdf (accessed on 15 December 2019).

agronomy

MDPI

Article

Unravelling the Endophytic Virome Inhabiting Maize Plant

Ayomide Emmanuel Fadiji, Onalenna Galeemelwe and Olubukola Oluranti Babalola *

Food Security and Safety Focus Area, Faculty of Natural and Agricultural Sciences, North-West University, Mmabatho 2735, South Africa

* Correspondence: olubukola.babalola@nwu.ac.za; Tel.: +27-18-389-2568

Abstract: Endophytes are well-known for their symbiotic interaction with plants and their ability to promote plant growth by producing various metabolites. The most well-studied endophytes are bacteria and fungi. For generations, viruses were misnamed, and their symbiotic associations were ambiguous. Recent advances in omics techniques, particularly next-generation sequencing, have given rise to novel developments in the mutualistic relationships that exist between plants and viruses. Endogenous viruses have received a lot of attention in the animal world, but limited information exists on their functions and importance to plants. Therefore, endophytic viral populations inhabiting the root of a maize plant were assessed in this study for the first time using shotgun metagenomics. Complete DNA was extracted and sequenced using shotgun metagenomics from the maize roots in farming sites where organic fertilization (FZ), inorganic fertilization (CZ), and maize planted with no fertilization (NZ) are being practised in an experimental field. Our results identified 2 orders namely: Caudovirales (67.5%) and Herpesvirales (28.5%) which dominated the FZ site, although they do not show any significant difference ($p > 0.05$) across the sites. At the class level *Microviridae*, *Phycodnaviridae*, *Podoviridae*, *Phycodnaviridae*, and *Poxviridae* dominated the FZ site. *Myoviridae* and *Podoviridae* were more abundant in the CZ site, while only Siphoviridae predominated the inorganic fertiliser site (NZ). Diversity analysis revealed that viral populations were more abundant in organic fertilization (FZ). Taken together, this research adds to our understanding of the symbiotic integration of endophytic viruses with maize plants and that their abundance is affected by farming practices. In addition, their potential can be exploited to solve a variety of agronomic issues.

Keywords: agricultural sustainability; farming practices; endophytes; metagenomics; viruses

Citation: Fadiji, A.E.; Galeemelwe, O.; Babalola, O.O. Unravelling the Endophytic Virome Inhabiting Maize Plant. *Agronomy* **2022**, *12*, 1867. https://doi.org/10.3390/agronomy12081867

Academic Editors: Raquel P. F. Guiné, António Dinis Ferreira, António Moitinho Rodrigues and Essaid Ait Barka

Received: 27 June 2022
Accepted: 3 August 2022
Published: 8 August 2022

1. Introduction

In the natural environment, plant health relies on an overabundance of interconnection between micro and macro-organisms. Endophytes are the diverse microbial group that lives in plant tissues in a mutualistic approach. It has been scientifically proven that endophytes can benefit their host through the mitigation of various agroclimatic conditions like broad abiotic and biotic stresses. Endophytic microbes are commonly acknowledged for the promotion of the growth of plants by metabolite creation, which enhances soil nutrients [1–3]. The root endophytes are of great significance because of their immediate imminent relationship with soil [4–6]. Various information exists on the interaction between the root of plants and endophytic micro-organisms [7–10]. Nonetheless, some viruses have been reported to be pathogenic in studies involving animals, plants, and humans. This discovery has presented a general negative image of viruses until recently. The introduction of next-generation sequencing and omic tools has helped improve our interest in understanding the symbiotic viral–host associations. The coexistence of viruses with the host plants and genomic association in an asymptomatic approach were accentuated by scientific studies [11,12]. Viral diverseness and symbiotic relationships require more scientific observation because most of them are unknown.

Viruses are ample and important biological individuals on earth. New research found that they are abundant in soil, desert, plant ecosystems, ocean, and the mammalian gut.

The broad misunderstanding linked with viral associations was highlighted by ecological surveys [11,12]. Viral interaction may also be mutualistic, but it cannot be symptomatic at all times. Interconnection between viruses and their hosts is conditional and dynamic in some instances. Symbiogenesis may easily be led by viral symbiosis through the genomic fusion of two individuals; this is where the evolution of new species takes place. The immense plethora of viral sequences in extant genomes can aid coevolution and symbiogenesis [13,14]. They play a crucial role in the diverseness of living things on the earth, including the coevolution of viruses and hosts. Based on studies on the diversity of plant virus biodiversity, viruses discovered in a larger number of plants were found to have an asymptomatic existence [15]. The influence of biotic and abiotic stresses is improved by viruses. The ability to survive under extreme temperatures as well as the disease-suppressive potentials of plants are some of the contributions of viruses to plant health. In Yellowstone National Park, the plant's ability to acclimatize to geothermal sectors with the capability to endure the increased level of temperature was discovered to be correlated with a novel fungus which consequently was affected by the virus [16]. The development of a nitrogen-fixing nodule, which is a situation where the quantity of nitrogen is sufficient in the soil as a means of energy conservation, can be restricted by *White Clover cryptic virus* [12].

The ribosomal RNA gene, which lacks a universal coding sequence, is discovered in all biological life making it difficult to analyse the diverseness of viruses. To unveil the enormous wealth of viral details from dissimilar environmental tests, metagenomic research using shotgun sequencing serves as a promising approach [15,17]. Sequences of viral endophyte were also contracted with the indigenous viral group of the soil to detect the particular relationship or straight transference coming to the root of the plant from the soil. A broad analysis was given to propose the likely useful impact of such a symbiotic relationship. This present study is crucial to discovering the wider expectation of an internal viral relationship. Having a better understanding of the viral interactions with plants and their abundance will help in addressing disease emergence in plants and identifying beneficial integration that might be used to treat a variety of agricultural concerns.

The majority of maize producers in South Africa utilise conventional farming practices and inorganic fertilisers to increase plant production. In addition to having negative environmental consequences, excessive usage of inorganic fertiliser also has negative effects on the seed quality, microbial populations, and increased lodging in the plant [18]. Similarly, chemical fertilisers are not economical and a non-renewable nutrient source for the plant [19]. Examining microbiological sources and organic farming, which have excellent properties for stimulating plant development and productivity, is urgently necessary.

Furthermore, it is uncommon to find a well-organised study on how various agricultural practices affect endophytic virome in maize roots. According to reports, the biggest population of endophytes is found in a plant's roots [20,21], which is why maize roots were chosen in this investigation. To the best of our knowledge, there has not been any research on how agricultural practices affect the composition and variety of endophytic viromes inhabiting the roots of maize plants using the shotgun metagenomic technique.

To acknowledge the enormous wealth of viral details from dissimilar environmental tests, metagenomic research employing shotgun sequencing has presented itself as one of the best approaches [15,17,22,23]. Therefore, we present the first study assessing the community structure of endophytic viromes in the roots of maize plants using the shotgun metagenomic techniques.

2. Materials and Methods

2.1. Seed Sourcing

The drought resistant WE 3127 maize seed used in this experiment was collected from North-West University School Farm, Molelwane, Mafikeng, North West Province, South Africa.

2.2. Root Sampling

Because of the experimental farmland's triangular form, each agricultural site was separated into 3 divisions for sampling purposes. The roots of the ten fresh plants in each division were selected randomly from the farming site and pooled to represent biological replicates with a total of 30 plants for each site. The maize roots were then uprooted for the experimental purpose at the fruiting stage of growth [24]. A total of ninety samples of the plant were assessed, with 3 replicates for the individual sampling site, indicating three regions. The collected samples were stored in ice, and then promptly taken to the laboratory for subsequent analysis.

2.3. Description of the Study Site and Experimental Design

Organic and inorganic experimental fields (approximately 500 acres) had been established for 15 years at the University School in Molelwane, North West of South Africa (25°47′25.24056″ S, 25°37′8.17464″ E). Shrubs and trees dominate this province. The average province temperature varies from 3–21 °C and 17–31 °C during winter and summer, respectively. The annual province's rainfall is about 360 mm. For a long period, the main crops planted at this experimental site were sorghum, maize, and soybean (maize-soybean-sorghum), with sorghum sown in 2019. The physicochemical parameters of the soil samples from the sampling sites were identical (66% silt, 22%, 12% clay, pH 6; 0.15% total N, 0.48% organic C, 101.5 ppm *p*, and 0.962 ppm) (Supplementary Table S1).

The two regimes of fertilization that were employed are organic fertilization (FZ) and inorganic fertilization (NZ) and have been existing for more than fifteen years, along with control with no fertiliser application (CZ). The quantity of the inorganic fertiliser that had been in use is 75 P_2O_5, 75 K_2O, and 150 N in kg ha^{-1}, while the organic fertiliser site had been applying cow manure with a 10,625 kg ha^{-1} dosage for more than 15 years following the international best practices [25], and the last site has never experienced the application of fertiliser. The WE 3127 seeds were planted on 3 sites, respectively, using a farming space measuring up to 10 m × 4 m, and was terminated during the summer of the year 2019. To avoid drought stress, all of the sites were irrigated as needed. Manual weed control was employed.

2.4. Surface Sterilization of Maize Roots

Soil particles that came with the roots of the plants from the experimental field were removed via sieving, the procedure outlined by [26], was employed for surface washing of the new roots. The roots were first immersed in 70% ethanol for 3 min. After which, they were washed for 5 min with a 2.5% sodium hypochlorite solution. They were then washed again with 70% ethanol for 30 s before being washed with distilled water that had been sterilised. To make sure that epiphytes were perfectly taken out and that the sterilization process was successfully done, the washed roots were chopped into little pieces and cultured on a yeast extract-mannitol medium (YEM) [27]. After 72 h, the Petri dishes were incubated at 30 °C and were then inspected for the growth of bacteria. The roots of maize plants from uncontaminated plates were selected for DNA extraction [28,29].

2.5. Extraction of DNA and Shotgun Sequencing

Using a sterilised knife, the maize roots were sliced into minute pieces and macerated using a Qiagen TissueLyser. Qiagen DNeasy Plant Mini Kit (Dusseldorf, Germany) was used to extract completed metagenome DNA from the root of the plant samples. The extracted DNA samples were then sent to the Molecular Research LP in Shallowater, TX, USA, where shotgun metagenomic sequencing was performed. The Nextera DNA Flex kit (Illumina, San Diego, CA, USA) was used to prepare the libraries, and the typical protocol was followed. The Life Technologies Qubit®dsDNA HS Assay Kit was employed for the determination of the actual DNA concentration in all of the samples. After the formation of the library, its final concentration was determined by employing the Qubit®dsDNA HS Assay Kit (Thermo Fisher Scientific, Waltham, MA, USA) and the Agilent 2100 Bioanalyzer.

The size of the library ranges from 683 to 877 bp, with an average of 731 bp. Library pooling was done with 0.6 nM ratios and paired-end sequencing was carried out using 300 cycles via the Illumina NovaSeq 6000 equipment.

2.6. Data Analysis

Sequences obtained for each metagenome were uploaded to the metagenomics rapid annotation online server (MG-RAST) [30], where QC of the raw sequences was performed, including the removal of adapter and low-quality reads using the Trimmomatic v 0.33 tool with default parameters [31]. Artificial sequences were removed, ambiguous bases ere filtered, a minimum read size was specified, and length filtering was all part of the quality control process. Following quality control, sequence annotation was performed using BLAT [32], against the M5NR database [33], which allows nonredundant database integration. The SEED database was used to categorise endophytic microbiomes, with characteristics including a 10^{-5} e-value cut-off and at least 60% similarity of the sequence to a subsystem. Sequences that failed that annotation test were not taken into consideration. However, because we focused on endophytic virome, we ignored sequences from bacteria, eukaryotes, archaea, and maize plants. The MG-RAST data normalization option was selected to reduce the impact of experimental error/noise. Each taxon's endophytic virome table was produced, and unclassified sequence reads were retained for statistical analysis. Furthermore, after an independent examination of the nine (9) sequences using MG-RAST, the relative abundance of the taxa in percentages was computed. For statistics, the mean values of the relative abundance of the three replicates for the experimental sites (CZ, FZ, and NZ) were employed. These standard sequences can be obtained in the PRJNA607664 NCBI SRA dataset.

2.7. Statistical Analyses

At the order level, the Shinyheatmap was used to plot a relative abundance graph of endophytic virome communities [34]. The Pielou evenness and Shannon diversity indices for all the sampling sites were analysed using PAST version 3.20 [35], and the indices across the farming sites were compared using the Kruskal–Wallis test. The beta diversity was defined using principal coordinate analysis (PCoA) based on a Kruskal–Wallis matrix, and the differences in community structure were assessed by employing the one-way analysis of similarities (ANOSIM) [35]. How the identified endophytic viral order was distributed among the maize plant fields was presented using Euclidean matrix-based principal component analysis (PCA).

3. Results

3.1. Metagenome Dataset and Quality Control

The sequence readings of samples were CZ (4839895527), FZ (2977205570), and NZ (48270695214), obtained for the three experimental sites. The sequenced reads for CZ were 334,259,767 with an average G + C content of 44%, FZ had 415,505,341 with an average G + C content of 44%, and NZ had 817,699,487 with an average G + C content of 49%, were obtained after quality-control analysis in MG-RAST. Sequences that mapped for identified proteins in the metagenome samples that passed the quality control assessment were 371,329, 325,439 and 643,141 for CZ, FZ, and NZ, respectively (Supplementary Table S2).

3.2. Community Structure and Abundance of Endophytic Virome Inhabiting Maize Root Samples

Two major viral orders identified in this experiment are the *Caudovirales* (67.5%) and *Herpesvirales* (68.5%) and are more abundant in the FZ site samples, although no significant difference ($p > 0.05$) exists across the sites (Figure 1). At the class level, Microviridae, *Phycodnaviridae*, *Podoviridae*, *Phycodnaviridae*, and *Poxviridae* dominated the FZ site (Figure 2). *Myoviridae* and *Podoviridae* were more abundant at the CZ site while *Siphoviridae* was found to be more abundant only in the site with the inorganic fertiliser (NZ), although, the difference across the experimental sites was found not to be significant ($p > 0.05$).

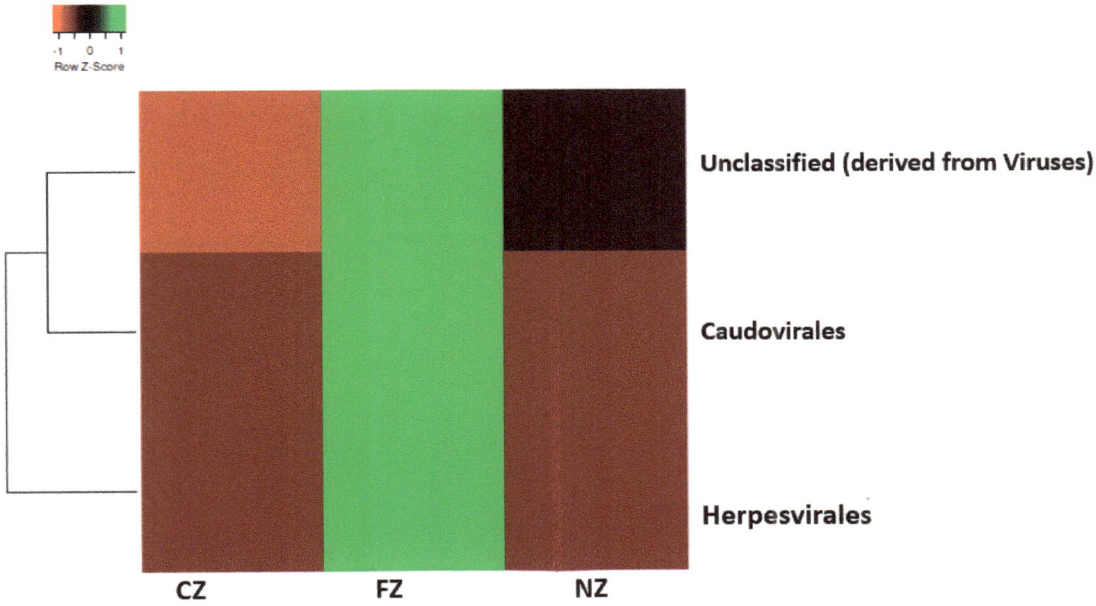

Figure 1. Heatmap of the order distribution of the notable endophytic virome from samples across the sites. The scale bar displays a colour-saturation gradient based on relative abundances that have been modified using the z-score for the endophytic virome.

Furthermore, at the genus level, unclassified Siphoviridae, Badnavirus, P2-like viruses, SPO1-like viruses, LUZ24-like viruses, N4-like viruses, Bpp 1-like viruses, Phi29-like viruses, T4-like viruses, T7-like viruses, L5-Like viruses, Lambda-like viruses and N15-like viruses dominated FZ sites. Chlamydiamicrovirus, unclassified Podoviridae, unclassified Myoviridae, and P22-like viruses were dominant in the CZ site, while unclassified Microviridae, Chlorovirus and T1-like viruses were found to be dominant in the NZ site (Figure 3). The PCA graph was employed in showing the virome distribution of the identified between the sites with the most abundant distribution observed in the organic farming site (FZ) (Figure 4).

3.3. Alpha (α) and Beta (β) Diversity of the Viral Endophytes across the Experimental Sites

The evenness and Shannon indexes derived for the order of the endophytic virome do not differ significantly ($p > 0.05$), while a significant difference ($p < 0.05$) was found at the genus level (Table 1). The virome community composition was analysed using PCoA with a Bray–Curtis dissimilarity matrix (Figure 5). The PCoA figure revealed that the FZ samples varied considerably from the CZ and NZ samples (Figure 5). ANOSIM revealed a significant difference (ANOSIM, $R = 0.67$, $p = 0.01$) in the diversity of the viral endophytes virome across the farming sites.

Table 1. Evenness and diversity assessment of endophytic virome across the sampling sites.

Level	Indices	CZ	FZ	NZ	*p*-Value
		Endophytic Virome			
Order	Shannon_H	0.59 ± 0.03	0.68 ± 0.14	0.48 ± 0.11	0.42
	Evenness_e^H/S	0.90 ± 0.03	0.66 ± 0.05	0.80 ± 0.14	
Genus	Shannon_H	1.77 ± 0.39	1.85 ± 0.22	1.60 ± 0.21	0.007
	Evenness_e^H/S	0.73 ± 0.23	0.85 ± 0.17	0.82 ± 0.17	

Mean ± SD (n = 3). p-values based on Kruskal–Wallis matrix test. NZ = samples from the inorganic experimental site, FZ = samples from the organic experimental site, and CZ = no fertiliser site/control samples.

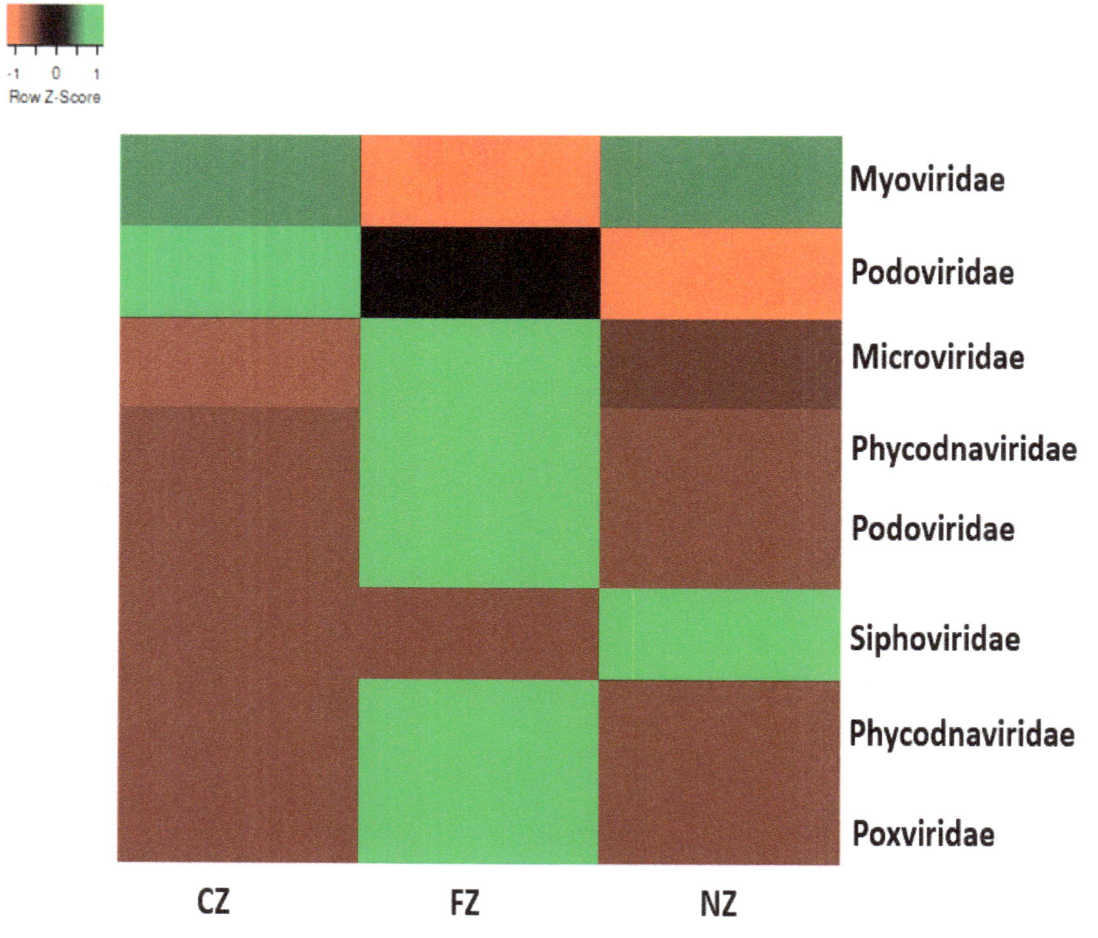

Figure 2. Heatmap of endophytic virome family. The scale bar displays a colour-saturation gradient based on relative abundances that have been modified using the z-score for the endophytic virome. NZ = samples from the inorganic experimental site, FZ = samples from the organic experimental site, and CZ = no fertiliser site/control samples.

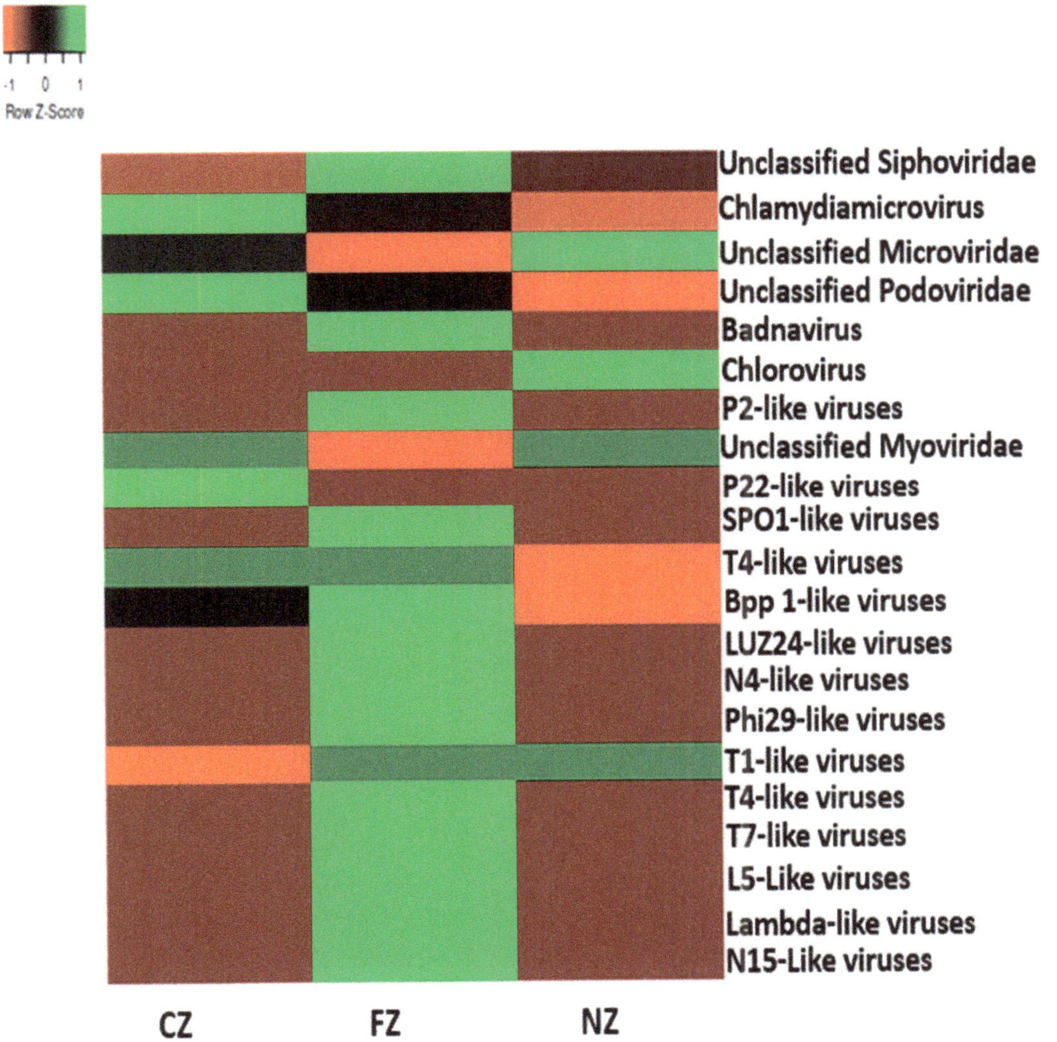

Figure 3. Heatmap of endophytic virome genus. The scale bar displays a colour-saturation gradient based on relative abundances that have been modified using the z-score for the endophytic virome. NZ = samples from the inorganic experimental site, FZ = samples from the organic experimental site, and CZ = no fertiliser site/control samples.

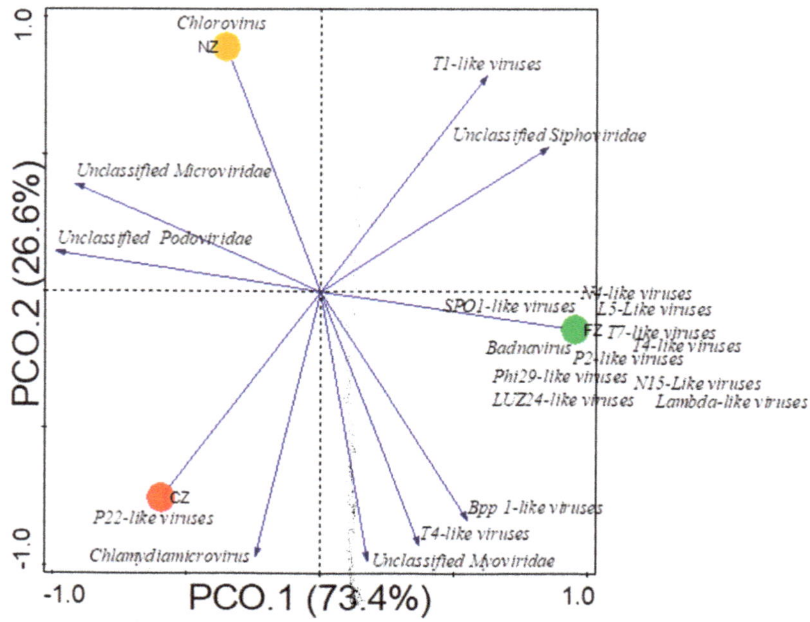

Figure 4. Principal component analysis graph of the mean metagenomes of endophytic virome. The effect of the metagenomes of the viral endophyte is shown by the vector arrow. Bray–Curtis dissimilarity matrix, axes 1 (73.4%), and 2 (26.6%) explained the variances.

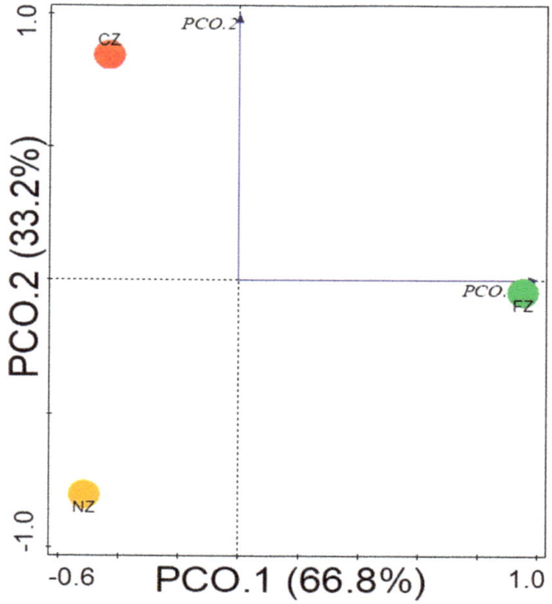

Figure 5. PCoA plot of the community composition of the endophytic virome in the experimental sites based on Bray–Curtis dissimilarities. FZ = samples from the organic experimental site, CZ = no fertiliser site/control samples, and NZ = samples from the inorganic experimental site.

4. Discussion

Farming techniques have a considerable effect on the abundance, diversity, and functions of microbial communities in the soil, and can thus be connected to increased crop output and development, as well as improve crop resistance to abiotic and biotic stress [36–38]. Using shotgun metagenomics, we investigated the effects of various farming practices on the community structure and abundance of endophytic virome inhabiting the root of maize grown under various fertiliser regimes. For years, viruses had a poor reputation and were mostly recognised for their capacity to spread disease. But more recently, symbiotic aspects using omics technologies have come into emphasis. The viral community discovered from maize root samples which are endogenous in origin were discussed in this study using the shotgun metagenomics. Interestingly, we also discovered phage virus genomes from the Caudovirales family, which may have spread from soil [39]. MG-RAST was used to examine the sequenced metagenome data collected. Genomes for the endophytic virome were identified, but plant-derived sequences were discarded.

Caudovirales and *Herpesvirales* were the major viral order in samples and the most predominant FZ site. This result agrees with an earlier study by Das, et al. [39], this may be a result of the application of organic fertiliser in the organic site, which might harbour more microbes. Caudovirales are a family of group-I viruses containing double-stranded DNA and an icosahedral head connected to a tail by a connector protein. The *Caulimoviridae* family provided the majority of unclassified sequences (Figure 1). *Caulimoviridiae* is a type of DNA virus with two strands of DNA. Their endogenous pararetroviral sequences have received a lot of attention (EPRVs). Natural integration into the host DNA has also been documented [40,41]. This natural interaction with the DNA of the host plant also points to a co-evolutionary relationship with the plant–virus pathosystem [42–44].

Also, at the class level *Siphoviridae*, *Microviridae*, *Phycodnaviridae*, *Podoviridae*, *Phycodnaviridae Poxviridae Myoviridae*, *Podoviridae*, and *Siphoviridae* were identified at the root of the maize plant. Most of the viral sequences discovered were comparable to those found in tea plants [39]. However, this study could imply that their relationship as an endogenous viral particle is well-known. Furthermore, at the genus level unclassified *Siphoviridae*, *Badnavirus*, P2-like viruses, SPO1-like viruses, Bpp 1-like viruses, LUZ24-like viruses, N4-like viruses, Phi29-like viruses, T4-like viruses, T7-like viruses, L5-like viruses, Lambda-like viruses, N15-Like viruses dominated the FZ site, *Chlamydiamicrovirus*, unclassified *Podoviridae*, unclassified *Myoviridae*, P22-like virueses dominated the CZ site, while unclassified *Microviridae*, *Chlorovirus*, and T1-like viruses were found to be dominant in the NZ site.

Badnavirus belongs to the *Caulimoviridae* family with the plant-associated bacilliform DNA virus. They have been reported to be a major pathogen of a variety of horticultural crops, including citrus, black pepper, cocoa, banana, taro, sugarcane, and yams [45,46]. Diseases of plants including root necrosis, leaf chlorosis, red vein banding in early leaves, tiny speckled pods, and the swelling of the stem/root followed by dieback are all caused by *Badnavirus* [39,45]. Several researchers have reported *Badnavirus* endogenous connection [47,48]. Endogenous recombination with the host genome, on the other hand, may not lead to infection in the host plant and can give protection against non-integrative counterparts [45]. Not many reports exist on its presence in maize plants. However, only one report of *Badnavirus* from the tea plant has been published, and it comes from Hao, et al. [49], who used metagenomic sequences of leaves and shoot samples.

The genus level was used for PCA due to the abundance of the virome at the genus level. The PCA graph revealed that each site has its unique viral genus, which accounts for 73.4% of the variance between all fertilization locations (Figure 4). The composition of sequences connected to each genus is reflected in the position of each endophytic virome; the vector arrows indicate the genus most heavily affected by the distribution. This information can be used to discover which viral genera are more prevalent at each sampling site when compared to others (Figure 4). In this investigation, viral genera were shown to be more prevalent in the FZ site than in other sites (Figure 4).

The Shannon and evenness indexes evaluated for each viral order revealed no significant differences ($p > 0.05$), while the result from the viral genus showed that they differ significantly ($p > 0.05$). Endophytic virome in maize grown in the organic farming site was more diverse and equally distributed than those in maize grown with inorganic or without fertiliser (Table 1). The result also agrees with the findings of Das et al. [39]. FZ's viral endophyte community structure differed from that of CZ and NZ, according to the PCoA plot (Figure 5). The endophytic virome in the root of maize plants differed significantly between sample sites, as seen by the Bray–Curtis dissimilarity matrix-based figure.

5. Conclusions

This is one of the foremost studies unravelling the diversity of endophytic virome inhabiting maize plants employing the shotgun metagenomics approach. This study gave a detailed taxonomic distribution of viral endophytes in maize roots and showed that farming practices have a significant effect on the abundance and diversity of these viromes. Endophytic viromes which were found to majorly dominate the roots of maize plants are *Caudovirales* and *Herpesvirales*. This report has added to our understanding of endogenous viruses, with a focus on the maize plant. There will be many more mutualistic viruses that need to be further studied to grasp their evolutionary significance. Understanding how viruses interact with plants and their diversity will help in communicating disease disclosure in plants and identifying the most favourable combination, which might be important in addressing a variety of agricultural issues. The findings of this study further add to our understanding of the virus–plant symbiotic connection.

Supplementary Materials: The following supporting information can be downloaded at: https://www.mdpi.com/article/10.3390/agronomy12081867/s1, Table S1: Physicochemical characteristics of the experimental field; Table S2: Analysis of sequenced data and diversity evaluation of the shotgun metagenomes of the maize plant from across the fertilizers sites.

Author Contributions: A.E.F. and O.G. handled the literature findings, carried out the laboratory and fieldwork, executed all necessary analyses, interpreted the results, and prepared the manuscript. O.O.B. initiated the next-generation sequence research, helped shape the research, verified the analytical methods, secured funds for the study, and commented on the manuscript at all stages. All authors have read and agreed to the published version of the manuscript.

Funding: The National Research Foundation of South Africa grants (UID123634 and UID132595).

Acknowledgments: A.E.F. gratefully acknowledged the North-West University for a Postdoctoral fellowship award. O.G. acknowledged the North-West University for the bursary that supported her studies. O.O.B. appreciates the National Research Foundation, South Africa, for the grants (UID123634 and UID132595) that have supported this research.

Conflicts of Interest: The authors declare no conflict of interest.

References

1. Chhabra, S.; Dowling, D.N. Endophyte-Promoted Nutrient Acquisition: Phosphorus and Iron. In *Functional Importance of the Plant Microbiome*; Doty, S., Ed.; Springer: Cham, Switzerland, 2017; pp. 21–42. [CrossRef]
2. Fadiji, A.E.; Ayangbenro, A.S.; Babalola, O.O. Shotgun metagenomics reveals the functional diversity of root-associated endophytic microbiomes in maize plant. *Curr. Plant Biol.* **2021**, *25*, 100195. [CrossRef]
3. Fadiji, A.E.; Ayangbenro, A.S.; Babalola, O.O. Unveiling the putative functional genes present in root-associated endophytic microbiome from maize plant using the shotgun approach. *J. Appl. Genet.* **2021**, *62*, 339–351. [CrossRef] [PubMed]
4. Fadiji, A.E.; Babalola, O.O. Elucidating mechanisms of endophytes used in plant protection and other bioactivities with multifunctional prospects. *Front. Bioeng. Biotechnol.* **2020**, *8*, 467. [CrossRef] [PubMed]
5. Fadiji, A.E.; Babalola, O.O. Exploring the potentialities of beneficial endophytes for improved plant growth. *Saudi J. Biol. Sci.* **2020**, *27*, 3622–3633. [CrossRef]
6. Fadiji, A.E.; Kanu, J.O.; Babalola, O.O. Metagenomic profiling of rhizosphere microbial community structure and diversity associated with maize plant as affected by cropping systems. *Int. Microbiol.* **2021**, *24*, 325–335. [CrossRef]
7. Gaiero, J.R.; McCall, C.A.; Thompson, K.A.; Day, N.J.; Best, A.S.; Dunfield, K.E. Inside the root microbiome: Bacterial root endophytes and plant growth promotion. *Am. J. Bot.* **2013**, *100*, 1738–1750. [CrossRef] [PubMed]

8. Borah, M.; Das, S.; Bora, S.S.; Boro, R.C.; Barooah, M. Comparative assessment of multi-trait plant growth-promoting endophytes associated with cultivated and wild Oryza germplasm of Assam, India. *Arch Microbiol* **2021**, *203*, 2007–2028. [CrossRef]
9. Mayerhofer, M.S.; Kernaghan, G.; Harper, K.A. The effects of fungal root endophytes on plant growth: A meta-analysis. *Mycorrhiza* **2013**, *23*, 119–128. [CrossRef]
10. Fadiji, A.E.; Ayangbenro, A.S.; Babalola, O.O. Organic Farming enhances the diversity and community structure of endophytic archaea and fungi in maize plant: A shotgun approach. *J. Soil Sci. Plant Nutr.* **2020**, *20*, 2587–2599. [CrossRef]
11. Roossinck, M.J. Plants, viruses and the environment: Ecology and mutualism. *Virology* **2015**, *479*, 271–277. [CrossRef]
12. Roossinck, M.J. Move over, bacteria! Viruses make their mark as mutualistic microbial symbionts. *J. Virol.* **2015**, *89*, 6532–6535. [CrossRef] [PubMed]
13. Roossinck, M.J. The good viruses: Viral mutualistic symbioses. *Nat. Rev. Microbiol.* **2011**, *9*, 99–108. [CrossRef] [PubMed]
14. Bao, X.; Roossinck, M.J. A life history view of mutualistic viral symbioses: Quantity or quality for cooperation? *Curr. Opin. Microbiol.* **2013**, *16*, 514–518. [CrossRef]
15. Roossinck, M.J.; Martin, D.P.; Roumagnac, P. Plant virus metagenomics: Advances in virus discovery. *Phytopathology* **2015**, *105*, 716–727. [CrossRef]
16. Márquez, L.M.; Redman, R.S.; Rodriguez, R.J.; Roossinck, M.J. A virus in a fungus in a plant: Three-way symbiosis required for thermal tolerance. *Science* **2007**, *315*, 513–515. [CrossRef] [PubMed]
17. Roossinck, M.J. The big unknown: Plant virus biodiversity. *Curr. Opin. Virol.* **2011**, *1*, 63–67. [CrossRef]
18. Xia, Y.; Sahib, M.R.; Amna, A.; Opiyo, S.O.; Zhao, Z.; Gao, Y.G. Culturable endophytic fungal communities associated with plants in organic and conventional farming systems and their effects on plant growth. *Sci. Rep.* **2019**, *9*, 1669. [CrossRef] [PubMed]
19. Scheiner, J.D.; Gutiérrez-Boem, F.H.; Lavado, R.S. Sunflower nitrogen requirement and 15N fertilizer recovery in Western Pampas, Argentina. *Eur. J. Agron.* **2002**, *17*, 73–79. [CrossRef]
20. Das, A.; Varma, A. Symbiosis: The Art of Living. In *Symbiotic Fungi. Soil Biology*; Varma, A., Kharkwal, A.C., Eds.; Springer: Berlin/Heidelberg, Germany, 2009; Volume 18, pp. 1–28. [CrossRef]
21. Sessitsch, A.; Hardoim, P.; Döring, J.; Weilharter, A.; Krause, A.; Woyke, T.; Mitter, B.; Hauberg-Lotte, L.; Friedrich, F.; Rahalkar, M. Functional characteristics of an endophyte community colonizing rice roots as revealed by metagenomic analysis. *Mol. Plant-Microbe Interact.* **2012**, *25*, 28–36. [CrossRef]
22. Fadiji, A.E.; Ayangbenro, A.S.; Babalola, O.O. Metagenomic profiling of the community structure, diversity, and nutrient pathways of bacterial endophytes in maize plant. *Antonie Van Leeuwenhoek* **2020**, *113*, 1559–1571. [CrossRef]
23. Babalola, O.O.; Fadiji, A.E.; Ayangbenro, A.S. Shotgun metagenomic data of root endophytic microbiome of maize (*Zea mays* L.). *Data Brief* **2020**, *31*, 105893. [CrossRef] [PubMed]
24. Xia, Y.; DeBolt, S.; Dreyer, J.; Scott, D.; Williams, M.A. Characterization of culturable bacterial endophytes and their capacity to promote plant growth from plants grown using organic or conventional practices. *Front. Plant Sci.* **2015**, *6*, 490. [CrossRef] [PubMed]
25. U.S. Department of Agriculture (USDA). *Organic Regulations*; USDA: Washington, DC, USA, 2014. Available online: https://www.ams.usda.gov/AMSv1.0/NOPOrganicStandards (accessed on 26 June 2022).
26. Liu, H.; Carvalhais, L.C.; Crawford, M.; Singh, E.; Dennis, P.G.; Pieterse, C.M.; Schenk, P.M. Inner plant values: Diversity, colonization and benefits from endophytic bacteria. *Front. Microbiol.* **2017**, *8*, 2552. [CrossRef] [PubMed]
27. Vincent, J.M. *A Manual for the Practical Study of the Root-Nodule Bacteria*; Wiley-Blackwell: Hoboken, NJ, USA, 1970; p. 440.
28. Correa-Galeote, D.; Bedmar, E.J.; Arone, G.J. Maize endophytic bacterial diversity as affected by soil cultivation history. *Front. Microbiol.* **2018**, *9*, 484. [CrossRef] [PubMed]
29. Marag, P.S.; Suman, A.; Gond, S. Prospecting endophytic bacterial colonization and their potential plant growth promoting attributes in hybrid maize (*Zea mays* L.). *Int. J. Curr. Microbiol. Appl. Sci.* **2018**, *7*, 1292–1304. [CrossRef]
30. Hong, C.E.; Kim, J.U.; Lee, J.W.; Bang, K.H.; Jo, I.H. Metagenomic analysis of bacterial endophyte community structure and functions in Panax ginseng at different ages. *3 Biotech* **2019**, *9*, 300. [CrossRef]
31. Bolger, A.M.; Lohse, M.; Usadel, B. Trimmomatic: A flexible trimmer for Illumina sequence data. *Bioinformatics* **2014**, *30*, 2114–2120. [CrossRef]
32. Kent, W.J. BLAT—the BLAST-like alignment tool. *Genome Res.* **2002**, *12*, 656–664.
33. Wilke, A.; Harrison, T.; Wilkening, J.; Field, D.; Glass, E.M.; Kyrpides, N.; Mavrommatis, K.; Meyer, F. The M5nr: A novel non-redundant database containing protein sequences and annotations from multiple sources and associated tools. *BMC Bioinform.* **2012**, *13*, 141. [CrossRef]
34. Khomtchouk, B.B.; Hennessy, J.R.; Wahlestedt, C. shinyheatmap: Ultra fast low memory heatmap web interface for big data genomics. *PLoS ONE* **2017**, *12*, e0176334. [CrossRef]
35. Carrell, A.A.; Frank, C. Bacterial endophyte communities in the foliage of coast redwood and giant sequoia. *Front. Microbiol.* **2015**, *6*, 1008. [CrossRef]
36. Kniss, A.R.; Savage, S.D.; Jabbour, R. Commercial crop yields reveal strengths and weaknesses for organic agriculture in the United States. *PLoS ONE* **2016**, *11*, e0165851. [CrossRef]
37. Gamboa, C.H.; Vezzani, F.M.; Kaschuk, G.; Favaretto, N.; Cobos, J.Y.G.; da Costa, G.A. Soil-root dynamics in maize-beans-eggplant intercropping system under organic management in a subtropical region. *J. Soil Sci. Plant Nutr.* **2020**, *20*, 1480–1490. [CrossRef]

38. Flores-Rentería, D.; Sánchez-Gallén, I.; Morales-Rojas, D.; Larsen, J.; Álvarez-Sánchez, J. Changes in the abundance and composition of a microbial community associated with land use change in a Mexican tropical rain forest. *J. Soil Sci. Plant Nutr.* **2020**, *20*, 1144–1155. [CrossRef]
39. Das, S.; Barooah, M.; Thakur, N. Endophytic Virome. *BioRxiv* **2019**, *602144*, 1–14. [CrossRef]
40. Harper, G.; Osuji, J.O.; Heslop-Harrison, J.P.; Hull, R. Integration of banana streak badnavirus into theMusagenome: Molecular and cytogenetic evidence. *Virology* **1999**, *255*, 207–213. [CrossRef] [PubMed]
41. Chen, S.; Kishima, Y. Endogenous pararetroviruses in rice genomes as a fossil record useful for the emerging field of palaeovirology. *Mol. Plant Pathol.* **2016**, *17*, 1317. [CrossRef]
42. Pahalawatta, V.; Druffel, K.; Pappu, H. A new and distinct species in the genus Caulimovirus exists as an endogenous plant pararetroviral sequence in its host, *Dahlia variabilis*. *Virology* **2008**, *376*, 253–257. [CrossRef]
43. Gnanasekaran, P.; Chakraborty, S. Biology of viral satellites and their role in pathogenesis. *Curr. Opin. Virol.* **2018**, *33*, 96–105. [CrossRef]
44. Gnanasekaran, P.; KishoreKumar, R.; Bhattacharyya, D.; Vinoth Kumar, R.; Chakraborty, S. Multifaceted role of geminivirus associated betasatellite in pathogenesis. *Mol. Plant Pathol.* **2019**, *20*, 1019–1033. [CrossRef]
45. Bhat, A.I.; Hohn, T.; Selvarajan, R. Badnaviruses: The current global scenario. *Viruses* **2016**, *8*, 177. [CrossRef] [PubMed]
46. Wilson, W.; Etten, J.L.V.; Allen, M. The Phycodnaviridae: The story of how tiny giants rule the world. *Curr. Top. Microbiol. Immunol.* **2009**, *328*, 4.
47. Hohn, T.; Richert-Pöggeler, K.R.; Staginnus, C.; Harper, G.; Schwarzacher, T.; Teo, C.H.; Teycheney, P.-Y.; Iskra-Caruana, M.-L.; Hull, R. Evolution of Integrated Plant Viruses. In *Plant Virus Evolution*; Roossinck, M., Ed.; Springer: Berlin/Heidelberg, Germany, 2008; pp. 53–81.
48. Staginnus, C.; Iskra-Caruana, M.-L.; Lockhart, B.; Hohn, T.; Richert-Poeggeler, K.R. Suggestions for a nomenclature of endogenous pararetroviral sequences in plants. *Arch. Virol.* **2009**, *154*, 1189–1193. [CrossRef] [PubMed]
49. Hao, X.; Zhang, W.; Zhao, F.; Liu, Y.; Qian, W.; Wang, Y.; Wang, L.; Zeng, J.; Yang, Y.; Wang, X. Discovery of plant viruses from tea plant (*Camellia sinensis* (L.) O. Kuntze) by metagenomic sequencing. *Front. Microbiol.* **2018**, *9*, 2175. [CrossRef] [PubMed]

agronomy

MDPI

Article

Effect of Controlled Atmospheres and Environmental Conditions on the Physicochemical and Sensory Characteristics of Sweet Cherry Cultivar Satin

Luís Pinto de Andrade [1,2,*], Abel Veloso [1,2], Christophe Espírito Santo [3,4], Pedro Dinis Gaspar [5,6], Pedro Dinho Silva [5,6], Mafalda Resende [3], Helena Beato [3], Cátia Baptista [3], Cristina Miguel Pintado [3], Luísa Paulo [3] and Maria Paula Simões [1,2]

1 IPCB-ESA—Instituto Politécnico de Castelo Branco, Escola Superior Agrária, Quinta da Senhora de Mércules, 6001-909 Castelo Branco, Portugal; abel.veloso@gmail.com (A.V.); mpaulasimoes@ipcb.pt (M.P.S.)
2 Instituto Politécnico de Castelo Branco, CERNAS—Centro de Estudos de Recursos Naturais, Ambiente e Sociedade, Av. Pedro Álvares Cabral 12, 6000-084 Castelo Branco, Portugal
3 CATAA—Centro de Apoio Tecnológico Agro Alimentar, Zona Industrial de Castelo Branco, Rua A, 6000-459 Castelo Branco, Portugal; cespiritosanto@cataa.pt (C.E.S.); mafalda.resende@cataa.pt (M.R.); helena.beato@cataa.pt (H.B.); catia.baptista@cataa-cei.pt (C.B.); cmiguel@cataa.pt (C.M.P.); luisa.paulo@cataa.pt (L.P.)
4 CFE—Centre for Functional Ecology, University of Coimbra, 3000-456 Coimbra, Portugal
5 Departamento de Engenharia Eletromecânica, UBI—Universidade da Beira Interior, Rua Marquês d'Ávila e Bolama, 6201-001 Covilhã, Portugal; dinis@ubi.pt (P.D.G.); dinho@ubi.pt (P.D.S.)
6 C-MAST—Centre for Mechanical and Aerospace Science and Technologies, Universidade da Beira Interior, Rua Marquês d'Ávila e Bolama, 6201-001 Covilhã, Portugal
* Correspondence: luispa@ipcb.pt

Citation: Pinto de Andrade, L.; Veloso, A.; Espírito Santo, C.; Dinis Gaspar, P.; Silva, P.D.; Resende, M.; Beato, H.; Baptista, C.; Pintado, C.M.; Paulo, L.; et al. Effect of Controlled Atmospheres and Environmental Conditions on the Physicochemical and Sensory Characteristics of Sweet Cherry Cultivar Satin. *Agronomy* **2022**, *12*, 188. https://doi.org/10.3390/agronomy12010188

Academic Editor: Milena Petriccione

Received: 16 December 2021
Accepted: 11 January 2022
Published: 13 January 2022

Abstract: Sweet cherry is a highly appreciated seasonal fruit with a high content of bioactive compounds; however, this highly perishable fruit has a relatively short shelf-life period. Here, we evaluated the evolution of the physicochemical and sensory qualities of sweet cherries (*Prunus avium* (L.) cv. Satin) under different storage conditions, namely at a Farmers' Organization (FO) and in a Research Centre (RC) under normal and four different conditions of controlled atmosphere for 49 days. Additional parameters were monitored, such as rotten fruit incidence and stem appearance. Temperature was the factor that most influenced the fruit quality changes over the study time. In fact, fruits stored at higher mean temperatures showed higher weight loss, higher variation in CIE-Lab colour parameters, higher firmness loss, and browner and more dehydrated stems and were less appealing to the consumer. Controlled atmosphere conditions showed a smaller decrease in CIE-Lab colour parameters and lower weight loss. The incidence of rotting was very low and was always equal or lower than 2% for all conditions. Thus, RC chamber conditions were able to sustain fruit quality parameters over 28 days under normal atmosphere conditions and 49 days under controlled atmosphere conditions.

Keywords: carbon dioxide; controlled atmosphere; refrigeration; sensory evaluation; sweet cherry

1. Introduction

Sweet cherry (*Prunus avium* L.) is a non-climacteric fruit with a significant content of nutrients and bioactive compounds, such as ascorbic acid, fibre, anthocyanins and carotenoids [1,2]. However, this highly perishable fruit has a short marketing period, extending from May in Southern Europe to August in Northern Europe [2,3].

In 2019, the three main global sweet cherry producers were Turkey (664,224 t), USA (321,420 t) and Chile (233,929 t). In the same year, Portugal produced 19,130 t of sweet cherry [4], mainly from the municipalities of Fundão, Covilhã and Belmonte, which are located in the Beira Interior region [5,6].

Sweet cherries deteriorate rapidly after being harvested, especially if stored at room temperature [7]. This may include changes in skin colour, peduncle dehydration and browning, pulp softening, decrease of acidity and rotting [8–11]. Fungal spoilage, especially from the genera *Botrytis*, *Monilia*, *Penicillium* and *Rhizopus*, is the most important reason for sweet cherry post-harvest losses [11–13].

The maturity of a fruit can be defined as the stage of development that is associated with the minimum acceptable quality for the consumer [14]. Soluble solids content (SSC), titrable acidity (TA), the ratio SSC/TA, skin colour and firmness are parameters that have been suggested to be used as indices to evaluate the maturity of sweet cherries [15–19]. More particularly, Crisosto et al. [15] report that, depending on the sweet cherry cultivar, a minimum of light red colour and/or 14% to 16% SSC is required for consumer acceptance.

Refrigeration is the most common technique used to extend fruit shelf life. Combined with refrigeration, controlled atmospheres (CAs) have been shown to help delay fruit quality decay. This technique uses low oxygen (O_2) and high carbon dioxide (CO_2) concentrations to lower the respiration rate and to stop or delay mould development [7,20,21], thus enabling fruit storage for longer time periods.

Specifically, CA conditions may help maintain adequate levels of acidity, brighter skin colour and greener stems [11,22]. High amounts of CO_2 have a fungistatic action (delay of fungal growth), preserving fruit quality and contributing to the extension of fruit shelf life [12,21]. Additionally, CA conditions induce metabolic changes in volatile compounds, phenolics and pigments [23].

Optimum values for O_2 and CO_2 concentrations vary among different studies. However, intervals are usually between 3% and 10% for O_2 concentration, and between 10% and 15% for CO_2 concentration, according to the review presented in Andrade et al. [24]. Very low concentration of O_2 and/or very high concentration of CO_2 may cause anaerobic fermentation, development of fruit injuries and off-flavours, leading to fruit spoilage [23].

In fact, O_2 concentrations under 1% may result in surface pitting and increase the risk of anaerobic fermentations occurring in the fruit tissues, which leads to the development of off-flavours [25,26], and CO_2 concentrations above 30% may induce skin discoloration and off-flavour occurrence [15].

This study intended to evaluate the effect of different storage conditions and different atmosphere compositions on the quality parameters of sweet cherry cv. Satin, namely loss of weight, CIE-Lab colour, firmness, soluble solids content, titrable acidity, sensorial variables and rotting incidence during a storage period of 49 days.

2. Materials and Methods

2.1. *Sweet Cherry Storage Conditions and Sampling*

The study was conducted over two years, 2019 and 2020, starting on the harvest day, on 12 June 2019 and on 16 June 2020, for a period of 28 to 49 days (Table 1), and weekly sampling analysis was performed. Sweet cherries came from the same orchard (Fundão, Portugal) to minimize the variability influence of environmental conditions, with the harvest date decided by the Farmers' Organization. However, in both years, the fruits used in the study were further selected to form homogeneous samples in terms of colour and weight.

Table 1. Description of the experimental design used in the sweet cherry cv. Satin conservation study: atmosphere conditions, storage time and location of the refrigeration chambers.

<table>
<tr><th>Treatment</th><th>Atmosphere</th><th>O_2 (%)</th><th>CO_2 (%)</th><th>Total Storage Time (Days)</th><th>Storage Location</th><th>Chamber</th></tr>
<tr><td>1</td><td rowspan="2">Normal</td><td rowspan="2">21</td><td rowspan="2">0.04</td><td rowspan="2">28</td><td>Farmers' Organization</td><td>A</td></tr>
<tr><td>2</td><td rowspan="5">Research Centre</td><td rowspan="3">B</td></tr>
<tr><td>3</td><td rowspan="4">Controlled</td><td rowspan="2">3</td><td>10</td><td rowspan="4">49</td></tr>
<tr><td>4</td><td>15</td></tr>
<tr><td>5</td><td rowspan="2">10</td><td>10</td><td rowspan="2">C</td></tr>
<tr><td>6</td><td>15</td></tr>
</table>

This experiment evaluated a total of 6 treatments (Table 1). Treatments 1 and 2 were under normal atmosphere (NA) conditions and treatments 3 to 6 were under controlled atmosphere (CA) conditions. The 6 treatments were organized in 3 refrigeration chambers, named A, B and C. Chamber A was located at FO and received treatment 1 (NA). Chambers B and C were located at RC and received treatments 2 to 6. Chamber B received treatments 2 (NA), 3 (3%O_2–10%CO_2) and 4 (3%O_2–15%CO_2), and chamber C received treatments 5 (10%O_2–10%CO_2) and 6 (10%O_2–15%CO_2). Sampling was performed once a week (every 7 days of storage), except for CA treatments where sampling was initiated after 14 days. The duration of the study was 28 days for NA (treatments 1 and 2) and 49 days for CA (treatments 3 to 6).

Temperature and relative humidity were monitored inside the refrigeration chambers using dataloggers (EL-USB-2-LCD+, Lascar Electronics). The 2 storage locations, namely FO (chamber A) and RC (chambers B and C) had different temperature conditions and similar relative humidity (Table 2).

Table 2. Temperature and relative humidity values in the three chambers of the sweet cherry cv. Satin conservation study. Chamber A was located in FO with treatment 1. Chambers B and C were located in RC with treatments 2 (NA), 3 (3%O_2–10%CO_2), 4 (3%O_2–15%CO_2), 5 (10%O_2–10%CO_2) and 6 (10%O_2–15%CO_2).

		Temperature (°C)			Relative Humidity (%)		
		A (FO)	B (RC)	C (RC)	A (FO)	B (RC)	C (RC)
2019	Mean ± Standard deviation	4.8 ± 2.7	2.0 ± 0.7	1.8 ± 0.6	91.1 ± 2.9	97.1 ± 1.2	97.0 ± 1.5
	Maximum	12.9	10.5	10.8	98.0	99.9	99.1
	Minimum	1.1	1.7	1.6	81.2	84.4	83.9
2020	Mean ± Standard deviation	6.7 ± 1.0	0.9 ± 0.4	1.6 ± 0.7	94.2 ± 3.3	98.7 ± 3.2	98.8 ± 3.3
	Maximum	12.0	6.8	10.7	100.0	100.0	100.0
	Minimum	4.5	0.3	0.6	81.4	60.2	58.8

The mean temperature in the FO (chamber A) was systematically higher than the mean temperatures in RC (chambers B and C), namely, for chamber A, 4.8 ± 2.7 °C (2019) and 6.7 ± 1.0 °C (2020), compared to chamber B, 2.0 ± 0.7 °C (2019) and 0.9 ± 0.4 °C (2020), and chamber C 1.8 ± 0.6 °C (2019) and 1.6 ± 0.7 °C (2020). In terms of relative humidity, all the chambers had an average higher than 90% during storage. According to Crisosto et al. [15], the optimum values of temperature and relative humidity for the storage of sweet cherry are, respectively, −0.5 ± 0.5 °C and 90–95%. All storage chambers were above that interval, yet the mean value of temperature from chamber A (FO) was significantly higher. The mean values found for the relative humidity were near (chamber A) or slightly above (chambers B and C) the recommended interval [15].

The sweet cherries used in the study were selected to be homogeneous both in colour and weight. For the physicochemical analysis, a total of 60 sweet cherries per treatment and sampling day were individually numbered and distributed in 3 small netted baskets (20 sweet cherries per basket). Each small basket was filled with 30 more cherries (to a total of 50 cherries), which were not used in the physicochemical analysis. The basket was then placed in the middle of a 5 kg tray with approximately 4500 cherries, simulating commercial storage conditions (Figure 1). The evaluation of stem aspects and rotting incidence was done with 50 sweet cherries per treatment and sampling day, corresponding to 1 full netted basket. These sweet cherries were not numbered and received only minimum handling to avoid unnecessary contamination.

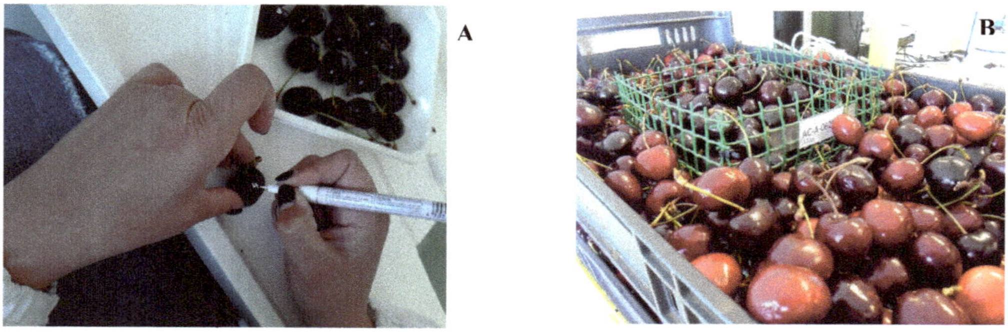

A B

Figure 1. Sweet cherries cv. Satin numbering (**A**) and netted basket with the numbered sweet cherries placed in the centre of a 5 kg commercial tray (**B**).

Fruit trays were stored on euro-sized pallets; CA treatments were covered and sealed with an LDPE plastic bag connected to a GAC 5000 unit (Fruitcontrol Equipment S.R.L.) to monitor and add gases when necessary (Figure 2). Gases used in the study were CO_2 (Biogon® C, E290, Linde), N_2 (Nitrogen 30, Sysadvance) and O_2 (air compressor, HYAC24-2, Hyundai).

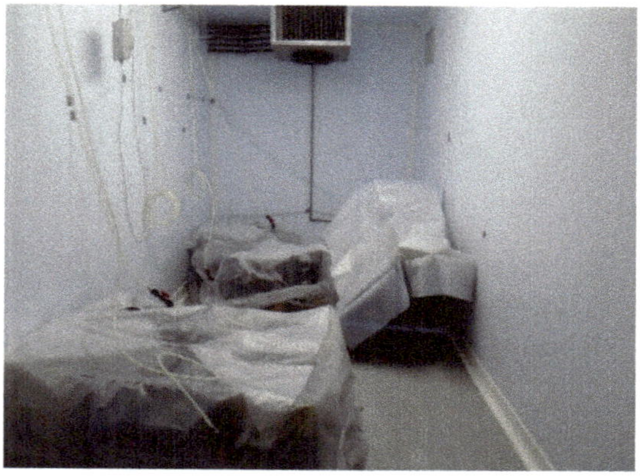

Figure 2. LDPE plastic bag cover of controlled atmosphere treatments and connection to the GAC 5000 unit for gas monitoring and control.

2.2. Fruit Quality Evaluation

2.2.1. Physicochemical Analysis

At day 0 (harvest), a sample of 60 fruits was used as a start reference based on the determination of weight, CIE-Lab colour parameters, firmness, soluble solids content (SSC) and titrable acidity. Additionally, all marked sweet cherries (60 cherries per treatment and sampling day) were individually evaluated for weight (digital scale, TE1502S, Sartorius) and colour, the 2 non-destructive parameters that can be monitored throughout the storage period.

Every sampling day, the weight, CIE-Lab colour parameters, firmness, SSC and titrable acidity were measured.

Weight loss (Δw) was expressed as a percentage of the initial weight. It was determined for each numbered sweet cherry by the ratio between the weight difference and the fruit weight on day 0. The weight difference corresponds to the difference between the weight on the sampling day and the weight on day 0.

Colour was evaluated on opposite sides of the numbered fruit with a tristimulus colorimeter (CR-400 Chroma Meter, Konica Minolta), the illuminant D65 and the software Colour DATA CM-S100w. CIE-Lab colour space was used. The differences in each colour parameter (ΔL, Δa and Δb) were determined between the sampling day and day 0.

SSC was determined using the refractometer PR-32 alpha, Atago. Results were expressed as °Brix.

Firmness was obtained by compressing the sample against a flat surface (TA-XTplus, Stable Microsystems), using a flat 75 mm diameter (P/75). The deformation rate was set to 5%, and the deformation speed to 1 mm/s. The results were expressed in newtons (N).

Titrable acidity was determined in the 3 sub-samples of 6–7 fruits (split of the 20 fruits) by potentiometric titration to pH 8.1, with a solution of NaOH 0.1 mol dm^{-3} using an automatic titrator (Titromatic 2S+3B, Crison) and the software TiCom. The results were expressed as the equivalent percentage of malic acid.

2.2.2. Rotten Fruit Evaluation

The incidence of rotten fruit was evaluated by direct observation based on the 50 fruits of the minimally handled tray. Results were expressed as a percentage.

2.2.3. Stem Visual Aspect

The visual stem aspect was determined using a scale of 1 to 5 (1—all brown and dry; 5—all green and turgid), created for this study (Figure 3).

Figure 3. Scale used to classify the aspects of stems, ranging from 1 (completely brown and dehydrated) to 5 (completely green and fresh).

2.2.4. Sensory Evaluation

The sensory evaluation was conducted in a sensory laboratory designed in accordance with ISO 8589 [27]. Parameters included the visual classification of the sweet cherry (aspect), firmness, juiciness, flavour and the overall sensory classification (global score). These parameters were evaluated by 10 consumers using a 9-point scale (1—dislike extremely, 5—neither like nor dislike, 9—like extremely). A minimum classification of 5 was defined as an acceptance indicator. The samples were presented randomly in Petri dishes, coded with a three digit number. Consumers were asked to rinse their palate with water after the evaluation of each sample.

2.2.5. Statistical Analysis

Statistical analysis was performed with SPSS 23 (IBM). For the physicochemical variables, the comparisons between means were performed with an Analysis of Variance (ANOVA), and the means were ordered using the Tukey HSD post hoc test. For the sensory variables, a Kruskal-Wallis test was used [28]. Statistical differences were accepted if $p < 0.05$.

3. Results and Discussion

3.1. Fruit Quality Evaluation

3.1.1. Physicochemical Analysis—Initial Conditions

Table 3 presents the initial values of weight, CIE-Lab colour variables, firmness, soluble solids content and titrable acidity. Colour-related variables, especially a^* and b^*, showed the highest variation between the 2 years, probably because of the natural heterogeneity of sweet cherry skin colour. Nevertheless, the values found in L^* were similar to what was reported by Harb et al. [29] for cv. Regina, more precisely, 27.9 and 29.0 in their studies conducted in 2001 and 2002, respectively. Firmness values, 3.17 N in 2019 and 3.85 N in 2020, were similar or slightly higher than those found by Paulo et al. [30] for cv. Sweetheart (2.9–3.3 N).

Table 3. Initial values of the physicochemical variables evaluated in the sweet cherry cv. Satin conservation study for each of the 2 years of analysis.

	Weight (g)	L^*	a^*	b^*	Firmness (N)	SSC (°Brix)	Titrable Acidity (% Malic Acid)
2019	9.36	33.35	24.25	8.20	3.17	16.92	0.50
2020	10.37	29.66	12.26	2.51	3.85	16.76	0.42

SSC—soluble solids content.

The values found for SSC, 16.92 °Brix (2019) and 16.76 °Brix (2020), were similar to the 16.1 °Brix reported by Costa [31], which indicates a high sugar content characteristic of this production region, where SSC frequently is over 20 °Brix. Simões et al. [32], using 12 cultivars, observed a minimum of 15.5 °Brix (cv. Earlise) and a maximum SSC of 24.4 °Brix (cv. Sweetheart). Remón et al. [22] reported 14.7 °Brix found in cv. Burlat, Akbudak et al. [33] reported an SSC of 13.8 °Brix for cv. 0900, and Ziraat and Dziedzic et al. [2] reported values between 14.8 °Brix and 16.6 °Brix for cv. Regina.

The acidity was 0.50% (2019) and 0.42% (2020), which is a lower value compared to other results. The values of titrable acidity are mainly determined by the cultivar, but there is also an influence of weather conditions. Costa [31] reported 0.77% for cv. Satin in the Beira Interior region. The titrable acidity levels found in the literature were also higher, more precisely, 0.66% for cv. Burlat [22], 0.7% for cv. 0900 Ziraat [33] and between 0.57% and 0.74% for cv. Regina [2].

3.1.2. Physicochemical Analysis—Weight Losses

Weight losses were higher in treatment 1 and significantly increased throughout the storage time, both in 2019 and in 2020 (Tables 4 and 5), reaching −8.83% in 2019 and −10.50% in 2020, after 28 days of storage. For treatment 2, weight loss reached −1.36% in 2019 and a residual value of −0.30% in 2020. This low value of 2020 was related to some condensation due to the proximity of an evaporator. Under CA conditions (treatments 3 to 6), it was always below −3.0% at the end of storage period of 49 days. In 2019, there were no significant differences in weight loss over the storage period for treatment 6 in 2019 and for treatments 4 and 5 in 2020. Globally, CA conditions correspond to lower weight losses, which is in accordance with Akbudak et al. [11,33], who reported higher weight losses under normal atmosphere (NA) conditions when compared to CA. However, the influence of different CA conditions on weight loss is not consensual among the reported studies. For example, Harb et al. [29] reported small or insignificant differences between the different treatments. Nevertheless, Akbudak et al. [11] pointed out that the water vapour accumulated inside the CA containers may have influenced the lower weight losses under these conditions. Sweet cherry low skin diffusion resistance and high surface/volume ratio increases the chances of weight loss [2,13].

Table 4. Comparison of physicochemical variables between treatments and sampling days in 2019 evaluated in the sweet cherry cv. Satin conservation study. Treatment 1 was located at the FO under normal atmosphere conditions. Treatments 2 to 6 were located at the Research Centre. Treatment 2 was under normal atmosphere conditions. The atmosphere compositions of treatments 3 to 6 were 3%O_2–10%CO_2, 3%O_2–15%CO_2, 10%O_2–10%CO_2 and 10%O_2–15%CO_2, respectively.

	Treat	7 d	14 d	21 d	28 d	35 d	42 d	49 d
Δ*w* (%)	1	$-2.95^{A} \pm 0.60$	$-4.92^{B-e} \pm 1.36$	$-7.18^{C-d} \pm 1.68$	$-8.83^{D-b} \pm 1.93$			
	2	$-0.66^{A} \pm 0.19$	$-0.64^{A-a} \pm 0.28$	$-0.92^{B-a} \pm 0.28$	$-1.36^{C-a} \pm 6.84$			
	3		$-1.51^{A-bc} \pm 0.60$	$-1.51^{A-b} \pm 0.47$	$-1.49^{A-a} \pm 0.50$	$-1.64^{AB-a} \pm 0.75$	$-1.92^{B-ns} \pm 0.91$	$-1.99^{B-b} \pm 0.99$
	4		$-1.22^{A-b} \pm 0.30$	$-1.75^{BC-bc} \pm 1.41$	$-1.52^{AB-a} \pm 0.60$	$-1.79^{BC-a} \pm 0.50$	$-2.09^{CD} \pm 0.70$	$-2.37^{D-b} \pm 0.95$
	5		$-1.76^{AB-c} \pm 0.44$	$-1.77^{ABC-bc} \pm 0.56$	$-2.28^{C-a} \pm 0.62$	$-1.95^{BC-a} \pm 0.96$	$-2.11^{BC} \pm 1.17$	$-1.41^{A-a} \pm 1.63$
	6		$-2.24^{NS-d} \pm 0.51$	$-2.20^{c} \pm 0.62$	$-2.32^{a} \pm 1.00$	$-2.64^{b} \pm 1.38$	-2.39 ± 0.95	$-2.36^{b} \pm 1.01$
Δ*L**	1	$0.43^{A} \pm 1.06$	$-1.21^{B-c} \pm 1.28$	$-3.06^{C-d} \pm 2.26$	$-1.23^{B-c} \pm 1.69$			
	2	$-0.57^{A} \pm 0.96$	$-1.27^{B-c} \pm 1.25$	$-2.06^{C-c} \pm 1.51$	$-0.26^{A-b} \pm 0.95$			
	3		$0.14^{D-a} \pm 1.44$	$1.68^{B-a} \pm 1.60$	$0.88^{C-a} \pm 0.65$	$1.32^{BC-ns} \pm 0.74$	$2.44^{A-ns} \pm 0.77$	$-0.17^{D-b} \pm 2.31$
	4		$-0.04^{D-a} \pm 0.98$	$0.08^{D-b} \pm 0.85$	$1.32^{BC-a} \pm 0.60$	$1.49^{B} \pm 0.87$	$2.60^{A} \pm 1.17$	$0.84^{C-a} \pm 1.00$
	5		$-0.23^{D-ab} \pm 0.80$	$-0.25^{D-b} \pm 1.26$	$0.82^{C-a} \pm 0.95$	$1.51^{B} \pm 0.76$	$2.63^{A} \pm 0.89$	$0.75^{C-a} \pm 0.90$
	6		$-0.79^{D-bc} \pm 1.47$	$-0.02^{C-b} \pm 0.78$	$0.82^{B-a} \pm 0.79$	$1.25^{B} \pm 1.54$	$2.30^{A} \pm 1.32$	$0.96^{B-a} \pm 0.98$
Δ*a**	1	$-2.48^{A} \pm 1.96$	$-4.18^{A-c} \pm 3.59$	$-10.46^{B-c} \pm 4.68$	$-10.66^{B-c} \pm 4.15$			
	2	$-1.74^{A} \pm 1.22$	$-2.70^{B-b} \pm 1.39$	$-4.01^{C-b} \pm 1.45$	$-6.54^{D-b} \pm 1.71$			
	3		$-0.83^{A-a} \pm 2.20$	$-1.20^{AB-a} \pm 3.22$	$-2.43^{BC-a} \pm 0.92$	$-3.39^{CD-ab} \pm 1.39$	$-4.74^{DE-ns} \pm 1.22$	$-5.27^{E-b} \pm 5.16$
	4		$-1.03^{A-a} \pm 1.18$	$-1.86^{B-a} \pm 0.84$	$-2.33^{C-a} \pm 1.14$	$-3.85^{C-b} \pm 1.31$	$-4.87^{D} \pm 1.55$	$-3.32^{C-a} \pm 1.63$
	5		$-1.33^{A-a} \pm 1.17$	$-1.75^{A-a} \pm 1.31$	$-2.49^{B-a} \pm 1.20$	$-2.87^{B-a} \pm 1.12$	$-4.22^{C} \pm 1.30$	$-2.49^{B-a} \pm 1.11$
	6		$-1.55^{A-a} \pm 1.25$	$-1.83^{A-a} \pm 1.11$	$-2.10^{A-a} \pm 0.97$	$-3.79^{BC-ab} \pm 1.42$	$-4.55^{C} \pm 2.30$	$-3.39^{B-a} \pm 1.75$
Δ*b**	1	$-0.36^{A} \pm 1.14$	$-1.59^{B-b} \pm 1.98$	$-4.39^{C-d} \pm 3.32$	$-3.31^{C-c} \pm 2.77$			
	2	$0.63^{A} \pm 0.66$	$-1.32^{B-b} \pm 1.24$	$-2.30^{C-c} \pm 1.52$	$-1.96^{C-b} \pm 1.31$			
	3		$0.23^{AB-a} \pm 1.37$	$0.49^{A-a} \pm 1.63$	$-0.52^{BC-a} \pm 0.75$	$-0.37^{ABC-b} \pm 0.88$	$-0.69^{C-b} \pm 0.99$	$-2.21^{D-b} \pm 3.29$
	4		$-0.14^{NS-a} \pm 0.65$	$-0.42^{b} \pm 0.50$	$-0.07^{a} \pm 0.64$	$-0.36^{b} \pm 1.01$	$-0.44^{ab} \pm 1.03$	$-0.41^{a} \pm 1.01$
	5		$-0.13^{AB-a} \pm 0.91$	$-0.44^{B-b} \pm 0.92$	$-0.35^{B-a} \pm 0.96$	$0.13^{A-a} \pm 0.66$	$-0.01^{AB-a} \pm 0.92$	$-0.08^{AB-a} \pm 0.89$
	6		$-0.23^{NS-a} \pm 0.67$	$-0.16^{ab} \pm 0.52$	$0.11^{a} \pm 0.69$	$-0.27^{ab} \pm 0.89$	$-0.32^{ab} \pm 1.56$	$-0.09^{a} \pm 0.98$
Firmness (N)	1	$2.68^{A} \pm 0.53$	$2.39^{B-c} \pm 0.54$	$2.22^{B-c} \pm 0.49$	$1.92^{C-d} \pm 0.53$			
	2	$3.42^{NS} \pm 0.61$	$3.57^{a} \pm 0.84$	$3.72^{a} \pm 0.82$	$3.75^{a} \pm 0.72$			
	3		$3.35^{B-ab} \pm 0.76$	$3.94^{A-a} \pm 0.87$	$3.72^{AB-a} \pm 0.85$	$3.58^{AB-b} \pm 0.76$	$3.97^{A-ns} \pm 1.06$	$3.73^{AB-ns} \pm 0.71$
	4		$3.58^{NS-a} \pm 0.81$	$3.65^{ab} \pm 0.87$	$3.59^{ab} \pm 0.91$	$3.77^{b} \pm 0.83$	3.81 ± 0.91	3.91 ± 0.75
	5		$3.32^{BC-ab} \pm 0.80$	$3.66^{AB-ab} \pm 0.82$	$3.23^{C-bc} \pm 0.69$	$3.76^{A-a} \pm 1.01$	$3.53^{ABC} \pm 0.88$	$3.48^{ABC} \pm 0.71$
	6		$3.15^{B-b} \pm 0.67$	$3.27^{AB-b} \pm 0.73$	$3.12^{B-c} \pm 0.73$	$3.56^{A-b} \pm 0.82$	$3.57^{A} \pm 0.96$	$3.64^{A} \pm 0.71$
SSC (°Brix)	1	$16.63^{B} \pm 1.81$	$17.60^{AB-a} \pm 2.46$	$17.79^{A-a} \pm 2.44$	$18.56^{A-a} \pm 2.73$			
	2	$16.27^{NS} \pm 1.27$	$16.33^{b} \pm 1.76$	$16.25^{b} \pm 2.07$	$16.70^{bc} \pm 2.16$			
	3		$16.70^{AB-ab} \pm 1.94$	$16.68^{AB-ab} \pm 1.67$	$17.33^{A-bc} \pm 2.05$	$16.16^{BC-b} \pm 2.47$	$15.94^{BC-ns} \pm 1.59$	$15.64^{C-ns} \pm 2.09$
	4		$17.06^{AB-ab} \pm 2.71$	$17.22^{A-ab} \pm 1.79$	$17.09^{AB-bc} \pm 2.92$	$16.55^{AB-b} \pm 2.49$	$16.40^{AB} \pm 1.59$	$15.92^{B} \pm 1.64$
	5		$16.90^{AB-ab} \pm 1.82$	$16.57^{B-b} \pm 2.07$	$16.16^{B-c} \pm 1.78$	$17.72^{A-a} \pm 2.37$	$16.43^{B} \pm 1.94$	$16.06^{B} \pm 2.07$
	6		$17.26^{AB-ab} \pm 2.66$	$17.81^{A-a} \pm 3.21$	$17.39^{AB-ab} \pm 1.91$	$16.18^{BC-b} \pm 2.27$	$16.45^{BC} \pm 1.91$	$15.62^{C} \pm 1.68$
Titrable acidity (% malic acid)	1	$0.36^{B} \pm 0.04$	$0.44^{AB-ns} \pm 0.02$	$0.45^{A-ns} \pm 0.02$	$0.43^{AB-ns} \pm 0.04$			
	2	$0.45^{NS} \pm 0.07$	0.47 ± 0.04	0.44 ± 0.01	0.47 ± 0.02			
	3		$0.43^{NS} \pm 0.06$	0.45 ± 0.04	0.44 ± 0.05	$0.39^{ns} \pm 0.03$	$0.43^{ns} \pm 0.03$	$0.41^{ns} \pm 0.03$
	4		$0.48^{NS} \pm 0.02$	0.50 ± 0.08	0.44 ± 0.05	0.41 ± 0.07	0.43 ± 0.05	0.42 ± 0.04
	5		$0.44^{NS} \pm 0.01$	0.46 ± 0.03	0.41 ± 0.06	0.43 ± 0.05	0.42 ± 0.02	0.36 ± 0.07
	6		$0.47^{NS} \pm 0.03$	0.47 ± 0.04	0.44 ± 0.06	0.39 ± 0.03	0.43 ± 0.01	0.40 ± 0.04

Different capital letters in the same row indicate statistical differences ($p < 0.05$) between sampling days. Different small letters in the same column indicate statistical differences ($p < 0.05$) between treatments. NS—non significant ($p \geq 0.05$) differences between sampling days, ns—non significant ($p \geq 0.05$) differences between treatments.

Comparing results from 2019 and 2020 for treatments 1 and 2, the higher weight losses that occurred in treatment 1 may have been caused by constant exposure to higher storage temperature. This was also noticed and reported by Dziedzic and co-authors [2].

Table 5. Comparison of physicochemical variables between treatments and sampling days in 2020, evaluated in the sweet cherry cv. Satin conservation study. Treatment 1 was located at FO under normal atmosphere conditions. Treatments 2 to 6 were located at RC. Treatment 2 was under normal atmosphere conditions. The atmosphere composition of treatments 3 to 6 was, respectively, 3%O_2–10%CO_2, 3%O_2–15%CO_2, 10%O_2–10%CO_2 and 10%O_2–15%CO_2.

	Treat	7 d	14 d	21 d	28 d	35 d	42 d	49 d
Δw (%)	1	$-3.30^{\text{A}} \pm 0.82$	$-6.06^{\text{B-c}} \pm 1.31$	$-7.22^{\text{C-d}} \pm 1.84$	$-10.50^{\text{D-d}} \pm 2.57$			
	2	$0.96^{\text{B}} \pm 1.24$	$2.12^{\text{A-a}} \pm 1.31$	$-1.08^{\text{D-a}} \pm 0.53$	$-0.30^{\text{C-a}} \pm 0.74$			
	3		$-1.48^{\text{A-b}} \pm 2.63$	$-1.44^{\text{A-ab}} \pm 1.11$	$-1.69^{\text{A-bc}} \pm 0.71$	$-1.66^{\text{A-ns}} \pm 1.15$	$-2.19^{\text{AB-ns}} \pm 1.55$	$-2.96^{\text{B-b}} \pm 2.03$
	4		$-1.41^{\text{NS-b}} \pm 0.48$	$-2.09^{\text{c}} \pm 0.90$	$-1.85^{\text{bc}} \pm 1.36$	-1.71 ± 1.26	-2.05 ± 1.90	$-2.10^{\text{ab}} \pm 1.95$
	5		$-2.05^{\text{NS-b}} \pm 1.51$	$-2.00^{\text{bc}} \pm 0.90$	$-1.58^{\text{b}} \pm 0.72$	-1.95 ± 1.84	-1.97 ± 1.35	$-1.72^{\text{a}} \pm 2.24$
	6		$-1.73^{\text{AB-b}} \pm 0.48$	$-1.71^{\text{A-bc}} \pm 1.05$	$-2.43^{\text{B-c}} \pm 1.52$	$-2.15^{\text{AB}} \pm 1.66$	$-2.33^{\text{AB}} \pm 1.46$	$-2.39^{\text{AB-ab}} \pm 1.63$
ΔL^*	1	$-0.46^{\text{A}} \pm 0.53$	$-0.70^{\text{AB-b}} \pm 0.90$	$-0.93^{\text{B-d}} \pm 0.63$	$-1.28^{\text{C-c}} \pm 0.65$			
	2	$-0.49^{\text{NS}} \pm 0.62$	$-0.41^{\text{ab}} \pm 0.75$	$-0.60^{\text{cd}} \pm 0.75$	$-0.75^{\text{b}} \pm 0.74$			
	3		$-0.25^{\text{A-a}} \pm 0.52$	$-0.61^{\text{B-cd}} \pm 0.59$	$-0.32^{\text{AB-a}} \pm 0.72$	$-0.18^{\text{A-ns}} \pm 0.72$	$-0.28^{\text{AB-ab}} \pm 0.62$	$-0.35^{\text{AB-ns}} \pm 0.65$
	4		$-0.26^{\text{AB-a}} \pm 0.78$	$-0.08^{\text{A-ab}} \pm 0.50$	$-0.34^{\text{AB-a}} \pm 0.68$	$-0.16^{\text{A}} \pm 0.91$	$-0.63^{\text{B-b}} \pm 1.07$	$-0.43^{\text{AB}} \pm 0.78$
	5		$-0.47^{\text{NS-ab}} \pm 0.65$	$-0.38^{\text{bc}} \pm 0.86$	$-0.31^{\text{a}} \pm 0.81$	-0.14 ± 0.75	$-0.37^{\text{ab}} \pm 0.74$	-0.41 ± 0.64
	6		$-0.19^{\text{AB-a}} \pm 0.72$	$0.07^{\text{A-a}} \pm 0.55$	$-0.38^{\text{BC-ab}} \pm 0.55$	$-0.25^{\text{ABC}} \pm 0.65$	$-0.24^{\text{ABC-a}} \pm 0.73$	$-0.53^{\text{C}} \pm 0.71$
Δa^*	1	$-1.72^{\text{A}} \pm 0.80$	$-3.31^{\text{B-b}} \pm 3.67$	$-5.18^{\text{C-c}} \pm 1.87$	$-5.60^{\text{C-c}} \pm 1.77$			
	2	$-0.62^{\text{A}} \pm 0.70$	$-1.52^{\text{B-a}} \pm 0.65$	$-2.09^{\text{C-b}} \pm 0.85$	$-2.86^{\text{D-b}} \pm 1.10$			
	3		$-0.78^{\text{A-a}} \pm 0.71$	$-1.50^{\text{B-ab}} \pm 1.02$	$-1.41^{\text{B-a}} \pm 1.39$	$-1.71^{\text{BC-ns}} \pm 0.89$	$-2.32^{\text{D-ns}} \pm 0.97$	$-2.17^{\text{C D-ab}} \pm 1.14$
	4		$-1.30^{\text{A-a}} \pm 0.80$	$-1.40^{\text{A-a}} \pm 1.03$	$-1.64^{\text{AB-a}} \pm 0.94$	$-2.21^{\text{BC}} \pm 1.45$	$-2.34^{\text{C}} \pm 1.32$	$-2.70^{\text{C-bc}} \pm 1.37$
	5		$-1.18^{\text{A-a}} \pm 0.98$	$-1.50^{\text{AB-ab}} \pm 1.14$	$-1.81^{\text{BC-a}} \pm 0.94$	$-1.99^{\text{BC}} \pm 1.16$	$-2.32^{\text{C}} \pm 1.06$	$-2.01^{\text{BC-a}} \pm 1.07$
	6		$-1.28^{\text{A-a}} \pm 0.93$	$-1.77^{\text{AB-ab}} \pm 0.90$	$-1.99^{\text{B-a}} \pm 1.21$	$-2.07^{\text{B}} \pm 0.88$	$-2.09^{\text{B}} \pm 1.23$	$-2.76^{\text{C-c}} \pm 1.22$
Δb^*	1	$-0.24^{\text{A}} \pm 0.32$	$-0.53^{\text{AB-b}} \pm 0.96$	$-0.94^{\text{C-b}} \pm 0.76$	$-0.84^{\text{BC-c}} \pm 0.62$			
	2	$-0.10^{\text{A}} \pm 0.27$	$-0.06^{\text{A-a}} \pm 0.38$	$-0.10^{\text{A-a}} \pm 0.33$	$-0.30^{\text{B-b}} \pm 0.60$			
	3		$0.10^{\text{A-a}} \pm 0.21$	$0.03^{\text{AB-a}} \pm 0.27$	$-0.04^{\text{B-a}} \pm 0.31$	$0.06^{\text{AB-ab}} \pm 0.24$	$-0.07^{\text{B-b}} \pm 0.27$	$0.02^{\text{AB-ns}} \pm 0.29$
	4		$0.04^{\text{NS-a}} \pm 0.19$	$0.08^{\text{a}} \pm 0.28$	$0.09^{\text{a}} \pm 0.24$	$0.16^{\text{ab}} \pm 0.29$	$0.15^{\text{a}} \pm 0.37$	0.07 ± 0.26
	5		$0.00^{\text{NS-a}} \pm 0.28$	$0.03^{\text{a}} \pm 0.32$	$0.04^{\text{a}} \pm 0.38$	$0.03^{\text{b}} \pm 0.41$	$0.05^{\text{ab}} \pm 0.29$	0.15 ± 0.32
	6		$-0.04^{\text{B-a}} \pm 0.30$	$0.05^{\text{AB-a}} \pm 0.28$	$0.09^{\text{AB-a}} \pm 0.32$	$0.17^{\text{A-a}} \pm 0.23$	$0.14^{\text{A-a}} \pm 0.35$	$0.13^{\text{A}} \pm 0.34$
Firmness (N)	1	$3.01^{\text{NS}} \pm 0.75$	$3.02^{\text{d}} \pm 0.61$	$2.72^{\text{d}} \pm 0.68$	$2.91^{\text{b}} \pm 0.91$			
	2	$4.65^{\text{NS}} \pm 1.33$	$4.94^{\text{a}} \pm 0.92$	$4.63^{\text{ab}} \pm 1.12$	$4.94^{\text{a}} \pm 1.04$			
	3		$3.90^{\text{C-bc}} \pm 0.82$	$4.23^{\text{BC-bc}} \pm 0.83$	$4.41^{\text{ABC-a}} \pm 1.16$	$4.77^{\text{A-ns}} \pm 1.07$	$4.38^{\text{ABC-b}} \pm 1.08$	$4.66^{\text{AB-ns}} \pm 0.93$
	4		$4.28^{\text{B-b}} \pm 1.03$	$4.73^{\text{AB-a}} \pm 1.08$	$4.64^{\text{AB-a}} \pm 0.93$	$4.80^{\text{AB}} \pm 1.07$	$5.02^{\text{A-a}} \pm 1.38$	$4.62^{\text{AB}} \pm 0.90$
	5		$4.19^{\text{A-b}} \pm 1.21$	$4.42^{\text{A-ab}} \pm 0.95$	$4.43^{\text{A-a}} \pm 1.03$	$4.62^{\text{A}} \pm 1.18$	$4.25^{\text{A-b}} \pm 0.92$	$4.37^{\text{A}} \pm 0.90$
	6		$3.58^{\text{C-c}} \pm 0.81$	$3.85^{\text{BC-c}} \pm 0.80$	$4.71^{\text{A-a}} \pm 0.98$	$4.53^{\text{A}} \pm 1.03$	$4.24^{\text{AB-b}} \pm 0.91$	$4.38^{\text{A}} \pm 1.13$
SSC (°Brix)	1	$17.11^{\text{NS}} \pm 2.47$	$17.66^{\text{a}} \pm 2.01$	$17.30^{\text{a}} \pm 1.89$	$17.81^{\text{a}} \pm 2.13$			
	2	$16.23^{\text{A}} \pm 1.40$	$15.44^{\text{B-c}} \pm 1.67$	$16.75^{\text{A-abc}} \pm 1.86$	$16.00^{\text{AB-b}} \pm 1.61$			
	3		$16.78^{\text{NS-ab}} \pm 1.94$	$16.61^{\text{abc}} \pm 1.70$	$16.50^{\text{b}} \pm 2.24$	$16.55^{\text{ns}} \pm 1.81$	$16.37^{\text{ns}} \pm 2.01$	$16.97^{\text{a}} \pm 2.48$
	4		$16.06^{\text{NS-bc}} \pm 2.21$	$16.02^{\text{c}} \pm 1.65$	$16.30^{\text{b}} \pm 1.94$	16.61 ± 2.37	16.46 ± 1.74	$15.95^{\text{ab}} \pm 1.75$
	5		$16.67^{\text{NS-ab}} \pm 2.09$	$16.32^{\text{bc}} \pm 2.02$	$16.00^{\text{b}} \pm 1.91$	16.37 ± 1.79	16.38 ± 1.94	$16.36^{\text{ab}} \pm 2.09$
	6		$17.12^{\text{A-ab}} \pm 2.52$	$17.06^{\text{A-ab}} \pm 2.05$	$16.64^{\text{AB-b}} \pm 2.29$	$16.71^{\text{AB}} \pm 1.94$	$16.76^{\text{AB}} \pm 1.86$	$15.66^{\text{B-b}} \pm 2.42$
Titrable acidity (% malic acid)	1	$0.48^{\text{A}} \pm 0.04$	$0.40^{\text{B-ab}} \pm 0.02$	$0.41^{\text{B-ab}} \pm 0.04$	$0.38^{\text{B-ab}} \pm 0.04$			
	2	$0.37^{\text{AB}} \pm 0.03$	$0.38^{\text{AB-b}} \pm 0.03$	$0.40^{\text{A-b}} \pm 0.02$	$0.36^{\text{B-b}} \pm 0.03$			
	3		$0.42^{\text{A-ab}} \pm 0.04$	$0.41^{\text{A-ab}} \pm 0.03$	$0.38^{\text{AB-ab}} \pm 0.02$	$0.36^{\text{B-ns}} \pm 0.04$	$0.35^{\text{B-ns}} \pm 0.03$	$0.34^{\text{B-ns}} \pm 0.03$
	4		$0.44^{\text{A-a}} \pm 0.04$	$0.39^{\text{AB-b}} \pm 0.01$	$0.39^{\text{AB-ab}} \pm 0.04$	$0.36^{\text{BC}} \pm 0.05$	$0.32^{\text{C}} \pm 0.03$	$0.31^{\text{C}} \pm 0.03$
	5		$0.44^{\text{A-a}} \pm 0.04$	$0.41^{\text{AB-ab}} \pm 0.03$	$0.37^{\text{BC-ab}} \pm 0.02$	$0.35^{\text{C D}} \pm 0.03$	$0.33^{\text{C D}} \pm 0.04$	$0.32^{\text{D}} \pm 0.03$
	6		$0.44^{\text{A-a}} \pm 0.04$	$0.44^{\text{A-a}} \pm 0.04$	$0.40^{\text{AB-a}} \pm 0.03$	$0.37^{\text{BC}} \pm 0.03$	$0.36^{\text{BC}} \pm 0.03$	$0.34^{\text{C}} \pm 0.03$

Different capital letters in the same row indicate statistical differences ($p < 0.05$) between sampling days. Different small letters in the same column indicate statistical differences ($p < 0.05$) between treatments. NS—non-significant ($p \geq 0.05$) differences between sampling days, ns—non-significant ($p \geq 0.05$) differences between treatments.

3.1.3. Physicochemical Analysis—CIE-Lab Colour Parameters

The range of ΔL^* values was higher in 2019 than in 2020 (Tables 4 and 5). In the first year, ΔL^* values were between −3.06 (treatment 1, 21 days) and 2.63 (treatment 5, 42 days). However, in 2020 the values were spread between −1.28 (treatment 1, 28 days) and 0.07 (treatment 6, 21 days).

CA conditions resulted mostly in positive values for ΔL^* in 2019 and negative values in 2020. Apart from a global decrease in the values of this variable found in 2019 and 2020 for treatment 1, the other treatments did not show any consistent evolution pattern either in 2019 or in 2020 (Tables 4 and 5), which was contrary to the more prevailing decrease pattern reported in other studies [7,22,33].

The treatments under CA conditions showed lower differences in the L^* value than the treatments under NA between 14 and 28 days of storage (in 2019, Table 4) or at 28 days of storage (in 2020, Table 5), indicating lower differences in the brightness of the fruits. Akbudak et al. [33], working with cv. 0900 Ziraat, also reported lower colour variation in sweet cherries stored under CA conditions. In a similar way, Yang et al. [7] reported that the sweet cherries cv. Lapins under CA conditions which, in their case, included argon (Ar) instead of nitrogen (N_2) as the inert gas (5%O_2–10%CO_2–85%Ar) had higher L^* values than the cherries under NA conditions.

A decreasing tendency was found in Δa^* for all treatments, pointing to a^* as the main affected colour parameter under storage (Tables 4 and 5), reaching −10.66 (in 2019) and −5.60 (in 2020) in treatment 1. Higher variation of a^* was observed in NA compared to CA, which is consistent with the results of Akbudak et al. [33].

In 2019 (Table 4) and 2020 (Table 5), b^* values showed a decreasing pattern throughout time for NA treatments (treatments 1 and 2) and showed a higher decrease in the b^* value (higher Δb^* in modulus) than the CA treatments, similar to what was reported by Akbudak et al. [33]. Δb^* values for CA conditions were always <1 for all treatments and remained stable through time for treatment 4 (2019 and 2020), treatment 6 (2019) and treatment 5 (2020), which means that CA conditions were better at preserving the colour characteristic than NA conditions.

The temperature difference between treatments 1 and 2 may have influenced the CIE-Lab colour parameter evolution, since treatment 1 showed a higher decrease in values than treatment 2 (Tables 4 and 5).

3.1.4. Physicochemical Analysis—Firmness

The fruit firmness from treatment 1 was lower than other treatments after 14 days (Tables 4 and 5), reaching 1.92 N (2019) and 2.91 N (2020). This might have been influenced by different temperature conditions since (a) treatment 2 fruit firmness did not statistically differ (in most cases) from CA fruit firmness treatments (3 to 6) and (b) the fruit firmness of treatment 1 was systematically lower than treatment 2, both under NA conditions. Dziedzic et al. [2] also found lower firmness values for 'Regina' sweet cherries stored under NA at 8 °C than at 2 °C in both years presented (12.2 N vs. 14.1 N in 2011 and 17.1 N vs. 18.9 N in 2012). However, since these authors used a different probe (8 mm) for the firmness determination, the absolute values cannot be directly compared with ours. Nevertheless, the trends between treatments remain valid.

In this study, no differences in fruit firmness between atmosphere compositions were found. However, in most studies, CA treatments showed similar or higher firmness than NA treatments. For example, Dziedzic et al. [2], using cv. Regina, reported mean values of 21.2 N when stored at 2 °C under 3%O_2–5%CO_2 and 18.9 N for NA conditions in their study performed in 2012. However, in the study performed the year before, also presented in the same article, Dziedzic et al. [2] reported similar mean values between CA (14.2 N) and NA (14.1 N) treatments.

Yang et al. [7] also reported a sharper decrease in fruit firmness from 'Lapins' sweet cherries stored under NA when compared to the fruits stored under CA conditions (5%O_2–10%CO_2–85%Ar). After 63 days of storage, the firmness of the fruits under NA was approximately 4 N, while the firmness was 5.5 N under CA treatment.

Treatment 6 (2019) showed an increase of firmness (reaching 3.64 N at 49 days). Similar evolution was also reported by Tian et al. [9] for 'Lapins' sweet cherries stored under CA (5%O_2–10%CO_2).

Globally fruit firmness was higher in 2020. At harvest, fruit firmness was 3.85 N in 2020 compared to 3.17 N in 2019 and remained very stable under CA conditions, inside the interval of 3.58 to 5.02 N in 2019 and 3.15 to 3.97 N in 2020 (Tables 4 and 5).

3.1.5. Physicochemical Analysis—Soluble Solids Content

SSC mean values at day 0 were 16.92 °Brix and 16.76 °Brix (Table 3), higher than the minimum indicated by Crisosto et al. (1996) for acceptable quality (14–16 °Brix).

Neither clear nor systematic evolution patterns throughout time were found (Tables 4 and 5). Treatment 1 showed, at the end of storage period (28 days), the highest values for SSC: 18.56 °Brix in 2019 and 17.81 °Brix in 2020. Our results did not show any relationship between SSC and CA conditions. The results found in the literature regarding the sweet cherry SSC evolution are not consensual. Dziedzic et al. [2] reported an increase of SSC for 'Regina' sweet cherries, higher at 8 °C than at 2 °C, which may be explained by the sugar concentration due to water loss [2] or by the conversion of cell wall polysaccharides into sugar, since sweet cherry starch amount is low [33]. On the other side, Yang et al. [7] reported a decrease in SSC of 'Lapins' sweet cherries, which was lower under CA conditions (5%O_2–10%CO_2).

3.1.6. Physicochemical Analysis—Titrable Acidity

Titrable acidity (TA) is very important for sweet cherry sensorial properties [12]. Starting from a low value of 0.50% (2019) and 0.40% (2020), no significant differences were found for titrable acidity evolution in 2019 (Table 4). In 2020, a decreasing tendency was found for all treatments through storage time (Table 5), which is similar to what was reported by other authors [2,7,12,22,29,33,34]. The use of organic acids as substrates for physiological processes may be one reason for TA decrease [2,33]. However, no differences were found in TA content between CA conditions for >35 d (Table 4). Akbudak et al. [11,33], working with cv. 0900 Ziraat, found higher values for TA in the CA treatments (5%O_2–5% to 25%CO_2) compared to NA conditions, which was not observed in our results.

3.1.7. Visual Observation—Rotten Fruits

Sweet cherries are highly perishable and susceptible to fungal fruit decay. The observation of rotten fruits was low in all results and was equal to or lower than 3% (Table 6). In both years, the first observed rotten fruits occurred after 21 days of storage, sooner than the 30 days reported by Akbudak et al. [11]. Due to a low percentage of rotten fruits, no relationship between the percentage of rotten fruits and atmosphere composition was found, contrary to what was reported by Akbudak et al. [33], who found a lower percentage of rotten fruits in the higher CO_2 treatments (20% CO_2 and 25% CO_2).

Table 6. Percentage of rotten sweet cherries cv. Satin for each year, treatment and storage time. Before 21 days of conservation, no rotten fruits were found. Treatment 1 was located at FO under normal atmosphere conditions. Treatments 2 to 6 were located at RC. Treatment 2 was under normal atmosphere conditions. The atmosphere composition of treatments 3 to 6 was, respectively, 3%O_2–10%CO_2, 3%O_2–15%CO_2, 10%O_2–10%CO_2 and 10%O_2–15%CO_2. Treatments 1 and 2 were evaluated until the 28th day. Treatments 3 to 6 were evaluated until the 49th day. The absence of values indicates 0% rotten sweet cherries.

	Treatment	21 d	28 d	35 d	42 d	49 d
2019	1	3%	2%			
	2		3%			
	3					2%
	4	2%	2%			
	5					
	6					
2020	1	3%	2%			
	2					
	3		2%			
	4					
	5					3%
	6			2%		

N = 50 fruits per treatment and sampling day.

3.1.8. Stem Aspect

Green and non-shrivelling stems are considered a freshness characteristic by consumers [2] and are very important in retail. Stem browning is related to the fact that water evaporates more easily from stems than from fruits [11].

All treatments showed a generic decrease in the mean classification values through the storage time (Figure 4). Treatment 1 had the lowest mean values: one scale point lower on average. This indicates a more severe stem dehydration and browning related to higher storage temperatures.

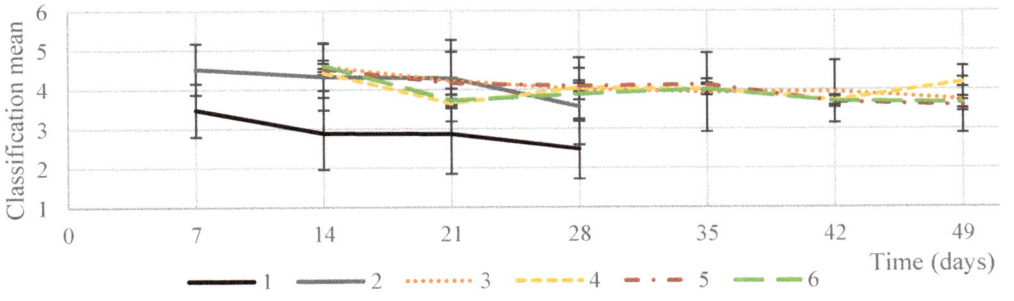

Figure 4. Stem aspect evolution (N = 50) expressed as mean values of the visual classification. Classification (Figure 3, Materials and methods) ranged from 1 (completely brown and dehydrated stem) to 5 (completely green and fresh stem). Treatment 1 was located at FO under normal atmosphere conditions. Treatments 2 to 6 were located at RC. Treatment 2 was under normal atmosphere conditions. The atmosphere compositions of treatments 3 to 6 were, respectively, 3%O_2–10%CO_2, 3%O_2–15%CO_2, 10%O_2–10%CO_2 and 10%O_2–15%CO_2.

CA conditions clearly maintained stems in the same classification throughout storage time. Thus, the browning and shrivelling of stems were kept at a minimum.

Differences were not observed between treatment 2 and CA treatments until 21 days (Figure 2). This indicates that, in our study, NA conditions preserved stem freshness until 21 days and did not differ considerably from CA conditions in the stem aspects. This observation differs from other authors [2,11,23,34]. Nevertheless, Stow et al. [35], working with sweet cherry cultivars Lapins and Colney, reported that, though the percentage of green stems was higher in the treatments with lower O_2 content (0.5% and 1%), it was similar for the control, NA conditions and the treatment with higher CO_2 content (10%), which, in their study, was conjugated with atmosphere O_2 content (21%).

3.1.9. Sensory Evaluation

In 2019, no differences between treatments were found in any sensory variable. In 2020, differences were found only at 28 days of storage for aspect, flavour and global score, and treatment 1 showed the lowest classification (Figure 5). Moreover, at 28 days of storage, the median of aspect scores in treatment 1 was below the minimum defined for sensory acceptance (5 points), which represents the negative influence of a higher storage temperature and not atmosphere composition, since no significant differences were found between treatment 2 (NA) and treatments 3 to 6 (CA).

In 2019, no differences were found between sampling days in any sensory variable, with the exception of aspect and flavour. The medians of the scores given to these two variables were lower at 28 days than at the other sampling days. Nevertheless, all medians of scores were equal or higher than the minimum defined for sensory acceptance (5).

In 2020 there was a global decrease in medians from all CA treatments in the variables of juiciness, flavour and global score. This decrease also occurred for firmness, except for treatment 3. Despite this, and similar to what was found in 2019, the medians of the scores at the end of the study (49 days) were all equal to or higher than the minimum established for sensory acceptance (5).

Figure 5. Results from the sensory evaluation of sweet cherries cv. Satin in 2020 after 14 days, 21 days and 28 days. The evaluation followed a 9-point scale ranging from 1 (dislike extremely) to 9 (like extremely). Treatment 1 was located at FO under normal atmosphere conditions. Treatments 2 to 6 were located at RC. Treatment 2 was under normal atmosphere conditions. The atmosphere compositions of treatments 3 to 6 were 3%O_2–10%CO_2, 3%O_2–15%CO_2, 10%O_2–10%CO_2 and 10%O_2–15%CO_2, respectively.

The influence of atmosphere composition in sweet cherry sensory characteristics, more precisely, in flavour, appear to be variable across different cultivars. Wang and Vestrheim [34] reported that atmosphere composition did not significantly influence the flavour of cv. Van, Sam and Stella. Nevertheless, according to the same authors, cv. Kristin, Huldra and Emperor Francis showed higher flavour scores under CA conditions.

4. Conclusions

The mean chamber temperature of the Farmers' Organization was higher than the chambers of the Research Centre, allowing differential treatment conditions. Higher mean storage temperature was correlated to higher weight loss, higher variation in colour parameters, especially a^* parameter, lower firmness, and more dehydrated and browner stems, resulting in lower sensorial fruit classifications, particularly concerning aspect, flavour and global score. Controlled atmosphere treatments maintain fruit quality, namely the colour parameters, fruit firmness, and sensorial elements. Additionally, at the end of our study (49 days), the classifications of all sensory parameters were above the limit defined for sensory acceptance. In CA conditions, fruit weight loss was 2% to 3% after 49 days of storage. Few rotten fruits, less than 2%, were observed throughout the storage period. No clear distinction could be found between CA treatments using sweet cherry cv. Satin.

Author Contributions: L.P.d.A.—conceptualization, methodology, supervision, writing—review and editing. A.V.—conceptualization, data collection and analysis, writing—original draft. C.E.S.—conceptualization, supervision, data collection, writing—review and editing. P.D.G.—conceptualization, methodology, supervision, writing—review and editing. P.D.S.—conceptualization, writing—review and editing. M.R.—data collection, writing—review and editing. H.B.—data collection. C.B.—data collection. C.M.P.—conceptualization, methodology, supervision, writing—review and editing. L.P.—data collection. M.P.S.—conceptualization, methodology, supervision, writing—review and editing. All authors have read and agreed to the published version of the manuscript.

Funding: This research work is part of the project "PrunusPÓS—Otimização de processos de armazenamento, conservação em frio e embalamento inteligente no pós-colheita de pêssego e cereja", No. PDR2020-101-031695 (PDR 2020), Portugal 2020. The participation was supported by national founds through the FCT—Foundation for Science and Technology, I.P., under the project UIDB/00681/2020.

Data Availability Statement: The data presented in this study are available upon request.

Acknowledgments: The authors would like to thank the Farmers' Organization, who allowed the use of the storage chambers and the respective monitoring of conditions.

Conflicts of Interest: The authors declare no conflict of interest.

References

1. Mccune, L.M.; Kubota, C.; Stendell-Hollis Cynthia, N.R.; Thomson, A.A. Cherries and Health: A Review. *Crit. Rev. Food Sci. Nutr.* **2010**, *51*, 1–12. [CrossRef] [PubMed]
2. Dziedzic, E.; Błaszczyk, J.; Kaczmarczyk, E. Postharvest properties of sweet cherry fruit depending on rootstock and storage conditions. *Folia Hortic.* **2017**, *29*, 113–121. [CrossRef]
3. Blando, F.; Oomah, B.D. Sweet and sour cherries: Origin, distribution, nutritional composition and health benefits. *Trends Food Sci. Technol.* **2019**, *86*, 517–529. [CrossRef]
4. FAO. FAOSTAT. 2021. Available online: http://www.fao.org/faostat/en/#home (accessed on 11 January 2022).
5. Gonçalves, A.C.; Campos, G.; Alves, G.; Garcia-Viguera, C.; Moreno, D.A.; Silva, L.R. Physical and phytochemical composition of 23 Portuguese sweet cherries as conditioned by variety (or genotype). *Food Chem.* **2021**, *335*, 127637. [CrossRef]
6. Dias, C. A Fileira Da Cereja Da Cova Da Beira. Master's Thesis, Polytechnic Institute of Castelo Branco, School of Agriculture, Castelo Branco, Portugal, 2012.
7. Yang, Q.; Zhang, X.; Wang, F.; Zhao, Q. Effect of pressurized argon combined with controlled atmosphere on the postharvest quality and browning of sweet cherries. *Postharvest Biol. Technol.* **2019**, *147*, 59–67. [CrossRef]
8. Wang, L.; Zhang, H.; Jin, P.; Guo, X.; Li, Y.; Fan, C.; Wang, J.; Zheng, Y. Enhancement of storage quality and antioxidant capacity of harvested sweet cherry fruit by immersion with β-aminobutyric acid. *Postharvest Biol. Technol.* **2016**, *118*, 71–78. [CrossRef]
9. Tian, S.P.; Jiang, A.L.; Xu, Y.; Wang, Y.S. Responses of physiology and quality of sweet cherry fruit to different atmospheres in storage. *Food Chem.* **2004**, *87*, 43–49. [CrossRef]
10. Dugan, F.M.; Roberts, R.G. Pre-harvest fungal colonization affects storage life of bing cherry fruit. *J. Phytopathol.* **1997**, *145*, 225–230. [CrossRef]
11. Akbudak, B.; Tezcan, H.; Eris, A. Evaluation of messenger plant activator as a preharvest and postharvest treatment of sweet cherry fruit under a controlled atmosphere. *Int. J. Food Sci. Nutr.* **2009**, *60*, 374–386. [CrossRef]
12. Serradilla, M.J.; del Villalobos, M.C.; Hernández, A.; Martín, A.; Lozano, M.; de Córdoba, M.G. Study of microbiological quality of controlled atmosphere packaged "Ambrunés" sweet cherries and subsequent shelf-life. *Int. J. Food Microbiol.* **2013**, *166*, 85–92. [CrossRef]
13. Serrano, M.; Martínez-Romero, D.; Castillo, S.; Guillén, F.; Valero, D. The use of natural antifungal compounds improves the beneficial effect of MAP in sweet cherry storage. *Innov. Food Sci. Emerg. Technol.* **2005**, *6*, 115–123. [CrossRef]
14. Crisosto, C.H. Stone fruit maturity indices: A descriptive review. *Postharvest News Inf.* **1994**, *5*, 65N–68N.
15. Crisosto, C.H.; Mitcham, E.J.; Kader, A.A. *Cherry Recommendations for Maintaining Postharvest Quality*. Perishables Handling #86. 1996. Available online: http://postharvest.ucdavis.edu/Commodity_Resources/Fact_Sheets/ (accessed on 11 January 2022).
16. Guyer, D.E.; Sinha, N.K.; Chang, T.S.; Cash, J.N. Physicochemical and sensory characteristics of selected Michigan sweet cherry (*Prunus avium* L.) cultivars. *J. Food Qual.* **1993**, *16*, 355–370. [CrossRef]
17. Drake, S.R.; Elfving, D.C. Indicators of Maturity and Storage Quality of 'Lapins' Sweet Cherry. *HortTechnology* **2002**, *12*, 687–690. [CrossRef]
18. Pereira, S.; Silva, V.; Bacelar, E.; Guedes, F.; Silva, A.P.; Ribeiro, C.; Gonçalves, B. Cracking in sweet cherry cultivars early bigi and lapins: Correlation with quality attributes. *Plants* **2020**, *9*, 1557. [CrossRef]
19. Çalhan, O.; Onursal, C.E.; Guneyli, A.; Eren, I.; Demirtas, I. Determination of Optimum Harvest Date of Sweet Cherry cv. Lapins Grown in Isparta. *Turk. J. Agric. Nat. Sci.* **2014**, *2*, 1905–1910.
20. Thompson, A.K. *Controlled Atmosphere Storage of Fruits Vegetables*, 2nd ed.; CABI: Wallingford, UK, 2010; ISBN 978-1-84593-646-4.
21. De Vries-Paterson, R.M. Fungistatic Effects of Carbon Dioxide in a Package Environment on the Decay of Michigan Sweet Cherries by Monilinia fructicola. *Plant Dis.* **1991**, *75*, 943. [CrossRef]
22. Remón, S.; Ferrer, A.; López-Buesa, P.; Oria, R. Atmosphere composition effects on Burlat cherry colour during cold storage. *J. Sci. Food Agric.* **2004**, *84*, 140–146. [CrossRef]
23. Chockchaisawasdee, S.; Golding, J.B.; Vuong, Q.V.; Papoutsis, K.; Stathopoulos, C.E. Sweet cherry: Composition, postharvest preservation, processing and trends for its future use. *Trends Food Sci. Technol.* **2016**, *55*, 72–83. [CrossRef]
24. Andrade, L.P.; Nunes, J.; Simões, M.P.; Morais, D.; Cannavarro, C.; Espírito Santo, C.; Gaspar, P.D.; Silva, P.D.; Resende, M.; Caseiro, C.; et al. Experimental study of the consequences of controlled atmosphere conservation environment on cherry characteristics. In Proceedings of the 25th IIR International Congress of Refrigeration, Montreal, QC, Canada, 24–30 August 2019; pp. 3059–3066. [CrossRef]
25. Kader, A.A. Regulation of fruit physiology by controlled/modified atmospheres. *Acta Hortic.* **1995**, *398*, 59–70. [CrossRef]
26. Beaudry, R.M. Responses of horticultural commodities to low oxygen: Limits to the expanded use of modified atmosphere packaging. *HortTechnology* **2000**, *10*, 491–500. [CrossRef]
27. ISO. *ISO 8589:2007—Sensory Analysis—General Guidance for the Design of Test Rooms*; ISO: Geneva, Switzerland, 2007.
28. Maroco, J. *Análise Estatística com Utilização do SPSS [in Portuguese]*, 3rd ed.; Edições Sílabo: Lisbon, Portugal, 2011.

29. Harb, J.; Streif, J.; Saquet, A. Impact of controlled atmosphere storage conditions on storability and consumer acceptability of sweet cherries "Regina". *J. Hortic. Sci. Biotechnol.* **2003**, *78*, 574–579. [CrossRef]
30. Paulo, L.; Resende, M.; Nunes, A.; Pintado, C.M.; Antunes, P. Quality parameters, total phenolic content and antioxidant activity in different maturation stages of 'Sweetheart' cherry. In Proceedings of the XIV Congresso Nacional de Ciencias Hortícolas, Orihuela, Spain, 3–5 June 2015; pp. 453–457.
31. Da Costa, F.M.M. Avaliação das Características Agronómicas da Cerejeira 'De Saco' na Região da Cova da Beira. Master's Thesis, University of Lisbon, Lisbon, Portugal, 2006.
32. Simões, M.P.; Stuburic, I.; Kamenjak, K. Qualidade das cerejas na região da Beira Interior. In Proceedings of the 3° Simpósio Nacional de Fruticultura, Vila Real, Portugal, 4–5 December 2014; pp. 213–218, ISBN 978-972-8936-16-7.
33. Akbudak, B.; Tezcan, H.; Eris, A. Determination of controlled atmosphere storage conditions for "0900 Ziraat" sweet cherry fruit. *Acta Hortic.* **2008**, *795*, 855–860. [CrossRef]
34. Wang, L.; Vestrheim, S. Controlled Atmosphere Storage of Sweet Cherries (*Prunus avium* L.). *Acta Agric. Scand. Sect. B Plant Soil Sci.* **2002**, *52*, 136–142. [CrossRef]
35. Stow, J.R.; Jameson, J.; Senner, K. Storage of cherries: The effects of rate of cooling, store atmosphere and store temperature on storage and shelf-life. *J. Hortic. Sci. Biotechnol.* **2004**, *79*, 941–946. [CrossRef]

agronomy

MDPI

Review

Harnessing the Known and Unknown Impact of Nanotechnology on Enhancing Food Security and Reducing Postharvest Losses: Constraints and Future Prospects

Ayomide Emmanuel Fadiji [1], Doctor Mziwenkosi Nhlanhla Mthiyane [1,2], Damian C. Onwudiwe [3] and Olubukola Oluranti Babalola [1,*]

1 Food Security and Safety Focus Area, Faculty of Natural and Agricultural Sciences, North-West University, Private Mail Bag X2046, Mmabatho 2735, South Africa; 33462623@mynwu.ac.za (A.E.F.); doc.mthiyane@nwu.ac.za (D.M.N.M.)
2 Department of Animal Science, School of Agricultural Sciences, Faculty of Natural and Agricultural Sciences, North-West University, Private Bag X2046, Mmabatho 2735, South Africa
3 Material Science Innovation and Modelling (MaSIM) Research Focus Area, Faculty of Natural and Agricultural Science, North-West University, Private Bag X2046, Mmabatho 2735, South Africa; damian.onwudiwe@nwu.ac.za
* Correspondence: olubukola.babalola@nwu.ac.za

Abstract: Due to the deterioration of natural resources, low agricultural production, significant postharvest losses, no value addition, and a rapid increase in population, the enhancement of food security and safety in underdeveloped countries is becoming extremely tough. Efforts to incorporate the latest technology are now emanating from scientists globally in order to boost supply and subsequently reduce differences between the demand and the supply chain for food production. Nanotechnology is a unique technology that might increase agricultural output by developing nanofertilizers, employing active pesticides and herbicides, regulating soil features, managing wastewater and detecting pathogens. It is also suitable for processing food, as it boosts food production with high market value, improves its nutrient content and sensory properties, increases its safety, and improves its protection from pathogens. Nanotechnology can also be beneficial to farmers by assisting them in decreasing postharvest losses through the extension of the shelf life of food crops using nanoparticles. This review presents current data on the impact of nanotechnology in enhancing food security and reducing postharvest losses alongside the constraints confronting its application. More research is needed to resolve this technology's health and safety issues.

Keywords: agriculture; food processing; food safety; nanoformulation; nanosensors

Citation: Fadiji, A.E.; Mthiyane, D.M.N.; Onwudiwe, D.C.; Babalola, O.O. Harnessing the Known and Unknown Impact of Nanotechnology on Enhancing Food Security and Reducing Postharvest Losses: Constraints and Future Prospects. *Agronomy* **2022**, *12*, 1657. https://doi.org/10.3390/agronomy12071657

Academic Editors: Raquel P. F. Guiné, António Dinis Ferreira and António Moitinho Rodrigues

Received: 23 June 2022
Accepted: 6 July 2022
Published: 12 July 2022

1. Introduction

One of the most challenging concerns of the 21st century is ensuring food security for the world's rapidly growing population. According to projections, the global food demand will increase from 59 to 98% by 2050, with a population of 9 billion [1,2]. Although the world's population is growing, most especially in developing nations, the local and international supply of food is being disrupted by the use of bio-resources for chemical manufacturing, production of energy, low-value addition, high postharvest losses, poor marketing systems, and inefficient distribution and other similar factors [3]. Farmers globally have concentrated most of their efforts on improving the yield of crops via extensive and intensive agriculture by implementing novel technologies and ideas [4,5]. Precision farming in combination with the application of nano-modified stimulants has boosted the current attempts. Agricultural efficiency, secure water usage, food quality, soil improvement and food distribution in outlets are all essential aspects of food security that nanotechnology research can improve [6].

It is critical to develop new technologies that will enhance product output while reducing food waste in order to preserve a nation's sustainable living standards and improve food security. Nanotechnology is notable for providing foods with exceptional physical quality in addition to incrementing nutrient bioavailability. Recently, most research studies have concentrated on expanding the use of nanotechnology in food processing and agricultural production [7,8]. Nutraceutical distribution, packing, food processing, serviceability, and quality control are all areas of nanotechnology research that are experiencing an increase in intellectual property, patents and publications [9].

Nanotechnology is among the most promising alternatives for increasing food availability and developing new products related to water application, food, environment, agriculture, energy, electronics, and medicine. It is a rapidly growing field with novel food research and exclusive agricultural applications [10]. Growth enhancement and decreasing postharvest expenditures through improved outcomes and assistance by advanced scientific studies involving biotechnology and nanotechnology in foodstuffs may be the ideal solution [11]. In agriculture, nanomaterials are being used to discourage the dependency on pesticides in crop production, limit the loss of nutrients in fertilization and boost crop output via nutrient and pest control [12].

Smart nutrition delivery, chemical pollutants, bioseparation of proteins, quick monitoring of biological, nutraceutical nanoencapsulation, solubilization, and distribution are examples of new themes approached by nanotechnology and involved in food security that might be greatly improved [13,14]. The application of nanocarrier techniques to reinforce bioactive components in order to adjust their biological accessibility and resistance against a variety of environmental or chemical variations is referred to as food nanotechnology [15]. It improves food dependability by enhancing sensory features such as color, texture, and taste [16]. It may also increase the capture and biological administration of nutraceuticals and medicine systems [17]. Food companies benefit from nanotechnology as a unique food packaging supply with improved mechanical and antibacterial properties [18,19]. Other advantages of the nanotechniques include the development of nanosensors for trace elements detection, monitoring the state of foodstuffs during storage, transportation and encapsulation of food modifiers or additional components [20]. There is an increased need for nanoparticles in food biotechnology, food processing, functional food creation, food packaging, detection of pathogens in food, food safety and prolonged shelf-life of food driven by nanotechnology-based applications [21]. Nanomaterials are very good in the enhancement of food security and in promoting the growth of the food production industry. Depositing food processing equipment (through biofilm coating), membranes, sieves, nanofabricated filters, catalytic agents, nanosized adsorbents and nanocomposite-based are all research areas that could contribute to the development of food processing [22]. Using nanoparticles in food packaging is reported to reduce the amount of time it takes for items to be packed, as well as the use of valuable raw resources and waste generation [13].

Nanotechnology has a promising potential for developing improved and novel products. Many scientific groups, however, remain skeptical of the public health problems linked with products emanating from nanotechnology, which requires further research [23,24]. The purpose of this review is to discuss the essentials of nanotechnology, as well as its applications in food process technology, postharvest, and packaging. It also explores its role in enhancing food security, as well as the challenges associated with its use in the agricultural and food systems.

2. Approaches to Nanotechnology

Nanomaterials can be synthesized by either a "bottom-up" or a "top-down" approach. On a commercial scale, nanomaterials are mostly produced using the "top-down" strategy, in which bulk precursors are reduced to nano size using nanolithography, precision engineering or milling techniques [13]. The milling process is used to secure flour having small size particles and high water-holding capacity. By reducing the size of green tea, the top-down technique can improve its antioxidant capabilities [25]. A study revealed that green tea powder with a particle size of 1000 nm has a stronger capacity to digest nutrients, resulting in an increased ability of oxygen removal by the enzyme dismutase and, hence, increased antioxidant activity [26]. A similar top-down strategy is homogenization, which is often employed in the dairy industry for the size reduction of globules, laser applications and vaporization linked to chilling [27]. A surface area with superior qualities allows achieving food material with the functions desired for the needed purpose.

The top-down technique is related to nanotechnology-based devices, which are often controlled with external forces to produce the necessary parameters and particle initiation from greater levels with stuffing and fragmenting, to obtain the required size [28]. In the bottom-up approach, atoms are turned into materials of nanoscale dimensions, and this may involve some complicated processes. The matching of attracting and repulsive interactions among paired molecules employed as components for making effective supramolecular assemblies is essential for self-assembly [26]. Self-assembly or clustering of casein micelles, for example, results in the creation of carbohydrate-binding proteins. The arrangement of casein micelles, the structures generated in protein–polysaccharide liposomes, and their aggregates are examples of nanostructures that are self-assembled in food. According to research, the bottom-up strategy allows for the creation of nanostructures with fewer faults and a more uniform chemical arrangement [29].

3. Nanotechnology in Agriculture

Agriculture is important in food production through the cultivation of a variety of crops and the rearing of livestock. Most developing countries regard it as the backbone of their economies, and it plays a major role in their progress and development. As the global population expands, so does the need for more food, and food engineers and scientists are devising innovative ways to boost agricultural productivity [30]. Agricultural nanotechnology has spent the last several years focusing on applications and research to address environmental and agriculture challenges for crop enhancement and increased production. In relation to undernutrition and the need to reduce hunger and child mortality, agricultural nanotechnology appears to be quite promising in underdeveloped nations [31]. Germany, China, the USA, France, Brazil, Korea and India are among the developed and rising nations that have shown a great interest in employing nanomaterials for agricultural purposes, as evidenced by the increased number of related patents and publications [31].

Nanotechnology may be used to refurbish the agricultural sector as a prospective tool; it also helps in studying the biochemical routes of crops by changing conservative approaches for assessing environmental difficulties and its application towards improved production [32]. Nanotechnology, when compared to eco-friendly agricultural biotechnology, demonstrates the potential for a greater and faster influence on all elements of the agro-value chain, resulting in synchronized benefits to the public, moral, legal, and environmental repercussions [33]. The techniques employed for traditional agricultural have been transformed through the use of nanoscale agrochemicals such as nanopesticides, nanofertilizers, nanoformulations, and nanosensors in agriculture (Figure 1).

In agriculture, nanotechnology has a variety of uses, which include the treatment of wastewater, quality improvement of contaminated soil, and the increment of crop yield through the use of sensors to detect diseases [22,34]. Nanobiosensors, for instance, are examples of nanotools that support the highly technological development of agricultural farms, while also aligning with the usage of nanotools for farm management accuracy and the control of agricultural inputs [35]. Nanopores carrying zeolite for delayed discharge and better efficiency, nanosensors for the measurement of soil quality, and smooth herbicide delivery mechanisms are some of the beneficial impacts of nanotechnology use in agricultural development [36].

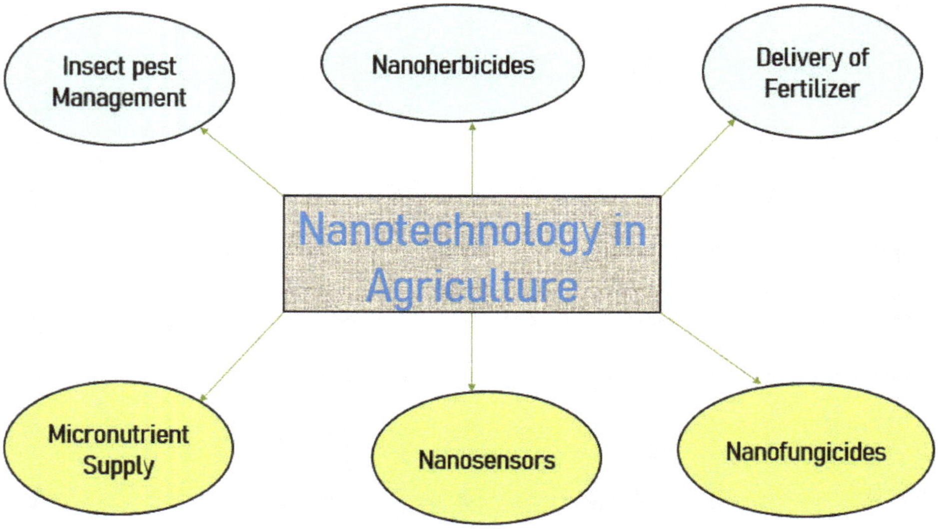

Figure 1. Schematic diagram of the potential applications of nanotechnology in agriculture.

Nano-forms of alumino-silicates, silver, silica, and carbon are examples of nanoparticles used for the monitoring of plant diseases. The application of nanomaterials in agricultural practice has been reported to decrease pesticide usage by ensuring a steady supply of energetic molecules. It enhances nutrient waste reduction during the application of fertilizer sand increases harvests by ensuring a better management of nutrients and water [31]. The responses of different cultivars of rice to engineered nanoparticles have also been studied at various stages of development and under various conditions [37].

Pests and diseases of plants have been reported to cause 20–40% of crop losses each year throughout the world [38]. Pest control in current farming techniques is mainly reliant on the use of pesticides such as herbicides, fungicides and insecticides. It is critical to produce cost-effective and very active insecticides that are eco-friendly. Pesticides may benefit from emerging concepts such as nanotechnology, which can reduce toxicity, improve shelf-life, and increase the solubility of poorly water-soluble pesticides, all of which might have a good impact on the environment [39]. Other studies also reported the importance of nanotechnology in agriculture, which majorly regards disease control and safety [31,33].

Pesticides and herbicides developed through nanotechnology help plants receive steady agricultural chemicals and supply nutrients in regulated proportion [40]. Nanoparticles may potentially be useful in the management of viruses, insects and pests that infect hosts [41]. For the manufacture of nano-insecticides, some polysaccharides such as polyesters, starch, chitosan, and alginates have been investigated [39]. In general, nanoparticles that can be used for plant protection may act in two ways: (a) they directly protect crops and (b) they act as carriers when commonly used pesticides are sprayed [42]. The application of nanoparticles in food production and plant protection, on the other hand, remains mostly unexplored [12].

4. Role of Nanotechnology in Postharvest Loss Reduction

More than 40% of food losses (cereals, fruit, pulses, oil crops, vegetables, fish, roots and tubers, dairy and meat) occur at the distribution and trade stages in developed countries, whereas more than 40% of losses recorded for foodstuffs occur at the processing and postharvest stages in developing countries [43]. Due to microbial attack, most of freshly harvested high-moisture crops are not well preserved, and may soon decay. Nanotechnology is a newer and more improved technology that can assist in reducing postharvest losses by creating an active packing element with a low amount of bioactive compounds, enhanced mechanical and gas capabilities, and a minimized influence on the quality of fruits and vegetables. [44].

Edible coatings are applied on food in the form of a liquid, usually by immersing the agricultural product in a material-providing solution created via a structural medium (protein, lipid, carbohydrate or a mixture). They prevent untreated foods from deteriorating by reducing respiration, preventing dehydration, preserving volatile fragrance molecules, inhibiting microbial development and refining textural qualities [45]. Nano-coatings with edible properties deposited on a variety of foods create a barrier to moisture and gas exchange, while also delivering enzymes, colors, antioxidants, browning-resistant chemicals, and tastes which may extend the shelf life of foods produced synthetically [45]. The method allows for the creation of nanoscale coatings with a thickness of up to 5 (5) nm [46]. Thin films and edible coatings are commonly used on horticulture products. Cost, functional characteristics, availability, mechanical properties (tension and elasticity), photosensitive properties (opacity and brilliance), the fence effect vs. gas flow, the structural barrier to the movement of water, sensory appropriateness, and microbes are all factors that influence their utilization [47].

To manage the postharvest activities of newly harvested items, numerous edible coatings that are within nanoscale dimensions are applied to food crops. Silver nanoparticles have lately generated a lot of interest because of their antibacterial qualities, which are important for the processing of food. The application of PVP-dependent Ag nanoparticles on asparagus significantly slowed down microbial growth and weight loss and reduce changes in the color of the skin [48]. A similar study found that edible gelatin-derived coverings containing nanocrystalline cellulose extended significantly the shelf-life of strawberries [49]. In comparison to other treatments, a chitosan supplement nano-silica coating increased the physiological and physicochemical value of longan fruit at room temperature by effectively generating an exceptional semi-permeable layer [50]. Furthermore, coatings with alginate or a lysozyme-based nanolaminate [51] and chitosan film-based nano-SiO_2 [52] were reported to preserve the value of fresh foods throughout a long storage. In addition, nano-ZnO coating prevented microbial decay and preserved the postharvest quality of selected fruits throughout storage [53].

5. Impact of Nanotechnology in Food Processing

The world's largest food industries are looking for new ways to improve food's value, safety and nutritional characteristics. To improve market price, production and quality in the food industry, newer technologies are necessary. Nanotechnology is useable in diverse ways in the food industry, including nanoencapsulation, nanoparticle-based food additives, nanosensors, nano-packing, and nanoparticle-based smart distribution systems [54]. Its uses include encapsulation, biopolymer matrix, emulsions, related colloids and simple solutions, all of which provide effective delivery methods. Nanotechnology is gaining attention in industrial food processing, notably for the encapsulation of flavors or the enhancement of odor, food texture modification or value increase, and as innovative viscosity- or gelatin-enhancing agents [55]. Food nanotechnology focuses on the creation of structures at the nanometer-scale having unique qualities that are often used for various applications, e.g., as delivery systems, food interaction surfaces with distinct superficial properties, food characterization tools, sensor technology, nanocomposite coatings and microfluidic instruments [10]. Figure 2 depicts the multiple applications of food technology.

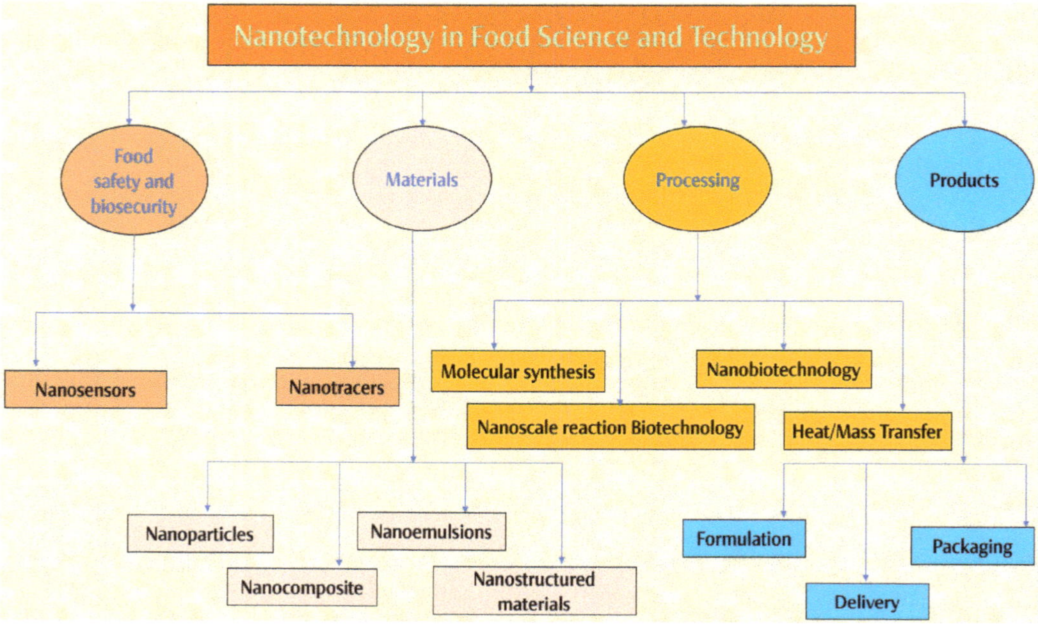

Figure 2. Application of nanotechnology in food science and technology.

At the nanoscale (water/oil system), nanoemulsions are formed by minute emulsion droplets with diameters of less than 100 nm [56]. The generation of scattered phases requires more mechanical energy, such as a high-pressure homogenizing step or the use of a sonication process or microfluidizer. The application of nanoemulsions brings about a reduction in the involvement of stabilizers by protecting the food from splitting and breaking, leading to a considerable reduction in the fat quantity required [9]. Numerous nanoemulsions often appear to be transparent optically and possess numerous technical advantages when the mixing of a liquid is involved [12]. Nanoemulsions' end products often appear creamy, just like other foodstuffs, with no alterations in flavor or taste [9]. They are used to improve drinks, oils taste, sweeteners, salad dressings and other processed foods. Nanomaterials are often involved in the processing of food as anti-caking agents, food additives, antimicrobial agents, transporters for nutrient delivery, filling factors for mechanical power increment and stability of packing material, as well as to improve the assessment of food security and safety through food nanosensing [57]. This is true for dietary additions as seen for nutraceuticals, which are produced with improved stability and bioavailability [55].

Nanoparticles are naturally present in a wide range of food items. Several meals include proteins that have globular structures with a size ranging from 10 to 100 nm, while others possess polymers that are linear, with a thickness of 1 nm, and mostly consist of lipids and carbohydrates. Casein, a nanoscale protein, is found in milk, while meat contains protein filaments that are classified as nanomaterials [12]. A number of food firms have been focusing on increasing food safety, production efficiency and nutritional characteristics in the beverage and food industry, which is a high-finance industry across the world [27]. Increased bioavailability and antibacterial capabilities, improved sensory qualities and guided distribution of substances with superior bioactivity are outstanding advantages of using nanoparticles in meatpacking and processing [58]. Nanotechnology is used to create motivating nano-protocols, fabricate ecologically friendly procedures and smart nano-packaging, manufacture goods with the best texture and flavor and produce low-calorie drinks and food items to improve the health of humans [59]. It can use instruments such as atomic force microscopy (AFM) for faster diagnosis of component shortfalls and enhance the development of nanosensors that can be used to identify infections in food [54]. Atomic force microscopy is a powerful tool for studying food assemblages and chemical interactions that occur at the nanoscale level [59]. It is also a very useful tool for studying fine food assemblages and chemical interactions at the nanoscale level [45,59].

Food science, like other scientific fields, uses modern nanotechnologies to improve consumers' quality of life through updated food formulations and packaging, innovative ingredient synthesis and process monitoring to produce healthier, safer, precise food systems, with high quality and long shelf lives [27]. Food security, functionality, processing and economic issues regarding distribution and efficiency are developing key links between nanotechnologies and food systems [54]. Nanofiltration, modification and absorption of nanoencapsulation, nanoscale enzyme-based reactors, mass and heat transfer and nanofabrication are multipurpose uses of nanotechnology in food processing. In the pharmaceutical industry, nanofiltration is very effective in purifying medications and as a necessary step in removing certain solutes. It is equally employed for the treatment of water and dairy products in a bid to improve the quality of products by eliminating salt from lactose [9]. Nanofabrication through mass and heat transfer improves package heat resistance. Nanoscale enzyme reactors are used to change food systems so to improve flavor and nutritional value and provide a variety of health benefits. As a result of their assistance in scattering food due to large surface-to-volume relationships, nanomaterials result in greater enzyme-mediated systems (to enhance economy, activity and shelf life) in comparison with macroscale amended goods [60]. The action of lipase in nanotubes was reported to be 70% greater when compared with that of conventional lipases. For instance, nano-SiO_2 particles considerably hydrolyzed olive oil, leading to increased reusability, adaptability and stability at extreme temperatures (65 °C) [61].

To extend the shelf life of food items, nanoencapsulation is also very important. This technique is commonly used to improve flavor, preserve food and provide cooking balance. Nanoceramic derived from nanocapsules in a pot-like form may be used for absorption modification to reduce the time used for cooking and that spent on oil. It further reduces trans fatty acids by using plant oil instead of hydrogenated oil and, finally, leads to safer nanofood production through nanocapsules used for the distribution of nutrients in food for enhanced absorption. Nanoencapsulation can hide taste and odors, manage food interactions with effective ingredients, regulate the release of dynamic agents, ensure accessibility at a specific time intervals and protect them from biological, heat, moisture, or chemical interferences. Its action displays similarities with those of other ingredients present in the system [21,62]. Metallic oxides such as titanium dioxide (TiO_2) and silica (SiO_2) are often used in food preparation as coloring additives. Wasted food nanoparticles based on SiO_2 nanomaterials are used to transfer odors or fragrances to foodstuffs [63].

6. Impact of Nanotechnology on Food Packaging

One of the important roles of nanotechnology is the protection of food from physical injury and quality deterioration. Food packaging should be passive, safe, low-cost, readily reusable or disposable, stable in transportation and storage and resistant to physical abuse. The content and kind of packing materials have an impact on food quality. Packaging materials with lightweight, heat resistance and strength, among other qualities, could be obtained using nanomaterials. Food packaging obtained through nanotechnology has been widely reported [64,65]. The global nano-based packaging beverage and food industry was predicted to reach USD 4.13 billion in 2008, rising to USD 7.3 billion by 2014, with an 11.65% estimated annual growth rate [55].

Active packing applications are used on a variety of metal and metal oxide nanoparticles. In food packaging, TiO_2 and silver (Ag) are extremely valuable [66]. Metal and metal oxide nanoparticles, as well as nanocomposites for food packaging, are used as antimicrobials in active packaging [29]. Due to its semiconducting qualities and improved electrical, photosensitive and optical properties, TiO_2 is commonly used as a stain, catalytic substrate and adsorbent material [67]. Rice storage at 70% relative humidity and a temperature of 37 °C enhanced food characteristics by combining dispersed and antibacterial Ag/TiO_2 nanoparticles with polyethene used as a packing agent [68]. Antimicrobial potential, oxygen transfer, enzyme mobilization and information on the level of vulnerability to degradation-related factors are advantages linked to nanoparticles used in food packaging [69]. When compared to traditional packing materials, material for nano-packaging generated a superior sensory quality [70]. A nano-polymer including zinc oxide and polylactic acid nanoparticles was used to produce a highly functional material for food packaging [45]. A coating with zinc oxide-treated semolina protein was evaluated for the packaging of food and produces a significant reduction in oxygen absorption as well as increased resistance to heat [71].

In food packaging, nanoclay, hydrated alumina–silicate and silicates have been used as layers [62]. The blending of clay with silicate and polymers is a promising nanocomposite contender for food packaging with excellent qualities [70]. Organoclay nanoparticles employed as antibacterial material have also been studied for their usefulness in food packaging [72]. As a result of the large surface area of nanoparticles and their increased activity, cellulose is considered a supporting material in a variety of nanomaterials and is being used in a variety of applications [70]. As a natural polymer, it is extremely strong, ecofriendly, easily recyclable and affordable [73]. Food packaging using nano-composite films modified with nano-cellulose and nano-chitosan-added films has shown an improvement in clearance, elongation to breaking, food protection features and tensile strength [45]. In comparison to chitosan alone, nano-bio-composited films made with polylactides–chitosan were proven to be acceptable for food packaging, with significant and long-term antioxidant prospects [74]. Packaging and paper coatings made of nanocomposites of copper and cellulose have also been reported to have antimicrobial activities [75].

Due to their intrinsic antibacterial properties, nanometer-thick chitosan films have been widely used [76]. Ionic binding is used to make chitosan nanoparticles via the electrostatic interaction of the positive amino sides of chitosan with polyanions as cross-linkers [70]. A chitosan/gelatin-based nanocomposite containing Ag nanoparticles is utilized in the packaging of food and appears to be a very effective protective packaging material for prolonging the lifespan of red grapes up to 14 days [77]. Gold–chitosan and silver–chitosan nanocomposite coatings are used as effective antibacterial agents against Gram-negative bacteria (*P. aeruginosa*), yeast (*C. albicans*), Gram-positive bacteria (*S. aureus*), and fungi (*A. niger*) in food packaging [78].

7. Nanosensors in Food Security

Carbon nanotubes are hollow carbonaceous materials composed of atomic groups arranged in a hexagonal pattern [79]. Nanotubes are used in alumina, medical instruments, food processing equipment and sports equipment. This is due to their ability to tolerate high temperatures and to their flexible and sturdy nature [54]. Inorganics, bio-microtubules, carbon, viral proteins, porins, amyloid proteins, carbohydrates, synthetic polymers, lactalbumin, lipids, DNA28 and other organics are often used to create nanotubular textures [45].

Carbon nanotubes increase food packaging's mechanical qualities [80]. Some polymers have had their tensile strength increased by using polyamides and carbon nanotubes [81]. Because of their lower cost, easy methodology, and acute detection property, carbon nanotube-based biosensors have been used to detect microorganisms, hazardous substances and other metabolites in food and drinks [82]. Carbon nanotubes made of TiO_2 have been shown to have increased disinfecting ability against *B. cereus* spores [83]. TiO_2 nanoparticles doped with silver showed improved their bactericidal effects against *E. coli* [81].

Carbon nanotubes' potent antibacterial properties often lead to the mass destruction of pathogens [80]. Partial hydrolysis of α-lactalbumin (milk proteins) resulted in the formation of nanotubes [13]. According to several studies, α-lactalbumin nanotubes have improved viscosity and hardness and therefore can be used as a thickening agent [84]. According to this study, α-lactalbumin nanotubes can be used as gelation agents, regulating viscosity, encapsulating agents with steady taste and drug delivery agents in pharmaceuticals and foods [85]. Nanotubes have also been studied in contemporary agriculture. Various nanoparticles have been reported to infiltrate the cell walls of plants. Carbon nanotubes have been reported to infiltrate tomato seeds and alter their growth and sprouting [86]. This is because of the greater water intake promoted by carbon nanotubes, which improves the plant's performance. The cell wall was also penetrated by gold-derived mesoporous silica nanoparticles, which aided DNA penetration [87].

8. Nanosensors in Agriculture and Food

The utilization of biosensors in conjunction with better microfluidics technology, nanomaterials, and molecular biology has huge implications for crop yield. They may also be used to track microbes' activities in the soil and anticipate the occurrence of soil diseases. The primary premise behind using a biosensor to examine soil is to determine how the action of negative and positive bacteria in soil is influenced by variations in oxygen consumption during respiration. They also provide several alternatives for detecting pollutants and their obstructive effects by utilizing novel nanomaterial characteristics [30].

For azelaic acid and methyl salicylate, biosensors for detecting nitrate concentrations in plants alongside markers for the identification of infected plants have been reported [88]. Biosensors were used to monitor infections emanating from *P. digitatum* in citrus [89]. Nanosensors and smart delivery systems are used in precision farming to aid in the efficient use of natural agricultural resources such as water, chemicals and nutrients. Applications include geographic systems, remote detection and satellite monitoring tools that can aid the detection of pests' activities in crops or of signs of stress, such as drought [90]. The use of separate sensors linked to real-time GPS tracking is anticipated to be crucial in nanotechnology-assisted instruments [91]. Nanosensors may be strategically placed across a field to monitor crop development and soil factors.

Nanosensors are gaining global attention due to their increasingly important impact in the food sector due to their rapid response capabilities in detecting microorganisms, harmful compounds and gases in packaged foods. Nanobiosensors have been proven to detect infections in processing facilities, so to inform clients and suppliers about food safety [92]. They have also been used to check for impurities, mycotoxins, pollutants and microorganisms in foodstuffs [93]. There have been reports on allergens detected by using nanoparticles and biosensor tools, near commercialization [94]. These technologies can also help in the detection of temperature, expiration date and time history.

9. Challenges and Future Prospects of Nanotechnology in Food Security

Nanotechnology is widely used in agriculture, industry and food production, as previously stated. Nanomaterials are associated with a host of safety problems linked to the fact that they might penetrate cells and remain in the system due to their tiny sizes [70,95]. Increased application of nanotechnology in agricultural operations and production of food is of significant concern to a broad portion of the society due to several antagonistic effects of different nanoparticles [30]. Despite the protection properties and quality of bulk substances being obvious, nanoscale complements consistently show different features when compared to macroscale complements [69]. Nanotechnology poses a concern owing to the use of tiny nanoparticle with big surface areas that are readily dispersed, may penetrate the cells and reach far-flung places of the body, posing a risk of toxicity [9]. Nanomaterials have the potential to react with biological specimens due to their size resemblance to DNA [30].

Environmental conditions can result in the degradation of nanocomposites, leading to the release of incorporated nanoparticles from polymeric materials into the environment [73]. For example, the food packaging material low-density polyethylene loses strength after exposure to environmental factors such as ozone or UV light under humid circumstances [64]. Low-density polyethylene samples oxidized under UV radiation or ozone undergo significant thermal, physical and structural changes [64]. Nanofertilizers and pesticides are used in agriculture and often spread into the soil, water and environment, leading to serious health-endangering consequences for farmers [27]. The development and yield of the plants might be affected by the build-up of nanoparticles in the soil, which may later accumulate in human tissues when plants are consumed [95].

When nanoparticles are discharged into the agro-environment, they quickly undergo a series of changes. One of the main concerns regards the unpredicted effects of nanoparticles on the human body. Nanoparticles can produce oxidative damage and unwanted responses; wasted nanoparticles may be hazardous [64,96]. Nanotoxicity is primarily mediated by the massive creation of free radicals, which cause oxidative stress in the cells, rendering them incapable of performing normal redox-regulated biological functions [29]. In humans, the breakdown of nano-clay found in low-density polyethylene clumps can result in alveolar basal epithelial cell cancer [77]. The antibacterial mechanism of Ag has been the subject of several investigations. However, being a notable heavy metal, it may cause toxicity in the body by denaturing enzymes and proteins when present in large concentrations; its danger should be calculated [66]. Research on TiO_2 and Ag nanoparticles, as well as on carbon nanotubes, revealed that they could penetrate the bloodstream and accumulate in organs due to their insoluble nature [70]. When TiO_2 is consumed as a food additive, it can induce oxidative stress, which causes inflammation, and genotoxicity, which causes chromosomal instability [97].

Inhalation, cutaneous contact and ingestion are different ways through which nanoparticles enter the body. The use of a large number of nanoparticles in food packaging might be a source of worry, as their leakage could lead to contamination of the environment and food materials [64]. However, there is a shortage of data regarding nanoparticles' migration from packaging materials into food, as well as their long-term toxicological effects [29].

The selection of green nanofillers in studies involving nanocomposites is critical for animal, human and environmental safety. Furthermore, concentration, particle size, molecular weight, diffusivity and certain compounds' stability in polymer blends, pH value, temperature, viscosity and polymeric structure, contact duration, food composition and mechanical pressure are critical factors to be considered [45,70]. It is, therefore, important that studies should be channeled towards determining the precise number of nanoparticles discharged into the environment, their buildup in plants and their influence on human health [95].

Generally, for the safe application of nanoparticles in the food industry, comprehensive guidelines, rules and regulatory systems are essential. For instance, in the USA, the Food and Drug Administration (FDA) monitors food packaging and nanofoods, whereas, in Europe, the European Union regulates nanotechnology-based food additives [98]. However, most nations that produce nanomaterials lack adequate and specific nanotechnology regulations [45]. For a legal nanotechnological application, comprehensive government legislation and guidelines, as well as a thorough procedure for impact and toxicological screening are required [98].

10. Conclusions

Nanotechnology is a relatively young but rapidly developing technology that has applications in a wide range of areas including food, agriculture, medicine, various industries and human activities across the world. It is a remarkable phenomenon that nanostructures and nanoparticles improve different qualities of the systems in which they are used because of their high surface area, reduced size and impressive properties such as high catalytic nature. Nanotechnology plays a major role in improving food security, especially in agriculture. This technology has the potential of improving crop yield through the provision of an effective insect, pest, microbial and weed management that has high economic value, safety and security. It further helps in ensuring food safety, monitoring food processing, stability, food modification and shelf life, sensing, extension, and food loss reduction. With improved safety, packaging materials, and stability, nanotechnology also reduces postharvest losses. Metallic and metal oxide nanoparticles such as Au, Zn, Ag, TiO_2, MgO, ZnO and SiO_2, which are often employed in food processing, might cause health problems due to their ease of penetration into the cells, causing undesirable responses in many animals and human organs, as well as plant parts. Future studies should focus on reducing the hazards posed by nanocomposite and nanoparticles attacks by employing greener synthesis and looking for simple and less expensive techniques for removing and degrading existing nanomaterials from the sites where they were deposited.

Author Contributions: A.E.F. and O.O.B. conceived the ideas, collected the data and developed the manuscript. D.M.N.M., D.C.O. and O.O.B. provided professional input and critiqued the work. All authors have read and agreed to the published version of the manuscript.

Funding: The National Research Foundation of South Africa grants (UID123634 and UID132595).

Acknowledgments: AEF is grateful to North-West University for a Postdoctoral fellowship. OOB appreciates the National Research Foundation, South Africa, for the grants UID123634 and UID132595 that supported herresearch endeavor.

Conflicts of Interest: The authors declare no conflict of interest.

References

1. Elemike, E.E.; Uzoh, I.M.; Onwudiwe, D.C.; Babalola, O.O. The role of nanotechnology in the fortification of plant nutrients and improvement of crop production. *Appl. Sci.* **2019**, *9*, 499. [CrossRef]
2. Duro, J.A.; Lauk, C.; Kastner, T.; Erb, K.-H.; Haberl, H. Global inequalities in food consumption, cropland demand and land-use efficiency: A decomposition analysis. *Glob. Environ. Change* **2020**, *64*, 102124. [CrossRef]
3. Barrett, C.B. Overcoming global food security challenges through science and solidarity. *Am. J. Agric. Econ.* **2021**, *103*, 422–447. [CrossRef]
4. Ajilogba, C.F.; Babalola, O.O.; Nikoro, D.O. Nanotechnology as Vehicle for Biocontrol of Plant Diseases in Crop Production. In *Food Security and Safety: Africa's Perspective*; Babalola, O.O., Ed.; Springer: Cham, Switzerland, 2021; pp. 709–724. [CrossRef]
5. Fadiji, A.E.; Babalola, O.O. Elucidating mechanisms of endophytes used in plant protection and other bioactivities with multifunctional prospects. *Front. Bioeng. Biotechnol.* **2020**, *8*, 467. [CrossRef]
6. Ashraf, S.A.; Siddiqui, A.J.; Abd Elmoneim, O.E.; Khan, M.I.; Patel, M.; Alreshidi, M.; Moin, A.; Singh, R.; Snoussi, M.; Adnan, M. Innovations in nanoscience for the sustainable development of food and agriculture with implications on health and environment. *Sci. Total Environ.* **2021**, *768*, 144990. [CrossRef]
7. Peters, R.J.; Bouwmeester, H.; Gottardo, S.; Amenta, V.; Arena, M.; Brandhoff, P.; Marvin, H.J.; Mech, A.; Moniz, F.B.; Pesudo, L.Q. Nanomaterials for products and application in agriculture, feed and food. *Trends Food Sci. Technol.* **2016**, *54*, 155–164. [CrossRef]
8. Abobatta, W.F. Nanotechnology application in agriculture. *Acta Sci. Agric.* **2018**, *2*, 99–102.
9. Dasgupta, N.; Ranjan, S.; Mundekkad, D.; Ramalingam, C.; Shanker, R.; Kumar, A. Nanotechnology in agro-food: From field to plate. *Food Res. Int.* **2015**, *69*, 381–400. [CrossRef]
10. Sadeghi, R.; Rodriguez, R.J.; Yao, Y.; Kokini, J.L. Advances in nanotechnology as they pertain to food and agriculture: Benefits and risks. *Annu. Rev. Food Sci. Technol.* **2017**, *8*, 467–492. [CrossRef]
11. Yadollahi, A.; Arzani, K.; Khoshghalb, H. The role of nanotechnology in horticultural crops postharvest management. In Proceedings of the Southeast Asia Symposium on Quality and Safety of Fresh and Fresh-Cut Produce, Bangkok, Thailand, 3–5 August 2009; Volume 875, pp. 49–56.
12. Prasad, R.; Bhattacharyya, A.; Nguyen, Q.D. Nanotechnology in sustainable agriculture: Recent developments, challenges, and perspectives. *Front. Microbiol.* **2017**, *8*, 1014. [CrossRef]
13. Sozer, N.; Kokini, J.L. Nanotechnology and its applications in the food sector. *Trends Biotechnol.* **2009**, *27*, 82–89. [CrossRef] [PubMed]
14. Ravichandran, R. Nanotechnology applications in food and food processing: Innovative green approaches, opportunities and uncertainties for global market. *Int. J. Green Nanotechnol. Phys. Chem.* **2010**, *1*, P72–P96. [CrossRef]
15. Reza Mozafari, M.; Johnson, C.; Hatziantoniou, S.; Demetzos, C. Nanoliposomes and their applications in food nanotechnology. *J. Liposome Res.* **2008**, *18*, 309–327. [CrossRef]
16. Kalita, D.; Baruah, S. The impact of nanotechnology on food. In *Nanomaterials Applications for Environmental Matrices*; Ronaldo, F.N., Odair, P.F., Vicente, O.S.N., Eds.; Elsevier: Amsterdam, The Netherlands, 2019; pp. 369–379.
17. Jafari, S.M.; McClements, D.J. Nanotechnology approaches for increasing nutrient bioavailability. *Adv. Food Nutr. Res.* **2017**, *81*, 1–30.
18. Rossi, M.; Passeri, D.; Sinibaldi, A.; Angjellari, M.; Tamburri, E.; Sorbo, A.; Carata, E.; Dini, L. Nanotechnology for food packaging and food quality assessment. *Adv. Food Nutr. Res.* **2017**, *82*, 149–204.

19. Mustafa, F.; Andreescu, S. Nanotechnology-based approaches for food sensing and packaging applications. *RSC Adv.* **2020**, *10*, 19309–19336. [CrossRef]
20. Chaudhry, Q.; Scotter, M.; Blackburn, J.; Ross, B.; Boxall, A.; Castle, L.; Aitken, R.; Watkins, R. Applications and implications of nanotechnologies for the food sector. *Food Addit. Contam.* **2008**, *25*, 241–258. [CrossRef]
21. Singh, T.; Shukla, S.; Kumar, P.; Wahla, V.; Bajpai, V.K.; Rather, I.A. Application of nanotechnology in food science: Perception and overview. *Front. Microbiol.* **2017**, *8*, 1501. [CrossRef]
22. Bartolucci, C. Nanotechnologies for agriculture and foods: Past and future. In *Nanotechnology in Agriculture and Food Science*; Axelos, M.A., Van De Voorde, M., Eds.; John Wiley & Sons: Hoboken, NJ, USA, 2017; pp. 1–14.
23. Díaz-Soler, B.M.; Martínez-Aires, M.D.; López-Alonso, M. Potential risks posed by the use of nano-enabled construction products: A perspective from coordinators for safety and health matters. *J. Clean. Prod.* **2019**, *220*, 33–44. [CrossRef]
24. Kamarulzaman, N.A.; Lee, K.E.; Siow, K.S.; Mokhtar, M. Public benefit and risk perceptions of nanotechnology development: Psychological and sociological aspects. *Technol. Soc.* **2020**, *62*, 101329. [CrossRef]
25. Shibata, T. Method for Producing Green Tea in Microfine Powder. U.S. Patent 6,416,803, 9 July 2002.
26. Sanguansri, P.; Augustin, M.A. Nanoscale materials development—A food industry perspective. *Trends Food Sci. Technol.* **2006**, *17*, 547–556. [CrossRef]
27. Roohinejad, S.; Greiner, R. Nanoscience: Relevance for agriculture and the food sector. In *Nanotechnology in Agriculture and Food Science*; Axelos, M.A., Van De Voorde, M., Eds.; John Wiley & Sons: Hoboken, NJ, USA, 2017; pp. 347–362.
28. Sangeetha, J.; Thangadurai, D.; Hospet, R.; Purushotham, P.; Karekalammanavar, G.; Mundaragi, A.C.; David, M.; Shinge, M.R.; Thimmappa, S.C.; Prasad, R. Agricultural Nanotechnology: Concepts, Benefits, and Risks. In *Nanotechnology*; Springer: Singapore, 2017; pp. 1–17.
29. Pathakoti, K.; Manubolu, M.; Hwang, H.-M. Nanostructures: Current uses and future applications in food science. *J. Food Drug Anal.* **2017**, *25*, 245–253. [CrossRef] [PubMed]
30. Baruah, S.; Dutta, J. Nanotechnology applications in pollution sensing and degradation in agriculture: A review. *Environ. Chem. Lett.* **2009**, *7*, 191–204. [CrossRef]
31. Gogos, A.; Knauer, K.; Bucheli, T.D. Nanomaterials in plant protection and fertilization: Current state, foreseen applications, and research priorities. *J. Agric. Food Chem.* **2012**, *60*, 9781–9792. [CrossRef]
32. Prasad, R.; Kumar, M.; Kumar, V. *Nanotechnology: An Agricultural Paradigm*; Springer Nature: Singapore, 2017; p. 371.
33. Sastry, K.; Rashmi, H.; Rao, N. Nanotechnology patents as R&D indicators for disease management strategies in agriculture. *J. Intellect. Prop. Rights* **2010**, *15*, 197–205.
34. Singh, H.; Sharma, A.; Bhardwaj, S.K.; Arya, S.K.; Bhardwaj, N.; Khatri, M. Recent advances in the applications of nano-agrochemicals for sustainable agricultural development. *Environ. Sci. Processes Impacts* **2021**, *23*, 213–239. [CrossRef] [PubMed]
35. Sivarethinamohan, R.; Sujatha, S. Unlocking the potentials of using nanotechnology to stabilize agriculture and food production. *AIP Conf. Proc.* **2021**, *2327*, 20022.
36. Chinnamuthu, C.; Boopathi, P.M. Nanotechnology and agroecosystem. *Madras Agric. J.* **2009**, *96*, 17–31.
37. Wang, Y.; Deng, C.; Rawat, S.; Cota-Ruiz, K.; Medina-Velo, I.; Gardea-Torresdey, J.L. Evaluation of the effects of nanomaterials on rice (*Oryza sativa* L.) responses: Underlining the benefits of nanotechnology for agricultural applications. *ACS Agric. Sci. Technol.* **2021**, *1*, 44–54. [CrossRef]
38. Flood, J. The importance of plant health to food security. *Food Secur.* **2010**, *2*, 215–231. [CrossRef]
39. Mali, S.C.; Raj, S.; Trivedi, R. Nanotechnology a novel approach to enhance crop productivity. *Biochem. Biophys. Rep.* **2020**, *24*, 100821.
40. Duhan, J.S.; Kumar, R.; Kumar, N.; Kaur, P.; Nehra, K.; Duhan, S. Nanotechnology: The new perspective in precision agriculture. *Biotechnol. Rep.* **2017**, *15*, 11–23. [CrossRef] [PubMed]
41. Khot, L.R.; Sankaran, S.; Maja, J.M.; Ehsani, R.; Schuster, E.W. Applications of nanomaterials in agricultural production and crop protection: A review. *Crop Prot.* **2012**, *35*, 64–70. [CrossRef]
42. Worrall, E.A.; Hamid, A.; Mody, K.T.; Mitter, N.; Pappu, H.R. Nanotechnology for plant disease management. *Agronomy* **2018**, *8*, 285. [CrossRef]
43. FAO. *The State of Food and Agriculture. Moving Forward on Food Loss and Waste Reduction*; FAO: Rome, Italy, 2019.
44. Flores-López, M.L.; Cerqueira, M.A.; de Rodríguez, D.J.; Vicente, A.A. Perspectives on utilization of edible coatings and nano-laminate coatings for extension of postharvest storage of fruits and vegetables. *Food Eng. Rev.* **2016**, *8*, 292–305. [CrossRef]
45. Neme, K.; Nafady, A.; Uddin, S.; Tola, Y.B. Application of nanotechnology in agriculture, postharvest loss reduction and food processing: Food security implication and challenges. *Heliyon* **2021**, *7*, e08539. [CrossRef]
46. Sekhon, B. Food nanotechnology—An overview. *Nanotechnol. Sci. Appl.* **2010**, *3*, 1–15.
47. Falguera, V.; Quintero, J.P.; Jiménez, A.; Muñoz, J.A.; Ibarz, A. Edible films and coatings: Structures, active functions and trends in their use. *Trends Food Sci. Technol.* **2011**, *22*, 292–303. [CrossRef]
48. An, J.; Zhang, M.; Wang, S.; Tang, J. Physical, chemical and microbiological changes in stored green asparagus spears as affected by coating of silver nanoparticles-PVP. *LWT-Food Sci. Technol. Soc.* **2008**, *41*, 1100–1107. [CrossRef]
49. Fakhouri, F.; Casari, A.; Mariano, M.; Yamashita, F.; Mei, L.I.; Soldi, V.; Martelli, S. Effect of a gelatin-based edible coating containing cellulose nanocrystals (CNC) on the quality and nutrient retention of fresh strawberries during storage. *IOP Conf. Ser. Mater. Sci. Eng.* **2014**, *64*, 012024. [CrossRef]

50. Shi, S.; Wang, W.; Liu, L.; Wu, S.; Wei, Y.; Li, W. Effect of chitosan/nano-silica coating on the physicochemical characteristics of longan fruit under ambient temperature. *J. Food Eng.* **2013**, *118*, 125–131. [CrossRef]
51. Medeiros, B.G.; Souza, M.P.; Pinheiro, A.C.; Bourbon, A.I.; Cerqueira, M.A.; Vicente, A.A.; Carneiro-da-Cunha, M.G. Physical characterisation of an alginate/lysozyme nano-laminate coating and its evaluation on 'Coalho'cheese shelf life. *Food Bioprocess Technol.* **2014**, *7*, 1088–1098. [CrossRef]
52. Yu, Y.; Zhang, S.; Ren, Y.; Li, H.; Zhang, X.; Di, J. Jujube preservation using chitosan film with nano-silicon dioxide. *J. Food Eng.* **2012**, *113*, 408–414. [CrossRef]
53. Sogvar, O.B.; Saba, M.K.; Emamifar, A.; Hallaj, R. Influence of nano-ZnO on microbial growth, bioactive content and postharvest quality of strawberries during storage. *Innov. Food Sci. Emerg. Technol.* **2016**, *35*, 168–176. [CrossRef]
54. Rashidi, L.; Khosravi-Darani, K. The applications of nanotechnology in food industry. *Crit. Rev. Food Sci. Nutr.* **2011**, *51*, 723–730. [CrossRef]
55. Duncan, T.V. Applications of nanotechnology in food packaging and food safety: Barrier materials, antimicrobials and sensors. *J. Colloid Interface Sci.* **2011**, *363*, 1–24. [CrossRef]
56. Anton, N.; Vandamme, T.F. Nano-emulsions and micro-emulsions: Clarifications of the critical differences. *Pharm. Res.* **2011**, *28*, 978–985. [CrossRef]
57. Ezhilarasi, P.; Karthik, P.; Chhanwal, N.; Anandharamakrishnan, C. Nanoencapsulation techniques for food bioactive components: A review. *Food Bioprocess Technol.* **2013**, *6*, 628–647. [CrossRef]
58. Ramachandraiah, K.; Han, S.G.; Chin, K.B. Nanotechnology in meat processing and packaging: Potential applications—A review. *Asian-Australas. J. Anim. Sci.* **2015**, *28*, 290. [CrossRef]
59. Katouzian, I.; Jafari, S.M. Nano-encapsulation as a promising approach for targeted delivery and controlled release of vitamins. *Trends Food Sci. Technol.* **2016**, *53*, 34–48. [CrossRef]
60. Yu, L.; Banerjee, I.A.; Gao, X.; Nuraje, N.; Matsui, H. Fabrication and application of enzyme-incorporated peptide nanotubes. *Bioconjugate Chem.* **2005**, *16*, 1484–1487. [CrossRef] [PubMed]
61. Bai, Y.-X.; Li, Y.-F.; Yang, Y.; Yi, L.-X. Covalent immobilization of triacylglycerol lipase onto functionalized nanoscale SiO_2 spheres. *Process Biochem.* **2006**, *41*, 770–777. [CrossRef]
62. Weiss, J.; Takhistov, P.; McClements, D.J. Functional materials in food nanotechnology. *J. Food Sci.* **2006**, *71*, R107–R116. [CrossRef]
63. Dekkers, S.; Krystek, P.; Peters, R.J.; Lankveld, D.P.; Bokkers, B.G.; van Hoeven-Arentzen, P.H.; Bouwmeester, H.; Oomen, A.G. Presence and risks of nanosilica in food products. *Nanotoxicology* **2011**, *5*, 393–405. [CrossRef]
64. Han, C.; Zhao, A.; Varughese, E.; Sahle-Demessie, E.J.N. Evaluating weathering of food packaging polyethylene-nano-clay composites: Release of nanoparticles and their impacts. *NanoImpact* **2018**, *9*, 61–71. [CrossRef] [PubMed]
65. Noorbakhsh-Soltani, S.; Zerafat, M.; Sabbaghi, S. A comparative study of gelatin and starch-based nano-composite films modified by nano-cellulose and chitosan for food packaging applications. *Carbohydr. Polym.* **2018**, *189*, 48–55. [CrossRef]
66. Li, L.; Zhao, C.; Zhang, Y.; Yao, J.; Yang, W.; Hu, Q.; Wang, C.; Cao, C. Effect of stable antimicrobial nano-silver packaging on inhibiting mildew and in storage of rice. *Food Chem.* **2017**, *215*, 477–482. [CrossRef]
67. Prakash, J.; Sun, S.; Swart, H.C.; Gupta, R.K. Noble metals-TiO2 nanocomposites: From fundamental mechanisms to photocatalysis, surface enhanced Raman scattering and antibacterial applications. *Appl. Mater. Today* **2018**, *11*, 82–135. [CrossRef]
68. Zhao, L.; Li, F.; Chen, G.; Fang, Y.; An, X.; Zheng, Y.; Xin, Z.; Zhang, M.; Yang, Y.; Hu, Q. Effect of nanocomposite-based packaging on preservation quality of green tea. *Int. J. Food Sci. Technol.* **2012**, *47*, 572–578. [CrossRef]
69. De Azeredo, H.M. Nanocomposites for food packaging applications. *Food Res. Int.* **2009**, *42*, 1240–1253. [CrossRef]
70. Sharma, C.; Dhiman, R.; Rokana, N.; Panwar, H. Nanotechnology: An untapped resource for food packaging. *Front. Microbiol.* **2017**, *8*, 1735. [CrossRef] [PubMed]
71. Jafarzadeh, S.; Alias, A.; Ariffin, F.; Mahmud, S. Characterization of semolina protein film with incorporated zinc oxide nano rod intended for food packaging. *Pol. J. Food Nutr. Sci.* **2017**, *67*, 183–190. [CrossRef]
72. Fasihnia, S.H.; Peighambardoust, S.H.; Peighambardoust, S.J. Nanocomposite films containing organoclay nanoparticles as an antimicrobial (active) packaging for potential food application. *J. Food Processing Preserv.* **2018**, *42*, e13488. [CrossRef]
73. Moustafa, H.; Youssef, A.M.; Darwish, N.A.; Abou-Kandil, A.I. Eco-friendly polymer composites for green packaging: Future vision and challenges. *Compos. Part B Eng.* **2019**, *172*, 16–25. [CrossRef]
74. Basu, A.; Kundu, S.; Sana, S.; Halder, A.; Abdullah, M.F.; Datta, S.; Mukherjee, A. Edible nano-bio-composite film cargo device for food packaging applications. *Food Packag. Shelf Life* **2017**, *11*, 98–105. [CrossRef]
75. Pinto, R.J.; Daina, S.; Sadocco, P.; Neto, C.P.; Trindade, T. Antibacterial activity of nanocomposites of copper and cellulose. *BioMed Res. Int.* **2013**, *2013*, 280512. [CrossRef]
76. Rieger, K.A.; Eagan, N.M.; Schiffman, J.D. Encapsulation of cinnamaldehyde into nanostructured chitosan films. *J. Appl. Polym. Sci.* **2015**, *132*, 41739. [CrossRef]
77. Kumar, S.; Shukla, A.; Baul, P.P.; Mitra, A.; Halder, D. Biodegradable hybrid nanocomposites of chitosan/gelatin and silver nanoparticles for active food packaging applications. *Food Packag. Shelf Life* **2018**, *16*, 178–184. [CrossRef]
78. Youssef, A.M.; Abdel-Aziz, M.S.; El-Sayed, S.M. Chitosan nanocomposite films based on Ag-NP and Au-NP biosynthesis by Bacillus subtilis as packaging materials. *Int. J. Biol. Macromol.* **2014**, *69*, 185–191. [CrossRef]
79. Scott, N. Nanotechnology and animal health. *Rev. Sci. Et Tech.* **2005**, *24*, 425. [CrossRef]

80. Abdelmonem, A.M. Application of carbon-based nanomaterials in food preservation area. In *Carbon Nanomaterials for Agri-Food and EnvironmentalApplications*; Abd-Elsalam, K.A., Ed.; Elsevier: Amsterdam, The Netherlands, 2020; pp. 583–593.
81. Kim, J.Y.; Han, S.I.; Hong, S. Effect of modified carbon nanotube on the properties of aromatic polyester nanocomposites. *Polymer* **2008**, *49*, 3335–3345. [CrossRef]
82. Singh, T.; Jyoti, K.; Patnaik, A.; Singh, A.; Chauhan, R.; Chandel, S. Biosynthesis, characterization and antibacterial activity of silver nanoparticles using an endophytic fungal supernatant of *Raphanus sativus*. *J. Genet. Eng.* **2017**, *15*, 31–39. [CrossRef] [PubMed]
83. Krishna, V.; Pumprueg, S.; Lee, S.-H.; Zhao, J.; Sigmund, W.; Koopman, B.; Moudgil, B. Photocatalytic disinfection with titanium dioxide coated multi-wall carbon nanotubes. *Process Saf. Environ. Prot.* **2005**, *83*, 393–397. [CrossRef]
84. Graveland-Bikker, J.; De Kruif, C. Unique milk protein based nanotubes: Food and nanotechnology meet. *Trends Food Sci. Technol.* **2006**, *17*, 196–203. [CrossRef]
85. Ipsen, R.; Otte, J. Self-assembly of partially hydrolysed α-lactalbumin. *Biotechnol. Adv.* **2007**, *25*, 602–605. [CrossRef]
86. Khodakovskaya, M.; Dervishi, E.; Mahmood, M.; Xu, Y.; Li, Z.; Watanabe, F.; Biris, A.S. Carbon nanotubes are able to penetrate plant seed coat and dramatically affect seed germination and plant growth. *ACS Nano* **2009**, *3*, 3221–3227. [CrossRef]
87. Torney, F.; Trewyn, B.G.; Lin, V.S.-Y.; Wang, K. Mesoporous silica nanoparticles deliver DNA and chemicals into plants. *Nat. Nanotechnol.* **2007**, *2*, 295–300. [CrossRef] [PubMed]
88. Griesche, C.; Baeumner, A.J. Biosensors to support sustainable agriculture and food safety. *Trends Anal. Chem.* **2020**, *128*, 115906. [CrossRef]
89. Chalupowicz, D.; Veltman, B.; Droby, S.; Eltzov, E. Evaluating the use of biosensors for monitoring of *Penicillium digitatum* infection in citrus fruit. *Sens. Actuators B Chem.* **2020**, *311*, 127896. [CrossRef]
90. Sekhon, B.S. Nanotechnology in agri-food production: An overview. *Nanotechnol. Sci. Appl.* **2014**, *7*, 31. [CrossRef]
91. Davari, M.; Bayat Kazazi, S.; Akbarzadeh Pivehzhani, O. Nanomaterials: Implications on agroecosystem. Nanotechnology. In *Nanotechnology*; Prasad, R., Kumar, M., Kumar, V., Eds.; Springer: Singapore, 2017; pp. 59–71.
92. Cheng, M.M.-C.; Cuda, G.; Bunimovich, Y.L.; Gaspari, M.; Heath, J.R.; Hill, H.D.; Mirkin, C.A.; Nijdam, A.J.; Terracciano, R.; Thundat, T. Nanotechnologies for biomolecular detection and medical diagnostics. *Curr. Opin. Chem. Biol.* **2006**, *10*, 11–19. [CrossRef] [PubMed]
93. Bratovčić, A.; Odobašić, A.; Ćatić, S.; Šestan, I. Application of polymer nanocomposite materials in food packaging. *Croat. J. Food Sci. Technol.* **2015**, *7*, 86–94. [CrossRef]
94. Warriner, K.; Reddy, S.M.; Namvar, A.; Neethirajan, S. Developments in nanoparticles for use in biosensors to assess food safety and quality. *Trends Food Sci. Technol.* **2014**, *40*, 183–199. [CrossRef]
95. Rajput, V.; Minkina, T.; Mazarji, M.; Shende, S.; Sushkova, S.; Mandzhieva, S.; Burachevskaya, M.; Chaplygin, V.; Singh, A.; Jatav, H. Accumulation of nanoparticles in the soil-plant systems and their effects on human health. *Ann. Agric. Sci.* **2020**, *65*, 137–143. [CrossRef]
96. Narei, H.; Ghasempour, R.; Akhavan, O. Toxicity and safety issues of carbon nanotubes. In *Carbon Nanotube-Reinforced Polymers: From Nanoscale to Macroscale*; Rafiee, R., Ed.; Elsevier: Amsterdam, The Netherlands, 2018; pp. 145–171.
97. Oleszczuk, B.-W.E.S.D. Effects of titanium dioxide nanoparticles exposure on human health—A review. *Biol. Trace Elem. Res.* **2020**, *193*, 118–129.
98. Nile, S.H.; Baskar, V.; Selvaraj, D.; Nile, A.; Xiao, J.; Kai, G. Nanotechnologies in food science: Applications, recent trends, and future perspectives. *Nano-Micro Lett.* **2020**, *12*, 1–34. [CrossRef]

agronomy

MDPI

Article

Chemical and Physical Properties of Some Hazelnut Varieties Grown in Portugal

Ana Cristina Ferrão [1], Raquel P. F. Guiné [1], Elsa Ramalhosa [2], Arminda Lopes [3], Cláudia Rodrigues [4], Hugo Martins [4], Roberto Gonçalves [4] and Paula M. R. Correia [1,*]

1 CERNAS-IPV Research Centre, Polytechnic Institute of Viseu, 3504-510 Viseu, Portugal; aferrao@esav.ipv.pt (A.C.F.); raquelguine@esav.ipv.pt (R.P.F.G.)
2 CIMO, Polytechnic Institute of Bragança, 5300-253 Bragança, Portugal; elsa@ipb.pt
3 Direcção Regional de Agricultura e Pescas do Centro, 3504-504 Viseu, Portugal; arminda.lopes@drapc.gov.pt
4 Agrarian School, Polytechnic Institute of Viseu, 3500-506 Viseu, Portugal; cjoana_97@hotmail.com (C.R.); hugomartinsagro@hotmail.com (H.M.); robertofilipe1994@gmail.com (R.G.)
* Correspondence: paulacorreia@esav.ipv.pt

Abstract: Hazelnuts are one of the most appreciated nuts worldwide due to their unique organoleptic and nutritional characteristics. The present work intended to analyse several physical and chemical properties of different hazelnut varieties grown in Portugal, namely Tonda de Giffoni, Grada de Viseu, Segorbe, Longa de Espanha, Butler, Gunslebert, and Negreta. In general, the results revealed statistically significant differences between the varieties under study. The Gunslebert had more elongated hazelnuts and with heavier shelled fruits, while the kernels of the Grada de Viseu revealed to be heavier. Grada de Viseu was harder in the shell, Gunslebert had a harder core, and Segorbe was more resistant to fracture. Fat was the more representative component for all varieties and in some cases the values of moisture and water activity were over the recommended amount ($\geq$0.62). Tonda de Giffoni was the variety with the highest induction time, indicating the highest oxidation stability. Moreover, discriminant analysis revealed that the variables more important to distinguish the varieties were protein (λ = 0.007) and water activity (λ = 0.010). The results of this study help to better understand the differences between some hazelnut varieties that are cultivated in Portugal, which gives important hints for all players in the hazelnut sector.

Keywords: chemical properties; hazelnuts; physical properties; specific extinction coefficients

Citation: Ferrão, A.C.; Guiné, R.P.F.; Ramalhosa, E.; Lopes, A.; Rodrigues, C.; Martins, H.; Gonçalves, R.; Correia, P.M.R. Chemical and Physical Properties of Some Hazelnut Varieties Grown in Portugal. *Agronomy* **2021**, *11*, 1476. https://doi.org/10.3390/agronomy11081476

Academic Editor: Luca Espen

Received: 21 June 2021
Accepted: 22 July 2021
Published: 25 July 2021

1. Introduction

Hazelnut (*Corylus avellana* L.) is the fruit of the hazel tree and belongs to the *Betulaceae* family. Due to their organoleptic and nutritional characteristics, hazelnuts are well appreciated worldwide. This fruit is traditionally consumed as a whole nut (raw or roasted) but hazelnut kernels can also be used in bakery products, chocolates, dairy products, snacks, and other confectionary products [1,2].

Hazelnuts are a good source of proteins, monounsaturated fatty acids (mainly oleic acid), carbohydrates, fibre, minerals, vitamins (such as vitamin E), phytosterols, and also phenolic antioxidants [3–5]. Its consumption is associated with several health benefits such as the prevention of cardiovascular diseases for example [6].

In Portugal, the dried fruit sector is of great importance in the economic sustainability of rural regions, as well as in the fight against social desertification, as it continues to be based on a regional tradition, is represented by family businesses, and is derived from smaller orchards [7]. In 2018, hazelnut production in Portugal was equal to 240 tons [8]. According to a study performed by Ferrão et al. [9], the main hazelnut varieties cultivated in Portugal are Grada de Viseu, Segorbe, Fertile de Coutard, Butler and Negreta, being Tonda de Giffoni one of the most productive varieties.

There are different standards to measure hazelnuts' quality in terms of dimensions, aspect, and hidden defects [10]. Furthermore, hazelnut properties vary according to the cultivar [11]. Knowledge of hazelnuts' physical and chemical properties is crucial as they are necessary to determine numerous parameters of extreme importance to guarantee the quality of the product. Moreover, the determination of these parameters also leads to the decision regarding what type of care should be given to the product; for example, regarding whether or not it should undergo any transformation before it is made available to the final consumer [12]. Food acceptability is influenced by different factors including food intrinsic properties such as taste, texture, aroma, and appearance [13,14]. According to Lopes et al. [4], hazelnuts' quality is not only related to its nutritional and chemical composition but also to some visual aspects such as the absence of broken cores and presentation of a clear and uniform colour. Thus, to guarantee the food quality of the products, it is of utmost importance to ensure that they are handled and stored under proper conditions [4,15]. According to their characteristics, the different varieties may be more suitable for industry or fresh consumption. The varieties of Grada de Viseu, Longa de Espanha, Butler, and Segorbe are more appropriate for fresh consumption, while the variety Negreta is more suitable for industry. Conversely, the varieties Tonda de Giffoni and Segorbe are suitable for both industry and fresh consumption [16]. Despite the importance of this theme, there are still only few studies on the physical and chemical properties of hazelnut varieties grown in Portugal.

Therefore, the aim of this study was to evaluate the physical and chemical properties of different hazelnut varieties cultivated in Portugal.

2. Materials and Methods

2.1. Hazelnut Samples

The hazelnut samples used in this study were from different varieties, namely Tonda de Giffoni, Grada de Viseu, Segorbe, Longa de Espanha, Butler, Gunslebert, and Negreta, collected in Viseu in central Portugal. Three sets of 1 kg of each hazelnut fruit were harvested in 2019 at the maturity state and then stored (whole hazelnut) until the analysis at 5 °C to guarantee the safety and quality of the product.

2.2. Colour Evaluation

The colour was examined using the colourimeter Konica Minolta CR-400 that evaluates the Cartesian coordinates CIE L*, a*, and b *. The L* coordinate quantifies the lightness variation, ranging from 0 (black) to 100 (white). The a* and b* represent chromaticity coordinates, varying from −60 to +60. The coordinate a* changes from red (+a) to green (−a) and the coordinate b* varies from yellow (+b) to blue (−b) [4,12,17]. For this analysis, 30 hazelnuts of each variety were used and measurements of different parts of the shell (2 on the brown shell and 2 on the hilum) and the kernels (2 on the skin and 2 on the inner core) of each variety were recorded. The measurements on the skin were made on the whole surface of the hazelnuts.

2.3. Analysis of Texture

Texture measurements were conducted using the texturometer TA.XT.Plus (Stable Micro Systems, Godalming, Surrey, UK). For these analyses, 30 fruits from each variety were submitted to two independent tests, namely shell crushing and core cutting. Both tests used a 500 N load cell.

The test performed for shell crushing was a measurement of force in compression using a flat P75 probe (diameter of 75 mm) that compressed the sample against the base of the texturometer. The speeds of the pre-test, test, and post-test were 1.0 mm/s in all cases. The distance was 6 mm and the trigger force considered was 0.2 N. From the test performed for shell crushing, was obtained a curve of force (N) versus distance (mm) which allowed calculating the hardness (the value of the force that corresponded to the crushing of the shell).

For the test of cutting the core, the probe used was the Blade Set HDP/BS (Warner-Bratzler). The pre-test and test speeds were 1.0 mm/s, while the post-test speed was 10.0 mm/s. For this test, the distance was 30 mm and the trigger force was 0.15 N. The obtained curve of force vs. distance allowed for the calculation of two textural properties: hardness (force at first peak) and friability (distance to first peak). Hardness was measured both for shell and unshelled hazelnuts, while friability was only measured for the core.

2.4. Density Evaluation

Density was assessed as described by Guiné et al. [12] and two types of density were measured: the true density and the apparent density. To assess the apparent density, two samples of 100 g of hazelnuts were weighted: one with the shell and one without the shell. Then, they were placed in two separate beakers: one with a capacity of 500 mL (for the shelled hazelnuts) and the other with a capacity of 250 mL (for the hazelnut kernels) in order to determine the apparent density of both samples. The apparent density was calculated by dividing the mass by the volume measured. In the case of true density, 250 mL of water was placed in two 500 mL beakers. To one beaker, 100 g of hazelnuts with the shells were added, while to the other beaker, 100 g of hazelnuts without the shell were added. In both cases, the initial volume and final volume was measured. The true density was calculated as the ratio between the mass and the differences observed in the volume. All the measurements were conducted in triplicates for each of the hazelnut samples.

2.5. Biometric Evaluation

For the biometric analysis, 50 hazelnuts of each variety were used and the measurements were performed using a calliper rule with a precision scale. The analyses were conducted for the whole fruit and for the corresponding kernels, and the numbers of hollow fruits were also registered.

Different biometric parameters were evaluated, namely width (wider equatorial zone), height (distance between centres), and thickness (narrow equatorial zone perpendicular to the latter). These parameters were evaluated for the whole hazelnuts and the kernels [18]. With the values obtained, it was possible to calculate the shape ratio and compression ratio as follows [4]:

$$\text{Shape ratio} = \frac{(\text{width} + \text{thickness})}{2 \times \text{height}} \tag{1}$$

$$\text{Compression ratio} = \frac{\text{width}}{\text{thickness}} \tag{2}$$

The kernel percentage was calculated as (kernel weight/nut weight) $\times$ 100 [19].

2.6. Chemical Analyses

To perform the chemical analyses regarding moisture, protein (% Nx5.23), fat, fibre, and ash, standard procedures were followed completely [20] and only the kernels were analysed after the milling and homogenization of the varieties were evaluated. The procedure was conducted in fresh hazelnuts.

Water activity was determined at a constant temperature using the Hygroscope Rotronic. The Rancimat method for oils and fats is an accelerated aging test and was used to evaluate the oxidation stability of hazelnut fat. [21]. This method was performed using the Rancimat equipment model 743, Metrohm from Herisau/Switzerland.

2.7. Specific Extinction Coefficients Analysis

The analysis of primary and secondary oxidation of the fat extracted from hazelnut kernels was evaluated using a spectrophotometer (T80 uv-vis spectrophotometer, PG instruments Ltd., Lutterworth, UK) and according to the procedure described at the Annex IV of the Regulation (EU) 2015/1833 [22]. For the determination of the spectrophotometric indices, each sample of hazelnut fat (0.6 g) was placed in a 10 mL flask and filled up to 10 mL with isooctane as this the stock solution. Then, two dilutions were made from the stock solution: one of 1:20 for the measurement at 232 nm and the other of 1:2 for the measurements at 264 nm, 268 nm, and 272 nm. In all cases, isooctane was added until 10 mL was reached.

The specific extinction coefficients (K) were calculated as follows:

$$K(\lambda) = \frac{\text{absorbance measured at wavelength } \lambda \times \text{ dilution factor}}{\text{weight of the sample } \times \text{ path length of the quartz cell (mm)}} \quad (3)$$

For each sample, two fat extractions were performed and then the procedure was conducted in duplicates for each extraction.

2.8. Statistical Analysis

The data were analysed using basic descriptive tools such as the mean value and standard deviation of a set of replication measurements. Additionally, one-way ANOVA was used to compare the means of three or more samples and to assess the differences between them using the post-hoc test Tukey HSD (Honestly Significant Difference); this test consists of a multiple comparison process performed in a single step and identifies which means are significantly different from others. The data were further submitted to a discriminant function analysis for the categorical variable VARIETY as the dependent variable and the 43 ratio variables (accounting for all the measured properties) as the independent variables. The analysis was performed using the stepwise method with Mahalanobis distance and the criteria was the probability of F. The tests of equality of co-variance matrices (Box's M test) and group means were performed to verify the assumptions for the utilization of this analysis.

For all tests, a level of significance of 5% ($p < 0.05$) was considered and the data analysis was performed with the Statistical Software for Social Sciences (SPSS) software from IBM Inc. (Armonk, NY, USA, version 26).

3. Results and Discussion

3.1. Physical Properties

3.1.1. Biometric Characteristics

Biometric evaluation is important not only to differentiate hazelnuts' varieties but also to decide their utilization and design for the processing equipment. For example, in the food industry, spherical nuts are preferred because they are easily processed [23]. Table 1 presents the values obtained for the weight of the shelled hazelnuts and kernels. In terms of the shelled hazelnuts, the fruits of Gunslebert were heavier on average (3.89 ± 0.64 g), while the fruits of Negreta were lighter on average (2.23 ± 0.37 g). In terms of the kernels, Grada de Viseu were considered the heavier fruit (1.70 ± 0.31 g), while Negreta was considered the lightest of all the varieties (1.12 ± 0.24 g). In a study performed by Lopes et al. [4], it was also found that Negreta appears among the lightest varieties.

Table 1. Weight of the hazelnut fruits and kernels (mean ± standard deviation).

Sample	Fruit Weight [1] (g)	Kernel Weight [1] (g)	Kernel Percentage [1] (%)
Grada de Viseu	3.82 ± 0.65 [e]	1.70 ± 0.31 [d]	44.14 ± 6.24 [a]
Tonda de Giffoni	3.16 ± 0.33 [c]	1.50 ± 0.18 [bc]	46.19 ± 4.86 [a]
Segorbe	2.56 ± 0.43 [b]	1.21 ± 0.22 [a]	57.27 ± 9.17 [b]
Longa de Espanha	2.68 ± 0.54 [b]	1.23 ± 0.34 [a]	67.68 ± 10.33 [c]
Butler	3.51 ± 0.42 [d]	1.58 ± 0.28 [cd]	53.16 ± 12.39 [a]
Gunslebert	3.89 ± 0.64 [e]	1.65 ± 0.20 [b]	44.23 ± 9.07 [a]
Negreta	2.23 ± 0.37 [a]	1.12 ± 0.24 [a]	63.91 ± 12.05 [c]
p-value	<0.05	<0.05	<0.05

[1] Mean values in the same column with the same letter are not statistically different ($p < 0.05$).

It was observed that the differences encountered between some of the varieties were statistically significant ($p < 0.05$).

Kernel percentage is an important parameter as it identifies which part is edible [18]. The results demonstrated that, regarding kernel percentage, the varieties can be grouped into three different groups: the first includes the varieties Grada de Viseu (44.14 ± 6.24%), Gunslebert (44.23 ± 9.07%), Tonda de Giffoni (46.19 ± 4.86%), and Butler (53.16 ± 12.39%); the second includes Segorbe (57.27 ± 9.17%); and the third group includes Negreta (63.91 ± 12.05%) and Longa de Espanha (67.68 ± 10.33). In previous studies [18,24,25], it was found that kernel percentages were lower (< 40.00%) than the values obtained in the present study.

As presented in Table 2, height varies from 1.39 ± 0.18 cm (Negreta) to 1.76 ± 0.14 cm (Longa de Espanha), with significant differences between some of the varieties ($p < 0.05$). It is important to highlight that height and thickness have the same variation profile for each variety.

Table 2. Biometric measurements of the hazelnut kernels (mean ± standard deviation).

Sample	Height [1] (cm)	Width [1] (cm)	Thickness [1] (cm)	Shape Ratio [1]	Compression Ratio [1]
Grada de Viseu	1.65 ± 0.11 ^{b}	1.60 ± 0.12 ^{e}	1.42 ± 0.15 ^{e}	0.92 ± 0.09 ^{c}	1.13 ± 0.12 ^{a}
Tonda de Giffoni	1.45 ± 0.10 ^{a}	1.58 ± 0.09 ^{e}	1.42 ± 0.09 ^{e}	1.04 ± 0.09 ^{d}	1.11 ± 0.07 ^{a}
Segorbe	1.43 ± 0,12 ^{a}	1.36 ± 0.11 ^{c}	1.22 ± 0.11 ^{c}	0.91 ± 0.09 ^{c}	1.12 ± 0.12 ^{a}
Longa de Espanha	1.76 ± 0.14 ^{c}	1.23 ± 0.18 ^{b}	1.11 ± 0.17 ^{b}	0.67 ± 0.08 ^{a}	1.13 ± 0.23 ^{a}
Butler	1.69 ± 0.13 ^{bc}	1.45 ± 0.13 ^{d}	1.31 ± 0.13 ^{d}	0.82 ± 0.08 ^{b}	1.23 ± 0.13 ^{a}
Gunslebert	1.67 ± 0.19 ^{b}	1.10 ± 0.21 ^{a}	1.01 ± 0.21 ^{a}	0.64 ± 0.14 ^{a}	1.10 ± 0.15 ^{a}
Negreta	1.39 ± 0.18 a	1.08 ± 0.12 ^{a}	1.00 ± 0.12 ^{a}	0.78± 0.30 ^{b}	1.09 ± 0.08 ^{a}
p-value	<0.05	<0.05	<0.05	<0.05	0.643

[1] Mean values in the same column with the same letter are not statistically different ($p < 0.05$).

As for width, values ranged from 1.08 ± 0.12 cm for Negreta to 1.60 ± 0.12 cm for Grada de Viseu, again with significant differences between the varieties ($p < 0.05$). Thickness assumed values between 1.00 ± 0.12 cm (Negreta), 1.42 ± 0.09 cm (Tonda de Giffoni), and 1.42 ± 0.15 cm (Grada de Viseu). The variety Gunslebert is one of the smaller and thicker varieties, unlike Tonda de Giffoni that is one of the widest and thickest varieties, which translates into a higher shape ratio (1.04 ± 0.09 cm). As for the compression ratio, all the varieties had values higher than 1 and Butler had the highest compression ratio (1.23 ± 0.13 cm). Although there were significant differences between some varieties in the shape ratio, the same did not occur with the compression ratio to which all the varieties had similar values. According to Lopes et al. [4], higher values of compression ratio indicate that these fruits are more asymmetric and values close to 1 indicate that these fruits are more rounded in the equatorial zone. Nut size is an important parameter as it provides indication about the kernel size. According to Yao and Mehlenbacher [26], kernels diameters ranging from 11 to 13 mm are the most suitable for kernel market.

3.1.2. Density

Knowledge about physical properties such as porosity in addition to apparent and true densities are very important in order to project storage facilities [27]. The mean values of the apparent and true densities with and without the shell are presented in Table 3. By analysing the results obtained, generally the values of the fruits without the shell were higher than those with the shell, which is expected considering there exists an important amount of void between the shell and the core. The results also demonstrated that for all varieties, true density was higher than the apparent density both for shelled and unshelled hazelnuts which can be explained by the fact that the round form of the elements does not allow a great compaction [12]. It was also observed that there were statistically significant differences between the values of densities among the samples under study ($p < 0.05$). Similar results were obtained in the study performed by Guiné et al. [12].

Table 3. Hazelnuts' density (mean ± standard deviation).

Sample	True Density (g/mL)		Apparent Density (g/mL)	
	With Shell [1]	Without Shell [1]	With Shell [1]	Without Shell [1]
Grada de Viseu	1.39 ± 0.05 cd	1.35 ± 0.09 ab	0.35 ± 0.01 ab	0.42 ± 0.01 ab
Tonda de Giffoni	1.17 ± 0.23 d	1.41 ± 0.12 a	0.33 ± 0.01 a	0.42 ± 0.01 ab
Segorbe	1.09 ± 0.09 bc	1.19 ± 0.07 c	0.47 ± 0.04 c	0.45 ± 0.02 bc
Longa de Espanha	1.56 ± 0.12 e	1.63 ± 0.07 a	0.31 ± 0.01 a	0.40 ± 0.01 a
Butler	0.79 ± 0.05 a	0.97 ± 0.03 b	0.40 ± 0.01 b	0.47 ± 0.02 c
Gunslebert	0.73 ± 0.02 a	0.96 ± 0.04 a	0.33 ± 0.010 a	0.40 ± 0.01 a
Negreta	0.91 ± 0.03 ab	1.07 ± 0.04 c	0.47 ± 0.01 c	0.47 ± 0.01 c
p-value	<0.05	<0.05	<0.05	<0.05

[1] Mean values in the same column with the same letter are not statistically different ($p < 0.05$).

3.1.3. Colour

Colour assessment is crucial because it is one of the most important food intrinsic properties affecting consumers' food choices. In fact, a change in food colour can lead to a rejection by consumers [28]. In Table 4, the colour coordinates for the shell, hilum, skin, and kernel of the hazelnuts studied are presented. Regarding the lightness of the shell (L* value), the variety with the highest value was Butler (49.34 ± 2.71) (Table 4), indicating that the shell of those hazelnuts were clear when compared to the others. Conversely, the blackest varieties were of Grada de Viseu (42.56 ± 3.05) and Negreta (45.29 ± 2.04). As for the values of a* on the shell, it was observed that for all varieties, the predominant colour was red with two distinct groups: one with a greater intensity of red (varieties Tonda de Giffoni, Negreta, Grada de Viseu, Butler, and Longa de Espanha) and the other with a lesser intensity (varieties Butler and Segorbe). In the case of the yellowness of the shell (b*), there were more differences found among the varieties and Butler had the highest value (31.77 ± 2.01).

Table 4. Colour coordinates (mean ± standard deviation).

		Grada de Viseu	Tonda de Giffoni	Segorbe	Longa de Espanha	Butler	Gunslebert	Negreta
Shell [1]	L*	42.56 ± 3.05 a	46.43 ± 2.19 bc	46.73 ± 2.47 c	45.65 ± 3.03 bc	49.34 ± 2.71 d	46.46 ± 2.04 bc	45.29 ± 2.04 b
	a*	19.11 ± 2.49 b	19.51 ± 2.28 b	16.33 ± 2.03 a	18.31 ± 2.48 b	18.61 ± 1.64 b	16.64 ± 3.26 a	19.23 ± 2.62 b
	b*	22.61 ± 3.78 a	28.43 ± 3.43 c	24.52 ± 3.29 ab	28.54 ± 4.25 c	31.77 ± 2.01 d	26.24 ± 4.58 b	24.61 ± 3.93 b
Hilum [1]	L*	49.79 ± 2.97 a	54.01 ± 6.28 b	50.28 ± 2.38 a	50.49 ± 3.10 a	48.99 ± 2.81 a	48.00 ± 6.95 a	50.49 ± 5.58 a
	a*	10.60 ± 1.35 b	9.37 ± 1.53 a	9.36 ± 1.14 a	9.18 ± 3.10 a	12.59 ± 1.66 c	8.75 ± 1.35 a	8.75 ± 1.40 a
	b*	23.02 ± 1.71 d	23.19 ± 2.75 d	21.22 ± 1.61 c	19.98 ± 1.64 b	24.71 ± 2.39 e	18.59 ± 3.38 a	20.67 ± 1.50 bc
Skin [1]	L*	43.75 ± 4.08 a	50.21 ± 5.07 d	48.76 ± 3.25 cd	46.36 ± 4.70 b	49.84 ± 3.62 d	49.35 ± 3.53 cd	47.46 ± 3.63 bc
	a*	15.56 ± 1.33 b	14.40 ± 1.32 a	15.55 ± 1.65 b	15.55 ± 1.54 b	17.08 ± 1.30 d	14.73 ± 1.33 a	15.83 ± 1.23 b
	b*	23.65 ± 2.17 a	25.00 ± 2.11 a	25.37 ± 2.66 a	25.59 ± 2.47 a	26.98 ± 1.84 a	25.26 ± 1.38 a	26.63 ± 1.84 a
Kernel [1]	L*	72.35 ± 4.27 cd	75.00 ± 5.26 ef	69.71 ± 5.21 bc	64.94 ± 6.48 a	68.78 ± 3.34 b	74.08 ± 4.85 de	77.50 ± 4.04 f
	a*	2.45 ± 0.94 b	2.30 ± 1.23 ab	2.73 ± 1.24 b	3.82 ± 1.78 c	2.91 ± 0.74 b	2.29 ± 0.97 ab	1.73 ± 0.83 a
	b*	25.24 ± 2.28 ab	24.59 ± 3.21 a	25.67 ± 2.04 ab	26.18 ± 2.32 b	29.72 ± 3.79 c	25.43 ± 2.50 ab	25.29 ± 1.98 ab

[1] Mean values in the same row with the same letter are not statistically different ($p < 0.05$).

As presented in Table 4, the hilum had lower values for a* when compared to the shell, indicating that the hilum is less red. In general, there were no differences among the varieties regarding the values of L* with the exception of Tonda de Giffoni that was clearer when compared to the others. As for the values of a* on the hilum, the variety with the highest value was Butler (12.59 ± 1.66), followed by Grada de Viseu (10.60 ± 1.35). Regarding the coordinate b*, it was also possible to distinguish two varieties, namely Grada de Viseu with the lowest value (18.59 ± 3.38) and Butler with the highest (24.71 ± 2.39).

Regarding the skin (Table 4), Grada de Viseu was the variety with blackest skin (43.75 ± 4.08) and both Tonda de Giffoni and Butler were the varieties with the lightest skin (50.21 ± 5.07 and 49.84 ± 3.62, respectively). Regarding the a* coordinate, the values ranged between 14.40 ± 1.32 (Tonda de Giffoni) and 17.08 ± 1.30 (Butler). The ANOVA test did not show significant differences among the samples for the coordinate b* (p = 0.418).

As presented in Table 4, the values of the coordinates L* and a* of the kernel are highly different from those of the shell and the skin. L* presented higher values (between 64.94 ± 6.48 and 77.50 ± 4.04 for Longa de Espanha and Negreta, respectively), indicating that the hazelnut kernels are whiter. For all samples, the values of a* were lower than the ones obtained for the shell, hilum, and skin, indicating that in the kernel, the red colour is less intense. The highest value was 3.82 ± 1.78 for the Longa de Espanha sample. The values of b* were all positive, indicating the presence of the yellow colour. Guiné et al. [29] reported a value for b* equal to 31.57, thus indicating that the kernels they studied were slightly more yellow than those in the present study. In another study by Mexis and Kontominas [30], colour parameters were found for hazelnuts of 30.45 for lightness (L*), 6.84 for redness (a*), and 8.70 for yellowness (b*), notably values that are considerably lower than those found in the present study. Lopes et al. [4] also evaluated the colour of the varieties present in this study with the exception of Longa de Espanha and observed different values than those presented in Table 4. These differences are perhaps due to the fact that hazelnuts' properties are dependent on the year and place of harvest, among other factors [31].

3.1.4. Texture Characteristics

The textural characteristics evaluated were hardness and friability. Hardness is related to the mechanical strength necessary to cause the crushing of a product [3] that in the case of hazelnuts is linked to the mechanical strength necessary to crush the outer shell, an important consideration as it ensures the physical integrity of the product, as well as ensures its ability to support the mechanical stress involved in the process of packing and transportation. However, if hardness is too high, it may represent a difficulty for industrial workers who need to remove the shell to process the hazelnuts' kernels. The friability measures the tendency of the products to fracture [32,33]. Table 5 presents the mean values obtained for the hazelnuts' textural properties. The results demonstrated that the hardness of the shell varied considerably according to variety, with Grada de Viseu having a very high value of hardness (389.91 ± 94.31 N) and Longa de Espanha having a lower value (82.56 ± 22.22 N). The hardness of the core also exhibited very high variability from 11.94 ± 3.20 N (Segorbe) to 79.00 ± 32.17 N (Gunslebert). The different values between the samples indicates that the forces necessary to cause the crushing of the kernel and to perform the hazelnut cut are dependent on the variety. In a study performed by Ghirardello et al. [34], it was found that the rupture force for hazelnut kernels was 91.83 ± 20.91 N, indicating that those hazelnut fruits were harder than the samples evaluated in this study. Thus, these results could be very important for producers and the food industry sector.

Table 5. Hazelnuts' textural properties (mean ± standard deviation).

Sample	Hardness of the Shell [1] (N)	Hardness of the Core [1] (N)	Friability of the Core [1] (mm)
Grada de Viseu	389.91 ± 94.31 [d]	68.15 ± 15.23 [bc]	5.15 ± 1.45 [ab]
Tonda de Giffoni	292.87 ± 65.26 [c]	56.53 ± 11.12 [b]	4.42 ± 1.25 [a]
Segorbe	201.85 ± 46.97 [b]	11.94 ± 3.20 [a]	11.21 ± 2.75 [c]
Longa de Espanha	82.56 ± 22.22 [a]	15.45 ± 3.70 [a]	10.50 ± 2.70 [c]
Butler	189.85 ± 53.42 [b]	15.11 ± 4.10 [a]	11.09 ± 3.13 [c]
Gunslebert	235.42 ± 56.21 [b]	79.00 ± 32.17 [c]	7.27 ± 4.69 [b]
Negreta	293.28 ± 47.41 [c]	57.39 ± 14.42 [b]	5.60 ± 3.73 [ab]
p-value	<0.05	<0.05	<0.05

[1] Mean values in the same column with the same letter are not statistically different ($p < 0.05$).

Regarding the friability of the core (Table 5), fruits of the varieties Segorbe, Longa de Espanha, and Butler were less susceptible to fracture (supporting a higher crushing distance before fracture: 11.21 ± 2.75 mm, 10.50 ± 2.70 mm, and 11.09 ± 3.13 mm, respectively), while the variety Tonda de Giffoni exhibited the lowest value of friability and therefore can be fractured more easily (4.42 ± 1.25 mm). This could be also important when the main objective of producers and the food industry is to maintain the whole core.

3.2. Chemical Properties

The results obtained for the chemical analyses of the hazelnuts' fruits under study are presented in Table 6. Moisture content varied from 4.77 ± 0.27 g/100 g (Negreta) to 9.78 ± 1.32 g/100 g (Longa de Espanha). According to Silva et al. [35], hazelnut moisture content in the natural state is about 4% to 6%. As can be observed, some of the analysed samples presented values higher than the above-mentioned limit that can compromise the storage time and quality of the fruits. The ANOVA test showed statistically significant differences among the varieties ($p < 0.05$) and this could be due to the intrinsic differences among the varieties, among drying conditions, and handling techniques.

Table 6. Hazelnuts' chemical properties (mean ± standard deviation as determined for the kernel).

Sample	Moisture [1] (g/100 g)	Water Activity [1]	Fat [1] (g/100 g)	Ash [1] (g/100 g)	Fibre [1] (g/100 g)	Protein [1] (g/100 g)	Induction Period (h)
Grada V.	7.18 ± 0.38 ^{bc}	0.73 ± 0.01 ^{c}	46.04 ± 1.53 ^{a}	2.59 ± 0.14 ^{a}	10.58 ± 0.77 ^{c}	17.56 ± 0.40 ^{e}	23.76 ± 2.16 ^{b}
Tonda G.	5.98 ± 0.15 ^{ab}	0.70 ± 0.01 ^{b}	72.29 ± 4.16 ^{b}	2.45 ± 0.18 ^{a}	11.53 ± 0.82 ^{c}	12.15 ± 0.27 ^{b}	24.21 ± 4.74 ^{b}
Segorbe	8.80 ± 0.38 ^{d}	0.80 ± 0.03 ^{f}	66.13 ± 1.75 ^{b}	3.08 ± 0.08 ^{b}	n.d. [2]	14.09 ± 0.34 ^{de}	21.58 ± 2.25 ^{b}
Longa E.	9.78 ± 1.32 ^{d}	0.77 ± 0.01 ^{d}	72.50 ± 3.79 ^{b}	2.61 ± 0.15 ^{a}	8.37 ± 0.34 ^{b}	17.43 ± 0.53 ^{c}	20.96 ± 1.70 ^{ab}
Butler	8.45 ± 0.21 ^{cd}	0.80 ± 0.01 ^{e}	69.07 ± 4.28 ^{b}	3.10 ± 0.05 ^{b}	11.11 ± 0.22 ^{c}	15.43 ± 0.75 ^{d}	22.47 ± 1.76 ^{b}
Gunslebert	5.67 ± 0.31 ^{ab}	0.60 ± 0.01 ^{a}	71.61 ± 2.50 ^{b}	3.80 ± 0.23 ^{c}	5.75 ± 0.41 ^{a}	11.41 ± 0.28 ^{b}	17.74 ± 0.21 ^{ab}
Negreta	4.77 ± 0.27 ^{a}	0.59 ± 0.01 ^{a}	70.06 ± 0.18 ^{b}	3.05 ± 0.14 ^{b}	5.30 ± 0.46 ^{a}	10.02 ± 0.27 ^{a}	14.02 ± 0.39 ^{a}
p-value	<0.05	<0.05	<0.05	<0.05	<0.05	<0.05	<0.05

[1] Mean values in the same column with the same letter are not statistically different ($p < 0.05$). [2] Not determined.

The water activity values ranged from 0.59 ± 0.01 (Negreta sample) to 0.80 ± 0.01 and 0.80 ± 0.03 (Butler and Segorbe samples, respectively). It is considered that the limit for fungal activity corresponds to a value of water activity equal to 0.62, indicating that with higher values, all microbial activity ceases [36]. Nevertheless, in the evaluated samples, that upper limit was exceeded and therefore some samples might be more susceptible to microbial or mould deterioration.

It is important to highlight that fat was the major component for all varieties, ranging from 46.04 ± 1.53 (Grada de Viseu) to 72.50 ± 3.79 (Tonda de Giffoni). However, considering the statistical analysis, the only result for fat that was significantly different is the result presented for Grada de Viseu with a low value, indicating that the other varieties presented have a similar fat content. These results are similar to those obtained by Oliveira et al. [5]. According to previous studies, hazelnut kernel composition is composed of 60% fat, 15% crude proteins, 4% ash, and 4% moisture on average, and these values are not very different from those obtained in the present study (Table 6) [5,37].

The ash, fibre, and protein contents showed significant differences between samples ($p < 0.05$). Ash was highest for the Gunslebert sample (3.80 ± 0.23 g/100 g) and lowest for the Tonda de Giffoni sample (2.45 ± 0.18 g/100 g). Oliveira et al. [5] found higher values of ash for hazelnuts in the range of 4.2–5.2 g/100 g. Moreover, the results illustrated that there are three different groups in terms of the ash content: the first is for the varieties Tonda de Giffoni, Grada de Viseu, and Longa de Espanha (2.45 ± 0.18, 2.59 ± 0.14, and 2.61 ± 0.15 g/100 g, respectively); the second for Negreta, Segorbe, and Butler (3.05 ± 0.14, 3.08 ± 0.08, and 3.10 ± 0.05 g/100 g, respectively); and the third for the variety Gunslebert (3.80 ± 0.23 g/100 g). Fibre varied from 5.30 ± 0.46 g/100 g (Negreta) and 5.75 ± 0.41 g/100 g (Gunslebert) to 11.53 ± 0.82 g/100 g (Tonda de Giffoni), 11.11 ± 0.22 g/100 g (Butler), and 10.58 ± 0.77 g/100 g (Grada de Viseu). According to

Tunçil [38], hazelnut is rich in dietary fibre, recognized for its potential to improve bowel function. In his work, the author reported a fibre content value of 17.78 g/100 g for a natural hazelnut, considering this lowered as hazelnuts were roasted. He also reported a very high value of fibre in the hazelnut skin (69.78 g/100 g). Regarding the protein content, it varied from 10.02 ± 0.27 g/100 g (Negreta) to 17.56 ± 0.40 g/100 g (Grada de Viseu). Oliveira et al. [5] reported values of protein in the range of 14.8–15.7 g/100 g and hence some of the samples evaluated in this work had higher values while others had a lower protein content.

It is very important to assess the oxidation stability of a sample because lipid oxidation is one of the most critical factors that affects the shelf-life and preservation capacity of food products [12,39]. The results of the fat oxidation test (Rancimat) presented in Table 6 demonstrated that the variety with the highest induction time was Tonda de Giffoni (24.21 ± 4.74 h), indicating a higher oxidation stability when compared to the other varieties. Again, there were some statistically significant differences found among the varieties under study.

3.3. Specific Extinction Coefficients

The specific extinction coefficients are used as indicators of oil oxidation, providing information about its quality and preservation state [40]. The parameter K_{232} is indicator of primary oxidation during oxidation processes and it is correlated with the formation of conjugated dienes of polyunsaturated fatty acids [37,41–44]. Conversely, K_{270} indicates the level of conjugated trienes that is representative of the secondary oxidation [45]. Furthermore, the resulting products of the secondary oxidation (aldehydes and ketones) also absorbs at wavelengths of 262, 268, and 274 nm [46].

Table 7 presents the results for the specific extinction coefficients at 232, 264, and 268 e 272 nm. The values for K_{232} varied between 1.66 ± 0.06 (Negreta) and 4.06 ± 0.24 (Segorbe), indicating that Negreta exhibited a lower presence of primary oxidation products than the other hazelnut varieties. Furthermore, significant differences were found among the varieties under study for all the determined parameters. The results also demonstrated that in all cases the values for Segorbe were considered higher and were statistically different from the other samples, indicating that this variety had a higher degree of oxidation that translates into a worse state of conservation. In the study of Pannico et al. [45], it was stated that K_{232} is considered the most important lipid oxidation parameter and that values higher than 2 are attributed to hazelnuts with taste defects, while values above 2.5 are considered rancid. According to Karoui et al. [47], an inappropriate storage of fruits or incorrect procedures of oil extraction usually correspond to an increase of K_{232}, while K_{270} increases when the oil is not fresh and results are from a previous harvest.

Table 7. Hazelnuts' specific extinction coefficients at 232, 264, and 268 e 272 nm (mean ± standard deviation).

Sample	K_{232} [1]	K_{264} [1]	K_{268} [1]	K_{272} [1]
Grada de Viseu	2.09 ± 0.01 [b]	0.09 ± 0.00 [ab]	0.09 ± 0.00 [ab]	0.09 ± 0.00 [ab]
Tonda de Giffoni	2.16 ± 0.14 [b]	0.07 ± 0.01 [a]	0.07 ± 0.01 [a]	0.07 ± 0.01 [a]
Segorbe	4.06 ± 0.24 [d]	0.22 ± 0.02 [d]	0.22 ± 0.02 [d]	0.21 ± 0.02 [d]
Longa de Espanha	3.23 ± 0.06 [c]	0.17 ± 0.00 [c]	0.17 ± 0.00 [c]	0.17 ± 0.00 [c]
Butler	2.36 ± 0.06 [b]	0.11 ± 0.00 [b]	0.12 ± 0.00 [b]	0.13 ± 0.03 [bc]
Gunslebert	2.42 ± 0.27 [b]	0.08 ± 0.00 [ab]	0.08 ± 0.00 [ab]	0.08 ± 0.00 [ab]
Negreta	1.66 ± 0.06 [a]	0.10 ± 0.04 [ab]	0.10 ± 0.04 [ab]	0.10 ± 0.04 [ab]
p-value	<0.05	<0.05	<0.05	<0.05

[1] Mean values in the same column with the same letter are not statistically different ($p < 0.05$).

3.4. Discriminant Function Analysis

The discriminant function analysis was performed for variety as dependent variable and considering as independent variables all the properties analysed in the hazelnuts. The results illustrated that only some variables had a significant Wilks' lambda according to the p-value of the F test ($p < 0.05$): namely L* of the shell, height, width, compression ratio, kernel height, kernel width, kernel thickness, kernel shape ratio, true density of the whole fruit with the shell, true density of the whole fruit without the shell, apparent density of the whole fruit with the shell, apparent density of the whole fruit without the shell, moisture content, ash, fat, fibre, protein, water activity, and specific extinction coefficient at 232 nm (K_{232}). The variables with a lower value of Wilks' lambda and therefore those more important for the discriminant analysis were protein ($\lambda = 0.007$) and water activity ($\lambda = 0.010$). The results for the correlation matrix indicated some very high values of the correlations including three values equal to or higher than 0.990. The results of discriminant function analysis are shown in Table 8 and they correspond to four functions. The first function with the highest eigenvalue explains 65.2% of variance, followed by F2 explaining 29.1%, which indicates these two functions account for most of the variance explained. Therefore, the other two functions are residual. Nevertheless, all four functions are significant ($p < 0.05$). The value of the Wilks' lambda is considerably lower for F1 that confirms its bet discriminant ability, although F2 also has a very low value of lambda. The canonical correlation coefficients that measure the strength of the association between each function and the dependent variable (VARIETY) are very high for all four functions, although notably highest for F1 and F2 as compared to F3 or F4.

Table 8. Results of discriminant function analysis.

Function	Eigenvalue	% Variance Explained	Canonical Correlation	Wilks' Lambda	p-Value
F1	601.7	65.2	0.999	1.31×10^{-8}	<0.05
F2	269.2	29.1	0.998	7.92×10^{-6}	<0.05
F3	43.08	4.7	0.989	0.002	<0.05
F4	9.600	1.0	0.952	0.094	<0.05

From the 43 independent variables used, only five were considered in the analysis as they had greater discriminant ability (water activity, protein, ash, b*of the shell, and apparent density with the shell). According to the standardized canonical discriminant function coefficients in which the largest absolute values correspond to variables with greater discriminating capacity, variable b*of the shell demonstrated greater discriminant ability in function F1 (4.821) and protein was more discriminating in F2 and F3 (0.874 and 0.863, respectively), while for function F4 the most discriminating variable is ash (1.079).

The results of the structure matrix (Table 9) show that the greater discriminant ability in functions F1 and F3 is provided by the variable of apparent density with the shell (0.199 and 0.546, respectively), while in F2 it is provided by the variable of protein (0.680) and in F4 it is provided by the variable of ash (0.986).

Table 9. Structure matrix accounting only for the five variables included in the analysis.

Variables	F1	F2	F3	F4
Water activity	0.143	0.546 *	−0.362	−0.382
Protein	0.028	0.680 *	0.479	−0.318
Ash	−0.039	−0.096	0.127	0.986 *
b*of the shell	−0.007	0.002	−0.014	0.124 *
Apparent density with the shell	0.199	−0.280	0.546 *	−0.515

* Largest absolute correlation between each variable and any discriminant function.

Figure 1 shows the distribution of the different variety groups according to the most relevant discriminant functions (F1 and F2). These results indicate that varieties Tonda de Giffoni, Longa de Espanha, and Segorbe are more similar considering the proximity of their centroids, while varieties Gunslebert, Negreta, Grada de Viseu, and Butler are different from all others.

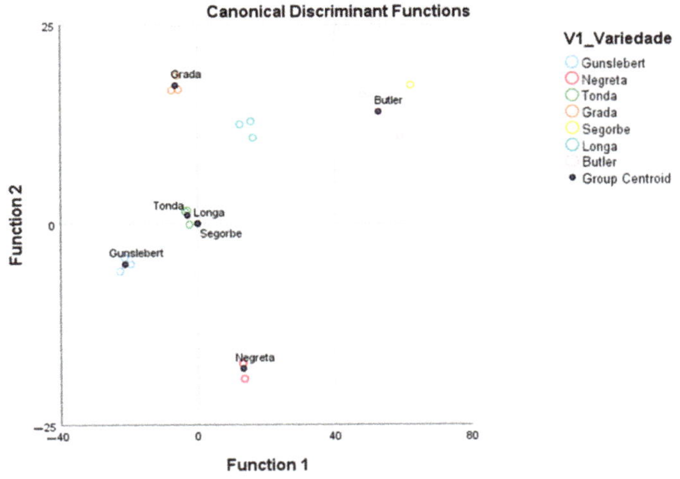

Figure 1. Discrimination of the seven varieties studied according to the two most important functions: F1 and F2.

The model includes four equations built on the basis of the canonical discriminant function coefficients as follows:

$$F1 = -399.6 + 395.7 \times aw - 4.2 \times Prt + 11.6 \times Ash + 345.6 \times DaS - 0.99 \times bS \quad (4)$$

$$F2 = -60.7 + 60.4 \times aw + 2.8 \times Prt + 1.6 \times Ash - 51.4 \times DaS - 0.06 \times bS \quad (5)$$

$$F3 = -0.123 - 85.2 \times aw + 2.8 \times Prt + 1.7 \times Ash + 51.2 \times DaS - 0.12 \times bS \quad (6)$$

$$F4 = -38.9 + 32.3 \times aw - 0.25 \times Prt + 6.1 \times Ash + 2.1 \times DaS + 0.06 \times bS \quad (7)$$

where aw is water activity, Prt is protein, DaS is the apparent density with the shell, and bS is the b* of the shell.

4. Conclusions

This study provided results that allowed for the comparison different hazelnut varieties. Regarding the biometric characteristics, it was concluded that Gunslebert had the heavier fruits while Grada de Viseu had the heavier kernels. In addition, Gunslebert was more elongated and Butler was more rounded in the equatorial zone. For all varieties, the apparent density was lower than the true density.

The results also established the expectable ranges for each colour parameters in the shell, hilum, skin, and kernel, with Butler exhibiting a clearer shell and Tonda de Giffoni exhibiting a clearer hilum and skin. In contrast, Negreta exhibited a clearer kernel. As for texture, Grada de Viseu had a harder shell while Gunslebert had a harder core. Additionally, varieties Segorbe, Longa de Espanha, and Butler were more resistant to fracture.

Regarding the chemical properties, as expected, the major component was fat, followed by protein, and then fibre. Very different values were found for the chemical components according to variety. Negreta exhibited the lowest amount of primary oxidation products, corresponding to a lower K_{232} value. The values of moisture content were over the recommended values in some cases (varieties Grada de Viseu, Segorbe, Longa de Espanha, and Butler) and the desirable values for water activity were surpassed in some samples (Grada de Viseu, Tonda de Giffoni, Segorbe, Longa de Espanha, and Butler). This indicates that these varieties are more susceptible to mould and microbial deterioration that represents a problem in the hazelnuts' conservation and storage.

The discriminant analysis illustrated that the variables of protein and water activity were the most important parameters. The distribution of the different variety groups according to the most relevant discriminant functions revealed that varieties Tonda de Giffoni, Longa de Espanha, and Segorbe were more similar, while varieties Gunslebert, Negreta, Grada de Viseu, and Butler were different from all others.

These findings are important for Portuguese producers, retailers, and industry sectors because they are useful to better understand the characteristics of different hazelnuts' varieties, they can be a tool to facilitate the choices for all players in the hazelnut sector, and they can provide context in view of the particular future usages intended for hazelnuts.

Author Contributions: Conceptualization, P.M.R.C. and R.P.F.G.; methodology, P.M.R.C. and R.P.F.G.; validation, P.M.R.C., R.P.F.G. and A.C.F.; formal analysis, A.C.F., E.R., C.R., H.M., R.G. and R.P.F.G.; investigation, A.C.F., C.R., H.M. and R.G.; resources, P.M.R.C. and A.L.; data curation, P.M.R.C., R.P.F.G. and A.C.F.; writing—original draft preparation, A.C.F. and R.P.F.G.; writing—review and editing, P.M.R.C. and R.P.F.G.; supervision, P.M.R.C.; project administration, P.M.R.C.; funding acquisition, P.M.R.C. All authors have read and agreed to the published version of the manuscript.

Funding: This study was supported by the National Funds through the FCT-Foundation for Science and Technology, I.P., within the scope of the projects Ref[a] UIDB/00681/2020 and UID/AGR/00690/2019.

Institutional Review Board Statement: Not applicable.

Informed Consent Statement: Not applicable.

Data Availability Statement: The data are available from the corresponding author upon request.

Acknowledgments: The authors would like to thank the CERNAS Research Centre, the Polytechnic Institute of Viseu, and the CIMO (Instituto Politécnico de Bragança) for their support. Thanks also to PDR2020 Program through project ValNuts (PDR2020-101-030759 and PDR2020-101-030756).

Conflicts of Interest: The authors declare no conflict of interest.

References

1. Petriccione, M.; Ciarmiello, L.F.; Boccacci, P.; De Luca, A.; Piccirillo, P. Evaluation of 'Tonda Di Giffoni' Hazelnut (*Corylus avellana* L.). *Clones. Sci. Hortic.* **2010**, *124*, 153–158. [CrossRef]
2. Ramalhosa, E.; Delgado, T.; Estevinho, L.; Pereira, J.A. Chapter 73—Hazelnut (*Corylus avellana* L.) Cultivars and Antimicrobial Activity. In *Nuts and Seeds in Health and Disease Prevention*; Preedy, V.R., Watson, R.R., Patel, V.B., Eds.; Academic Press: San Diego, CA, USA, 2011; pp. 627–636. ISBN 978-0-12-375688-6.
3. Guiné, R.P.F.; Almeida, C.F.F.; Correia, P.M.R.; Mendes, M. Modelling the Influence of Origin, Packing and Storage on Water Activity, Colour and Texture of Almonds, Hazelnuts and Walnuts Using Artificial Neural Networks. *Food Bioprocess Technol.* **2015**, *8*, 1113–1125. [CrossRef]
4. Lopes, A.; Matos, A.; Guiné, R. Evaluation of morphological and physical characteristics of hazelnut varieties. *Millenium-J. Educ. Technol. Health* **2016**, 13–24. [CrossRef]
5. Oliveira, I.; Sousa, A.; Morais, J.S.; Ferreira, I.C.F.R.; Bento, A.; Estevinho, L.; Pereira, J.A. Chemical Composition, and Antioxidant and Antimicrobial Activities of Three Hazelnut (*Corylus avellana* L.) Cultivars. *Food Chem. Toxicol.* **2008**, *46*, 1801–1807. [CrossRef]
6. Perna, S.; Giacosa, A.; Bonitta, G.; Bologna, C.; Isu, A.; Guido, D.; Rondanelli, M. Effects of Hazelnut Consumption on Blood Lipids and Body Weight: A Systematic Review and Bayesian Meta-Analysis. *Nutrients* **2016**, *8*, 747. [CrossRef]
7. Cabo, P.; Ramalhosa, E.; Bento, A. *Frutos Secos de Casca Rija Em Portugal: Produção e Mercados*; Sociedade de Ciências Agrárias de Portugal: Lisboa, Portugal, 2019; pp. 23–24.
8. Instituto Nacional de Estatística. *Estatísticas Agrícolas: 2018*; INE: Lisboa, Portugal, 2019.
9. Ferrão, A.C.; Guiné, R.; Rodrigues, M.; Droga, R.; Correia, P. Post-Harvest Characterization of the Hazelnut Sector. *Rev. Mill.* **2020**, *6e*, 11–20. [CrossRef]
10. Di Caro, D.; Liguori, C.; Pietrosanto, A.; Sommella, P. Quality Assessment of the Inshell Hazelnuts Based on TD-NMR Analysis. *IEEE Trans. Instrum. Meas.* **2020**, *69*, 3770–3779. [CrossRef]
11. Rovira, M.; Hermoso, J.F.; Romero, A.J. Performance of Hazelnut Cultivars from Oregon, Italy, and Spain, in Northeastern Spain. *HortTechnology* **2017**, *27*, 631–638. [CrossRef]
12. Guiné, R.; Rodrigues, C.; Correia, P.; Ramalhosa, E. Evaluation of Some Physical and Chemical Properties of Hazelnuts. In Proceedings of the FABE 2019: Food and Biosystems Engineering Conference, Crete Island, Greece, 30 May–2 June 2019.
13. Costell, E.; Tárrega, A.; Bayarri, S. Food Acceptance: The Role of Consumer Perception and Attitudes. *Chemosens. Percept.* **2009**, *3*, 42–50. [CrossRef]

14. Murray, J.M.; Baxter, I.A. SENSORY EVALUATION | Food Acceptability and Sensory Evaluation. In *Encyclopedia of Food Sciences and Nutrition*, 2nd ed.; Caballero, B., Ed.; Academic Press: Oxford, UK, 2003; pp. 5130–5136; ISBN 978-0-12-227055-0.
15. Guiné, R.P.F.; Almeida, C.F.F.; Correia, P.M.R. Influence of Packaging and Storage on Some Properties of Hazelnuts. *Food Measure* **2015**, *9*, 11–19. [CrossRef]
16. Silva, A.P.; Santos, F.A.; Santos, A.S.; Sousa, V.; Lopes, A.; Mota, B.S.; Leme, P.; Carvalho, J.; Borges, O.; Ribeiro, R.M.; et al. *Variedades de Aveleira*; Projecto AGRO 162; 2004. Available online: https://www.drapc.gov.pt/base/documentos/variedades_aveleira.pdf (accessed on 22 July 2021).
17. Gómez-Polo, C.; Muñoz, M.P.; Lorenzo Luengo, M.C.; Vicente, P.; Galindo, P.; Martín Casado, A.M. Comparison of the CIELab and CIEDE2000 Color Difference Formulas. *J. Prosthet. Dent.* **2016**, *115*, 65–70. [CrossRef]
18. Correia, P.; Rodrigues, C.; Filipe, A.; Guiné, R. Evaluation of the Biometric Characteristics of Hazelnuts. In Proceedings of the FABE 2019: Food and Biosystems Engineering Conference, Crete Island, Greece, 30 May–2 June 2019.
19. Ozturk, S.C.; Ozturk, S.E.; Celik, I.; Stampar, F.; Veberic, R.; Doganlar, S.; Solar, A.; Frary, A. Molecular Genetic Diversity and Association Mapping of Nut and Kernel Traits in Slovenian Hazelnut (*Corylus avellana*) Germplasm. *Tree Genet. Genomes* **2017**, *13*, 16. [CrossRef]
20. AOAC. *Official Methods of Analysis of AOAC International*; Association of Official Analytical Chemists: Rockville, MD, USA, 2019.
21. Metrohm, A.G. *Oxidation Stability of Oils and Fats-Rancimat Method*; Application Bulletin 204/2 e; Metrohm International: Herisau, Switzerland, 2018.
22. Regulation (EU) 2015/1833. *Characteristics of Olive Oil and Olive-Residue Oil and on the Relevant Methods of Analysis*; European Union: Brussels, Belgium, 2015.
23. Król, K.; Gantner, M.; Piotrowska, A. Morphological Traits, Kernel Composition and Sensory Evaluation of Hazelnut (*Corylus avellana* L.) Cultivars Grown in Poland. *Agronomy* **2019**, *9*, 703. [CrossRef]
24. Miletić, R.; Žikić, M.; Mitić, N.; Nikolić, R. Pomological Characteristics of Superior Selections of European Filbert-*C. Avellana* L.- and Turkish Hazel-*C. Colurna* L.-. *Genetika* **2005**, *37*, 103–111. [CrossRef]
25. Mohammadzedeh, M.; Fattahi, R.; Zamani, Z.; Khadivi-Khub, A. Genetic Identity and Relationships of Hazelnut (*Corylus avellana* L.) Landraces as Revealed by Morphological Characteristics and Molecular Markers. *Sci. Hortic.* **2014**, *167*, 17–26. [CrossRef]
26. Yao, Q.; Mehlenbacher, S. Distribution of Quantitative Traits in Hazelnut Progenies. *Acta Hortic.* **2001**, 143–162. [CrossRef]
27. Altuntas, E.; Erkol, M. Physical Properties of Shelled and Kernel Walnuts as Affected by the Moisture Content. *Czech. J. Food Sci.* **2010**, *28*, 547–556. [CrossRef]
28. Spence, C. On the Psychological Impact of Food Colour. *Flavour* **2015**, *4*, 21. [CrossRef]
29. Guiné, R.; Almeida, C.; Correia, P. Properties of Hazelnut as Influenced by Packaging and Storage. In Proceedings of the 9th Baltic Conference on Food Science and Technology, Jelgava, Latcia, 8–9 May 2014; p. 142.
30. Mexis, S.F.; Kontominas, M.G. Effect of γ-Irradiation on the Physicochemical and Sensory Properties of Hazelnuts (*Corylus Avellana* L.). *Radiat. Phys. Chem.* **2009**, *78*, 407–413. [CrossRef]
31. Barreira, J.C.M.; Casal, S.; Ferreira, I.C.F.R.; Peres, A.M.; Pereira, J.A.; Oliveira, M.B.P.P. Chemical Characterization of Chestnut Cultivars from Three Consecutive Years: Chemometrics and Contribution for Authentication. *Food Chem. Toxicol.* **2012**, *50*, 2311–2317. [CrossRef]
32. Almeida, C. *Efeitos Das Condições de Conservação Nas Características de Frutos Secos. Trabalho de Estágio*; Escola Superior Agrária de Viseu: Viseu, Portugal, 2013.
33. Gharibzahedi, S.M.T.; Mousavi, S.M.; Hamedi, M.; Khodaiyan, F. Comparative Analysis of New Persian Walnut Cultivars: Nut/Kernel Geometrical, Gravimetrical, Frictional and Mechanical Attributes and Kernel Chemical Composition. *Sci. Hortic.* **2012**, *135*, 202–209. [CrossRef]
34. Ghirardello, D.; Zeppa, G.; Rolle, L.; Gerbi, V.; Contessa, C.; Valentini, N.; Botta, R.; Griseri, G. Effect of Different Storage Conditions on Hazelnut Quality. *Acta Hortic.* **2014**, *1052*, 315–318. [CrossRef]
35. Silva, A.P.; Santos, F.A.; Santos, A.S.; Lopes, A.; Assunção, A.; Carvalho, J.; Borges, O.; Ribeiro, R.M.; Fernandes, S.T.; Santos, T. *A Avelã Na Sua Alimentação*; O Incremento da Produtividade da Aveleira em Portugal; João Azevedo Editor: Mirandela, Portugal, 2003.
36. Guiné, R.P.F. *Unit Operations for the Food Industry. Volume I: Thermal Processing & Nonconventional Technologies*; LAP Lambert Academic Publishing: Saarbrücken, Germany, 2013.
37. Pannico, A.; Cirillo, C.; Giaccone, M.; Scognamiglio, P.; Romano, R.; Caporaso, N.; Sacchi, R.; Basile, B. Fruit Position within the Canopy Affects Kernel Lipid Composition of Hazelnuts. *J. Sci. Food Agric.* **2017**, *97*, 4790–4799. [CrossRef] [PubMed]
38. Tunçil, Y.E. Dietary Fibre Profiles of Turkish Tombul Hazelnut (Corylus Avellana L.) and Hazelnut Skin. *Food Chem.* **2020**, *316*, 126338. [CrossRef]
39. Mateos, R.; Uceda, M.; Aguilera, M.P.; Escuderos, M.E.; Beltrán Maza, G. Relationship of Rancimat Method Values at Varying Temperatures for Virgin Olive Oils. *Eur. Food Res. Technol.* **2006**, *223*, 246–252. [CrossRef]
40. Malvis, A.; Šimon, P.; Dubaj, T.; Sládková, A.; Ház, A.; Jablonský, M.; Sekretár, S.; Schmidt, Š.; Kreps, F.; Burčová, Z.; et al. Determination of the Thermal Oxidation Stability and the Kinetic Parameters of Commercial Extra Virgin Olive Oils from Different Varieties. Available online: https://www.hindawi.com/journals/jchem/2019/4567973/ (accessed on 3 February 2021).
41. Casal, S.; Malheiro, R.; Sendas, A.; Oliveira, B.P.P.; Pereira, J.A. Olive Oil Stability under Deep-Frying Conditions. *Food Chem. Toxicol.* **2010**, *48*, 2972–2979. [CrossRef] [PubMed]

42. Kıralan, M.; Bayrak, A. Oxidative and Antiradical Stabilities of Two Important Virgin Olive Oils from Ayvalik and Memecik Olive Cultivars in Turkey. *Int. J. Food Prop.* **2013**, *16*, 649–657. [CrossRef]
43. Knezović, Z.; Trgo, M.; Stipišić, A.; Sutlović, D. Impact of metals from the environment on chemical changes in olive oil. In Proceedings of the 15th Ružička Days-Book of Abstracts, Vukovar, Croatia, 11–12 September 2014.
44. Salek, R.N.; Burešová, I.; Kráčmar, S.; Lorencová, E.; Zálešáková, L.; Dabash, V. Evaluation of Selected Physicochemical Parameters of Extra Virgin Olive Oil Commercialized in the Czech Market and Stored during a Period of 5 Months. *Potravin. Slovak J. Food Sci.* **2017**, *11*, 664–672. [CrossRef]
45. Pannico, A.; Schouten, R.E.; Basile, B.; Woltering, E.J.; Cirillo, C. Non-Destructive Detection of Flawed Hazelnut Kernels and Lipid Oxidation Assessment Using NIR Spectroscopy. *J. Food Eng.* **2015**, *160*, 42–48. [CrossRef]
46. Sinha, N.K. *Handbook of Food Products Manufacturing*; John Wiley & Sons: Hoboken, NJ, USA, 2007; ISBN 978-0-470-11354-7.
47. Karoui, I.J.; Ayari, J.; Ghazouani, N.; Abderrabba, M. Physicochemical and Biochemical Characterizations of Some Tunisian Seed Oils. *OCL* **2020**, *27*, 29. [CrossRef]

agronomy

MDPI

Article

Agri-Food Waste as a Method for Weed Control and Soil Amendment in Crops

Paula Lorenzo [1,*], Rosa Guilherme [2,3], Sara Barbosa [3,4], António J. D. Ferreira [3,4] and Cristina Galhano [1,3,4]

1 Centre for Functional Ecology (CFE)—Science for People & the Planet, Department of Life Sciences, University of Coimbra, 3000-456 Coimbra, Portugal; cicgalhano@esac.pt
2 Direção Regional de Agricultura e Pescas do Centro, Av. Fernão Magalhães, 465, 3000-177 Coimbra, Portugal; rosa.guilherme@drapc.gov.pt
3 CERNAS—Centro de Estudos de Recursos Naturais, Ambiente e Sociedade, Escola Superior Agrária de Coimbra, Instituto Politécnico de Coimbra, Bencanta, 3045-601 Coimbra, Portugal
4 Coimbra Agriculture School, Polytechnic of Coimbra, Bencanta, 3045-601 Coimbra, Portugal; sara_catarina_barbosa@hotmail.com (S.B.); aferreira@esac.pt (A.J.D.F.)
* Correspondence: paulalorenzo@uc.pt

Abstract: The continued and extensive use of synthetic herbicides to control weeds to maximize crop yield is no longer sustainable, as it results in negative impacts on the environment and human health. Innovative sustainable and resilient food production systems should preserve resources and environmental health by incorporating alternative natural herbicides, recycling waste, and favoring a circular economy. The present work assesses the value of different organic waste (*Urtica dioica* residues, *Vicia faba* pods, spent coffee grounds, and corn cobs) as bioherbicides and fertilizers in different seasons through pot and field two-year sequential experiments. Pot assays revealed that *V. faba* pods, spent coffee grounds, and corn cob waste showed the best inhibitory effect, which were subsequently evaluated in the Spring–Summer and Autumn crops. In the field, spent coffee grounds reduced the biomass of total naturally-emerged weeds and stimulated crop growth under scarce rainfall and warm days. However, its effect varied under different environmental conditions. Spent coffee grounds can partially control weeds in the field, which valorizes them as a bioherbicide and boosts sustainable agriculture.

Keywords: bioherbicides; corn cobs; organic manures; spent coffee grounds; *Vicia faba* pods

Citation: Lorenzo, P.; Guilherme, R.; Barbosa, S.; Ferreira, A.J.D.; Galhano, C. Agri-Food Waste as a Method for Weed Control and Soil Amendment in Crops. *Agronomy* **2022**, *12*, 1184. https://doi.org/10.3390/agronomy12051184

Academic Editor: Victor Galea

Received: 5 April 2022
Accepted: 12 May 2022
Published: 14 May 2022

1. Introduction

The widespread use of synthetic herbicides and fertilizers on agricultural lands to maximize yield and feed an increasing population has resulted in air, soil, and water pollution [1–3]. The overuse of herbicides has additionally resulted in an increase in herbicide-resistant weeds [4]. This reduces the effectiveness of the long-term use of herbicides, compromising crops yield [5]. Environmental contamination by synthetic pesticides results in human diseases [6,7] and to the depletion of non-human biota populations [6]. This has led to an increasing demand for organic products free of synthetic herbicides, able to provide sustainable and safe alternatives, leading to a rapid increase of organic farming, which contributes to environmental protection [8,9]. In the last ten years, the European Union's organic farming area increased by 70% [9]. Maintaining agricultural production based on synthetic agrochemicals has proven to be environmentally and socially unsustainable in the long term. As a result, both the United Nations, through its 2030 Agenda, and the European Union, with the European Green Deal, are deeply committed to ensuring the use of sustainable and resilient food production systems that preserve ecosystems for future generations. The European Commission aims to significantly and urgently reduce the dependence on pesticides and fertilizers by promoting a circular bioeconomy and integrated pest management to achieve sustainable food production [10]. Conducting sustainable agriculture involves abandoning old risky practices and adopting new, safer farming strategies that preserve the environment [11].

Herbicides can be partially replaced by natural phytotoxic compounds or organic mulches to control weeds [12–14]. Natural compounds are considered safer than synthetic herbicides, since they have shorter environmental half-lives and rarely contain halogenated atoms [15,16]. Similarly, renewable manures made with organic waste can be a source of nutrients for crops [17,18], reducing the use of synthetic fertilizers. Organic amendments also improve soil fertility and favor water retention [19,20]. Additionally, the use of biopesticides and biofertilizers contributes to the recovery and maintenance of soil quality and health, which are essential in agriculture [21].

Valuing waste by finding new potential uses for it, while reducing the use of synthetic agrochemicals in agriculture, embodies a holistic and sustainable alternative that prevents environmental contamination, provides new raw materials, and favors zero waste, by turning waste into by-products. This contributes towards the implementation of the circular economy strategy and the Waste Directive approved by the Council of the European Union on 30 May 2018 [22]. In Portugal, three types of common organic waste include spent coffee grounds (CG), *Vicia faba* L. (broad bean) pods, and corn cobs. Coffee is the most popular beverage in Portugal with a consumption average rate per capita of 4.15 kg/year[1] [23], providing a considerable daily amount of CG. However, this waste is marginally recycled as a mushroom substrate or adsorbent for cationic dyes in sewage water [24]. Biodiesel production from CG was also evaluated, but it does not seem to be worthwhile since the production costs outweigh the energy revenues [24]. *Vicia faba* is a popular crop in Portugal. Data from 2017 pointed out that 2892 tons of broad beans were produced in the country that year, with the fresh pods being thrown away to landfills. Corn (*Zea mays* L.) is the main cereal crop in Portugal for grain production [25]. Corn cob waste is generally burned [26]. Recent studies evaluated its potential as a raw building material for thermal insulation [26,27] and a source of xylan to produce nanoxylan particles with antileishmanial and antifungal activities [28].

Despite the abundance of CG, *V. faba* pods, and corn cob waste, their potential as new agricultural resources has not been broadly explored. Extensive literature has focused on CG properties and its potential uses [29,30]. For agriculture purposes, CG may be a nutrient source, since it increases the content of organic matter, carbon, total nitrogen, available potassium and phosphorous, and reduced C/N ratio in soils [31,32]. This waste also increases moisture in sandy-loam soil [33]. The direct application of CG produced contradictory results on crop yields; while low doses can slightly affect the plant growth [33,34], high doses suppress the growth of a range of different crops [32,33], although the inhibitory effect can be alleviated through composting [34]. Crop-growth inhibition by CG seems to be related to the presence of phytotoxic compounds such as caffeine, tannins, and chlorogenic acids [31,35]. However, this negative effect might be useful to control weeds in sustainable agriculture, a relevant issue that was scarcely explored for CG [31]. In the case of *V. faba*, the literature highlighted both its phytotoxic and fertilizing properties. Interestingly, fresh flowering *V. faba* plants incorporated into the soil reduced weed biomass in maize crops by 50% in the short-term, but increased crop and weed biomass after the phytotoxic effects disappeared [36]. For *V. faba* pod waste in particular, Al-Chammaa et al. [37] found that sorghum plants almost doubled their biomass in response to soil amendments with this type of green manure, with the N from green manure being effectively used (up to 45%) by sorghum. Pod-waste extracts are known to have phytotoxic activity on three nematode species under in vitro and in vivo conditions [38]. Corn cobs are a source of antioxidant phenolic compounds that could be useful in food and pharmacological industries [39,40]. However, their use for agricultural purposes has not yet been explored.

This study presents a combined strategy to reduce waste and the overuse of synthetic agrochemicals. Therefore, the objective of this study is to explore the potential herbicidal effect of different agri-food waste incorporated into the soil as tools to control emerging weeds in Spring–Summer and Autumn crops, and simultaneously to evaluate the waste effect on crops' biomass and soil quality. We hypothesize that these waste would contribute towards conducting more sustainable practices in agriculture by reducing the use of synthetic herbicides and fertilizers, and favor a circular economy by providing new uses for overlooked agri-food byproducts.

2. Materials and Methods

2.1. Evaluated Waste

The flowering plants of *Urtica dioica* L. were weeded at the Escola Superior Agrária de Coimbra farm (ESAC, Coimbra, Portugal; 40°13′02″ N, 8°26′56.2″ W) on 29 May 2019, allowed to dry in a shaded placed until 3 June 2019, cut into 4–7 cm pieces, and stored in a paper box before use.

The fruitful plants of *Vicia faba* L. were harvested from crops in central Portugal on 3 June 2019 and on 29 May 2020, and immediately processed in the FRIP company (Gafanha da Nazaré, Portugal) 2–3 days before setting up the experiments. Fresh-oxidized pods were used in the Spring–Summer experiments to test their effect, but overprocessing was avoided. However, dry pods were used in the Autumn experiments because broad bean is only cultivated during the Winter–Spring season at this latitude. Fresh-oxidized pods were dried at 65 °C for preservation and stored at dark-room temperature before use.

The spent coffee grounds were obtained from local coffee shops. In October 2019, fresh one-day spent coffee grounds were collected and immediately used in the Autumn pot experiment (2019) as fresh waste. The amount of coffee grounds needed for the field experiments, set in 2020, was much higher. However, the availability of spent coffee grounds was sporadic due to the COVID-19 pandemic restrictions, which kept local coffee shops closed. Therefore, we decided to collect spent coffee grounds when available, dry them at room temperature to prevent fungal contamination, and mix them before setting up the experiments.

The corn cobs were obtained from local farmers in September 2019 and 2020, ground using a standard garden grinder to provide 3–8 cm particles, and stored at room temperature before use.

2.2. Waste Extracts for Pot Experiments

In Spring–Summer 2019, dry *U. dioica* and fresh pods of *V. faba* were used to obtain water extracts at 20% mixing 20 g of waste with 100 mL of distilled water for 24 h, at room temperature. These were periodically shaken. The extracts were then filtered through a gauze and immediately used.

In Autumn 2019, water extracts using *U. dioica* or *V. faba* (20%, 20 g dry waste 100 mL^{-1} distilled water) were obtained as described for the Spring–Summer extracts.

2.3. Crop Species for Field Evaluation

To evaluate the waste effect on crops, several species were selected to ensure different taxa and crop-edible plant parts. *Lactuca sativa* L. and *Raphanus sativus* var. *niger* (Mill.) J. Kern. were chosen to represent leaf-edible and bulb-edible crops under Spring–Summer conditions, while *Brassica rapa* subsp. *nipposinica* (L. H. Bailey) Hanelt (leaf-edible) and *Beta vulgaris* L., sp. Pl. (bulb-edible) were chosen to represent crops under the Autumn conditions. The seedlings of those species were obtained from local suppliers before conducting the field experiments.

2.4. Experimental Set-Up

Two trials in two consecutive years were conducted. First, the herbicidal effect of different waste on weeds naturally emerging from agricultural soils was evaluated in pots in Spring–Summer and Autumn. Agri-food waste showed the best herbicidal results, which is the reason why they were subsequently assayed in field plots to test their effect on weeds, crops, and physico-chemical soil properties during the Spring–Summer and

Autumn seasons. To facilitate their potential use by farmers, fresh waste were primarily used when available to avoid later overprocessing. However, the availability of some agri-food waste depended on crop timing. To standardize the procedures, fresh agri-food waste were dried for preservation and used when needed.

2.4.1. Pot Trials

Experiment 1—Bioherbicidal Effect of Different Waste on Emerging Spring–Summer Weeds

The first pot experiment was established outdoors at ESAC farm (Coimbra, Portugal; 40°12′47,3″ N, 8°27′05,1″ W) from 6 June to 18 July 2019. This region is characterized by a warm temperate climate with a dry summer—Csb (Köppen–Geiger climate classification, [41]), with very scarce rainfall and an average medium temperature of 20 °C for the experimental period (Figure S1: meteorological data were obtained from the ESAC meteorological station). The area is devoted to organic production that is heavily infested by several weed species, mainly *Cyperus* spp. The soil characteristics at the experimental site were the following in average: $pH_{(H_2O)}$ = 6.5, 1.3% organic matter (OM), 84 mg P_2O_5 kg^{-1} extractable phosphorus (P), 122 mg K_2O kg^{-1} extractable potassium (K), and 0.13% total Kjeldahl nitrogen (N). Fifteen-liter pots (29 cm diameter, 24 cm high) were filled with agricultural soil (Ap horizon) collected from this location and watered with 2 L of tap water. The pots were randomly assigned to one of the treatments: dry *U. dioica* waste (hereafter U), fresh *V. faba* waste (hereafter F), extract of dry *U. dioica* waste (hereafter UE), extract of fresh *V. faba* waste (hereafter FE), control without any addition of waste or extract as a negative control (hereafter C), and Cyperal herbicide (Benfuresato 40%) as a positive control (hereafter H). The treatments were applied in the doses described in Table 1. The control pots received water from the Mondego River (pH 7.34; electrical conductivity 0.10 mS cm^{-1}; [42]). The treatments were replicated five times and the pots were randomly distributed. The doses for fresh (28 Mg ha^{-1}) and dry (9 Mg ha^{-1}) waste applied in the present study were similar to those for phytotoxic mulches or manures commonly used in weed control [14,43,44]. The dose for aqueous treatments was the quantity of waste extracts needed to moisten the soil, while avoiding percolation. U and F were incorporated into the first 10 cm soil layer. EU, EF, and H were homogeneously applied on the top of the soil. After treatment application, the pots were covered with plastic bags for 24 h to avoid premature treatment leachate by rain on June 6th (Figure S1). Then, the bags were removed, and weeds grew under natural conditions. The pots were watered with the same quantity of tap water two or three times a week, avoiding percolation.

Table 1. Doses for treatments tested in Experiment 1.

Treatment	Abbreviation	Dose
River water/no waste (negative control) (L ha^{-1})	C	45,418.8
Vicia faba pod extract (L ha^{-1})	FE	45,418.8
Urtica dioca extract (L ha^{-1})	UE	45,418.8
Vicia faba pod waste (Mg ha^{-1})	F	28
Urtica dioca waste (Mg ha^{-1})	U	9
Herbicide (positive control) (L ha^{-1})	H	3

At the end of the experiment, the weeds from each pot were uprooted, classified into monocotyledons, dicotyledons, and *Cyperus* spp., typically the most problematic weed, and dried at 65 °C until a constant weight was achieved. Determinations included the aboveground biomass (g m^{-2}) of each weed group and total weeds.

Experiment 2—Bioherbicidal Effect of Different Waste on Emerging Autumn Weeds

A second pot experiment was conducted in the same place, from 15 October to 26 November 2019. Environmental conditions throughout the experiment were characterized by average values of 6 mm for rainfall and 14 °C of medium temperature (Figure S1:

data collected at ESAC). Fifteen-liter pots were filled with the same agricultural soil as for Experiment 1 and randomly assigned to these treatments: dry U, dry F, UE, FE, fresh spent coffee grounds waste (CG), dry corn cob waste (CC), C, and H. The treatments were applied in the doses described in Table 2 and were replicated five times. The control treatment received water from the Mondego River.

Table 2. Doses for treatments tested in Experiment 2.

Treatment	Abbreviation	Dose
River water/no waste (negative control) (L ha^{-1})	C	45,418.8
Vicia faba pod extract (L ha^{-1})	FE	45,418.8
Urtica dioca extract (L ha^{-1})	UE	45,418.8
Vicia faba pod waste (Mg ha^{-1})	F	9
Urtica dioca waste (Mg ha^{-1})	U	9
Spent coffee grounds waste (Mg ha^{-1})	CG	28
Corn cob waste (Mg ha^{-1})	CC	9
Herbicide (positive control) (L ha^{-1})	H	3

Waste and extract treatments were applied as described in Experiment 1. At the beginning of the experiment, the pots were covered by plastic bags for 48 h to avoid premature treatments leachate by rain from 17 October to 19 October (Figure S1). Then, the bags were removed, and the weeds grew normally. The pots were naturally rain-fed.

At the end of the experiment, the weed biomass as indicated for Experiment 1 and soil samples from the first 15 cm of the top soil layer in each pot were collected. The soil samples were dried at 30 °C, sieved through a 2 mm sieve (Soil Mill Pulverizette 8, Fritsch, Idar-Oberstein, Germany), and analyzed following Magalhães et al. [45] to determine the pH, OM, extractable P, extractable K, and total N. Briefly, the soil pH was determined using 1:5 soil volume/distilled water volume ratio. Organic matter was estimated by oxidizing organic carbon in 2 g soil samples at 600 °C (model SC-144 DR, LECO, St. Joseph, Michigan, United States) and multiplying the organic carbon value by the 1.724 factor. Extractable P and K were extracted in 2 g soil samples using the Egnér–Riehm method. P was calorimetrically determined at 650 nm. K was quantified using an atomic absorption spectrophotometer (model Analyst 300 with software AA Winlab, Perkin Elmer, Waltham, Massachusetts, United States) based on a calibration curve with K_2O standard solutions. Total N was determined in 2.5 g soil samples using the Kjeldahl method [45].

2.4.2. Field Evaluation

Pot experiments 1 and 2 indicated that F, CG, and CC treatments had the best herbicidal effect on the biomass of the total weeds. Therefore, these treatments were selected to evaluate their herbicidal effect under field conditions.

Experiment 3—Evaluation of Agri-Food Waste Effects on Spring–Summer Crops and Associated Weeds

The Spring–Summer field experiment was established on ESAC's agricultural farm (Coimbra, Portugal, 40°13′05.7″ N 8°26′54.1″ W) from 1 June to 27 July 2020, with the same climate as Experiment 1. This area was used to grow potatoes under organic farming and was characterized by the following soil properties: 7 ± 0.0 pH, 2.40 ± 0.10% OM, 319.33 ± 63.40 mg kg^{-1} P, 308.00 ± 43.03 mg kg^{-1} K, and 0.15 ± 0.01% N. The average values for rainfall and mean temperature during the experiment were 0.24 mm and 22 °C, respectively (Figure S2: data collected at ESAC).

An experimental area covering 319 m^2 was ploughed and earth-milled on 29 May 2020, and split into 16 6 m × 1.5 m plots, divided from each other by 1 m in all directions. The applied treatments were dry CG, dry CC, fresh F, and C, with four replicates each. The treatments were randomly distributed by plots at doses of 9 Mg ha^{-1}, 9 Mg ha^{-1}, and 28 Mg ha^{-1} for CG, CC, and F, respectively, and manually incorporated into the first 15 cm of the soil layer on 1 June 2020. Seedlings of *L. sativa* and *R. sativus* were immediately planted in each plot in three central intermixed rows with 0.25 m separating the rows: one central *R. sativus* row (0.75 m from margins) with 0.10 m between plants, and two *L. sativa* rows (0.5 m from margins) with 0.30 m between plants [46]. The plots were watered by a drip irrigation system (2.66 L s^{-1} flow rate per emitter) for 30 min, every 2–3 days [42].

The normalized difference vegetation index (NDVI) was recorded weekly by measuring plant spectral reflectance with a GreenSeeker® handheld sensor (NTech Industries Incorporation, Ukiah, CA, USA; [47]) for the first seven weeks after waste incorporation until stable values were reached. Readings were collected at a height of 1 m from the topsoil layer along each plot and covering the central area.

The aboveground weed biomass was sampled in four random frames (0.25 m × 0.25 m) per plot, excluding plot margins, at four and eight weeks after waste incorporation (on 29 June and 27 July 2020, respectively). The weeds were cut at ground level, separated into monocotyledons, dicotyledons and *Cyperus* spp., as the most problematic weed in the area, and dried at 65 °C for 72 h to determine the dry biomass for each weed group and total weeds in each plot.

Lactuca sativa and *Raphanus sativus* were harvested on 16 July 2020, when commercial size was reached. Twelve *L. sativa* and fifteen *R. sativus* plants per plot were uprooted in both rows following the one-yes/one-no or one-yes/two-no schemes, respectively, discarding plants at 0.50 m row margins. The crop plants were washed to remove soil, divided into leaves and roots, and dried at 65 °C until constant weight to obtain the leaf, root, and total biomass. Additionally, *L. sativa* leaves, *R. sativus* roots, and waste were powdered to 0.05 mm size particles and analyzed to determine the total content of macronutrients (phosphorous, P; calcium, Ca; magnesium, Mg; potassium, K; nitrogen, N) and micronutrients (copper, Cu; iron, Fe; zinc, Zn; manganese, Mn). Briefly, the crop or waste dry composite samples (0.5 g) were extracted with HCl after ignition at 480–500 °C for 14 h. The elements Cu, Zn, Mn, Fe, Ca, Mg, and K were assessed based on a calibration curve for each one, using an atomic absorption spectrophotometer (model Analyst 300, Perkin Elmer). Phosphorous was colorimetrically quantified, at 470 nm, using the vanadate–molybdate reagent in nitric acid solution and a calibration curve. The total N was determined using the Kjeldahl method [45]. Among CG, CC, and F waste, CG had the highest content of N, Cu, and Mn; F was rich in P, Ca, K, Zn, and Fe; and CC showed the lowest content of nutrients, except for Mg, which was the highest (Table S1).

The soil samples were collected at the same time as the weed samples. A composed-soil sample was obtained per plot mixing soil four subsamples collected from the top 30 cm of soil. The soil samples were air-dried, sieved through a 2 mm sieve, and analyzed to determine the pH, OM, P, K, and N, as described in Experiment 2.

Experiment 4—Evaluation of Agri-Food Waste Effect on Autumn Crops and Associated Weeds

In the Autumn, an additional field experiment was conducted in the ESAC farm located 87 m apart from Experiment 3, from 25 September to 2 December 2020. For this period, the rainfall averaged per day was 3 mm, with heavy rain concentrated from 19 to 28 October, achieving 44 mm in one day (Figure S2, data collected at ESAC). The average mean temperature was 15 °C (Figure S2).

The same plot preparation, experimental design, treatments, and waste doses used for Experiment 3 were established, except for F that was calculated based on the same fresh volume used in Experiment 3, resulting in 3.84 Mg dry weight ha^{-1}. Seedlings of *Brassica rapa* and *Beta vulgaris* were alternately planted in four pure rows separated by 0.30 m between rows. Two *B. rapa* and two *B. vulgaris* rows were planted with 0.20 and 0.15 m between plants, respectively [46]. The seedlings were watered with river water immediately after plantation, and then rain fed.

NDVI was recorded for the first six weeks, like in Experiment 3. Weed samplings were performed four and eight weeks after waste incorporation on 26 October and 23 November, respectively, as described in Experiment 3. The crops were harvested when they had achieved commercial size, e.g., on 26 October 26 for *B. rapa*, and on 2 December 2021 for *B. vulgaris*. Ten and thirteen plants of *B. rapa* and *B. vulgaris*, respectively, were sampled in respective lines following the one-yes/one-no scheme, avoiding the 0.50 m row margins. These plants were processed as indicated for the crops in Experiment 3 to obtain leaf, root, and total biomass. Chemical analyses of *B. rapa* leaves and *B. vulgaris* roots were conducted as described in Experiment 3.

Soil sampling and analyses followed the pattern of Experiment 3.

2.5. Statistical Analyses

For the pot experiments, analyses of variance via general linear models (LMs, *lm()* function, 'stats' package) or generalized linear models (GLMs with Gamma error and identity or log link, *glm()* function, 'stats' package) were conducted. These were performed separately for each season experiment to test the effect of treatment (Spring–Summer: U, F, EU, EF, C, H; Autumn: U, F, CC, CG, UE, FE, H) on the biomass of monocotyledons, dicotyledons, *Cyperus* spp., and total weed groups. The effect of treatments on soil parameters (pH, MO, K, N, P) in Autumn was assessed using LMs.

In the case of field experiments, analyses were separately performed for the Spring–Summer and Autumn experiments. Two-way repeated measured models via generalized linear mixed models (GLMMs, *lmer()* function, 'lme4' package) were conducted to test whether treatment (C, CG, F, CC; fixed factor), time (sampling week; repeated measures) and their interaction had an effect on NDVI. Plot was considered as a random factor. To explore the effect of treatment (C, CG, F, CC; fixed factor), time (week four and eight; repeated measures) and treatment × time interaction on biomass of different weed groups (monocotyledons, dicotyledons, *Cyperus* spp., and total weeds), two-way repeated measures GLMMs were used with sampling units nested in plots as a random factor. The effect of treatment (C, CG, F, CC; fixed factor) on crop biomass (leaves, roots, and total) was assessed via one-way linear mixed models (LMMs, *lmer()* function, 'lme4' package) with plot as a random factor for each crop. In crop chemicals analysis, individuals from the same plot were pulled together to obtain a composite sample per plot. Therefore, one-way LMs or GLMs with Gamma error and identity link were conducted to test for the effect of treatment (C, CG, F, CC) on the content of each nutrient (P, Ca, Mg, K, N, Cu, Zn, Fe, Mn) in each crop. Finally, a two-way LM was used to assess the effect of treatment (C, CG, F, CC) and time (week four and eight) and the interaction between these two factors on the soil parameters (pH, OM, P, K, N) for Spring–Summer data, while in Autumn, the effect of the same factors using two-way LMs in the case of soil pH, OM, K, and N and a two-way GLM (Gamma error and identity link) for soil P was evaluated.

Post hoc mean separations were conducted using the *lsmeans()* function ('lsmeans' package) by comparing the least-square means obtained in each model.

All statistical analyses were conducted in R [48]. The level of significance was set at $p \leq 0.05$.

3. Results

3.1. Bioherbicidal Effect on Emerging Common Weeds in Pots

For the Spring–Summer (Experiment 1), treatment significantly affected the biomass of all weed groups (Figure 1). The biomass of dicotyledons was reduced when treated with FE, F, and U relative to the control (Figure 1A). However, the biomass of monocotyledons and *Cyperus* spp. did not differ from the control under any waste treatment (Figure 1B and 1C, respectively). Considering total weeds, the plant biomass increased in UE and decreased by 46% and 39% in F and U, respectively, when compared to the control (Figure 1D). As expected, pots treated with herbicide (positive control) always showed the greatest inhibition (Figure 1A–D).

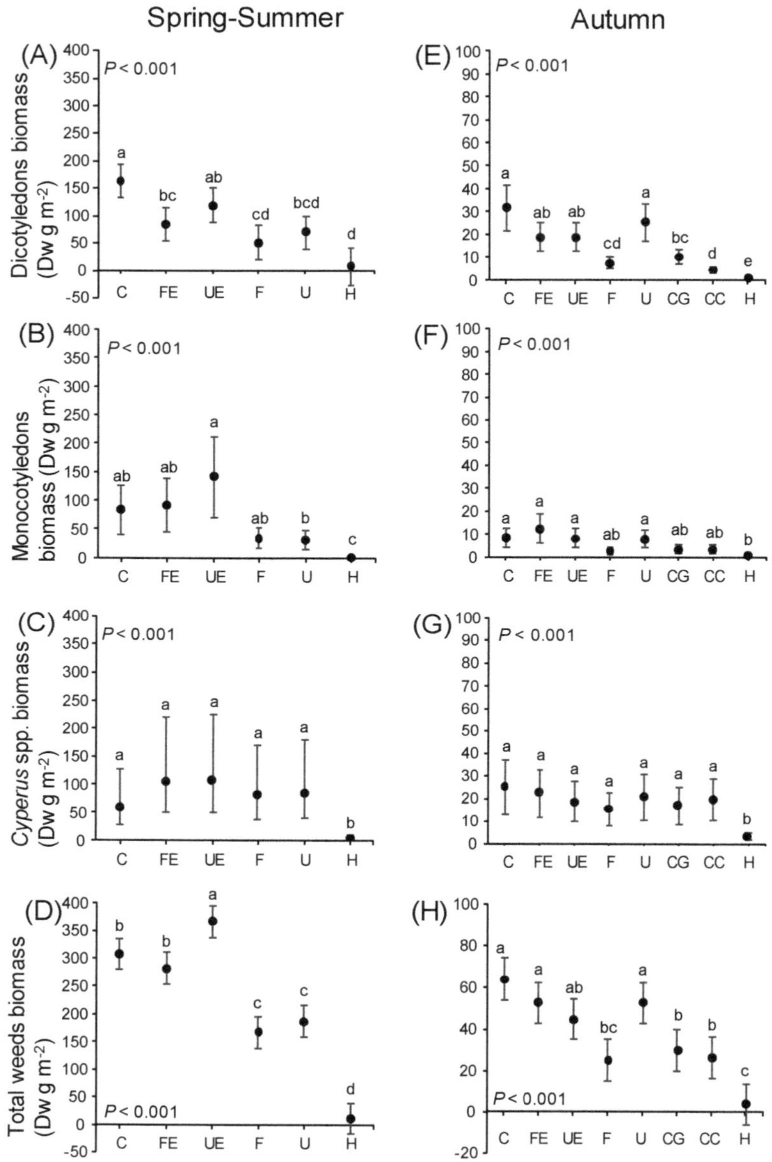

Figure 1. Effect of waste treatments on the biomass of dicotyledon weeds (**A**,**E**); monocotyledon weeds (**B**,**F**); the weed *Cyperus* spp. (**C**,**G**); and total weeds (**D**,**H**) emerged from pots filled with agricultural soil and placed outdoors in Spring–Summer (Experiment 1, 2019) and Autumn (Experiment 2, 2019). C = control, FE = *Vicia faba* pod extract, UE = *Urtica dioica* extract, F = *Vicia faba* pod waste, U = *Urtica dioica* waste, CG = spent coffee grounds waste, CC = corn cobs waste, H = positive control with herbicide. Dw = dry weight. Figures show least-square mean values (±confidence intervals). $n = 5$. For *Cyperus* spp. biomass in Spring–Summer, confidence intervals were back-transformed from the log scale. Different letters indicate statistical significance according to the lsmeans function at $p \leq 0.05$.

In Autumn (Experiment 2), treatment had a significant biomass impact for all weed groups (Figure 1). Dicotyledons showed reduced biomass in F, CG, and CC waste treatments (Figure 1E). As found in Spring–Summer, the biomass of monocotyledons and *Cyperus* spp. showed no significant differences between waste treatments and the control (Figure 1F and 1C, respectively). For total weeds, plant biomass was inhibited in F (61%), CG (54%), and CC (59%) pots compared to the control pots, being close to herbicide inhibition (Figure 1H). Again, herbicide showed the lowest weed biomass for all weed groups (Figure 1E–H). Regarding soil from pots in Autumn, treatment significantly affected pH, the content of OM, and N (Table 3). Soils treated with FE, UE, F, U, and CC had higher pH than control soils (Table 3). Soils with F and CG showed increased OM, while CG increased the content of N (Table 3).

Table 3. Values of soil properties obtained in soils mixed with different waste and collected at the end of Experiment 2 (Autumn pots, 2019). Least-square mean values (±confidence intervals) are shown. n = 5.

Treatment	pH	Organic Matter (%)	N (%)	P_2O_5 (mg kg^{-1})	K_2O (mg kg^{-1})
River water/no waste (negative control)	6.06 ± 0.17^{c}	2.70 ± 0.28^{c}	0.17 ± 0.02^{b}	422 ± 120	390 ± 160
Vicia faba pod extract	6.40 ± 0.17^{ab}	2.81 ± 0.28^{bc}	0.17 ± 0.02^{b}	358 ± 120	413 ± 160
Urtica dioca extract	6.44 ± 0.17^{ab}	2.80 ± 0.28^{bc}	0.17 ± 0.02^{b}	298 ± 120	346 ± 160
Vicia faba pod waste	6.41 ± 0.17^{ab}	3.15 ± 0.28^{b}	0.19 ± 0.02^{ab}	350 ± 120	576 ± 160
Urtica dioca waste	6.57 ± 0.17^{a}	3.12 ± 0.28^{bc}	0.19 ± 0.02^{ab}	390 ± 120	442 ± 160
Spent coffee grounds waste	6.28 ± 0.17^{bc}	3.86 ± 0.28^{a}	0.22 ± 0.02^{a}	309 ± 120	316 ± 160
Corn cob waste	6.46 ± 0.17^{ab}	2.91 ± 0.28^{bc}	0.16 ± 0.02^{b}	255 ± 120	373 ± 160
Herbicide (positive control)	6.32 ± 0.17^{abc}	2.73 ± 0.28^{bc}	0.17 ± 0.02^{b}	388 ± 120	333 ± 160
p	**<0.001**	**<0.001**	**<0.001**	0.116	0.052

Values in bold indicate significance at $p \leq 0.05$. Least-square mean values within a column and without statistical letters are not significantly different, when followed by different superscript letters statistical differences were found according to the lsmeans function at $p \leq 0.05$.

3.2. Waste Performance in Field Crops

In Spring–Summer (Experiment 3), time, treatment, and the time × treatment interaction had a significant effect on NDVI (Table S2). This parameter increased over time (Figure 2A). Plots treated with CG and CC registered lower NDVI than the control plots (Figure 2B). A significant reduction in NDVI was found for CG plots at weeks three and four and for CC at week four (Figure 2C). For the weed groups, treatment significantly affected the biomass of dicotyledons, monocotyledons, and total weeds (Table S2). Plots with F waste significantly increased dicotyledons' biomass (Figure 3A). However, plots receiving CG reduced the biomass of monocotyledons and total weeds 2- and 1.8-fold, respectively (Figure 3B,E). Time also significantly affected the biomass of all weed groups, being higher at week eight after waste incorporation (Tables S2 and S3). A treatment × time interaction was observed for dicotyledons' biomass and total weeds (Table S2). At week eight, plots receiving F waste had higher dicotyledons' biomass (Figure S3A). However, plots with CG reduced the biomass of all weed groups, although a significant reduction was only found for total weeds (43%), compared with the control (Figure S3D). In crops, the biomass of leaf- and bulb-edible plants was significantly affected by treatment (Figure 4A,B). This effect was influenced by the plot covariate for the leaf biomass of *L. sativa* and for leaf and total biomass of *R. sativus* (Figure 4A,B). CG significantly increased the leaf, root, and total biomass by 72, 109, and 76% in the case of *L. sativa* (Figure 4A) and by 57, 66, and 60% for *R. sativus* (Figure 4B). Treatment also affected the content of N, Fe, and Mn in *L. sativa* leaves and of P, Zn, Fe, and Mn in *R. sativus* roots (Table S4). In *L. sativa,* leaves from the CC treatment had a higher content of Fe than the control (Table S4). Plots treated with F had a higher content of N and Mn than CC and CG plots, respectively, but their content was not different from the control plots (Table S4). For *R. sativus* roots, the content of P an Zn increased in F plots (Table S4). Zn was additionally increased by CC (Table S4). However, CC reduced the Fe content (Table S4). Differences in the content of Mn occurred between F and CG and CC, but their values did not differ from the control (Table S4). For soil parameters, only the time factor significantly affected the content of OM and K, which decreased after eight weeks of waste incorporation (Table 4).

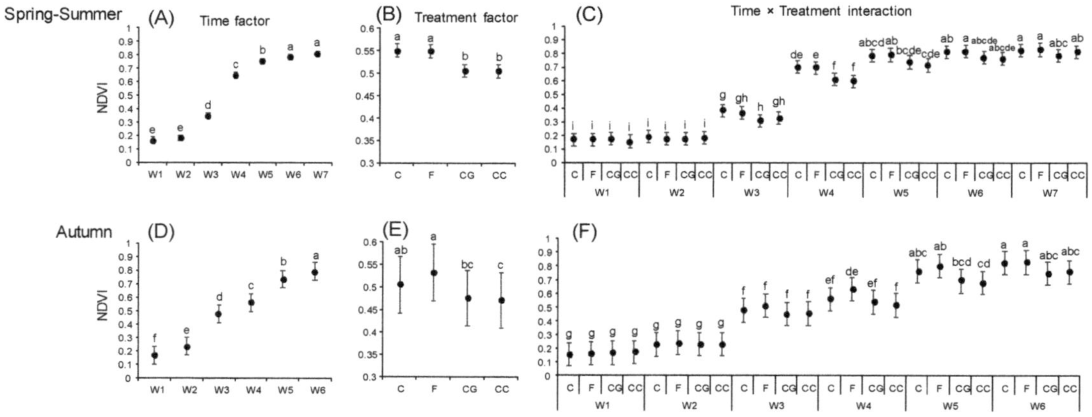

Figure 2. Effect of time (sampling week, (**A**,**D**)), treatment (different waste, (**B**,**E**)), and the interaction between these two factors (**C**,**F**)) on the normalized difference vegetation index (NDVI) recorded in field experiments under Spring–Summer and Autumn conditions (Experiment 3 and 4, 2020, respectively). W1, W2, ..., W7 = week 1, week 2, ..., week 7 after waste incorporation. C = control, F = *Vicia faba* pod waste, CG = spent coffee grounds waste, CC = corn cob waste. Figures show least-square mean values (±confidence intervals). $n = 4$. Different letters refer to statistical significance according to significant p values from Table S2 and lsmeans function at $p \leq 0.05$.

In Autumn (Experiment 4), the NDVI parameter was significantly affected by treatment, time, the interaction between these two factors and plot as covariate (Table S2). The values of NDVI were different among sampling weeks and increased over time (Figure 2D). This parameter was lower in CC compared to the control (Figure 2E). Nevertheless, considering treatment and time together, significant differences were registered between F and CC for weeks four and five (Figure 2F). Concerning weeds, treatment only significantly affected the biomass of *Cyperus* spp. (Table S2, Figure 3), which presented higher values for F compared to CC (Figure 3G). Time had a significant effect on the biomass of dicotyledons, *Cyperus* spp., and total weeds (Table S2). This effect was also influenced by plot for dicotyledons (Table S2). When affected by time, the biomass of the weed groups was higher at week eight (Table S3). The treatment × time interaction did not affect weed biomass (Table S2, Figure S3E–H). For crops, the leaf, root, and total biomass of leaf-edible *B. rapa* were significantly affected by treatment (Figure 4C). Leaf and total biomass were also influenced by plot as a covariate (Figure 4C). Plots treated with F and CG had smaller *B. rapa* plants than the control plots (Figure 4C). Treatment also affected the biomass of bulb-edible *B. vulgaris* plants, with an additional effect of the covariate plot on roots (Figure 4D). The leaf, root, and total biomass of *B. vulgaris* were reduced by 42–51% when grown in plots with CG or CC compared with the control plots (Figure 4D). Regarding plant nutrients, treatment only had a significant effect on K in both *B. rapa* and *B. vulgaris* (Table S4). *B. rapa* leaves from plants growing in F and CC plots showed a lower content of K than the control plots (Table S4). For *B. vulgaris*, the content of K was lower in CG and CC plots compared with the control (Table S4). Concerning the soil parameters, soils collected in F plots had significantly lower pH, N, P, and K than the control plots (Table 4). Time also significantly affected the content of N and P, presenting higher values at week four after waste incorporation (Table 4). Finally, the treatment × time interaction had a significant effect on K, which increased in CG plots at week four, but rapidly decreased in CG and F plots at week eight compared with the control plots (Table 4).

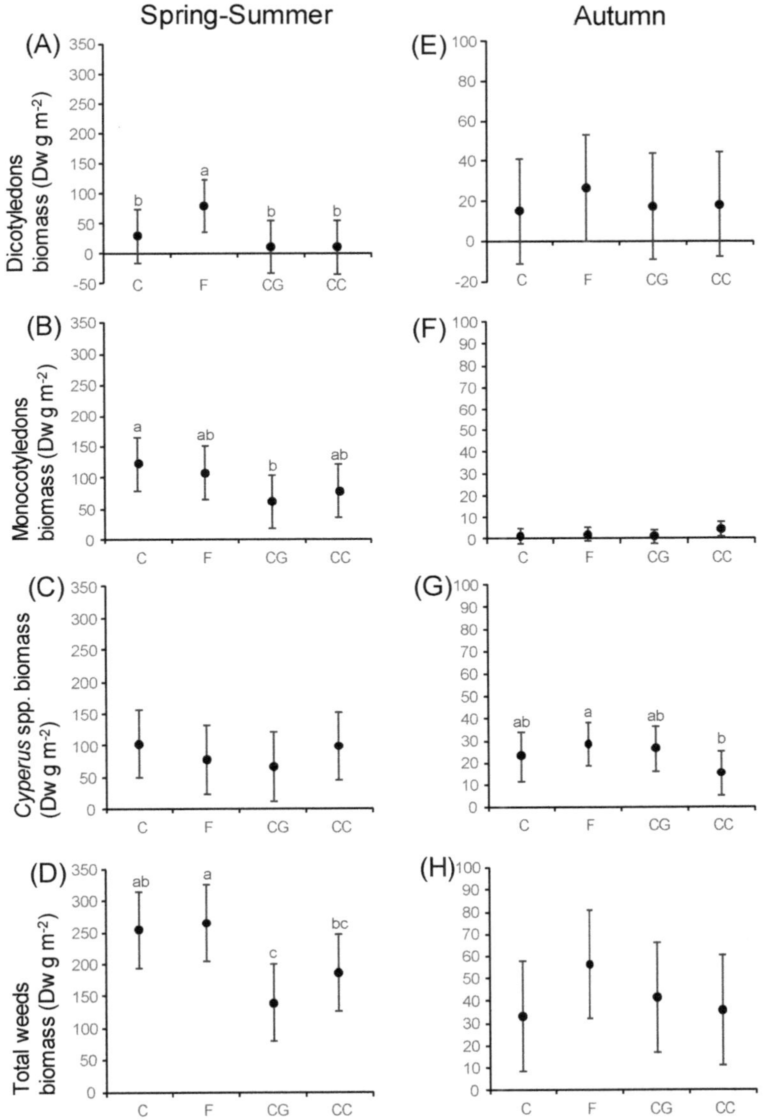

Figure 3. Effect of waste treatments on biomass of dicotyledon weeds (**A**,**E**); monocotyledon weeds (**B**,**F**); the weed *Cyperus* spp. (**C**,**G**); and total weeds (**D**,**H**) in experimental field plots in Spring–Summer (Experiment 3, 2020) and Autumn (Experiment 4, 2020). C = control, F = *Vicia faba* pod waste, CG = spent coffee grounds waste, CC = corn cob waste. Dw = dry weight. Figures show least-square mean values (±confidence intervals). $n = 16$. Least-square mean values without statistical letters are not significantly different. Different letters indicate statistical significance according to significant p values from Table S2 and lsmeans function at $p \leq 0.05$.

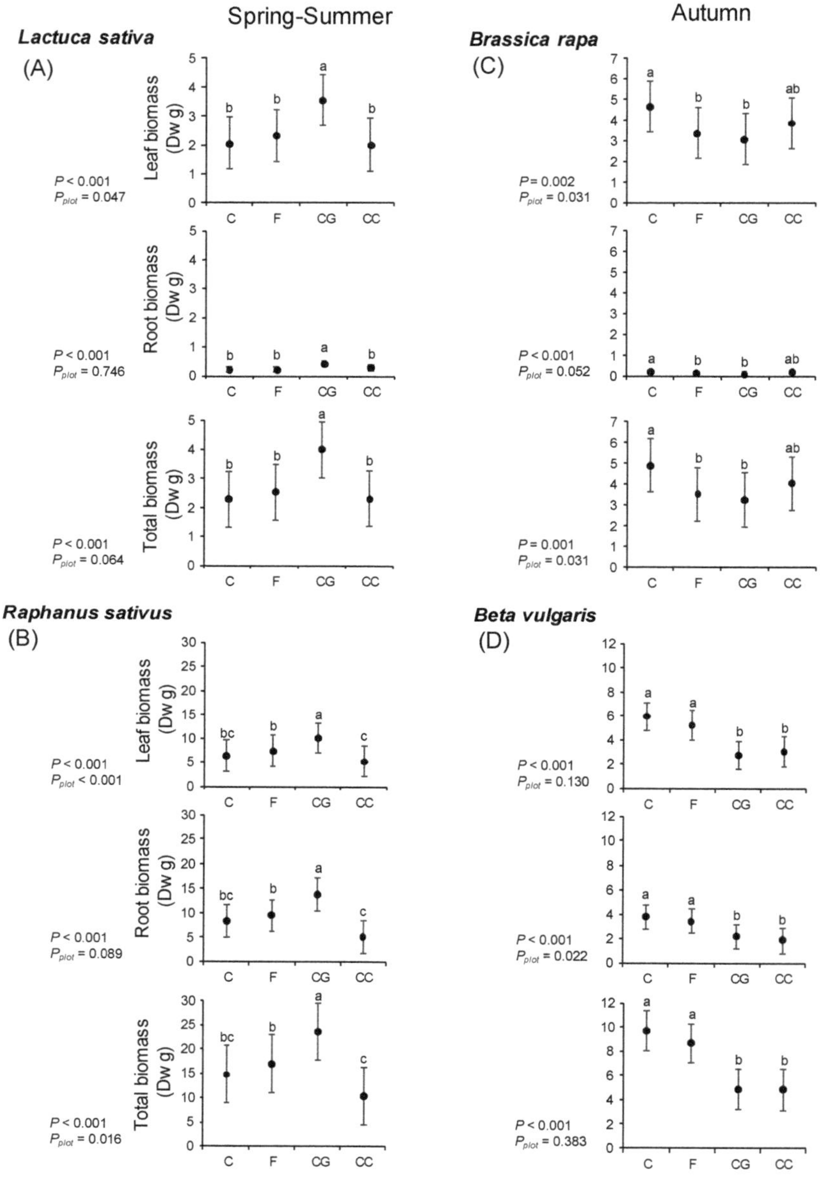

Figure 4. Effect of waste treatments on leaf, root, and total biomass of *Lactuca sativa* (**A**) and *Raphanus sativus* (**B**) grown in field plots in Spring–Summer (Experiment 3, 2020) and of leaf, root, and total biomass of *Brassica rapa* (**C**) and *Beta vulgaris* (**D**) grown in field plots during Autumn (Experiment 4, 2020). Upper figures = leaf-edible crops. Lower figures = bulb-edible crops. C = control, F = *Vicia faba* pod waste, CG = spent coffee grounds waste, CC = corn cobs waste. Dw = dry weight. plot = random factor. Figures show least-square mean values (±confidence intervals). n = 4–12, 10–15, 9–11, and 10–14 for *L. sativa*, *R. sativus*, *B. rapa*, and *B. vulgaris*, respectively. Different letters indicate statistical differences according to the lsmeans function at $p \leq 0.05$ after conducting linear mixed models.

Table 4. Effect of treatment (different waste), time (sampling week), and the interaction between them on soil properties from soils collected in the experimental field plots during Experiment 3 (Spring–Summer, 2020) and Experiment 4 (Autumn, 2020). Least-square mean values (±confidence intervals) are shown. $n = 4$.

Season	Source of Variation	Source Level	pH	Organic Matter (%)	N (%)	P_2O_5 (mg kg^{-1})	K_2O (mg kg^{-1})
Spring–Summer	Treatment (Tr)	C	6.79 ± 0.17	1.89 ± 0.30	0.123 ± 0.02	191 ± 120.0	227 ± 63.0
		F	6.90 ± 0.17	1.91 ± 0.30	0.129 ± 0.02	251 ± 120.0	265 ± 63.0
		CG	6.89 ± 0.17	2.06 ± 0.30	0.144 ± 0.02	218 ± 120.0	267 ± 63.0
		CC	6.96 ± 0.17	1.88 ± 0.29	0.130 ± 0.02	215 ± 120.1	295 ± 63.0
		p	0.285	0.643	0.137	0.817	0.259
	Time (t)	W4	6.92 ± 0.11	2.07 ± 0.19 ^{a}	0.133 ± 0.01	246 ± 75.0	309 ± 40.0 ^{a}
		W8	6.84 ± 0.11	1.80 ± 0.18 ^{b}	0.129 ± 0.01	192 ± 75.0	218 ± 39.0 ^{b}
		p	0.207	**0.021**	0.559	0.196	**0.001**
	Tr × t	C W4	6.83 ± 0.2	1.80 ± 0.46	0.123 ± 0.03	191 ± 188.0	271 ± 98.6
		F W4	6.95 ± 0.26	2.15 ± 0.46	0.139 ± 0.03	306 ± 188.2	335 ± 98.6
		CG W4	6.95 ± 0.26	2.23 ± 0.47	0.141 ± 0.03	234 ± 188.2	298 ± 98.9
		CC W4	6.97 ± 0.27	2.10 ± 0.46	0.130 ± 0.03	252 ± 188.2	334 ± 99.0
		C W8	6.75 ± 0.26	1.99 ± 0.47	0.122 ± 0.03	190 ± 188.0	184 ± 98.8
		F W8	6.85 ± 0.26	1.66 ± 0.46	0.119 ± 0.03	197 ± 188.0	196 ± 98.5
		CG W8	6.83 ± 0.27	1.89 ± 0.46	0.147 ± 0.03	202 ± 188.0	237 ± 99.0
		CC W8	6.95 ± 0.2	1.66 ± 0.46	0.130 ± 0.03	178 ± 188.5	256 ± 98.9
		p	0.950	0.139	0.528	0.884	0.670
Autumn	Treatment	C	6.41 ± 0.17 ^{a}	1.92 ± 0.21 ^{ab}	0.146 ± 0.01 ^{ab}	244 ± 69.0 ^{a}	304 ± 74.0 ^{a}
		F	6.15 ± 0.18 ^{b}	1.82 ± 0.21 ^{b}	0.126 ± 0.01 ^{c}	122 ± 34.2 ^{b}	191 ± 74.0 ^{b}
		CG	6.39 ± 0.18 ^{ab}	2.16 ± 0.21 ^{a}	0.161 ± 0.01 ^{a}	299 ± 103.0 ^{a}	247 ± 74.0 ^{ab}
		CC	6.36 ± 0.18 ^{ab}	2.04 ± 0.21 ^{ab}	0.141 ± 0.01 ^{bc}	186 ± 52.0 ^{ab}	203 ± 74.0 ^{ab}
		p	**0.032**	**0.023**	**<0.001**	**<0.001**	**0.023**
	Time	W4	6.29 ± 0.11	2.01 ± 0.13	0.149 ± 0.01 ^{a}	264 ± 54.0 ^{a}	248 ± 46.0
		W8	6.36 ± 0.11	1.95 ± 0.13	0.138 ± 0.01 ^{b}	161 ± 31.0 ^{b}	224 ± 46.0
		p	0.263	0.438	**0.036**	**0.009**	0.396
	Tr × t	C W4	6.36 ± 0.28	1.97 ± 0.33	0.158 ± 0.02	228 ± 99.0 ^{bc}	313 ±116.0
		F W4	6.06 ± 0.27	1.81 ± 0.33	0.129 ± 0.02	133 ± 57.6 ^{bcd}	196 ± 116.0
		CG W4	6.36 ± 0.27	2.27 ± 0.33	0.167 ± 0.02	512 ± 221.5 ^{a}	261 ± 115.4
		CC W4	6.38 ± 0.27	2.00 ± 0.33	0.142 ± 0.02	185 ± 80.1 ^{bc}	221 ± 115.0
		C W8	6.46 ± 0.28	1.87 ± 0.33	0.135 ± 0.02	261 ± 112.8 ^{ab}	294 ± 116.0
		F W8	6.24 ± 0.28	1.82 ± 0.33	0.123 ± 0.02	111 ± 48.0 ^{cd}	185 ± 115.2
		CG W8	6.42 ± 0.28	2.05 ± 0.33	0.154 ± 0.02	85 ± 36.8 ^{d}	233 ± 116.0
		CC W8	6.33 ± 0.28	2.08 ± 0.33	0.140 ± 0.02	187 ± 81.6 ^{bc}	185 ± 115.7
		p	0.677	0.537	0.448	**<0.001**	0.989

Values in bold indicate significance at $p \leq 0.05$. Least-square mean values within a column for each source of variation and without statistical letters are not significantly different; when accompanied by different superscript letters, statistical differences were found according to the lsmeans function at $p \leq 0.05$. C = control, F = *Vicia faba* pod waste, CG = spent coffee grounds waste, CC = corn cob waste. W4 = week 4 after waste incorporation, W8 = week 8 after waste incorporation.

4. Discussion

The Spring–Summer pot experiment revealed that U and F waste were efficient in reducing the biomass for almost all weed groups. However, U waste lost its inhibitory effect in the Autumn pot experiment, with F, CG, and CC being the most effective treatments in reducing the biomass of weeds emerging during that season. The inhibitory effect of these waste did not seem to be influenced by environmental conditions nor by a nutrient deficit. F waste showed exactly the same inhibitory pattern for dicotyledons, monocotyledons, and total weeds in Spring–Summer and in Autumn, and soil treated with this waste had higher OM content than the control soils. Similarly, soils mixed with CG were enriched with OM and N. No soil with F, CG, and CC showed nutrient or pH limitations compared to the control soils. This implies that the reduced growth found for weeds in these treatments are related to the release of phytotoxic compounds by the waste. In fact, CG inhibited the growth of different crops and weeds when directly applied to the soil, probably due to its natural toxic compounds [31–33,35]. A recent study on weed control using *V. faba* found that green manures of this plant reduced by half weed biomass in field maize crops, with the inhibitory effect occurring within 27 days after the incorporation of green manure [36]. After 27 days, the inhibition disappeared and the fertilizing effects were evident [36]. Water-soluble and volatile organic compounds released by *V. faba* probably play a role in reducing weed biomass [49]. For *V. faba* pods in particular, a herbicidal effect on weeds was scarcely studied. However, pod extracts of this species had nematocidal activity [38]. As far as we know, the inhibitory effect of CC waste or its composition in phytotoxic compounds is unknown. This could mean that toxic compounds were not abundant in this waste or were not released during the pot experiment duration (six weeks). One plausible explanation for the reduced weed growth found in CC pots may be related to the C/N ratio and the lower nutrient content of this raw waste compared to CG and F (Table S1). The C/N ratios for CG and F are around 23:1 and 18:1, respectively [33,37]. However, CC has a C/N ratio of 120:1 [50], which is very high, explaining its poor quality. Although the soil from CC pots did not show nutrient limitations relative to control soils, CC soils had lower OM and N content than CG soils. This could indicate that CC raw waste contains a poor nutrient stock that was likely firstly immobilized by the microbial biomass (i.e., N), leading to a lower nutrient availability for plants, which, in turn, hindered their growth [33].

As F, CG, and CC showed the best inhibitory effect on weed biomass in the Spring–Summer and Autumn pot experiments, consequently, their effectiveness was evaluated in real field crops in both seasons. Under field conditions, the results varied with the seasons. The CG waste reduced weed biomass and increased crop biomass in Spring–Summer, but had no effect on weeds and reduced crop biomass in Autumn. The NDVI parameter indicated that the negative effect caused by CG occurred at week three and four after waste incorporation. The inhibitory trend of this waste was similar in both seasons, but the significant effect was only found in Spring–Summer. The F and CC waste did not impact weed biomass regardless of season and crop biomass in Spring–Summer. However, F and CC inhibited the growth of *B. rapa* and *B. vulgaris* in Autumn. The different results found between the Spring–Summer and Autumn field trials may be firstly explained by different environmental conditions during the experiments. In Spring–Summer, warm days with scarce precipitation prevailed. Lower temperatures and rainy days were more frequent in Autumn. Heavier rain events, especially those concentrated in short periods, as found for the Autumn field trial, may contribute to soil nutrient leachate [51,52]. This could explain the lower content of all nutrients measured in the soils from F plots, which in turn may result in the lower biomass and K leaf content of *B. rapa* found in these plots. High rainfall concentrated at week four in the Autumn experiment could also lead to a phytotoxicity loss from CG by leaching toxic compounds, reducing its effectiveness. This may explain why no significant differences were found between CG and the control treatments in Autumn, despite having a similar inhibitory trend (NDVI) to the Spring–Summer field trial. The remaining phytotoxicity of CG seemed to be sufficient to affect the growth of Autumn crops, since no nutrient deficit was found in the CG plots. Additionally, contrasting behavior between the Spring–Summer and Autumn crops in the CG plots might be related to the different sensitivity of plant species to the same phytochemicals [53]. On the other hand, CC waste did not show a clear pattern for weeds or crops. However, the values of both weeds and crops biomass tended to be lower than in the control plots. This could be related to the high C/N ratio, as previously discussed [50]. Additional studies under field conditions should be conducted to clarify this issue.

The waste dose applied may also influence the results. Most studies assessing the value of CG as an organic amendment found that doses higher than 10% had a toxic effect on lettuce, broccoli, leek, radish, viola, and sunflower [32–34]. However, growing substrate containing up to 5% CG stimulated seed emergence of *Brassica* crops [54]. In our field trials, CG was used at 9 Mg ha^{-1}, which is lower than that previously reported [32] and agreed with those results, showing positive effects of CG at low densities [54]. Interestingly, we found that CG dose positively impacted the growth of both *L. sativa* and *R. sativus* and additionally reduced weed biomass when precipitation was scarce, and temperature was warm (Spring–Summer). However, this effect was not consistent under different environmental conditions (Autumn), probably because the dose used represents a threshold between low and medium-high doses under field conditions. The effect of CG at different doses in the field is an interesting topic that requires further research.

Cyperus spp. are among the most serious weed species in arable lands in warmer countries, substantially reducing crop yield, especially in organic agriculture [55]. They efficiently reproduce by seed and tubers and show tolerance to various herbicides [56]. For these reasons, *Cyperus* spp. have become particularly worrisome to farmers and efficient alternatives are urgently needed to control them. That is why we paid attention to evaluating the effect of the tested waste on this weed group. Unfortunately, the tested waste did not effectively control *Cyperus* weeds. However, in field trials, the lowest biomass value of *Cyperus* spp. was obtained from CG plots in Spring–Summer and CC plots in Autumn. As discussed above, the applied dose might be the key to obtaining effective results. Hardgrove and Livesley [33] found that amendments with increasing rates of CG decreased weed growth. Therefore, it can be suggested that higher doses of CG and CC than those applied in the present study may contribute to reducing *Cyperus* spp.; for that, an additional evaluation is required.

On the other hand, organic waste is expected to improve soil fertility. For example, Cervera-Mata et al. [32] showed that the addition of CG at 10% to two different soils increased the content of organic carbon, total N, and available K after 15, 30, 45, and 60 days. Similarly, Chrysargyris et al. [54] found that a substrate mixture including CG showed enriched content of N, P, K, and some micronutrients. In the case of F waste, Al-Chammaa et al. [37] proved that adding green manures of *V. faba* pods as soil amendment with, 15 days before plant sowing, provided a considerable amount of N requirements for *Sorghum* plants. These three studies were conducted in pots using a range of substrates (peat–perlite and field soil). In our study, a more basic pH (by F, CG, and CC) and improved OM content (by F and CG) and total N (by CG) were obtained when the effect of these wastes was evaluated in pots. Against expectations, field trials indicated that F, CG, and CC did not increase soil nutrients. Although we did not check for this, one plausible explanation might be related to the different decomposition rates occurring in the pots (e.g., higher temperature) that accelerates the process. Moreover, even if a similar decomposition rate occurred between pot and field trials, leaching or movement of nutrients from pots was likely to be hampered by a physical barrier.

5. Conclusions

Finding new alternatives to control weeds is a major task for the future of organic agriculture, as well as for environmental and human health preservation. Various organic waste can play an important role in replacing synthetic pesticides, which is why we assessed the role of *Urtica dioica* waste, *Vicia faba* pods, spent coffee grounds, and corn cob waste. The work presented here shows that the applied doses of organic waste can reduce the sprouting of weeds, but can also have a negative impact on cash crop yields.

The findings of this study show that, among all of the evaluated waste, the application of CG at 9 Mg ha^{-1} reduced the total biomass of naturally-emerged weeds, while stimulating crop growth, under scarce rainfall and warm days. The inhibitory effect is likely to occur between weeks three and four, after the incorporation of the waste into the soil. The inhibitory and stimulating effects of CG obtained under real conditions suggest that this waste could partially replace the use of synthetic agrochemicals and provide new evidence to boost a circular economy. However, the effectiveness of this waste seems to be limited by environmental conditions and probably by the applied dose. Despite these promising results, additional field research is required to corroborate the obtained effect of CG at different doses in the long term, covering several years.

Supplementary Materials: The following supporting information can be downloaded at: https://www.mdpi.com/article/10.3390/agronomy12051184/s1: Figure S1: Meteorological data registered during pot experiments in Spring–Summer (A, Experiment 1) and Autumn (B, Experiment 2) 2019; Figure S2: Meteorological data registered during field experiments in Spring–Summer (A,

Experiment 3) and Autumn (B, Experiment 4) 2020; Figure S3: Effect of the interaction between waste treatments and time on biomass of different weed groups (dicotyledons, monocotyledons, *Cyperus* spp., total weeds) in experimental filed plots in Spring–Summer (Experiment 3, 2020) and Autumn (Experiment 4, 2020); Table S1: Content of macro- and micronutrients of agri-food waste used in the field trials; Table S2: Repeated measures results obtained via generalized linear mixed models to test for the effect of treatment (different waste), time (sampling week), and the interaction between them on the normalized difference vegetation index (NDVI) of weeds and crops and on the biomass of different weed groups in experimental field plots under Spring–Summer and Autumn conditions (Experiment 3 and 4, 2020, respectively); Table S3: Effect of time on biomass of different weed groups (dicotyledons, monocotyledons, *Cyperus* spp., total weeds) in experimental field plots in Spring–Summer (Experiment 3, 2020) and Autumn (Experiment 4, 2020); Table S4: Effect of treatment (agri-food waste) on the macro- and micronutrients content of *Lactuca sativa* leaves and *Raphanus sativus* roots grown in field plots in Spring–Summer (S-S) (Experiment 3, 2020) and of *Brassica rapa* leaves and *Beta vulgaris* roots grown in field plots during Autumn (A) (Experiment 4, 2020).

Author Contributions: Conceptualization, P.L., R.G. and C.G.; methodology, P.L., R.G., S.B. and C.G.; formal analysis, P.L.; investigation, P.L., R.G., S.B. and C.G.; resources, P.L., R.G., A.J.D.F. and C.G.; data curation, P.L., R.G., S.B. and C.G.; writing—original draft preparation, P.L.; review and editing, R.G., S.B., A.J.D.F. and C.G.; supervision, P.L.; additional funding acquisition, P.L., R.G., A.J.D.F. and C.G. All authors have read and agreed to the published version of the manuscript.

Funding: This work was supported by the Portuguese Foundation for Science and Technology—FCT/MEC and ERDF, within the PT2020 Partnership Agreement, and COMPETE 2020, grant number UID/BIA/04004/2019 and FCT-UIDB/681/2020. P.L. was funded by FCT, grant SFRH/BPD/88504/2012 and contract IT057-18-7248.

Data Availability Statement: The data that support the findings of this study are available from the corresponding author, P.L., upon request.

Acknowledgments: The authors thank the Frip company, Dona Olga Bar, Abadia Bar, Novo Pinguin, Pedro Moreira, the local farmers, Francisco Farias, and the ESAC students for the supplied waste or field and technical support.

Conflicts of Interest: The authors declare no conflict of interest.

References

1. Cycon, M.; Mrozik, A.; Piotrowska-Seget, Z. Bioaugmentation as a strategy for the remediation of pesticide-polluted soil: A review. *Chemosphere* **2017**, *172*, 52–71. [CrossRef] [PubMed]
2. De Corato, U. Towards new soil management strategies for improving soil quality and ecosystem services in sustainable agriculture: Editorial overview. *Sustainability* **2020**, *12*, 9398. [CrossRef]
3. Khan, M.A.; Costa, F.B.; Fenton, O.; Jordan, P.; Fennell, C.; Mellander, P.E. Using a multi-dimensional approach for catchment scale herbicide pollution assessments. *Sci. Total Environ.* **2020**, *747*, 141232. [CrossRef] [PubMed]
4. Duke, S.O.; Heap, I. Evolution of weed resistance to herbicides: What have we learned after 70 years? In *Biology, Physiology and Molecular Biology of Weeds*; Jugulam, M., Ed.; CRC Press: Boca Raton, FL, USA, 2017; pp. 63–86.
5. Green, J.M.; Owen, M.D.K. Herbicide-resistant crops: Utilities and limitations for herbicide-resistant weed management. *J. Agric. Food Chem.* **2011**, *59*, 5819–5829. [CrossRef]
6. Carvalho, F.P. Pesticides, environment, and food safety. *Food Energy Secur.* **2017**, *6*, 48–60. [CrossRef]
7. Fasusi, O.A.; Babalola, O.O. The multifaceted plant-beneficial rhizobacteria toward agricultural sustainability. *Plant Protect. Sci.* **2021**, *57*, 95–111.
8. European Commission (EC). *Regulation No 834/2007. Official Journal of the European Union on Organic Production and Labeling of Organic Products and Repealing Regulation (EEC)*; No 2092/91. OJL 189, 20.7.2007; European Commission (EC): Brussels, Belgium, 2007; pp. 1–23.
9. European Commission (EC). *Organic Farming in the EU. A Fast Growing Sector*; EU Agricultural Markets Briefs; European Commission (EC): Brussels, Belgium, 2019.
10. European Commission (EC). *Communication from the Commission to the European Parliament, the Council, the European Economic and Social Committee and the Committee of the Regions. A Farm to Fork Strategy for a Fair, Healthy and Environmentally-Friendly Food System*; Communication No 381/2020. 20.5.2020; European Commission (EC): Brussels, Belgium, 2020; pp. 1–20.
11. Semida, W.M.; Beheiry, H.R.; Sétamou, M.; Simpson, C.R.; Abd El-Mageed, T.A.; Rady, M.M.; Nelson, S.D. Biochar implications for sustainable agriculture and environment: A review. *S. Afr. J. Bot.* **2019**, *127*, 333–347. [CrossRef]

12. Lorenzo, P.; Reboredo-Durán, J.; Muñoz, L.; Freitas, H.; González, L. Herbicidal properties of the commercial formulation of methyl cinnamate, a natural compound in the invasive silver wattle (*Acacia dealbata*). *Weed Sci.* **2020**, *68*, 69–78. [CrossRef]
13. Puig, C.G.; Álvarez-Iglesias, L.; Reigosa, M.J.; Pedrol, N. *Eucalyptus globulus* leaves incorporated as green manure for weed control in maize. *Weed Sci.* **2013**, *61*, 154–161. [CrossRef]
14. Souza-Alonso, P.; Puig, C.G.; Pedrol, N.; Freitas, H.; Rodríguez-Echeverría, S.; Lorenzo, P. Exploring the use of residues from the invasive *Acacia* sp. for weed control. *Renew. Agric. Food Syst.* **2020**, *35*, 26–37. [CrossRef]
15. Duke, S.O.; Dayan, F.E.; Romagni, J.G.; Rimando, A.M. Natural products as sources of herbicides: Current status and future trends. *Weed Res.* **2000**, *40*, 99–111. [CrossRef]
16. Dayan, F.E.; Owens, D.K.; Duke, S.O. Rationale for a natural products approach to herbicide discovery. *Pest Manag. Sci.* **2012**, *68*, 519–528. [CrossRef] [PubMed]
17. Kadoglidou, K.; Kalaitzidis, A.; Stavrakoudis, D.; Mygdalia, A.; Katsantonis, D. A novel compost for rice cultivation developed by rice industrial by-products to serve circular economy. *Agronomy* **2019**, *9*, 553. [CrossRef]
18. Pellejero, G.; Palacios, J.; Vela, E.; Gajardo, O.; Albrecht, L.; Aschkar, G.; Chrorolque, A.; García-Navarro, F.J.; Jiménez-Ballesta, R. Effect of the application of compost as an organic fertilizer on a tomato crop (*Solanum lycopersicum* L.) produced in the field in the Lower Valley of the Río Negro (Argentina). *Int. J. Recycl. Org. Waste Agric.* **2021**, *10*, 145–155.
19. Diacono, M.; Montemurro, F. Long-term effects of organic amendments on soil fertility. A review. *Agron. Sustain. Dev.* **2010**, *30*, 401–422. [CrossRef]
20. Zhou, H.; Chen, C.; Wang, D.; Arthur, E.; Zhang, Z.; Guo, Z.; Peng, X.; Mooney, S.J. Effect of long-term organic amendments on the full-range soil water retention characteristics of a Vertisol. *Soil Till. Res.* **2020**, *202*, 104663. [CrossRef]
21. Campos, E.V.; Proença, P.L.; Oliveira, J.L.; Bakshi, M.; Abhilash, P.C.; Fraceto, L.F. Use of botanical insecticides for sustainable agriculture: Future perspectives. *Ecol. Indic.* **2019**, *105*, 483–495. [CrossRef]
22. European Commission (EC). *Directive (EU) 2018/851 of the European Parliament and of the Council of 30 May 2018 Amending Directive 2008/98/EC on Waste*; OJL 150, 14.6.2018; European Commission (EC): Brussels, Belgium, 2018; pp. 109–140.
23. Oliveira, T.; Dias, R.R. Tendências Económico-Sociais No Consumo de Café em Portugal Para. 2021. Available online: https://www.academia.edu/16593649/O_Cafe_Perspectivas_e_Tende_ncias_Sociais_em_Portugal (accessed on 30 September 2021).
24. Caldeira, D.C.A. Valorização da Borra de Café: Otimização da Produção de Biodiesel por Catálise Enzimática. Ph.D. Thesis, Instituto Superior de Engenharia do Porto, Porto, Portugal, 2015.
25. AMPROMIS. Associąião Nacional dos Produtores de Milho e Sorgo. 2021. Available online: https://www.anpromis.pt/ (accessed on 25 August 2021).
26. Pinto, J.; Cruz, D.; Paiva, A.; Pereira, S.; Tavares, P.; Fernandes, L.; Varum, H. Characterization of corn cob as a possible raw building material. *Constr. Build. Mater.* **2012**, *34*, 28–33. [CrossRef]
27. Ramos, A.; Briga-Sá, A.; Pereira, S.; Correia, M.; Pinto, J.; Bentes, I.; Teixeira, C.A. Thermal performance and life cycle assessment of corn cob particleboards. *J. Build. Eng.* **2021**, *44*, 102998. [CrossRef]
28. Viana, R.L.S.; Fidelis, G.P.; Medeiros, M.J.C.; Morgano, M.A.; Alves, M.G.C.F.; Passero, L.F.D.; Pontes, D.L.; Theodoro, R.C.; Arantes, T.D.; Sabry, D.A.; et al. Green synthesis of antileishmanial and antifungal silver nanoparticles using corn cob xylan as a reducing and stabilizing agent. *Biomolecules* **2020**, *10*, 1235. [CrossRef]
29. Stylianou, M.; Agapiou, A.; Omirou, M.; Vyrides, I.; Ioannides, I.M.; Maratheftis, G.; Fasoula, D. Converting environmental risks to benefits by using spent coffee grounds (SCG) as a valuable resource. *Environ. Sci. Pollut. R.* **2018**, *25*, 35776–35790. [CrossRef] [PubMed]
30. McNutt, J.; He, Q. Spent coffee grounds: A review on current utilization. *J. Ind. Eng. Chem.* **2019**, *71*, 78–88. [CrossRef]
31. Yamane, K.; Kono, M.; Fukunaga, T.; Iwai, K.; Sekine, R.; Watanabe, Y.; Iijima, M. Field evaluation of coffee grounds application for crop growth enhancement, weed control, and soil improvement. *Plant Prod. Sci.* **2014**, *17*, 93–102. [CrossRef]
32. Cervera-Mata, A.; Pastoriza, S.; Rufián-Henares, J.A.; Párraga, J.; Martín-García, J.M.; Delgado, G. Impact of spent coffee grounds as organic amendment on soil fertility and lettuce growth in two Mediterranean agricultural soils. *Arch. Agron. Soil Sci.* **2018**, *64*, 790–804. [CrossRef]
33. Hardgrove, S.J.; Livesley, S.J. Applying spent coffee grounds directly to urban agriculture soils greatly reduces plant growth. *Urban For. Urban Green.* **2016**, *18*, 1–8. [CrossRef]
34. Gomes, T.; Pereira, J.A.; Ramalhosa, E.; Casal, S.; Baptista, P. Effect of fresh and composted spent coffee grounds on lettuce growth, photosynthetic pigments and mineral composition. In *VII Congreso Ibérico de Agroingeniería y Ciencias Horticolas*; Téllez, A., Masaguer, F., Sancho, I., Robinson, M., Altisent, M., Ballesteros, F., Strap, E., Eds.; SECH e SEAgIng: Madrid, Spain, 2014; pp. 1–5.
35. Batish, D.R.; Singh, H.P.; Kaur, M.; Kohli, R.K.; Yadav, S.S. Caffeine affects adventitious rooting and causes biochemical changes in the hypocotyl cuttings of mung bean (*Phaseolus aureus* Roxb.). *Acta Physiol. Plant.* **2008**, *30*, 401–405. [CrossRef]
36. Álvarez-Iglesias, L.; Puig, C.G.; Revilla, P.; Reigosa, M.J.; Pedrol, N. Faba bean as green manure for field weed control in maize. *Weed Res.* **2018**, *58*, 437–449. [CrossRef]
37. Al-Chammaa, M.; Al-Ain, F.; Kurdali, F. Growth, nitrogen and phosphorus uptake of sorghum plants as affected by green manuring with pea or faba bean shell pod wastes using N. *Open Agric. J.* **2019**, *13*, 133–145. [CrossRef]
38. Roshan-Bakhsh, A.; Pourjam, E.; Ayyari, M.; Pedram, M. Biocontrol properties of some agricultural waste extracts on three nematode species in *in vitro* and *in vivo* conditions. *Nematology* **2019**, *21*, 837–846. [CrossRef]

39. Hernández, M.; Ventura, J.; Castro, C.; Boone, V.; Rojas, R.; Ascacio-Valdés, J.; Martínez-Ávila, G. Uplc-esi-qtof-ms2-based identification and antioxidant activity assessment of phenolic compounds from red corn cob (*Zea mays* L.). *Molecules* **2018**, *23*, 1425. [CrossRef]
40. Gullón, P.; Eibes, G.; Lorenzo, J.M.; Pérez-Rodríguez, N.; Lú-Chau, T.A.; Gullón, B. Green sustainable process to revalorize purple corn cobs within a biorefinery frame: Co-production of bioactive extracts. *Sci. Total Environ.* **2020**, *709*, 136236. [CrossRef] [PubMed]
41. Kottek, M.; Grieser, J.; Beck, C.; Rudolf, B.; Rubel, F. World map of the Köppen-Geiger climate classification updated. *Meteorol. Z.* **2006**, *15*, 259–263. [CrossRef]
42. Guilherme, R.; Reboredo, F.; Guerra, M.; Ressurreição, S.; Alvarenga, N. Elemental composition and some nutritional parameters of sweet pepper from organic and conventional agriculture. *Plants* **2020**, *9*, 863. [CrossRef] [PubMed]
43. Abbas, T.; Nadeem, M.A.; Tanveer, A.; Ali, H.H.; Farooq, N. Role of allelopathic crop mulches and reduced doses of tank-mixed herbicides in managing herbicide-resistant *Phalaris minor* in wheat. *Crop Prot.* **2018**, *110*, 245–250. [CrossRef]
44. Farooq, M.; Nawaz, A.; Ahmad, E.; Nadeem, F.; Hussain, M.; Siddique, K.H. Using sorghum to suppress weeds in dry seeded aerobic and puddled transplanted rice. *Field Crop. Res.* **2017**, *214*, 211–218. [CrossRef]
45. Magalhães, M.C.; Cameira, M.C.; Pato, R.L.; Santos, F.; Bandeira, J. Biomassa florestal residual: Efeitos da sua remoção na qualidade do solo. *Rev. Ciências Agrárias* **2011**, *34*, 205–217.
46. Almeida, D. *Manual de Culturas Hortícolas*; Coleção: Guias Práticos; Presença: Lisboa, Portugal, 2006; Volume II, p. 400.
47. Ali, A.M.; Thind, H.S. A framework for refining nitrogen management in dry direct-seeded rice using GreenSeeker™ optical sensor. *Comput. Electron. Agric.* **2015**, *110*, 114–120. [CrossRef]
48. R Development Core Team. *R: A Language and Environment for Statistical Computing*; R Foundation for Statistical Computing: Vienna, Austria, 2015.
49. Álvarez-Iglesias, L. *Vicia faba* L. for Weed Control: From Biochemical Evidences to Field Application. Ph.D. Thesis, Universidade de Vigo, Vigo, Spain, 2015.
50. Osunde, M.O.; Olayinka, A.; Fashina, C.D.; Torimiro, N. Effect of carbon-nitrogen ratios of lignocellulosic substrates on the yield of mushroom (*Pleurotus pulmonarius*). *OALib. J.* **2019**, *6*, 1–8. [CrossRef]
51. Miranda, J.; Costa, L.M.D.; Ruiz, H.A.; Einloft, R. Composição química da solução de solo sob diferentes coberturas vegetais e análise de carbono orgânico solúvel no deflúvio de pequenos cursos de água. *Rev. Bras. Cienc. Solo* **2006**, *30*, 633–647. [CrossRef]
52. Urbaniak, M.; Gągała, I.; Szewczyk, M.; Bednarek, A. Leaching of PCBs and nutrients from soil fertilized with municipal sewage sludge. *Bull. Environ. Contam. Tox.* **2016**, *97*, 249–254. [CrossRef]
53. Kobayashi, K. Factors affecting phytotoxic activity of allelochemicals in soil. *Weed Biol. Manag.* **2004**, *4*, 1–7. [CrossRef]
54. Chrysargyris, A.; Antoniou, O.; Xylia, P.; Petropoulos, S.; Tzortzakis, N. The use of spent coffee grounds in growing media for the production of Brassica seedlings in nurseries. *Environ. Sci. Pollut. Res.* **2021**, *28*, 24279–24290. [CrossRef] [PubMed]
55. Peerzada, A.M. Biology, agricultural impact, and management of *Cyperus rotundus* L.: The world's most tenacious weed. *Acta Physiol. Plant.* **2017**, *39*, 270. [CrossRef]
56. Riemens, M.M.; van der Weide, R.Y.; Runia, W.T. *Nutsedge, Biology and Control of Cyperus rotundus and Cyperus esculentus, Review of a Literature Survey*; Plant Research International B.V.: Wageningen, The Netherlands, 2008.

agronomy

MDPI

Review

Sustainable Approach to Weed Management: The Role of Precision Weed Management

António Monteiro [1,2,*] and Sérgio Santos [1]

[1] Agrarian Superior School, Polytechnic Institute of Viseu, Quinta da Alagoa, 3500-606 Viseu, Portugal; sergioalvesstos@gmail.com

[2] CERNAS, Research Centre for Natural Resources, Environment and Society, Polytechnic Institute of Viseu, Polytechnic Campus, 3500-606 Viseu, Portugal

* Correspondence: amonteiro@esav.ipv.pt

Abstract: In the last few decades, the increase in the world's population has created a need to produce more food, generating, consequently, greater pressure on agricultural production. In addition, problems related to climate change, water scarcity or decreasing amounts of arable land have serious implications for farming sustainability. Weeds can affect food production in agricultural systems, decreasing the product quality and productivity due to the competition for natural resources. On the other hand, weeds can also be considered to be valuable indicators of biodiversity because of their role in providing ecosystem services. In this sense, there is a need to carry out an effective and sustainable weed management process, integrating the various control methods (i.e., cultural, mechanical and chemical) in a harmonious way, without harming the entire agrarian ecosystem. Thus, intensive mechanization and herbicide use should be avoided. Herbicide resistance in some weed biotypes is a major concern today and must be tackled. On the other hand, the recent development of weed control technologies can promote higher levels of food production, lower the amount of inputs needed and reduce environmental damage, invariably bringing us closer to more sustainable agricultural systems. In this paper, we review the most common conventional and non-conventional weed control strategies from a sustainability perspective, highlighting the application of the precision and automated weed control technologies associated with precision weed management (PWM).

Keywords: agricultural production; sustainability; weed management; herbicide resistance; weed control technologies

Citation: Monteiro, A.; Santos, S. Sustainable Approach to Weed Management: The Role of Precision Weed Management. *Agronomy* **2022**, *12*, 118. https://doi.org/10.3390/agronomy12010118

Academic Editor: David Clements

Received: 16 November 2021
Accepted: 31 December 2021
Published: 4 January 2022

1. Introduction

The world population has rapidly exceeded seven billion and is expected to reach nine billion by 2050 [1]. Current crop production levels are not adequate to feed the growing population, and meeting this anticipated demand could be a huge challenge for humanity [2]. Climate change, the scarcity of arable land and water resources and the threat from diseases, pests and weeds are additional issues [3] that make the pressure on agricultural systems greater than ever before [4], with implications, in the short and long term, for sustainability, for the planet and for the quality of life of living beings. Weeds have been a persistent problem in agriculture since its beginning [5]. Weeds hinder the growth of crops by competing with the plants for water, nutrients and sunlight, which results in large losses in crop production. Most weeds are either controlled mechanically through specific cultivation practices or with the application of herbicides [6]. However, intensive mechanization increases soil erosion [7], leading to a loss of fertility. The use of herbicides contaminates the soil, water, food and air, causing diseases in humans and animals [8], creating the phenomena of herbicide resistance and unbalancing ecosystems. From this perspective, biodiversity plays a preponderant role in the provision of ecosystem services in agricultural systems. Agrobiodiversity can have a direct effect on services when increased crop diversity increases food resources, or when cover crop diversity increases plant biomass, improving water quality and lowering runoff [9]. However, agrobiodiversity and services, such as pollination, improved soil structures and natural pest control, are increasingly threatened by the massive elimination of weeds and wild plants, as well as due to species' toxification by agrochemical inputs [10]. Weeds perform a range of ecosystem functions in terms of soil quality and biodiversity support, which can help to sustain agroecosystem productivity in the long term [11]. Thus, as sustainable agriculture has the capacity to save natural resources for the future and develop farms with less cost, a transition to sustainable weed control is necessary for a variety of environmental, social and economic reasons [12]. Sustainable weed management comprises a suite of weed management options, including integrated weed management (IWM) [13], which is based on the employment of a multiplicity of weed control strategies [1]. IWM aims to optimize crop production and increase grower profit through the concerted use of preventive strategies, scientific knowledge, management skills, monitoring procedures and the efficient use of control practices [14]. In this context, a wide and rapidly expanding range of new technologies have been developed and implemented in agricultural practices, which also play a key role in progress towards economically and environmentally sustainable weed management [6]. Precision weed management leads to a reduction of inputs without decreasing weed control effectiveness [1]. Studies and experiments have shown significant potential savings and technical progress in sensing, weeding and spraying technologies. Some of these technologies have been commercially exploited [6]. Thus, the current paper aims to describe conventional and non-conventional weed control strategies from a sustainability perspective, showing the value of the application of different innovations associated with PWM. The remainder of this paper is organized as follows: in Section 2, we present the search strategy chosen to carry out the literature review. In Section 3 we discuss the weed concept in the context of population ecology and the sustainability of agricultural systems. Section 4 introduces weed management techniques and discusses the conventional and non-conventional methods and their limitations. Section 5 addresses the topic of precision weed control and the current technological trends it encompasses. Finally, in Section 6 we present our conclusions regarding the potential role of precision weed control in taking integrated weed management to another level.

2. Search Strategy

The main motivation for carrying out this review lies in the fact that precision weed management plays a key role in farming production and sustainability. A literature analysis for the review involved a keyword-based search mainly for conference and/or journal articles. The ScienceDirect and IEEE Xplore scientific research databases, as well as the Google Scholar web scientific search engine, were selected to perform this review. We collected the information about the currently available weed management techniques, summarizing several methods for this agricultural practice. To achieve the objective proposed for this research, 119 studies were collected using the search terms "weed management" and "precision weed" without any restrictions for the years considered or language. Other keywords were also used, such as "sustainable weed management", "non-chemical weed management" or "weed technology". After excluding the duplicated studies (39 total), 80 articles were analyzed.

3. Weeds: Threat or Benefit?

Weeds are certainly as old as agriculture, and from the beginning, farmers recognized that the presence of those unsown plant species interfered with the growth of the crop they intended to produce [15]. The term "weed" can be defined, from an agronomical point of view, as any plant not intentionally sown or propagated by the farmer that requires management to avoid any interference with crop or livestock production [16]. Indeed, "volunteer crops", such as buckwheat, rye, Japanese millet, corn or soybean, can become weeds when they self-seed and emerge in another part of the crop rotation when they are no longer wanted [17].

Weeds are one of the main factors restricting food production in farming systems all over the world. By growing simultaneously with crops, weeds decrease productivity and even the quality of the harvested product, whether due to competition for water, sunlight, nutrients and space, allelopathy [18] or parasitism [17,19]. Furthermore, weeds are the costliest category of agricultural threats, causing more than 45% of loss in yields of field crops, when compared to crop diseases (25%) or insect pests (20%) [12]. Yield losses caused by weeds depend on several factors, such as weed emergence time, weed density, type of weeds and crops, etc. Uncontrolled weeds can result in 100% yield loss [20]. In addition, weeds interact with other biological agents of the ecosystem, acting as a host for insects and pathogens (fungi and bacteria), which can cause serious damage to crop plants [13]. Moreover, weeds decrease land value (especially perennial and parasitic weeds), and interfere with water management (e.g., increased losses through evapotranspiration, reduced water flow in irrigation ditches, etc.) [17]. Serious weed problems develop when a susceptible crop, a large weed seed bank in the soil (including both true seeds and vegetative propagules of perennial weeds), and a favorable environment for weed growth occur together [16]. All weeds' interactions with human goals represent a permanent constraint to human activities and this justifies the employment of control tactics aimed at killing or managing as many weeds as possible. However, the complete eradication of actual (emerged plants) and potential (seed bank) weeds is unachievable [15]. The presence of weeds in the crop field also provides a series of agronomic and ecological benefits when they occur, especially at low densities [17].

Based on the ecological approach, weeds are considered to be important indicators of biodiversity, playing a key role in providing food or shelter for animal species. For example, a large proportion of the decline in farmland birds has been associated with a reduction in weed occurrence in arable crops [15]. Weeds with a deep and extensive root system can decrease soil erosion and mineral nutrient leaching, conserve soil moisture and improve soil structure [17]. Weeds can also be an indirect resource for predatory species, as, in fact, they could provide alternative food sources for the biotic agents that play a key role in pest control. A solid knowledge of the long-term dynamics of the weed population and how it is affected by different weed management strategies is critical for developing an optimum crop management strategy [15]. Agricultural practices that enhance diversity and ecosystem functioning offer, in general, potential improvements for sustaining agricultural productivity and farm livelihoods, as well as broader biodiversity and better ecosystem functioning. More diverse agroecosystems usually provide: (a) greater resilience; (b) less dependence on pesticides; (c) more reliable incomes and (d) better quality of life for the rural population [11].

4. Weed Management and the Need for a New Paradigm

Managing weeds has always been placed at the center of agricultural activity by farmers since ancient times [17]. The control of weeds is a big challenge in agriculture and in many cases a complex, controversial and also expensive problem to solve [10]. In fact, weed management accounts for nearly one third of the total cost of the production of field crops [12]. This agronomic practice goes beyond the control of existing weed problems and places greater emphasis on preventing weed reproduction, reducing weed emergence after crop planting and minimizing weed competition with the crop [21].

Currently, weed management in agricultural systems branches out into two distinct directions corresponding to different approaches. On the one hand is the widespread use of synthetic herbicides, while on the other, weed control is widely based on mechanical, cultural and physical methods [17]. Mechanical methods are generally inefficient, while herbicides have a negative impact on the ecosystem. In this regard, mechanical and chemical weed control has disadvantages that will probably impede their effectiveness for future weed management. Thus, weed management requires an integrated approach that minimizes the drawbacks of mechanical and chemical weed control [22]. Indeed, there is a great need for a new weed management paradigm in modern agriculture that is based on ecological principles and non-conventional weed management approaches [23]. Sustainable weed control for the crop can significantly influence the operation of machinery, the reduction of pest habitats (e.g., for voles) and make contributions to satisfactory economic benefits through the quality of harvested products [24], as required by the market [25].

IWM plays a key role in the weed management of the advanced cropping systems of developed countries, especially in the European Union, while it is still poorly adopted in developing countries [17]. A combined use of different weed control methods (agronomic, physical, mechanical and chemical) within a system, rather than relying on a single method [20], is required in IWM [26]. This strategy is important for reducing the selection pressure for the development of resistance to any single method of weed control [20]. Furthermore, the use of non-chemical weed tactics in minor crops is important due to the scarce availability of chemical compounds [26]. Unlike traditional processes, IWM integrates many agro-ecological practices, such as the role of conservation tillage and crop rotation on weed seed bank dynamics, the ability to predict the critical period of weed interference and its competition with crops, and the specific critical levels of crop/weed interaction [22]. In a broader context of IWM, emerging technologies have the potential to change the current approach to weed control and help significantly reduce environmental impacts, such as herbicide resistance or drift and the high cost of inputs and labor, without decreasing weed control efficacy. Several methods are being developed to observe and detect weeds so that control measures can be applied wherever and whenever they are needed. This paradigm shift is based on an interdisciplinary work to harness powerful technology tools and use them to control weeds [1]. From this perspective, we will present in the next section the PWM's contributions to weed control, which could be considered to be an important upgrade in IWM.

4.1. Conventional Weed Control Strategies

4.1.1. Herbicide Control

The use of herbicides, also known as chemical substance applications, is at the top of weed control methods. Herbicides can improve production efficiency, facilitate reduced tillage production systems [21] and require less costs and human effort [17]. Herbicides were introduced in agriculture mainly to combat the weeds that compete with crops for nutrients and sunlight. Other common uses in the farm are to eradicate invasive plant species or undesirable plants for livestock farms [10]. A large number of herbicides have been produced and are currently under development for field crops. Herbicides can be classified according to the chemical family, the time of application (preplant, pre-emergence and postemergence), the mechanism of action, their formulation, the site of uptake and their selectivity [17]. In a similar manner to other pesticides, herbicides' active ingredients are biologically active compounds. They are designed to pass through membranes and diffuse into the interior of living cells to exert the desirable toxic action [10]. The use of herbicides should be as minimal and as efficient as possible in order to eliminate the negative environmental impacts, which can bring them a step closer to agricultural sustainability [27]. In this sense, global changes and herbicide policies in Europe compel farmers to reduce their herbicide use (Directive 2009/128/CE, 2009) in order to limit their impact on human health and the environment. In response, farmers need to replace herbicides with a combination of multiple, mostly preventive and partially efficient practices [28].

4.1.2. Mechanical Control

Controlling weeds via mechanical means is challenging and requires the combination of different weeding techniques and cultivation strategies to achieve economically acceptable weed control levels [29]. Mechanical weed control destroys weeds or reduces their competitive ability via physical means [30]. In addition, mechanical weeding can provide effective weed management even when other methods are not possible and can outperform them in some situations [31]. The choice of mechanical weeding method depends in part on practical considerations, such as the crop, soil type, price, operating costs and labor requirements [10].

Mechanical weed control is strongly associated with cultivating tillage [27,32]. Various forms of tillage are used for mechanical weed control; however, cultivating tillage is the key component in mechanical control in growing crops [30]. Cultivating tillage is carried out after crop sowing/planting to control weeds, and consists of shallow tillage with a variety of control tools, such as the tine harrow, rotary hoe or row cultivator. It includes whole-field cultivation, inter-row cultivation and intra-row cultivation [32], which are used to cultivate with the same intensity both on and between the crop rows, are used only between rows in row crops and are used to remove weeds from the row [21], respectively. Tillage has some beneficial factors, such as the capability of decomposing soil organic matter through soil disturbance, soil aeration, improved soil moisture status and the improved accessibility of organic residues to decomposers, all while being less labor intense [24]. Reduced tillage can also lead to indirect environmental benefits, including reductions in water pollution through pesticides and fertilizer runoff [33].

Mowing and cutting also have a large impact on weed control [32]. These operations are commonly used in turf, in rights of way, in vineyards, in orchards, in pastures and in forage crops [21]. These techniques are seldom efficient enough to obtain total weed control. Cutting and mowing weeds reduces their leaf area, slows their growth and decreases or prevents seed production [34]. However, weed control by cutting or mowing may be complicated due to the adaptation mechanisms to continuous defoliation of some weeds, especially perennial ones in grasslands. A high frequency of cutting of these weeds is required to achieve high degrees of weed control [32].

4.2. *Non-Conventional Weed Control Strategies*

4.2.1. Mulching

Covering the soil with plant residues/wastes or synthetic materials, commonly referred as "mulching", is one of the most popular management methods which can decrease weed issues, either by preventing weed seed germination altogether, or by blocking the growth of emerging seedlings [10]. It also promotes the sustainable management of water and biodiversity [19]. The additional advantage of mulching includes the management of temperature fluctuations and improved physical, chemical and biological characteristics of the soil. Mulches are available in distinct ways, including natural mulches, such as straw, sawdust, weeds, paper and plant residues, and synthetic mulches, such as plastic [25]. Materials, such as black polyethylene, have been used for weed control in a range of agricultural production systems [12], namely in horticultural crops (e.g., strawberry, tomato, eggplant, muskmelon, watermelon, etc.) [26]. Plastic mulches have been developed that filter out photosynthetically active radiation but let through infrared light to warm the soil. These infrared-transmitting mulches have been shown to be effective at controlling weeds [12]. Mulching is usually more effective against annual weeds rather than perennial weeds (e.g., *Cyperus* spp., *Elymus repens* (L.) Gould., *Cynodon dactylon* (L.), *Sorghum halepense* (L.) Pers.) because they greatly perforate plastic [16].

4.2.2. Cover Crops and Living Mulches

The adoption of a cover crops strategy can improve farm sustainability. A cover crop is any living ground cover that is planted into or after a main crop and then commonly eliminated before the next crop plantation [35]. Cover crops suppress weeds by occupying their ecological niche and competing for resources, while their soil surface residues inhibit weeds through physical, biotic and allelopathic interactions. The cover crop species can inhibit weed seed germination through the deposition of allelochemical compounds, which may be secreted both from living plants and decaying cover crop residues. Those species can be grown in rotation at times when crops are not being grown or simultaneously during part or all of the commercial growing season [34]. Cover crops enhance soil quality and carbon sequestration, facilitate machinery passage and increase microbial, vegetal and animal biodiversity [36]. However, the great benefits of cover crops as weed mitigators are usually associated with high cover crop biomass or rapid soil cover. If biomass and residues are scarce or decompose within a short period of time, herbicide use might be needed, depending on weed pressure. In this sense, choosing the best-adapted vegetation cover species is critical [37]. In addition, the choice of proper termination methods for cover crops can influence weed suppression capacity. Rolling-crimping and flail-mowing are effective mechanical methods for cover crop termination. Flail-mowing results in small fragments of material that decompose faster and are less persistent as mulch. Rolling flattens the cover down to form a mulch layer that decomposes slowly, when compared with flailing, and can provide more complete ground cover [38].

Living mulches are cover crops sown previously or at the same time as the main crop and maintained as a living ground cover throughout the growing season. If the living mulch is a perennial, it may be possible to maintain it without the need for reseeding every year [35]. Living mulches can decrease nutrient leaching, especially of nitrates, along with the absorption of carbon and nitrogen [39,40], and provide efficient control of soil erosion, build up the organic matter for better soil structure, and provide a habitat for beneficial insects [25]. Some conditions are essential to improve the efficiency of the living mulches, such as areas with fertile soils, a sufficient water supply and the absence of perennial weed species. In addition, living mulches should be used only with established crops, as the competition for water and nutrients is much greater in the early stages of plant growth [24].

4.2.3. Soil Solarization

Soil solarization is an eco-friendly and cost-saving process of soil disinfestation [41], compatible with organic and integrated crop management systems [42], that uses the sun's heat to control weeds [43]. This method consists of placing a cover, such as a black or transparent plastic, over the soil surface to trap solar radiation and promote an increase in soil temperature [34]. The plastic cover must stay on the soil surface for 4 to 12 weeks [22]. In order to be effective for weed control, the soil needs to be kept moist during that time and, for large areas, preferably under drip irrigation [40], and intense radiation throughout the day is required [34]. This process is particularly applicable for the Mediterranean climate and similar climates due to the occurrence of high air temperatures in the summer [44], and higher exposure to high-energy electromagnetic radiation. Soil solarization allows farmers to maintain a high soil temperature (>40b °C), which is enough to eliminate weed seeds, plants, insects and plant pathogens, such as nematodes and fungal diseases. The application of solarization is generally restricted to vegetable and minor crops (e.g., tomato, radish, lettuce, colewort, cucumber and pepper) under greenhouse cultivation, although it is considered to be effective also in open field conditions [26].

4.2.4. Thermal Weed Control

The thermal control of weeds is based on the use of fire, flaming, hot water, steam and freezing, which provide rapid weed control without leaving chemical residues in the soil and water. Moreover, thermal methods are selective towards the weeds, do not disturb the soil and, therefore, do not bring the buried seeds to the soil surface, as is the case with cultivation methods [34]. Flaming is the thermal method most commonly applied in organic and conventional farming systems, and relies on propane gas burners or, recently, renewable alternatives, such as hydrogen, to generate combustion temperatures up to 1900 °C [17], rapidly raising the temperature of the exposed plant tissues. Heat injury causes the destruction of plant membranes, which results in the loss of cell function. Eventually, the plants die or become severely weakened [43]. Dicotyledonous and young weed plants are more sensitive compared to developed plants and grass species [26]. Flaming is most effective at controlling erect and broad-leaved weeds in an early stage of growth, and it has been shown to be less effective in the control of grassy and prostrate weeds [25]. It should be noted that flaming should not be confused with burning, since plant tissues do not ignite, but heat rapidly up to the point of rupturing the cell membranes [17]. As an alternative to herbicides, the efficiency of flaming can be enhanced by its integration with tillage or mulching strategies [25]. Although thermal weed methods do not leave chemical residues in the soil and water, this approach uses large amounts of fossil fuels per unit area. The effectiveness of thermal means on weeds can be influenced by several factors, including temperature, exposure time and energy input [34].

4.2.5. Weed Control through Livestock Grazing

There is a readily available and under-exploited method that is fast proving very effective for weed control: livestock grazing. Incorporating grazing management into weed management plans has been recognized as one of the key components in successfully addressing weed problems [45]. Furthermore, in the agroforestry systems, for example, the combination of livestock with trees and shrubs provides multiple benefits, including biodiversity conservation and improved soil fertility [46]. Weed control with livestock grazing aims to manipulate patterns of defoliation to place a target plant at a competitive disadvantage relative to other plants in the community [45]. The efficacy and conservation benefits of targeted grazing can vary based on the timing, duration and intensity of grazing, as well as the grazing species [29]. In most cases, grazing does not eradicate a mature infestation of weeds. For successful weed control, grazing animals must be fenced into or off an area in order to adjust the grazing pressure. The ability to concentrate stock on weed infestations at some stages of growth or times of the year, and the ability to keep them off pasture or weeds at other times, is often the key to weed control [47]. Cattle, sheep and goats are ruminants and the most common animals used for weed control. Combining ruminant grazing with other weed management tools can offer an integrated approach that may be very cost effective [19,47]. In particular, sheep are a great tool for managing weed problems [43]. Increasingly, farmers are coming to view grazing sheep as an effective way to manage weeds and cover crops instead of chemicals, tillage or mowing. For example, sheep can replace the use of herbicides or mowing to manage floor vegetation in vineyards and orchards. Important considerations when grazing with sheep include the need for regular rotations, temporary fencing and protection against predators. In addition, sheep should not be able to eat the crop itself, which is essential for vineyard managers, for example [48].

4.3. Limitations of Conventional and Non-Conventional Weed Control Strategies

Apart from the advantages of using herbicides for weed control, there are also disadvantages, mainly due to limitations of the conventional spraying technologies [27]. Continuous use of the same group of herbicides over a period of time on the same piece of land leads to ecological imbalance in terms of weed shift, herbicide resistance in weeds and environmental pollution [12]. Indeed, the overuse of herbicides with the same mode of action may lead to the development of herbicide-resistant weed populations [32]. As a result, agricultural landscapes now tend to be dominated by a few weed species that are difficult to control and that provide a poor resource for farmland biodiversity [11]. For example, cutleaf evening primrose (*Oenothera laciniata* Hill) has become resistant to glyphosate and paraquat [22]. Herbicides can also have negative side effects, such as surface and ground water contamination, as well as leaving herbicide residues in the food chain [32,49]. In addition, chemical herbicides can substantially decrease the soil microbial communities and earthworm populations, and the persistent effects of weed suppression can lead to the reduction of nutrient availability and soil biodiversity [25].

In the same way, the excessive use of tillage results in substantial harmful effects on the soil quality parameters, including biological diversity, soil structure and water storage capacity. Tillage reduces the supply of carbon and nitrogen nutrients to microorganisms [25]. Soil erosion and soil degradation, inherent in tillage-based systems, increase the environmental pollution from agricultural chemical inputs, such as fertilizers and pesticides, compromising the sustainability of crop production and ecosystem services, as well as threatening global food security in the long run [15]. Moreover, the operation may face limitations owing to adverse weather conditions. There are also potential problems associated with minimum tillage or non-tillage. The bulk density and compaction of the topsoil increases, and the phytosanitary situation worsens with a higher spread of fungal diseases and the weed infestation of crops [50]. Furthermore, farmers using reduced tillage may choose to rely increasingly on herbicides and pesticides to deal with these threats [38] and, as a result, the phytotoxicity of the soil increases.

Ground cover methods, flaming or livestock grazing for weed control also have a few limitations. For example, mulching is cost intensive on a large scale, can promote changes in the soil due to the continuous use of the same mulching material and some of the organic mulches have allelopathic effects on crops [51]. In addition, many types of organic mulching, such as grass and straw, contain seeds which could allow weeds to grow and acidify the soil [52]. Cover crops incur expenses for novel equipment, more complicated management practices and time spent seeding and eliminating cover crops instead of managing cash crops [53]. Living mulches can reduce main crop growth and yield due to competition for water and nutrients, increase pest populations and the risk of diseases. Moreover, living mulches can also promote allelopathy [54]. Soil solarization induces high temperatures that can be lethal to bacteria and fungi. In some species, if the lethal temperature is not reached, dormancy can be broken, allowing an emergence of a new flush of weed seedlings. This can occur along the topsoil layer [22]. Solarization tends to result in a flush of nutrients which should be managed by immediately establishing the crop after plastic removal to prevent nutrient loss [40]. In a flaming strategy, fuel and water consumption can be high, and the flame has restrictions for use during the summer from a fire prevention standpoint. However, smaller, more portable units are now available and provide another tool for the spot control of escape weeds or around sheds and other pieces of infrastructure [26]. Finally, weed control via livestock grazing can cause damage to the soil structure and non-target species, lead to the spread of weed seeds in feces or on wool, hair or hooves, or even cause the loss of animal condition or liveweight [55].

Some of the limitations described above can be mitigated or even eliminated when technology associated with PWM is integrated. The use of the internet, the various types of sensors, artificial intelligence or machine learning can provide potential improvements to IWM. It may be said that we are entering a new era of agriculture, Agriculture 4.0, where precision is the rule [56].

5. Precision Weed Management

Smart farming technologies, such as smart sensors, remote sensing, air vehicles, satellites, the Internet of Things (IoT) technology, etc., are becoming increasingly common in modern agriculture to assist in optimizing agricultural production and minimizing the wastes and costs [57]. Precision farming or site-specific crop management is a concept based on sensing or observing and responding with management actions to spatial and temporal variability in crops. The "sensing" component of the concept is a fundamental element of precision farming [58], as is variable rate technology (VRT), which offers an effective way to protect the environment and increase economic benefits [59]. This technology works by integrating a variable rate control system with a sprayer for fertilizer, pesticide or herbicide applications. The application at a varied rate can be fundamentally based on maps or sensors [60]. Indeed, there are two main methods for implementing site-specific variable rate applications (VRA): map-based VRA, which adjusts the application rate of a crop production input based on the information contained in a digital map of field properties, and sensor-based systems that use data from real-time sensors to match inputs to the needs of the soil and crop [61]. From this perspective, precision farming technologies can provide many benefits for weed management practices [21]. As mentioned above, weeds are a persistent problem, and the continuing rise in numbers of herbicide-resistant biotypes reinforces the lesson that weed control technology (Table 1) must constantly advance to stay ahead of weed evolution and adaptation [2].

Table 1. Overview of precision weed control technologies.

Weed Control Technologies	Method	Remarks	Drawbacks	Ref.
UAV's	Combination of UAVs and GPS technologies	Fast and precise in situ remote sensing or survey operations. Excellent control in the presence of obstacles, no compaction and minimal labor involved.	These systems do not offer the same territorial coverage as satellites. Some technology literacy is required.	[13]
Hyperspectral imaging sensors	Hyperspectral imaging system coupled to a micro-spray heated oil application system	Less computationally intensive. Robust to visual occlusion of the leaf margin. Customizable spray application for various herbicides based on weed species.	Requires a multi-season calibration process.	[62]
Automatic weeders	Intra-row robotic weeder (Robovator)	Recognition of the crop row and the size difference between the crop and weed. Removes 95% of weeds.	The machine cannot distinguish between weed and crop. It can only distinguish between small and large plants.	[25,63]
Precision spray systems	Autonomous robot for precision spraying	Autonomously sprays targets with high accuracy.	N.A.	[63]
Weed sprayers	Machine vision weed spot-sprayer	Distinguishes weed leaves from maize plants with more than 90% accuracy.	N.A.	[64]

A wide range of weed sensing techniques have been studied since the beginning of the century. With large areas, the most cost-effective approach may be remote sensing to provide a farm, or a large area encompassing several farms, with maps of weed occurrence [6]. Remote sensing uses satellite or manned/unmanned aerial vehicles to collect data. Satellite-based remote sensing is well suited for surveying a large area and can help with large-scale crop yield monitoring. Satellite imagery lacks precision in assessing small areas, especially for weed detection, spatial distribution and herbicide injury evaluations. These tasks require high-resolution imagery, which is typically achieved through closer observations using manned/unmanned aerial or ground vehicles [65].

Unmanned aerial vehicles (UAVs) can be highly valuable, since they allow for site-specific weed management (SSWM), an improved weed management approach for the highly efficient and environmentally safe control of weed populations, enabling precise and continuous monitoring and mapping of weed infestation [13]. In addition, there are other advantages to using UAV technology. Indeed, UAVs provide helpful information for the precise application of amounts of water in the required field, contributing to water savings in agriculture [66]. Additionally, the use of UAVs for spraying and seeding purposes can prevent problems of subsoil compaction [67]. This technology also has the potential to minimize soil degradation, the loss of soil fertility and the subsequent contamination of water due to the excessive use of fertilizers, and can potentially save time by tremendously reducing inspection times [68]. The applications of UAVs have been increasing in forestry, rangeland ecology and agronomic cropping systems, among several other fields [65]. The combination of UAVs with advanced cameras and sensors, able to discern specific weeds, and global navigation satellite system (GNSS) or global positioning system (GPS) technologies, which provide geographical information for field mapping, can help in precisely monitoring large areas in a few minutes. Currently, UAVs stand out among the other remote sensing platforms, as they can fly at low altitudes, capture images with millimetric accuracy and provide data on demand in critical moments, which are not feasible with aerial or satellite platforms [69]. When compared to unmanned ground vehicles (UGVs), UAVs take less time to monitor or survey the crop field and have optimal control in the presence of any natural barrier, which is critical when working between crop rows [70]. Mainly three types of cameras are used for weed identification with UAVs: red, green and blue (RGB), multispectral and hyperspectral cameras, which can recognize weed patches with good accuracy depending on the flying altitude, camera resolution and UAV used. Therefore, the combined use of UAVs and image processing technologies may help to effectively control different weed species interfering with the crops with relevant environmental benefits [13].

Multispectral and hyperspectral imaging sensors mounted on UAVs have been used successfully to detect weeds and distinguish species. This kind of technology can provide valuable information that is not obtained by RGB cameras or not visible to the naked eye. In particular, hyperspectral imaging has been used more often to classify agricultural systems and vegetation because it has more bands compared to that of multispectral sensors [65]. In fact, the most powerful and, to date, the only method capable of robust, automated in-field discrimination of individual plant species is based upon hyperspectral imaging. The hyperspectral imaging concept has been demonstrated in the field with lettuce (*Lactuca sativa* L.) and tomato (*Lycopersicum esculentum* Mill.) crops, with between-species pixel-level recognition rates above 75% and crop vs. weed discrimination rates above 90% [43]. For example, Zhang et al. [62] developed a hyperspectral imaging system coupled to a micro-spray heated oil application system for weed control within the seed lines of early growth tomatoes. According to the authors, the hyperspectral imaging system correctly identified 95%, 94% and 99% of tomatoes, black night shade and pigweed, respectively. This technology is less computationally intensive than shape-based pattern recognition, it is robust to visual occlusion of the leaf margin and the species recognition ability can be used to customize the spray application of multiple herbicidal materials based upon the weed species. However, the requirement of a multi-season calibration process is a disadvantage of this method [37].

Automation technologies and mechatronics are likely to become more effective and commercially viable as future weed control strategies and they are already being used in industrialized countries with specific crops [71]. Typically, vegetable crops, such as broccoli, cabbage, field-grown flowers, herbs, lettuce, onion and tomato, among others, are hand weeded to achieve intra-row weed control. In this sense, the industry has responded to the need for automation of intra-row cultivators [2] as a viable alternative to hand weeding [47]. According to Peruzzi et al. [72], four kinds of intra-row robotic weeders are commercially available for precision weed-management systems: Robovator (Frank Poulsen Engineering Aps., Hvalsø, Denmark); Robocrop (Tillett and Hague Technology Ltd., Greenfield, Bedfordshire, England); IC-cultivator (Machinefabriek Steketee BV, Haringvliet, The Netherlands); and Remoweed (Costruzioni Meccaniche Ferrari, Guidizzolo MN, Italy). Robovator is considered to be the most effective intra-row weed-management system, and is used predominantly in organic farming [25]. The Robovator system is designed to detect the difference between the crop plant and weed based on the recognition of the crop row and the size difference between the crop and weed. With the Robovator, each row has a camera, and images from the cameras are processed to determine the position of the crop, and then the computer signals the actuator to operate at the proper location. The Robovator intelligent cultivator was evaluated with different types of crops [43]. With broccoli and lettuce, for example, the Robovator reduced hand-weeding time by 39% and 27%, respectively, compared with the standard cultivator [2]. Interest in automation of weed sprayers has been rising in recent decades [73]. Precision spraying is able to minimize the amount of herbicide needed on a given crop, compared with traditional broadcast sprayers that usually treat the entire field to control weed populations, which potentially results in unnecessary application to areas that do not require treatment. The application of herbicide in a specific location, i.e., where weeds occur, could reduce costs, the risk of crop damage and excess pesticide residue, as well as potentially reducing the environmental impact [27]. The effectiveness of precision spray systems is based on high levels of crop/weed differentiation, accurate spray prescription maps, the knowledge of the sprayer tip location relative to the target weed location, accurate herbicide placement and control of the spray drift [43]. For example, spot spraying systems provide potential savings with herbicide use, which can range from 5% to nearly 90% [6,74,75], depending mainly on the spatial and temporal distribution of weeds found in the treated fields. In addition, according to Jensen et al. [76], the detailed and resource-efficient approach of herbicide spraying with SSWM in smart farming can decrease herbicide consumption by 40% to 60%. An autonomous robot for precision spraying was developed by Søgaard and Lund [63]. According to the authors, the system was able to deliver lower doses (2.5 μL) autonomously by spraying targets with sub-centimeter accuracy. The system was further tested in field trials planted with oilseed rape as a test weed. In the study, the plant surface area was found to have a large effect on machine performance [27]. Additionally, Kargar and Shirzadifar [64] developed a machine vision weed spot sprayer for maize fields. The system used image segmentation and feature extraction to distinguish the grass leaves from maize plants with over 90% accuracy. As corn leaves are much wider than grass, the detection accuracy is increased. Lastly, H-Sensor (Agricon GmbH, Ostrau, Germany) and See and Spray (Blue River Technology, Sunnyvale, CA, USA) are commercial spraying systems that use artificial intelligence and are able to distinguish between crop plants and several weeds [27].

6. Conclusions

With the growth of the world population and the consequent need to ensure the supply of food by increasing agricultural production, there is a need for improved management of the world's agricultural resources while minimizing the negative impact on the environment. From an agronomic point of view, weeds are considered to be a threat with serious implications for agricultural efficiency, causing yield losses. However, from an ecological perspective, they can also be considered to be valuable indicators of biodiversity in the agrarian ecosystem, as well as providers of ecological services as a component of the agroecosystem. Weed management involves several methods. Nevertheless, a single method of control will not provide adequate long-term weed management, and instead often results in increasing resistance. Therefore, the need to integrate different weed control methods under a holistic approach is critical.

The use of herbicides creates imbalances in the ecosystem, even causing the resistance of some species to the continued use of these chemical agents. In addition, no less serious are the environmental problems they cause and their consequent threat to the well-being and health of animals and humans.

Thus, the sustainable management of the agricultural system, namely of weeds, is an important issue for the present and future of humanity. In addition to integrated management, the development of precision technologies inherent to weed control can be a valuable contribution to improved sustainability and agricultural yield. In this sense, we would suggest a more effective involvement of researchers and farmers with the integration of ecological and technological principles into weed management decision making.

Author Contributions: Conceptualization, A.M.; methodology, A.M. and S.S.; software, S.S. and A.M.; validation, A.M; formal analysis, A.M.; investigation, A.M. and S.S.; resources, A.M. and S.S.; data curation, S.S.; writing—original draft preparation, A.M. and S.S.; writing—review and editing, A.M.; visualization, S.S. and A.M.; supervision, A.M.; project administration, A.M.; funding acquisition, A.M. All authors have read and agreed to the published version of the manuscript.

Funding: This research was funded by the FCT—Foundation for Science and Technology, I.P., Portugal, within the scope of the project Ref. UIDB/00681/2020. The APC was funded by FCT—Ref. UIDB/00681/2020.

Institutional Review Board Statement: Not applicable.

Informed Consent Statement: Not applicable.

Data Availability Statement: Not applicable.

Acknowledgments: The authors would like to acknowledge support from the FCT—Foundation for Science and Technology, I.P., within the scope of the project Ref. UIDB/00681/2020, as well as the CERNAS Research Centre and the Polytechnic Institutes of Viseu for their support.

Conflicts of Interest: The authors declare no conflict of interest.

References

1. Young, S.; Pierce, J. *Automation: The Future of Weed Control in Cropping Systems*; Springer: Berlin/Heidelberg, Germany, 2014; pp. 249–259. [CrossRef]
2. Westwood, J.; Charudattan, R.; Duke, S.; Fennimore, S.; Marrone, P.; Slaughter, D.; Swanton, C.; Zollinger, R. Weed Management in 2050: Perspectives on the Future of Weed Science. *Weed Sci.* **2018**, *66*, 275–285. [CrossRef]
3. Wang, A.; Zhang, W.; Wei, X. A review on weed detection using ground-based machine vision and image processing techniques. *Comput. Electron. Agric.* **2019**, *158*, 226–240. [CrossRef]
4. Seelan, S.; Laguette, S.; Casady, G.M.; Seielstad, G.A. Remote sensing applications for precision agriculture: A learning community approach. *Remote Sens. Environ.* **2003**, *88*, 157–169. [CrossRef]
5. Fontanelli, M.; Frasconi, C.; Martelloni, L.; Pirchio, M.; Foster, R. Weed management in autumn fresh market spinach: A nonchemical alternative. *HortTechnology* **2015**, *25*, 177–184. [CrossRef]
6. Christensen, S.; Søgaard, H.T.; Kudsk, P.; Nørremark, M.; Lund, I.; Nadimi, E.S.; Jørgensen, R. Site-specific weed control technologies. *Weed Res.* **2009**, *49*, 233–241. [CrossRef]
7. Guccione, G.; Schifani, G. Technological innovation, agricultural mechanization and the impact on the environment: Sod seeding and minimum tillage. Prospettive e proposte mediterranee. *J. Econ. Agric. Environ.* **2001**, *3*, 29–36.
8. Ribas, P.; Matsumura, A. A química dos agrotóxicos: Impacto sobre a saúde e meio ambiente. The chemistry of pesticides: Impact on health and the environment. *Rev. Lib.* **2009**, *10*, 149–158.
9. Wood, S.A.; Karp, D.S.; DeClerck, F.; Kremen, C.; Naeem, S.; Palm, C.A. Functional traits in agriculture: Agrobiodiversity and ecosystem services. *Trends Ecol. Evol.* **2015**, *30*, 531–539. [CrossRef]
10. PAN—Pesticide Action Network Europe. *Alternative Methods in Weed Managment to Glyphosate*; Intergrated Weed Management. Pesticide Action Network Europe: Brussels, Belgium, 2017. Available online: https://www.pan-europe.info/sites/pan-europe.info/files/Report_Alternatives%20to%20Glyphosate_July_2018.pdf (accessed on 23 September 2021).
11. MacLaren, C.; Storkey, J.; Menegat, A. An ecological future for weed science to sustain crop production and the environment. A review. *Agron. Sustain. Dev.* **2020**, *40*, 24. [CrossRef]
12. Gnanavel, I. Eco-Friendly Weed Control Options for Sustainable Agriculture. *Sci. Int.* **2015**, *3*, 37–47. [CrossRef]
13. Esposito, M.; Crimaldi, M.; Cirillo, V. Drone and sensor technology for sustainable weed management: A review. *Chem. Biol. Technol. Agric.* **2021**, *8*, 18. [CrossRef]
14. Hartzler, R.; Buhler, D. Management of agricultural weeds. Ecologically Based Integrated Pest Management. In *Opender KoulGerrit and W. Cuperus*; CABI: Wallingford, UK, 2007; pp. 37–51. ISBN 9781845931636.
15. Radicetti, E.; Mancinelli, R. Sustainable Weed Control in the Agro-Ecosystems. *Sustainability* **2021**, *13*, 8639. [CrossRef]
16. Schonbeck, M. *Principles of Sustainable Weed Management in Organic Cropping Systems*, 3rd ed.; Workshop for Farmers and Agricultural Professionals on Sustainable Weed Management; Clemson University: Clemson, SC, USA, 2011.
17. Scavo, A.; Mauromicale, G. Integrated Weed Management in Herbaceous Field Crops. *Agronomy* **2020**, *10*, 466. [CrossRef]
18. Monteiro, A.; Souza, M.; Lins, H.; Teófilo, T.; Júnior, A.; Silva, S.; Mendonça, V. A new alternative to determine weed control in agricultural systems based on artificial neural networks (ANNs). *Field Crop. Res.* **2021**, *263*, 108075. [CrossRef]
19. ADVID—Associação para o Desenvolvimento da Viticultura Duriense. Diferentes Estratégias Para o Controlo de Infestantes em Viticultura de Encosta. Different Strategies for Weed Control in Hillside Viticulture. 2019. Available online: https://www.advid.pt/uploads/DOCUMENTOS/Subcategorias/Comunicacao/MANUAL_INFESTANTES_FINAL_WEB.pdf (accessed on 30 August 2021).
20. Chauhan, B.S. Grand Challenges in Weed Management. In *Frontiers in Agronomy*; Frontiers Media SA: Lausanne, Switzerland, 2020; Volume 1. [CrossRef]
21. Ghosheh, H. Precision Weed Management Research Advancement in the Near East Global Proliferation of Precision Agriculture and its Applications. 2010. Available online: https://www.ispag.org/proceedings/?action=abstract&id=137&title=Precision+Weed+Management+Research+Advancement+In+The+Near+East+&search=types (accessed on 7 September 2021).
22. Sims, B.; Corsi, S.; Gbehounou, G.; Kienzle, J.; Taguchi, M.; Friedrich, T. Sustainable Weed Management for Conservation Agriculture: Options for Smallholder Farmers. *Agriculture* **2018**, *8*, 118. [CrossRef]
23. Korresa, N.E.; Burgosa, N.; Travlos, I.; Vurro, M.; Gitsopoulos, T.; Varanasi, V.; Duke, S.; Kudsk, P.; Brabham, C.; Rouse, C.; et al. Chapter Six-New directions for integrated weed management: Modern technologies, tools and knowledge discovery. *Adv. Agron.* **2019**, *155*, 243–319. [CrossRef]
24. Hammermeister, A. Organic weed management in perennial fruits. *Sci. Hortic.* **2016**, *208*, 28–42. [CrossRef]
25. Mia, M.J.; Massetani, F.; Murri, G.; Neri, D. Sustainable alternatives to chemicals for weed control in the orchard—A Review. *Hortic. Sci.* **2020**, *47*, 1–12. [CrossRef]

26. Pannacci, E.; Lattanzi, B.; Tei, F. Non-chemical weed management strategies in minor crops: A review. *Crop. Prot.* **2017**, *96*, 44–58. [CrossRef]
27. Partel, V.; Kakarla, S.C.; Ampatzidis, Y. Development and evaluation of a low cost and smart technology for precision weed management utilizing artificial intelligence. *Comput. Electron. Agric.* **2019**, *157*, 339–350. [CrossRef]
28. Colas, F.; Cordeau, S.; Granger, S.; Jeuffroy, M.H.; Pointurier, O.; Queyrel, W.; Rodriguez, A.; Villerd, J.; Colbach, N. Co-development of a decision support system for integrated weed management: Contribution from future users. *Eur. J. Agron.* **2020**, *114*. [CrossRef]
29. Lehnhoff, E.A.; Rew, L.J.; Mangold, J.M.; Seipel, T.; Ragen, D. Integrated Management of Cheatgrass (*Bromus tectorum*) with Sheep Grazing and Herbicide. *Agronomy* **2019**, *9*, 315. [CrossRef]
30. Lingenfelter, D.; Curran, W. Weed Management in Pasture Systems. In *Agronomy Facts 62*; College of Agricultural Sciences, Penn State Extension: Pennsylvania, PA, USA, 2001. Available online: https://extension.psu.edu/weed-management-in-pasture-systems (accessed on 3 September 2021).
31. Hinds, J. A Whole-Farm Approach to Managing Pests. SARE-Sustainable Agriculture Research and Education. 2020. Available online: https://www.sare.org/wp-content/uploads/A_Whole-Farm_Approach_to_Managing_Pests.pdf (accessed on 16 January 2020).
32. Rueda-Ayala, V.; Rasmussen, J.; Gerhards, R. Mechanical Weed Control. In *Precision Crop Protection-the Challenge and Use of Heterogeneity*; Oerke, E.C., Gerhards, R., Menz, G., Sikora, R., Eds.; Springer: Dordrecht, The Netherlands, 2020. [CrossRef]
33. Van Acker, R.; Rahman, M.; Cici, S. Pros and Cons of GMO Crop Farming. Oxford Research Encyclopedia of Environmental Science. Available online: https://oxfordre.com/environmentalscience/view/10.1093/acrefore/9780199389414.001.0001/acrefore-9780199389414-e-217#acrefore-9780199389414-e-217-bibliography-0001 (accessed on 9 November 2021).
34. Radicetti, E. Ecological Weed Management. Ph.D. Thesis, Università degli Studi della Tuscia Di Viterbo, Viterbo, Italy, 2012. [CrossRef]
35. Hartwig, N.; Ammon, H. Cover crops and living mulches. *Weed Sci.* **2002**, *50*, 688–699. [CrossRef]
36. Sportelli, M.; Frasconi, C.; Fontanelli, M.; Pirchio, M.; Raffaelli, M.; Magni, S.; Caturegli, L.; Volterrani, M.; Mainardi, M.; Peruzzi, A. Autonomous Mowing and Complete Floor Cover for Weed Control in Vineyards. *Agronomy* **2021**, *11*, 538. [CrossRef]
37. Želazny, W.; Licznar-Malanczuk, M. Soil quality and tree status in a 12-year-old apple orchard under three mulch-based floor management systems. *Soil Tillage Res.* **2018**, *180*, 250–258. [CrossRef]
38. Wayman, S.; Cogger, C.; Benedict, C.; Burke, I.; Collins, D.; Bary, A. The influence of cover crop variety, termination timing and termination method on mulch, weed cover and soil nitrate in reduced-tillage organic systems. *Renew. Agric. Food Syst.* **2015**, *30*, 450–460. [CrossRef]
39. Büchi, L.; Wendling, M.; Amossé, C.; Jeangros, B.; Charles, R. Cover crops to secure weed control strategies in a maize crop with reduced tillage. *Field Crop. Res.* **2020**, *247*, 107583. [CrossRef]
40. Merfield, C. Organic Weed Management: A Practical Guide. 2002. Available online: https://researcharchive.lincoln.ac.nz/bitstream/handle/10182/4902/Merfield_organic_weed_2002.pdf?sequence=1&isAllowed=y (accessed on 18 August 2021).
41. Candido, V.; Addabbo, T.; Basile, M.; Castronuovo, D.; Miccolis, V. Greenhouse soil solarization: Effect on weeds, nematodes and yield of tomato and melon. In *Agronomy for Sustainable Development*; Springer/EDP Sciences/INRA: Berlin/Heidelberg, Germany, 2008; Volume 28, pp. 221–230. [CrossRef]
42. Peerzada, A.M.; Chauhan, B.S. Thermal weed control: History, mechanisms, and impacts. In *Non-Chemical Weed Control*, 1st ed.; Peerzada, A.M., Chauhan, B.S., Eds.; Academic Press: Cambridge, MA, USA, 2018; pp. 9–31. [CrossRef]
43. Fennimore, S.; Slaughter, D.; Siemens, M.; Leon, R.; Saber, M. Technology for Automation of Weed Control in Specialty Crops. *Weed Technol.* **2016**, *30*, 823–837. [CrossRef]
44. Oka, Y.; Shapira, N.; Fine, P. Control of root-knot nematodes in organic farming systems by organic amendments and soil solarization. *Crop. Prot.* **2007**, *26*, 1556–1565. [CrossRef]
45. Frost, R.; Launchbaugh, K. Prescription Grazing for Rangeland Weed Management A new look at an old tool. *Rangelands* **2003**, *25*, 43–47. [CrossRef]
46. Wilson, M.H.; Lovell, S.T. Agroforestry—The Next Step in Sustainable and Resilient Agriculture. *Sustainability* **2016**, *8*, 574. [CrossRef]
47. Melander, B.; Liebman, M.; Davis, A.S.; Gallandt, E.R.; Bàrberi, P.; Moonen, A.; Rasmussen, J.; van der Weide, R.; Vidotto, F. Non-Chemical Weed Management. In *Weed Research*; Hatcher, P.E., Froud-Williams, R.J., Eds.; Wiley: New York, NY, USA, 2017. [CrossRef]
48. Gonçalves, P.; Nóbrega, L.; Monteiro, A.; Pedreiras, P.; Rodrigues, P.; Esteves, F. SheepIT, an E-Shepherd System for Weed Control in Vineyards: Experimental Results and Lessons Learned. *Animals* **2021**, *11*, 2625. [CrossRef] [PubMed]
49. Lancaster, S.R. *Chemical Weed Control for Field Crops, Pastures, Rangeland, and Noncropland*; Kansas State University: Manhattan, KS, USA, 2021; p. 148. Available online: https://www.bookstore.ksre.ksu.edu/pubs/CHEMWEEDGUIDE.pdf (accessed on 23 September 2021).
50. Zhelezova, S.V.; Melnikov, A.V.; Ananiev, A.A. Pros and cons of no-till technology in a long-term field experiment on sod-podzolic soil. *IOP Conf. Ser. Earth Environ. Sci.* **2019**, *368*, 012055. [CrossRef]
51. Peera, P.; Debnath, S.; Maitra, S. Mulching: Materials, Advantages and Crop Production. In *Protected Cultivation and Smart Agriculture*; Sagar, M., Dinkar, J.G., Tanmoy, S., Eds.; New Delhi Publishers: New Delhi, India, 2020; pp. 55–66. [CrossRef]

52. Kader, M.A.; Singha, A.; Begum, M.A.; Jewel, A.; Khan, F.H.; Khan, N.I. Mulching as water-saving technique in dryland agriculture: Review article. *Bull. Natl. Res. Cent.* **2019**, *43*, 147. [CrossRef]
53. Hoorman, J.J. Using Cover Crops to Improve Soil and Water Quality. Ohio State University Sustainable Agriculture Fact Sheets. 2009. Available online: http://ohioline.osu.edu/factsheet/anr-57 (accessed on 22 August 2021).
54. Dabney, S.M.; Delgado, A.; Reeves, D.W. Using winter cover crops to improve soiland water quality. *Commun. Soil Sci. Plant Anal.* **2001**, *32*, 1221. [CrossRef]
55. Popay, I.; Field, R. Grazing Animals as Weed Control Agents. *Weed Technol.* **1996**, *10*, 217–231. [CrossRef]
56. Santos Valle, S.; Kienzle, J. Agriculture 4.0—Agricultural Robotics and Automated Equipment for Sustainable Crop Production. In *Integrated Crop Management*; FAO: Rome, Italy, 2020; Volume 24, Available online: http://www.fao.org/3/cb2186en/CB2186EN.pdf (accessed on 24 August 2021).
57. Monteiro, A.; Santos, S.; Gonçalves, P. Precision Agriculture for Crop and Livestock Farming—Brief Review. *Animals* **2021**, *11*, 2345. [CrossRef]
58. Zhang, Q. *Precision Agriculture Technology for Crop Farming*; CRC Press: Boca Raton, FL, USA, 2015.
59. Dammer, K.H.; Wartenberg, G. Sensor-based weed detection and applicationof variable herbicide rates in real time. *Crop. Prot.* **2007**, *26*, 270–277. [CrossRef]
60. Lima, A.; Mendes, K. Variable Rate Application of Herbicides for Weed Management in Pre- and Postemergence. In *Pests, Weeds and Diseases in Agricultural Crop and Animal Husbandry Production, Dimitrios Kontogiannatos*; Kourti, A., Mendes, K.F., Eds.; IntechOpen: London, UK, 2020. [CrossRef]
61. Forouzanmehr, E.; Loghavi, M. Design, development and field evaluation of a map-based variable rate granular fertilizer application control system. *Agric. Eng. Int. CIGR J.* **2012**, *14*, 255–264.
62. Zhang, Y.; Staab, E.S.; Slaughter, D.C.; Giles, D.K.; Downey, D. Precision automated weed control using Hyperspectral vision identification and heated oil. *ASABE Tech. Libr.* **2009**, 21–24. [CrossRef]
63. Søgaard, H.T.; Lund, I. Application accuracy of a machine vision-controlled robotic micro-dosing system. *Biosyst. Eng.* **2007**, *96*, 315–322. [CrossRef]
64. Kargar, B.; Shirzadifar, M. Automatic weed detection system and smart herbicide sprayer robot for corn fields. In Proceedings of the 2013 First RSI/ISM International Conference on Robotics and Mechatronics (ICRoM), Tehran, Iran, 13–15 February 2013; pp. 468–473.
65. Burgos, N.R.; Rouse, C.E.; Singh, V.; Salas-Perez, R.; Bagavathiannan, M. Technological Advances for Weed Management. In *APWSS 50th Anniversary Celebratory*; Chandrasena, N., Rao, A.N., Eds.; Asian-Pacific Weed Science Society (APWSS) Publication: Honolulu, HI, USA, 2017; pp. 65–79.
66. Gómez-Candón, D.; Virlet, N.; Labbé, S.; Jolivot, A.; Regnard, J.L. Field phenotyping of water stress at tree scale by UAV-sensed imagery: New insights for thermal acquisition and calibration. *Precis. Agric.* **2016**, *17*, 786–800. [CrossRef]
67. Rani, A.; Chaudhary, A.; Sinha, N.; Mohanty, M.; Chaudhary, R. Drone: The green technology for future agriculture. *Har. Dhara* **2019**, *2*, 3–6.
68. Ahmad, A.; Ordoñez, J.; Cartujo, P.; Martos, V. Remotely Piloted Aircraft (RPA) in Agriculture: A Pursuit of Sustainability. *Agronomy* **2021**, *11*, 7. [CrossRef]
69. Jiménez-Brenes, F.M.; López-Granados, F.; Torres-Sánchez, J.; Peña, J.M.; Ramírez, P.; Castillejo-González, I.L. Automatic UAV-based detection of Cynodon dactylon for site-specific vineyard management. *PLoS ONE* **2019**, *14*, e0218132. [CrossRef]
70. Hassanein, M.; El-Sheimy, N. An efficient weed detection procedure using low-cost UAV imagery system for precision agriculture applications. *Int. Arch. Photogramm. Remote. Sens. Spat. Inf. Sci.* **2018**, *42*, 181–187. [CrossRef]
71. King, A. Technology: The Future of Agriculture. *Nature* **2017**, *544*, S21–S23. [CrossRef]
72. Peruzzi, A.; Martelloni, L.; Frasconi, C.; Fontanelli, M.; Pirchio, M.; Raffaelli, M. Machines for non-chemical intra-row weed control: A review. *J. Agric. Eng.* **2017**, *48*, 57–70. [CrossRef]
73. Fernandez-Quintanilla, C.; Peña-Barragán, J.M.; Andújar, D.; Dorado, J.; Ribeiro, A.; López-Granados, F. Is the current state-of-the-art of weed monitoring suitable for site-specific weed management in arable crops? *Weed Res.* **2018**, *58*, 259–272. [CrossRef]
74. Riar, D.S.; Ball, D.A.; Yenish, J.P.; Burke, I.C. Light-activated, sensor-controlled sprayer provides effective postemergence control of broadleaf weeds in fallow. *Weed Technol.* **2011**, *25*, 447–453. [CrossRef]
75. Berge, T.; Goldberg, S.; Kaspersen, K.; Netland, J. Towards machine vision based site-specific weed management in cereals. *Comput. Electron. Agric.* **2012**, *81*, 79–86. [CrossRef]
76. Jensen, H.; Jacobsen, L.; Pedersen, S.; Tavella, E. Socioeconomic impact of widespread adoption of precision farming and controlled traffic systems in Denmark. *Precis. Agric.* **2012**, *13*, 661–677. [CrossRef]

agronomy

MDPI

Review

Integrated-Smart Agriculture: Contexts and Assumptions for a Broader Concept

Vítor João Pereira Domingues Martinho * and Raquel de Pinho Ferreira Guiné

Agricultural School (ESAV) and CERNAS-IPV Research Centre, Polytechnic Institute of Viseu (IPV), 3504-510 Viseu, Portugal; raquelguine@esav.ipv.pt
* Correspondence: vdmartinho@esav.ipv.pt

Abstract: The innovative technologies developed in the different fields of science (nanotechnology, artificial intelligence, genetic modification, etc.) opened new and infinite possibilities for the several stakeholders that carry out their activities in the different economic sectors. For agriculture, these new approaches are particularly relevant and may bring interesting contributions, considering the specificities of the sector, often dealing with contexts of land abandonment and narrow profit margins. Nonetheless, the question in these unstopped evolutions is about the interlinkages with sustainability. In this context, the objectives of this study are to highlight the main insights from the available scientific literature about the interrelationships between the new trends in the agriculture and the sustainability. To achieve these aims, a search on the Web of Science Core Collection (WoS) and Scopus databases was carried out, on 15 May 2021, for the topics 'smart agriculture' and 'sustainability'. A total of 231 documents (102 from WoS and 129 from Scopus) were obtained, remaining 155 documents after removing the duplicated, which were surveyed through systematic review following the PRISMA (Preferred Reporting Items for Systematic Reviews and Meta-Analyses) approach. As main insights, the concerns of the researchers with the impacts on the sustainability from the transformations in the farming organization are highlighted. On the other hand, it was shown the relevance and the new opportunities, including in terms of food supply, arising from the precision agriculture, agricultural intelligence, vertical/urban farming, circular economy, internet of things, and crowdfarming. We suggest the new and wider concept of 'integrated-smart agriculture', better than 'climate-smart agriculture'.

Keywords: agriculture; new technologies; sustainability; systematic review; PRISMA

Citation: Martinho, V.J.P.D.; Guiné, R.d.P.F. Integrated-Smart Agriculture: Contexts and Assumptions for a Broader Concept. *Agronomy* **2021**, *11*, 1568. https://doi.org/10.3390/agronomy11081568

Academic Editor: Domenico Ronga

Received: 4 July 2021
Accepted: 3 August 2021
Published: 5 August 2021

Publisher's Note: MDPI stays neutral with regard to jurisdictional claims in published maps and institutional affiliations.

1. Introduction

The technological progress opens several opportunities for the different socioeconomic sectors, including the farming sector, in a perspective of smart agriculture, but brings, also, various challenges [1] threats that may compromise the sustainability of the development process worldwide. One example of this paradigm is the internet of things (IoT). The concept of IoT refers to the digital interconnection of everyday objects with the Internet, connecting objects instead of than people, allowing the intercommunication of data between sensors and digital controllers, for example. IoT is a set of networks which connect things capable of processing and communicating data between each other [2]. Hence, the IoT may contribute to increase the agricultural efficiency in the use of soil, water and energy (some agricultural resources where an efficient management is crucial); however, it also creates new risks associated with confidentiality and integrity [2].

In any case, the agriculture is faced with new challenges, such as the population rising, the urban agglomeration, scarcity of resources, climate change, and waste management. These tasks call for pioneering solutions. The technological development may play here a relevant and determinant role, namely with innovative practices associated with, for example, the vertical/urban farming, seawater and desert farming [3], as well as smart agriculture [4]. The literature highlights the relevance of the smart agriculture concept and practices to increase the sustainability in the farms. Smart agriculture practices are envisaged as the most appropriate adaptation strategies that will allow achieving food security, while at the same time being able to mitigate climatic changes. This is achieved through the preservation of natural resources and sustainability of vital ecosystem services. Smart agriculture refers to a number of tools that help the management of agricultural resources and crop production. Smart agriculture is relating to the utilization of technologies such as the Internet of Things, sensors, geospatial location, robotics, and artificial intelligence.

Specifically, the artificial intelligence may be crucial for a more efficient use of the resources, but also, for a better disease and pest control, data analysis, productions' management and fill the gap between farmers and knowledge, allowing in this way higher productivities and competitiveness [5]. In general, the scientific community interlinks the artificial intelligence with sustainability and a more circular economy [6]. Potentialities of the artificial intelligence are referred for decades by the researchers [7]. The smart agriculture aims to promote more sustainability, increasing the farming productivity, dealing with the climate change implications and reducing the greenhouse gas emissions [8]. For these aims the artificial intelligence may bring interesting contributions, as well as, other new approaches and technologies, such as the IoT. In the current contexts of climate change [9], the innovation [10] is crucial to achieve sustainable development goals (SDGs) [11]. The agricultural institutions, namely the cooperatives, are fundamental to support the farmers in the process of agricultural digitalization [12] that is in course in the sector [13] in the different world countries. The cooperatives and the associations are important institutions in the farming sector, considering their proximity with the several agricultural stakeholders. To improve the sustainability in the agricultural sector through a smarter agriculture, the institutions (namely the associations and cooperatives) play a determinant role. The adaptability of farmers to the new technological demands might be difficult, especially for small dimension farms and family farmers. Hence the role of cooperatives might be of great help to connect them to the new technological demands. Other dimensions are related with the commitment of the farmers and the design of adjusted policy instruments to promote the adoption of innovative approaches in a sustainable perspective. Indeed, the agricultural policies and planning impact significantly the structure of the farms and the evolution of the sector, namely in the European Union through the CAP (Common Agricultural Policy). In fact, the CAP is the main driver of the European farming structures, including in the sustainability dimensions [14].

In parallel to this process of agricultural digitalization to promote smarter and sustainable farming practices, the public institutions should be able to implement strategies that increase the organization of the sector, improve the sills of the farmers and encourage the technological transfer from the scientific community to the farms and the sector. The organization of the sector and the skills of the farmers are, indeed, determinant for a more sustainable farming development [15]. These three dimensions will be crucial for an effective modernization of the agricultural sector in a perspective of a more sustainable development.

In this perspective, the main objective of this study is to highlight the main relationships between the smart agriculture (modern agricultural practices based on the new technologies, such as the IoT [16], that combine scientific research and innovation) and the sustainability (capacity to meet the present needs without compromise the future generations [17,18]). In practice the main question here is the following: what are the main relationships between the smart agriculture and the sustainability? Is the concept of 'climate-smart agriculture' (CSA) in general considered in the scientific literature sufficiently broader to capture the several interrelationships among the smart agriculture and the sustainability? CSA is a concept introduced by FAO (Food and Agricultural Organization) in 2010 and is known as the "triple win" approach [19]. CSA is an approach to the management of landscapes with the purpose of adapting crop and animal production to the climate changes originated by the human action on the planet. However, some researchers argue that the CSA concept is a narrow approach about the current farming contexts and a broader statement [20], involving interdisciplinary dimensions [21], and using recent sustainability indicators [22] is needed. For that, a systematic literature review was carried out based on the PRISMA approach and on a preliminary bibliometric analysis. The PRISMA statement consists on a checklist with several items and on a flow diagram with few phases [23]. PRISMA is used as a guiding methodology for conduction reviews, in many fields of science, however with some degree of subjectivity. In fact, the scientific research involves always decisions that depend on the authors' perspective. Nonetheless, in order to deal with this subjectivity the PRISMA statement suggests transparency in the description of the decisions made. The literature review reveals that there are not many studies (or none) that consider the topics here addressed jointly with the PRISMA approach or bibliometric analysis, highlighting the novelty of this research.

2. Materials and Methods

To achieve the objectives proposed for this research 231 studies were obtained from the Web of Science Core Collection [24] and from Scopus [25] for the topics 'smart agriculture' and 'sustainability' in a search carried out on 15 May 2021, without any restrictions for the years considered or language (in general the abstracts are in English). The identification of these topics was based on a preliminary survey of the literature related with these fields and on the findings of, for example, Ruiz-Real et al. [6]. Using the software Zotero [26] 76 duplicated documents were identified. It was considered other topics of search, such as, for example, 'smart farming', 'digital farming', or 'digital agriculture'. A quick search on the WoS (Core Collection), for example, shows that the interlinkages of these topics with the sustainability are more limited (a little more than 50 documents in total), highlighting the 'smart agriculture' as a wider term. It was considered the concept of 'precision agriculture'; however, the CSA is a broader approach that aims the optimization of the whole agricultural system [27]. It was used, also, the term 'sustain*' instead of 'sustainability', however, with this alternative documents were obtained about other issues that not properly about sustainability that is the main focus of this research. In fact, sustainability is not the same that sustainable development, for example. In addition, it could be used the topic "climate-smart agriculture", however this was already addressed by other studies [28]. There is always subjectivity in the choice of the topics of search. The PRISMA statement [23] about the search strategy suggests, namely, a clear explanation of the search terms used and the identification of the databases considered. In any case, these are suggestions that may be explored in future studies with other approaches. Indeed, the identification of directions for future research is an underlying objective of scientific studies.

After removing the duplicated studies 155 researches were considered to carry out a systematic review, following the PRISMA [23] approach schematised in Figure 1.

To better organise the literature review, namely in terms of sub-topics, the PRISMA approach was complement with a preliminary bibliometric analysis, following, for example, Martinho [29] for food marketing, or Nadaraja et al. [30] for sustainable agriculture, or Martinho for agri-food contexts [31]. The bibliometric analysis was performed with bibliographic data (full counting and 1 as the minimum number of occurrences of a keyword), considering co-occurrence as links and keywords as items, following the procedures of the VOSviewer [32,33] software. The bibliometric analysis as support for literature review was considered, for instance, by Martinho [14,15,34]. In addition a factor-cluster analysis was carried out to better identify groups to cluster the several documents with the topics addressed, following Stata [35–37] and Torres-Reyna procedures [38].

Tables 1 and 2 and Figure 2 show that the main keywords found in the documents related with the topics 'smart agriculture' and 'sustainability' may be clustered in the following groups (the name for these groups were identified considering the keywords associated at each cluster highlighted, namely in Table 2 and Figure 2): environmental impacts and climate change; new technologies and approaches; food supply and security; farming systems and crop management; multifunctionality and agricultural/rural development. In fact, the information in Table 2 is in agree with the results presented in Table 1 and confirms a group of keywords, in general, related with the new technologies and approaches (cluster 1), other group for the sustainability (cluster 2), for the crop management (cluster), for the food supply (cluster 4) and for the rural development (cluster 5). Figure 2 highlight, also, the relevance of dimensions associated with sustainability, food security, crops management, new technologies, and agricultural/rural development.

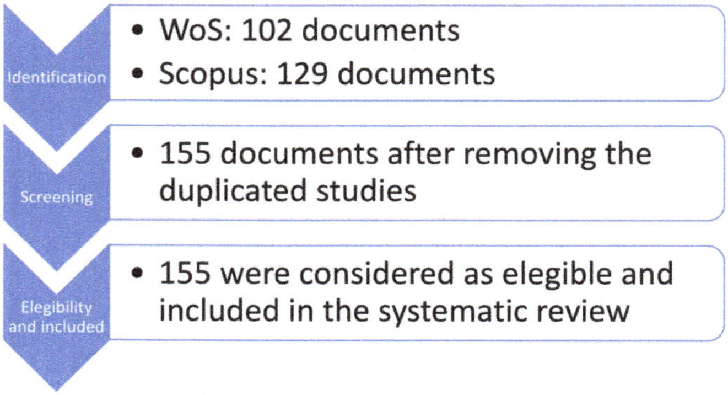

Figure 1. Steps to select the documents for systematic review, considering the PRISMA approach.

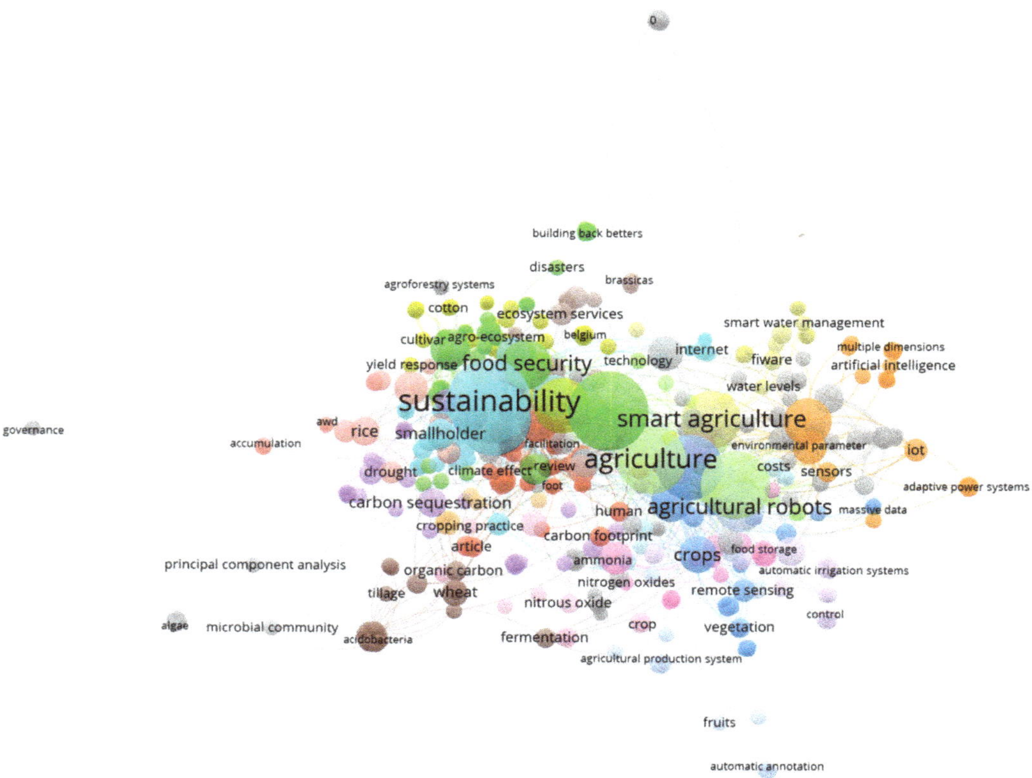

Figure 2. Network visualization map for the topics 'smart agriculture' and 'sustainability'.

Table 1. Top 50 keywords for the topics 'smart agriculture' and 'sustainability'.

Keyword	Occurrences	Average Publication Year
sustainability	49	2019
climate change	44	2019
climate-smart agriculture	39	2019
agriculture	33	2019
sustainable development	30	2019
smart agriculture	24	2020
agricultural robots	17	2020
food security	17	2018
internet of things	16	2020
alternative agriculture	12	2018
adaptation	11	2018
irrigation	11	2019
sustainable agriculture	11	2020
crops	10	2019
environmental sustainability	9	2019
adoption	8	2019
farming system	8	2020
greenhouse gases	8	2019
mitigation	8	2018
agroforestry	7	2018
conservation agriculture	7	2019
food supply	7	2018
adaptive management	6	2017
carbon sequestration	6	2019
cultivation	6	2018
resilience	6	2019
rice	6	2018
smallholder	6	2019
smart farming	6	2020
sustainable intensification	6	2019
wireless sensor networks	6	2019
agricultural development	5	2017
environmental technology	5	2018
greenhouse gas	5	2019
wheat	5	2018
agricultural ecosystem	4	2018
agricultural production	4	2018
agrometeorology	4	2019
agronomy	4	2019
biodiversity	4	2019
coffee	4	2019
crop production	4	2017
crop yield	4	2018
drought	4	2018
ecosystem services	4	2018
efficiency	4	2020
fertilizers	4	2019
impacts	4	2019
india	4	2019
innovation	4	2019

Table 2. Results for the factor-cluster analysis with the information from top 50 keywords (occurrences and average publication year).

Keyword	Cluster
sustainability	1
climate change	1
climate-smart agriculture	1
agriculture	1
sustainable development	1
smart agriculture	1
agricultural robots	1
internet of things	1
sustainable agriculture	1
farming system	1
smart farming	1
food security	2
alternative agriculture	2
adaptation	2
irrigation	2
environmental sustainability	2
mitigation	2
agroforestry	2
carbon sequestration	2
resilience	2
smallholder	2
agrometeorology	2
agronomy	2
biodiversity	2
coffee	2
fertilizers	2
impacts	2
india	2
crops	3
adoption	3
greenhouse gases	3
conservation agriculture	3
sustainable intensification	3
wireless sensor networks	3
greenhouse gas	3
efficiency	3
innovation	3
food supply	4
cultivation	4
rice	4
environmental technology	4
wheat	4
agricultural production	4
ecosystem services	4
adaptive management	5
agricultural development	5
agricultural ecosystem	5
crop production	5
crop yield	5
drought	5

3. Systematic Review

In this section will be carried out a systematic review organised considering the PRISMA approach and bibliometric analysis carried out on the previous section. In practice 231 documents were obtained from the WoS and Scopus (102 and 129, respectively) for a search carried out on 15 May 2021 for the topics 'smart agriculture' and 'sustainability'.

After removing the duplicated (76 studies) 155 documents were surveyed. To better show how the research questions and gaps were addressed, for a more sustainable development in the framework of the SDGs [39], the main insights are highlighted in Table 3. Each one of those findings will be presented deeper in the next subsection for literature review. In this perspective, this part will be structured in the following subsections: environmental impacts and climate change; new technologies and approaches; food supply and security; farming systems and crop management; multifunctionality and agricultural/rural development.

Table 3. Main insights from the systematic review.

References	Main Highlights
[2]	The new technologies and approaches are not exempt of risks and vulnerabilities
[40]	CSA approach is a promising solution for the sustainability
[41]	Some studies use the terminology of Environment-Smart Agriculture (ESA)
[19]	CSA is a concept presented by FAO in 2010, is known as the "triple win" approach
[42]	CSA practices improve the soil resilience and quality
[43]	The Internet of Things (IoT) and the Internet of Everything (IoE) may bring relevant added value for the farms
[44]	The wireless sensor network is an interesting tool to collect data
[45]	The biosensors are other techniques to collect information
[46]	Mobile applications, big data analytics and information systems, cloud computing, drones, blockchain, artificial intelligence
[47]	An efficient use of the agriculture resources, such as water, soil and energy, is crucial for competitiveness and food and security
[48]	Agriculture is one of the most vulnerable sectors to the global warming
[49]	The agricultural sector contributes with about a third of the anthropogenic GHG emissions worldwide
[50]	The eco-efficiency is the buzzword for the sustainability
[51]	The rice-wheat cropping systems concern particularly the researchers specifically in South Asia
[52]	Africa is another world region where it is important to promote cleaner farming systems
[53]	Sometimes the sustainable practices are misunderstood in these countries
[54]	In other cases and contexts there is not a convergent view about the CSA practices
[20]	CSA concept has a narrow perspective about the current farming contexts and a wider debate is needed
[55]	Rural development may benefit from the concept of smart villages
[56]	Sometimes is easier to convince the entrepreneurs than the policymakers
[57]	For an effective CSA implementation the farmers should be involved in the policy design process
[58]	Vocational training and the extension services may contribute for the adoption of the CSA practices
[59]	The European Union invested over the last years a significant part of its budget to promote CSA practices

3.1. Environmental Impacts and Climate Change

In the relationships between the agricultural sector, food security and the environmental impacts, on the framework of the climate change, the climate-smart agriculture (CSA) approach appears as a promising solution [40] for the sustainability [4] with the following three objectives: improve the resilience of the farming sector to the climate change; mitigate the greenhouse gas (GHG) emissions; and guarantee the food security [60]. The resilience will be the main challenge to deal with the climate change [61] and the transition to smart solutions is unstoppable [62]. The objective is to achieve sustainability, resilience, wellbeing, and development [63] with new approaches [64]. In fact, the agricultural sector suffers from several particularities that, often, compromise its competitiveness. The new approaches associated with the different dimensions of the smart agriculture may be an interesting contribution for the farm profitability and financial performance.

Some studies use the terminology of environment-smart agriculture to address the relationships between the agricultural activities and the environment [41]. CSA is a concept presented by FAO in 2010, relaunched by the Conference of Paris in 2015 and is known as the "triple win" approach [19]. The idea is to find solutions for the farms in order to minimise the global warming consequences [65]. The practices associated with CSA are recognized as intensive agricultural techniques compatible with a sustainable development [66]. They are included, also, in silvo-aquaculture [67] and integrated aquaculture [68] systems, in a context called as Agriculture 4.0 [69]. These practices are not understood and implemented in a similar way worldwide [70]. These findings should be considered and addressed properly by the diverse stakeholders, namely the policymakers, for an effective and adjusted implementation of a sustainable smart agriculture.

In these frameworks, the energy management, in an efficient way [71], is critical [72], as well as the soil use [73]. CSA practices improve the soil resilience and quality [42]. Nonetheless, for the policy and planning design, it is important to find metrics that allow to put together the three aims [74] and to bring more insights about this concept [75]. The agricultural practices have impacts on the environment and the climate change, but have, also, implications from the global warming, and this brings several challenges for the farmers and policymakers [76]. To deal with the new contexts faced by the agricultural sector will be needed robust policies [77] and institutions [78], including non-governmental [79], at local, regional, national, and international levels. Indeed, the public institutions and the cooperatives, for example, are crucial for a better organisation of the sector and to increase the compliance with the strategies designed for the agriculture.

3.2. *New Technologies and Approaches*

The Internet of Things (IoT) and the Internet of Everything (IoE) have had great impact on the farms [43] and are forms to improve the productivity in the use of several agricultural resources [80], namely the water, through approaches of smart irrigation [81] and precision agriculture [82]. The smart irrigation systems are important to collect and work environmental data [83]. The IoT allows to implement automated operations with reduced supervision [84] in the whole food chain [85] and agricultural production [86], including in greenhouse agriculture [87] control [88] and in diverse farming systems [89]. The water management is critical, where the IoT may contribute significantly for a more balanced use [90], as well as in the soil health assessment [91] and fertilization management [92], in a perspective of a more competitive agriculture [93]. The potable water will be one of the scarcest resources and here the new technologies will be determinant for a more efficient management.

The new technologies and approaches are, also, important methodologies to support the farmers in other agronomic practices, some of them are available in open source solutions [94]. In some circumstances the government supports are decisive for the new technologies' promotion, namely at an initial phase [95]. The wireless sensor network, for example, is an interesting tool to collect data about weather, soil, and plant conditions to provide the farmers with information to better manage the pest and disease control and the fertilizers' use [44]. The biosensors are other techniques to collect information about the condition of the plant and assess its exposure to biotic and abiotic stresses [45]. Namely, in Europe, there are various projects to create new approaches for the farms that include smart technologies. The several findings obtained by the scientific literature may have an important role as a basis of knowledge for these developments.

The continuous assessment of the plant resilience is determinant in a context of climate change and agricultural conditions transformation [96]. The specificities of the conditions faced by the agricultural practices in drylands systems are other contexts where the new technologies may offer relevant added value [97], specifically in Africa [98]. Mobile applications, big data analytics and information systems, cloud computing, drones, blockchain, artificial intelligence [46], remote sensing [99] are other terminologies referred in the scientific literature for a smarter agriculture. However, the new technologies and approaches related with the smart agriculture are not exempt of risks and vulnerabilities [2]. This is an important aspect that must be highlighted by the scientific literature. Indeed, the new technologies and approaches bring new solutions and perspectives, but are not exempt of negative consequences for the sector and for the farmers.

3.3. Food Supply and Security

An efficient use of the agriculture resources, such as water, soil and energy, is crucial to guarantee the competitiveness of the farming sector and consequently the food self-sufficiency and security [47]. Irrigation availability appears as the main driver of the farmers' decisions, jointly with farm labour, seasonality, climate, land, wildfires, and diseases and pest control [100]. An efficient water management is one of the most important practices in the farms [101] and is fundamental for a dynamic and competitive agricultural sector in an era of great challenges [102]. In fact, often, the water availability and management appear as the main concerns for the various stakeholders, specifically the farmers.

Agriculture is one of the most vulnerable sectors to the global warming, which, allied with the population growing, creates serious problems of food security worldwide [48]. Indeed, the demand for agricultural goods increases continuously and the availability of resources decreases [103]. The smart agriculture concept provides adjusted tools to better collect, transmit, select, and analyze data, in a perspective of smarter management [104], with big data [105], for more sustainable farms and, consequently, for a more balanced food supply [106]. The availability of data is crucial for an adjusted assessment of the present and future food supply and security scenarios [107] and here the smart approaches may bring important added value.

The problems related with food security are particularly worrisome in Africa, specifically between the small farmers [108], and these scenarios were worsened with the climate changes contexts [109]. More than 200 million have problems of sub nutrition and the perspective is for this scenario to become worse in the nest years and decades [110]. India is another context where the climate change and the food security bring new tasks for the farmers and policymakers [111]. The agricultural strategy instruments and the related organizations play a crucial role to guarantee food security, specifically where the risks are higher [112]. There are some specific contexts that deserve a special attention and this is highlighted by the scientific community.

3.4. Farming Systems and Crop Management

The effective contribution of the smart and sustainable agriculture for a more balanced development depends on the farming systems' management [113], on the agricultural practices implemented in the farms [114], namely those related with irrigation and fertilization [115], on the local conditions [116], specificities [117], and strategies [118]. Other particularities and solutions were highlighted by the researchers, such as the presented in Table 4.

Table 4. Particularities and solutions highlighted by the literature for a more balanced agricultural development.

References	Particularities and Solutions
[119]	Fodder banks
[120]	Fermentation of agricultural waste
[121]	Models to identify tomato ripeness
[122]	No-tillage, waste management, and agricultural diversification
[123]	Conservation agriculture
[124]	Based on conservation tillage systems
[125]	Nanotechnology
[126]	Including for carbon management in soil
[127]	Drought-tolerant seeds
[128]	Integrated pest control, combined crop-animal agriculture and organic composting
[129]	Fertilizer trees and shrubs
[130]	Terrace landscapes
[131]	Annual crops planted with coconuts
[132]	Agroforestry structures
[133]	Microalgae
[134]	Dambo cultivation
[135]	Valorisation of agro-food byproducts
[136]	Traditional agriculture
[137]	Integrated farming systems
[138]	'4R' approach (right source, right rate, right time, right place)
[139]	Agronomic rotations and cover cropping
[140]	"Positive Deviance" (identifying practices from farms with higher performance)
[141]	Genetic strategies
[142]	Vertical farming
[143]	In the cities
[144]	Crop residues management through principles of bioeconomy
[145]	Certification strategies

These solutions as CSA are not universal [146] and depend on the specificities of each context [147].

The agricultural sector contributes with about a third of the anthropogenic GHG emissions worldwide, for what is urgent to find innovative approaches to deal with these impacts [49]. The contexts of the environmental impacts are worse when considered the food industry sector [148]. The eco-efficiency is the buzzword for the sustainability of the farming systems [50] and for a more sustainable crop management [149], where the innovative irrigation adjuvants may have important roles [150].

The new contexts that appeared around the world in the recent years changed the paradigm of the economic sector organization and call for novel ways of farm management [151]. The rice-wheat cropping systems concern particularly the researchers [152], specifically in South Asia [51], because of its importance for the food security [153] and the problems associated with the soil quality, water scarcity, and availability of some production factors, such as labor [154]. Africa is another world region where it is important to promote cleaner farming systems [52], in a framework of CSA practices [155], and where several projects [156], and studies were carried out [157]. This considers the complexity of the African farming systems [158] and its vulnerability to the global warming [159]. The impacts of the climate change in the African countries are, in fact, problematic [160]. Namely because the difficulties in the pest control. However, sometimes the sustainable practices are misunderstood in these countries [53].

In other cases and contexts there is not a convergent view about the CSA practices between the different stakeholders [54]. The perceptions of the diverse actors about the smart agriculture have here their impacts [161]. It is important, namely through the extension services to assess and work the perceptions of the farmers about the smart practices in the farms, because this is decisive to achieve the objectives intended by the governments, including in terms of sustainability. In addition, some studies argue that the CSA concept has a narrow perspective about the current farming contexts and a wider debate is needed [20], involving interdisciplinary researchers [21] and using recent sustainability indicators, namely those related with the socioeconomic dimensions [22]. These findings support the thesis argued in this study that a broader concept is needed, in a more integrated perspective, to capture the several relationships between the smart agriculture and the sustainability.

In fact, innovative approaches are needed to deal with the multiple and conflicting domains related with the sustainable development [162] in the current realities [163]. The current impacts in the farming systems are implications of the global warming and come from socioeconomic changes, for what are needed integrated approaches [164]. Viticulture and the copper toxicity is another concern [165], as well as the cotton production [166], oil palm plantations [167], cocoa [168], and coffee production [169].

3.5. Multifunctionality and Agricultural/Rural Development

A better organised and planned farming sector may improve productivity and profit. The smart agriculture approaches may be determinant to obtain a more competitive and sustainable agricultural sector with advantages for the rural development in a context of smart villages [55], where the ecosystems are able to adapt to the new realities [170]. The terminology of "smart" came to stay, including for the cities [171]. The concept of climate-smart village and the associated practices are particularly relevant in countries where there are serious problems of food security, such as the India context [172].

The rural and agricultural policies are fundamental instruments to promote and encourage the smart agriculture practices, specifically in the developing countries, where the CSA adoption is low [173]. However, sometimes is easier to convince the entrepreneurs than the policymakers [56], and this may be, in certain circumstances, a real barrier to implement innovation [174]. The adoption of CSA practices is, indeed, a great challenge around the world for farmers and public institutions [175], including in the OECD countries [176]. For an effective CSA implementation the farmers should be involved in the policy design process, namely to be considered the local risks and particularities [57]. The involvement of the stakeholders in the decision processes, in a perspective of citizens/community engagement [177] or multiple engagements [178], improves the compliance with the strategies designed.

The participation of the women in the processes of decision may increase the adoption of CSA approaches, because of the concerns with the family food security. In addition, the vocational training and the extension services are other factors that may contribute for a more effective adoption of the CSA practices [58], as well as, educational programs [179] to prepare the next generations [180]. The extension services are crucial to advice the farmers [181] in the implementation of the CSA practices [182], nonetheless the particularities of these services in some countries may compromise their contributions [183].

In Vietnam, for example, the adoption of CSA practices by the rice farmers is influenced by the following variables: perceptions about the climate change, educational level, credit and capital availability, land tenure, farm size, availability of extension services, and access to markets [184]. In Nigeria the choice of adaptation practices is influenced by the following dimensions: age, land tenure, extension services, gender, farm size, assets, experience, and credit availability [185]. The determinants do adopt sustainable practices seem to be similar in these different parts of the world, where, for example, the land access and use has its relevance [186].

The European Union invested over the last years a significant part of its budget to promote research about the implementation of smart agriculture in the member-states, in the frameworks of the Horizon 2020 and the CAP (Common Agricultural Policy) [59]. The rural and agricultural institutions and their digitalisation is fundamental for a more integrated development [12], namely to avoid the land abandonment and the reduction in the number of small farms in the Europe context [187].

4. Discussion

There are not so many systematic reviews about the 'smart agriculture' and 'sustainability' topics and less (or none) considering, as complement, the bibliometric analysis, as highlighted by Martinho [31], showing the novelty of this research.

The technological development has been, over the past years, providing tools with applicability in the domain of agriculture and particularly aiming at more sustainable agricultural systems. This technological development is continuously evolving and the knowledge transfer must be, now more than ever, a reality seen as a way to improve agricultural production allied to sustainability goals. Hence new opportunities come from smart agriculture and IoT [43] as ways to improve agricultural efficiency and use of scarce resources such as soil, water, and energy. However, also some possible threats underlie the application of these new technologies, most especially those linked with ethics, since aspect such as confidentiality or integrity might be at risk [2].

Smart agriculture with an integrated view targeting sustainability can help face the challenges [104] derived from the urgent need to feed the growing world population, from inequality of population distribution around the globe or even the asymmetries coming from ultra-high density urban areas as opposed to depopulated rural areas, allied to the scarcity of natural resources, climate change demands, waste management needs, and circular economy promotion. Artificial intelligence and agricultural digitalization certainly contribute to improve productivity while reducing the input/output ratios, thus turning the systems target effective. Optimization of natural resources and of production factors together with cleaner pest control strategies, are on the verge of greener and highly efficient agricultural systems [100].

Climate-smart agriculture improves energy/water/soil management and increments resilience of the agricultural sector, contributing to guarantee food security and attenuating GHG emissions, thus finding a compromise between intensive agricultural production and global warming consequences [47]. Concepts of precision agriculture, smart irrigation and smart fertilization are possible owing to IoT and IoE. They allow to save resources, by for example collecting data from climate, soil, or plants to manage agricultural inputs, while improving productivity and reducing environmental impacts [80], either related with GHG emissions or the biosystem's ecology. Through controlled and automated farming systems it is possible to improve competitiveness [84], with positive socioeconomic impacts, specifically in rural communities. The challenges are particularly relevant in developing countries where food security is a major risk. Also, the sector of organic farming greatly benefits from these technologies, since in these farming systems better management is crucial for profitability. However, implementation of long term measures and a vast adoption of artificial intelligence and smart agriculture technologies faces some resistance of farmers, especially in some communities [173], much owing to lack of knowledge or lack of appropriate technical support. To this matter, engagement of women into these matters is seen as a major opportunity to better prepare the future generations.

The targets that smart agriculture tries to reach and respond to encompass not only the climate effects, but also other environmental factors as well as social and economic aspects linked with the life of farmers and rural communities all over the world. Hence, the concept of smart agriculture shall be expanded to include all these dimensions and we therefore suggest that an integrated-smart agriculture concept could be used [20].

The main insights discussed in this section show that the topics and the approaches considered, when put together, have novelty [70,74], namely the consideration of the bibliometric analysis for a systematic review in these topics. On the other, the objectives proposed for this study (What are the main relationships between the smart agriculture and the sustainability? Is the concept of 'climate-smart agriculture' (CSA) in general considered in the scientific literature sufficiently broader to capture the several interrelationships among the smart agriculture and the sustainability?) allowed to highlight the main contributions from the smart agriculture for the sustainability and the main limitations of this concept (smart agriculture). In addition, other approaches and topics of search were suggested for future research and highlighted topics already explored directly by the literature.

5. Conclusions

The main conclusions point out to the necessity to face the present and future challenges of the agricultural sector in a more effective way and making use of the technological possibilities that smart agriculture brings. The challenges linked to providing food to the growing world population while at the same time guaranteeing the sustainability of food supply chains must be addressed and looked at as new opportunities to use technology to the service of mankind and the planet. The efficient use of resources, the improvement of the input/output balance of agricultural systems, the mitigation of climate change effect or the socioeconomic impact over rural populations, all are, but not exclusively, part of this global approach. The conceptualization of an integrated approach where all aspects of the problem are included is urgent and therefore we are led to suggest that the somewhat limited concept of climate-smart agriculture could be expanded to a broader concept of 'integrated-smart agriculture'.

In terms of practical implications, the main insights obtained with this research suggest a more effective involvement of the stakeholders in the processes of decision and design of agricultural policy instruments. The participation of the women may be important, considering their concerns with the sustainability and the wellbeing of the family members. In these contexts, the contributions of the different institutions are determinant, specifically through services of extension to involve and support the farmers. Educational programs and vocational training courses are also determinant for an effective implementation of the Integrated-Smart Agriculture practices.

Table 5 presents the main policy recommendations and suggestions for future research from the findings obtain in this research.

Table 5. Policy recommendations, limitations, and future research.

	Policy and Future Studies Suggestions
Policy recommendations	- Design new policy instruments that promote the smart agriculture practices in a more integrated way, namely through a greater involvement of the women in the farms. - Create programs to standardize the perceptions about the smart agricultural approaches. - Promote courses to train and raise awareness of the farmers about the advantages and disadvantages of the smart approaches in the farms. - Involve deeper the interrelationships between the actors for a better understanding about the integrated-smart agriculture approaches.
Limitations	- The selection of the topics to search the documents in the scientific databases has always some subjectivity. The identification of the topics of search in this research is no exception. However, this aspect was discussed and presented suggestions to be addressed in future research with other approaches.
Future research suggestions	- For future research, it could be interesting to deeper survey the farmers about the main constraints to implement an effective plan for smart agriculture practices adoption. - These results of these studies will give new suggestions to take into account the local specificities and better design adjusted plans and policy instruments.

Author Contributions: Conceptualization, V.J.P.D.M. and R.d.P.F.G.; Methodology, V.J.P.D.M.; Formal analysis, V.J.P.D.M. and R.d.P.F.G.; Investigation, V.J.P.D.M. and R.d.P.F.G.; Writing—original draft preparation, V.J.P.D.M.; Writing—review and editing, R.d.P.F.G.; Funding acquisition, R.d.P.F.G. Both authors have read and agreed to the published version of the manuscript.

Funding: This work is funded by National Funds through the FCT—Foundation for Science and Technology, I.P., within the scope of the project Ref[a] UIDB/00681/2020.

Institutional Review Board Statement: Not applicable.

Informed Consent Statement: Not applicable.

Data Availability Statement: The data presented in this study are available on request from the corresponding author.

Acknowledgments: Furthermore, we would like to thank the CERNAS Research Centre and the Polytechnic Institute of Viseu for their support.

Conflicts of Interest: The authors declare no conflict of interest.

References

1. Xie, L.; Luo, B.; Zhong, W. How Are Smallholder Farmers Involved in Digital Agriculture in Developing Countries: A Case Study from China. *Land* **2021**, *10*, 245. [CrossRef]
2. Anand, P.; Singh, Y.; Selwal, A.; Alazab, M.; Tanwar, S.; Kumar, N. IoT Vulnerability Assessment for Sustainable Computing: Threats, Current Solutions, and Open Challenges. *IEEE Access* **2020**, *8*, 168825–168853. [CrossRef]
3. Clercq, M.D.; Vats, A.; Biel, A. *Agriculture 4.0: The Future of Farming Technology*; World Government Summit and Oliver Wyman: New York, NY, USA, 2018; p. 30.
4. Conteratto, C.; do Carmo Martinelli, G.; de Oliveira, L. Food Security, Smart Agriculture and Sustainability: The State of the Art in the Scientific Field. *RISUS* **2020**, *11*, 33–43. [CrossRef]
5. Eli-Chukwu, N.C. Applications of Artificial Intelligence in Agriculture: A Review. *Eng. Technol. Appl. Sci. Res.* **2019**, *9*, 4377–4383. [CrossRef]
6. Ruiz-Real, J.L.; Uribe-Toril, J.; Torres Arriaza, J.A.; de Pablo Valenciano, J. A Look at the Past, Present and Future Research Trends of Artificial Intelligence in Agriculture. *Agronomy* **2020**, *10*, 1839. [CrossRef]
7. Schroeder, K.; Kamel, A.; Sticklen, J.; Ward, R.; Ritchie, J.; Schulthess, U.; Rafea, A.; Salah, A. Guiding Object-Oriented Design Via the Knowledge Level Architecture: The Irrigated Wheat Testbed. *Math. Comput. Model.* **1994**, *20*, 1–16. [CrossRef]
8. Food and Agriculture Organization of the United Nations. Climate-Smart Agriculture. Available online: http://www.fao.org/climate-smart-agriculture/en/ (accessed on 29 June 2021).
9. Rosenzweig, C.; Mbow, C.; Barioni, L.G.; Benton, T.G.; Herrero, M.; Krishnapillai, M.; Liwenga, E.T.; Pradhan, P.; Rivera-Ferre, M.G.; Sapkota, T.; et al. Climate Change Responses Benefit from a Global Food System Approach. *Nat. Food* **2020**, *1*, 94–97. [CrossRef]
10. Herrero, M.; Thornton, P.K.; Mason-D'Croz, D.; Palmer, J.; Benton, T.G.; Bodirsky, B.L.; Bogard, J.R.; Hall, A.; Lee, B.; Nyborg, K.; et al. Innovation Can Accelerate the Transition towards a Sustainable Food System. *Nat. Food* **2020**, *1*, 266–272. [CrossRef]
11. Herrero, M.; Thornton, P.K.; Mason-D'Croz, D.; Palmer, J.; Bodirsky, B.L.; Pradhan, P.; Barrett, C.B.; Benton, T.G.; Hall, A.; Pikaar, I.; et al. Articulating the Effect of Food Systems Innovation on the Sustainable Development Goals. *Lancet Planet. Health* **2021**, *5*, e50–e62. [CrossRef]
12. Manuel Ciruela-Lorenzo, A.; Rosa Del Aguila-Obra, A.; Padilla-Melendez, A.; Jose Plaza-Angulo, J. Digitalization of Agri-Cooperatives in the Smart Agriculture Context. Proposal of a Digital Diagnosis Tool. *Sustainability* **2020**, *12*, 1325. [CrossRef]
13. Vorotnikov, I.L.; Ukolova, N.V.; Monakhov, S.V.; Shikhanova, J.A.; Neyfeld, V.V. Economic Aspects of the Development of the "Digital Agriculture" System. *Sci. Pap. Ser. Manag. Econ. Eng. Agric. Rural Dev.* **2020**, *20*, 633–638.
14. Martinho, V.J.P.D. Exploring the Topics of Soil Pollution and Agricultural Economics: Highlighting Good Practices. *Agriculture* **2020**, *10*, 24. [CrossRef]
15. Martinho, V.J.P.D. Agricultural Entrepreneurship in the European Union: Contributions for a Sustainable Development. *Appl. Sci.* **2020**, *10*, 2080. [CrossRef]
16. Bu, F.; Wang, X. A Smart Agriculture IoT System Based on Deep Reinforcement Learning. *Futur. Gener. Comp. Syst.* **2019**, *99*, 500–507. [CrossRef]
17. Sverdrup, H.; Svensson, M.G.E. Defining sustainability. In *Developing Principles and Models for Sustainable Forestry in Sweden*; Sverdrup, H., Stjernquist, I., Eds.; Managing Forest Ecosystems; Springer: Dordrecht, The Netherlands, 2002; Volume 5, pp. 21–32. ISBN 978-94-015-9888-0.
18. Kuhlman, T.; Farrington, J. What Is Sustainability? *Sustainability* **2010**, *2*, 3436–3448. [CrossRef]
19. Gras, C.; Cáceres, D.M. Technology, Nature's Appropriation and Capital Accumulation in Modern Agriculture. *Curr. Opin. Environ. Sustain.* **2020**, *45*, 1–9. [CrossRef]
20. Taylor, M. Climate-Smart Agriculture: What Is It Good For? *J. Peasant Stud.* **2018**, *45*, 89–107. [CrossRef]

21. Torquebiau, E.; Rosenzweig, C.; Chatrchyan, A.M.; Andrieu, N.; Khosla, R. Identifying Climate-Smart Agriculture Research Needs. *Cah. Agric.* **2018**, *27*. [CrossRef]
22. Van Wijk, M.T.; Merbold, L.; Hammond, J.; Butterbach-Bahl, K. Improving Assessments of the Three Pillars of Climate Smart Agriculture: Current Achievements and Ideas for the Future. *Front. Sustain. Food Syst.* **2020**, *4*, 148. [CrossRef]
23. Liberati, A.; Altman, D.G.; Tetzlaff, J.; Mulrow, C.; Gøtzsche, P.C.; Ioannidis, J.P.A.; Clarke, M.; Devereaux, P.J.; Kleijnen, J.; Moher, D. The PRISMA Statement for Reporting Systematic Reviews and Meta-Analyses of Studies That Evaluate Health Care Interventions: Explanation and Elaboration. *PLoS Med.* **2009**, *6*, e1000100. [CrossRef] [PubMed]
24. Web of Science. Web of Science (Core Collection). Available online: https://apps.webofknowledge.com/WOS_GeneralSearch_input.do?product=WOS&search_mode=GeneralSearch&SID=E5O6jHu4Udf12OXx4mG&preferencesSaved= (accessed on 15 May 2021).
25. Scopus. Scopus Database. Available online: https://www.scopus.com/search/form.uri?display=basic#basic (accessed on 15 May 2021).
26. Zotero. Zotero Software. Available online: https://www.zotero.org/ (accessed on 15 May 2021).
27. Mizik, T. Climate-Smart Agriculture on Small-Scale Farms: A Systematic Literature Review. *Agronomy* **2021**, *11*, 1096. [CrossRef]
28. Totin, E.; Segnon, A.C.; Schut, M.; Affognon, H.; Zougmore, R.B.; Rosenstock, T.; Thornton, P.K. Institutional Perspectives of Climate-Smart Agriculture: A Systematic Literature Review. *Sustainability* **2018**, *10*, 1990. [CrossRef]
29. Martinho, V.J.P.D. Food Marketing as a Special Ingredient in Consumer Choices: The Main Insights from Existing Literature. *Foods* **2020**, *9*, 1651. [CrossRef] [PubMed]
30. Nadaraja, D.; Lu, C.; Islam, M.M. The Sustainability Assessment of Plantation Agriculture—A Systematic Review of Sustainability Indicators. *Sustain. Prod. Consum.* **2021**, *26*, 892–910. [CrossRef]
31. Martinho, V.J.P.D. Agri-Food Contexts in Mediterranean Regions: Contributions to Better Resources Management. *Sustainability* **2021**, *13*, 6683. [CrossRef]
32. Van Eck, N.J.; Waltman, L. VOSviewer Manual: Manual for VOSviewer version 1.6.16, 2020. Center for Science and Technology Studies, Leiden University. Available online: https://www.vosviewer.com/documentation/Manual_VOSviewer_1.6.16.pdf (accessed on 15 May 2021).
33. VOSviewer. VOSviewer Software. Available online: https://www.vosviewer.com/ (accessed on 15 May 2021).
34. Martinho, V.J.P.D. Interrelationships between Renewable Energy and Agricultural Economics: An Overview. *Energy Strategy Rev.* **2018**, *22*, 396–409. [CrossRef]
35. StataCorp. *Stata 15 Base Reference Manual*; Stata Press: College Station, TX, USA, 2017.
36. StataCorp. *Stata Statistical Software: Release 15*; StataCorp LLC: College Station, TX, USA, 2017.
37. Stata. Stata: Software for Statistics and Data Science. Available online: https://www.stata.com/ (accessed on 28 May 2021).
38. Torres-Reyna, O. Getting Started in Factor Analysis (Using Stata). Available online: http://www.princeton.edu/~{}otorres/Stata/Factor (accessed on 28 May 2021).
39. United Nations. THE 17 GOALS. Sustainable Development. Available online: https://sdgs.un.org/goals (accessed on 29 June 2021).
40. Brohm, K.-A.; Klein, S. The Concept of Climate Smart Agriculture—A Classification in Sustainable Theories. *Int. J. Qual. Res.* **2020**, *14*, 291–302. [CrossRef]
41. Noor-E-Sabiha; Rahman, S. Environment-Smart Agriculture and Mapping of Interactions among Environmental Factors at the Farm Level: A Directed Graph Approach. *Sustainability* **2018**, *10*, 1580. [CrossRef]
42. Choudhary, M.; Jat, H.S.; Datta, A.; Sharma, P.C.; Rajashekar, B.; Jat, M.L. Topsoil Bacterial Community Changes and Nutrient Dynamics Under Cereal Based Climate-Smart Agri-Food Systems. *Front. Microbiol.* **2020**, *11*, 1812. [CrossRef]
43. Idoje, G.; Dagiuklas, T.; Iqbal, M. Survey for Smart Farming Technologies: Challenges and Issues. *Comput. Electr. Eng.* **2021**, *92*. [CrossRef]
44. Chacho, P.; Rivera, P.; Maza, J.; Icaza, D.; Arias, P.; Diaz, J.; Amon, X.; Sumba, N.; Pallmay, D.; Coyago, C. Wireless system for the control and real time monitoring of agricultural areas of Arenillas canton in Ecuador. In Proceedings of the 2019 IEEE CHILEAN Conference on Electrical, Electronics Engineering, Information and Communication Technologies (CHILECON), Valparaiso, Chile, 29–31 October 2019; Institute of Electrical and Electronics Engineers Inc.: New York, NY, USA, 2019.
45. Coppedè, N.; Janni, M.; Bettelli, M.; Maida, C.L.; Gentile, F.; Villani, M.; Ruotolo, R.; Iannotta, S.; Marmiroli, N.; Marmiroli, M.; et al. An in Vivo Biosensing, Biomimetic Electrochemical Transistor with Applications in Plant Science and Precision Farming. *Sci. Rep.* **2017**, *7*, 16195. [CrossRef]
46. El Bilali, H.; Bottalico, F.; Ottomano Palmisano, G.; Capone, R. Information and communication technologies for smart and sustainable agriculture. In Proceedings of the IFMBE Proceedings; Brka, M., Omanovic-Miklicanin, E., Karic, L., Falan, V., Toroman, A., Eds.; Springer: Berlin/Heidelberg, Germany, 2020; Volume 78, pp. 321–334.
47. Agrawal, S.; Jain, A. Sustainable Deployment of Solar Irrigation Pumps: Key Determinants and Strategies. *Wiley Interdiscip. Rev. Energy Environ.* **2019**, *8*, e325. [CrossRef]
48. Ahmed, M.; Asif, M.; Sajad, M.; Khattak, J.Z.K.; Ijaz, W.; Wasaya, A.; Chun, J.A. Could Agricultural System Be Adapted to Climate Change? A Review. *Aust. J. Crop Sci.* **2013**, *7*, 1642–1653.
49. Arakelyan, I.; Wreford, A.; Moran, D. Can agriculture be climate smart. In *Building a Climate Resilient Economy and Society: Challenges and Opportunities*; Edward Elgar Publishing Ltd.: Cheltenham, UK, 2017; pp. 115–131. ISBN 9781785368455.

50. Luo, X.-S.; Muleta, D.; Hu, Z.; Tang, H.; Zhao, Z.; Shen, S.; Lee, B.-L. Inclusive Development and Agricultural Adaptation to Climate Change. *Curr. Opin. Environ. Sustain.* **2017**, *24*, 78–83. [CrossRef]
51. Bhatt, R.; Kukal, S.S.; Busari, M.A.; Arora, S.; Yadav, M. Sustainability Issues on Rice-Wheat Cropping System. *Int. Soil Water Conserv. Res.* **2016**, *4*, 64–74. [CrossRef]
52. Branca, G.; Braimoh, A.; Zhao, Y.; Ratii, M.; Likoetla, P. Are There Opportunities for Climate-Smart Agriculture? Assessing Costs and Benefits of Sustainability Investments and Planning Policies in Southern Africa. *J. Clean. Prod.* **2021**, *278*, 123847. [CrossRef]
53. Findlater, K.M.; Kandlikar, M.; Satterfield, T. Misunderstanding Conservation Agriculture: Challenges in Promoting, Monitoring and Evaluating Sustainable Farming. *Environ. Sci. Policy* **2019**, *100*, 47–54. [CrossRef]
54. Long, T.B.; Blok, V.; Poldner, K. Business Models for Maximising the Diffusion of Technological Innovations for Climate-Smart Agriculture. *Int. Food Agribus. Manag. Rev.* **2017**, *20*, 5–23. [CrossRef]
55. Adesipo, A.; Fadeyi, O.; Kuca, K.; Krejcar, O.; Maresova, P.; Selamat, A.; Adenola, M. Smart and Climate-Smart Agricultural Trends as Core Aspects of Smart Village Functions. *Sensors* **2020**, *20*, 5977. [CrossRef]
56. Faling, M.; Biesbroek, R. Cross-Boundary Policy Entrepreneurship for Climate-Smart Agriculture in Kenya. *Policy Sci.* **2019**, *52*, 525–547. [CrossRef]
57. Chandra, A.; McNamara, K.E.; Dargusch, P. The Relevance of Political Ecology Perspectives for Smallholder Climate-Smart Agriculture: A Review. *J. Polit. Ecol.* **2017**, *24*, 821–842. [CrossRef]
58. Aryal, J.P.; Farnworth, C.R.; Khurana, R.; Ray, S.; Sapkota, T.B.; Rahut, D.B. Does Women's Participation in Agricultural Technology Adoption Decisions Affect the Adoption of Climate-Smart Agriculture? Insights from Indo-Gangetic Plains of India. *Rev. Dev. Econ.* **2020**, *24*, 973–990. [CrossRef]
59. Blau, J. Europe Moves toward Sustainable Farming. *Res. Technol. Manag.* **2016**, *59*, 3–6.
60. Acosta-Alba, I.; Chia, E.; Andrieu, N. The LCA4CSA Framework: Using Life Cycle Assessment to Strengthen Environmental Sustainability Analysis of Climate Smart Agriculture Options at Farm and Crop System Levels. *Agric. Syst.* **2019**, *171*, 155–170. [CrossRef]
61. Oerther, D.B. From disaster to development: Finance provides a platform to empower technology for resilience to climate change. In *Proceedings of the Procedia Engineering*; Vidan, A., Shoag, D., Eds.; Elsevier Ltd.: New York, NY, USA, 2016; Volume 159, pp. 267–271.
62. Tripathy, A.S.; Sharma, D.K. Image processing techniques aiding smart agriculture. In *Modern Techniques for Agricultural Disease Management and Crop Yield Prediction*; IGI Global: Hershey, PA, USA, 2019; pp. 23–48. ISBN 9781522596349.
63. Venkatramanan, V.; Shah, S. Climate smart agriculture technologies for environmental management: The intersection of sustainability, resilience, wellbeing and development. In *Sustainable Green Technologies for Environmental Management*; Springer: Singapore, 2019; pp. 29–51. ISBN 9789811327728.
64. Venkatramanan, V.; Shah, S.; Prasad, R. *Global Climate Change and Environmental Policy: Agriculture Perspectives*; Springer: Singapore, 2019; p. 435. ISBN 9789811395703.
65. Pareek, A.; Dhankher, O.P.; Foyer, C.H. Mitigating the Impact of Climate Change on Plant Productivity and Ecosystem Sustainability. *J. Exp. Bot.* **2020**, *71*, 451–456. [CrossRef] [PubMed]
66. Adelhart Toorop, R.; Ceccarelli, V.; Bijarniya, D.; Jat, M.L.; Jat, R.K.; Lopez-Ridaura, S.; Groot, J.C.J. Using a Positive Deviance Approach to Inform Farming Systems Redesign: A Case Study from Bihar, India. *Agric. Syst.* **2020**, *185*. [CrossRef]
67. Bosma, R.H.; Nguyen, T.H.; Siahainenia, A.J.; Tran, H.T.P.; Tran, H.N. Shrimp-Based Livelihoods in Mangrove Silvo-Aquaculture Farming Systems. *Rev. Aquac.* **2016**, *8*, 43–60. [CrossRef]
68. Tran, N.; Cao, Q.L.; Shikuku, K.M.; Phan, T.P.; Banks, L.K. Profitability and Perceived Resilience Benefits of Integrated Shrimp-Tilapia-Seaweed Aquaculture in North Central Coast, Vietnam. *Mar. Policy* **2020**, *120*, 104153. [CrossRef]
69. Furstenau, L.B.; Sott, M.K.; Kipper, L.M.; Machado, E.L.; Ricardo Lopez-Robles, J.; Dohan, M.S.; Cobo, M.J.; Zahid, A.; Abbasi, Q.H.; Imran, M.A. Link Between Sustainability and Industry 4.0: Trends, Challenges and New Perspectives. *IEEE Access* **2020**, *8*, 140079–140096. [CrossRef]
70. Fusco, G.; Melgiovanni, M.; Porrini, D.; Ricciardo, T.M. How to Improve the Diffusion of Climate-Smart Agriculture: What the Literature Tells Us. *Sustainability* **2020**, *12*, 5168. [CrossRef]
71. Streimikis, J.; Miao, Z.; Balezentis, T. Creation of Climate-Smart and Energy-Efficient Agriculture in the European Union: Pathways Based on the Frontier Analysis. *Bus. Strategy Environ.* **2021**, *30*, 576–589. [CrossRef]
72. Bhattacharyya, P.; Pathak, H.; Pal, S. *Energy Management for Climate-Smart Agriculture*; Green Energy and Technology; Springer Science and Business Media Deutschland GmbH: Berlin, Germany, 2020; pp. 73–84.
73. Bhattacharyya, P.; Pathak, H.; Pal, S. *Soil Management for Climate-Smart Agriculture*; Green Energy and Technology; Springer Science and Business Media Deutschland GmbH: Berlin, Germany, 2020; pp. 41–56.
74. Arenas-Calle, L.N.; Whitfield, S.; Challinor, A.J. A Climate Smartness Index (CSI) Based on Greenhouse Gas Intensity and Water Productivity: Application to Irrigated Rice. *Front. Sustain. Food Syst.* **2019**, *3*, 105. [CrossRef]
75. Bhatasara, S.; Nyamwanza, A. Sustainability: A Missing Dimension in Climate Change Adaptation Discourse in Africa? *J. Integr. Environ. Sci.* **2018**, *15*, 83–97. [CrossRef]
76. Alhassan, A.-R.M.; Yang, C.; Ma, W.; Li, G. Influence of Conservation Tillage on Greenhouse Gas Fluxes and Crop Productivity in Spring-Wheat Agroecosystems on the Loess Plateau of China. *PeerJ* **2021**, *9*, e11064. [CrossRef] [PubMed]

77. Nhamo, N.; Chikoye, D. Smart Agriculture: Scope, Relevance, and Important Milestones to Date. In *Smart Technologies for Sustainable Smallholder Agriculture: Upscaling in Developing Countries*; Elsevier Inc.: New York, NY, USA, 2017; pp. 1–20. ISBN 9780128105221.
78. Paul, B.K.; Groot, J.C.J.; Birnholz, C.A.; Nzogela, B.; Notenbaert, A.; Woyessa, K.; Sommer, R.; Nijbroek, R.; Tittonell, P. Reducing Agro-Environmental Trade-Offs through Sustainable Livestock Intensification across Smallholder Systems in Northern Tanzania. *Int. J. Agric. Sustain.* **2020**, *18*, 35–54. [CrossRef]
79. Yamoah, F.A.; Kaba, J.S.; Amankwah-Amoah, J.; Acquaye, A. Stakeholder Collaboration in Climate-Smart Agricultural Production Innovations: Insights from the Cocoa Industry in Ghana. *Environ. Manag.* **2020**, *66*, 600–613. [CrossRef]
80. Emira, S.S.A.; Youssef, K.Y.; Abouelatta, M. Adaptive power system for IoT-based smart agriculture applications. In Proceedings of the 2019 15th International Computer Engineering Conference (ICENCO), Cairo, Egypt, 29–30 December 2019; Institute of Electrical and Electronics Engineers Inc.: New York, NY, USA, 2019; pp. 126–131.
81. Kadar, H.H.; Sameon, S.S.; Rafee, P.A. Sustainable water resource management using IOT solution for agriculture. In Proceedings of the 2019 9th IEEE International Conference on Control System, Computing and Engineering (ICCSCE), Penang, Malaysia, 29 November–1 December 2019; Institute of Electrical and Electronics Engineers Inc.: New York, NY, USA, 2019; pp. 121–125.
82. Kamienski, C.; Soininen, J.-P.; Taumberger, M.; Dantas, R.; Toscano, A.; Cinotti, T.S.; Maia, R.F.; Neto, A.T. Smart Water Management Platform: IoT-Based Precision Irrigation for Agriculture. *Sensors* **2019**, *19*, 276. [CrossRef]
83. Adenugba, F.; Misra, S.; Maskeliūnas, R.; Damaševičius, R.; Kazanavičius, E. Smart Irrigation System for Environmental Sustainability in Africa: An Internet of Everything (IoE) Approach. *Math. Biosci. Eng.* **2019**, *16*, 5490–5503. [CrossRef]
84. Islam, N.; Rashid, M.M.; Pasandideh, F.; Ray, B.; Moore, S.; Kadel, R. A Review of Applications and Communication Technologies for Internet of Things (Iot) and Unmanned Aerial Vehicle (Uav) Based Sustainable Smart Farming. *Sustainability* **2021**, *13*, 1821. [CrossRef]
85. Symeonaki, E.G.; Arvanitis, K.G.; Piromalis, D.D. Current Trends and Challenges in the Deployment of IoT Technologies for Climate Smart Facility Agriculture. *Int. J. Sustain. Agric. Manag. Inform.* **2019**, *5*, 181–200. [CrossRef]
86. Symeonaki, E.G.; Arvanitis, K.G.; Piromalis, D.D. Cloud computing for IoT applications in climate-smart agriculture: A review on the trends and challenges toward sustainability. In *Innovative Approaches and Applications for Sustainable Rural Development, Haicta 2017*; Theodoridis, A., Ragkos, A., Salampasis, M., Eds.; Springer International Publishing Ag: Cham, Switzerland, 2019; pp. 147–167. ISBN 978-3-030-02312-6.
87. Tripathy, P.K.; Tripathy, A.K.; Agarwal, A.; Mohanty, S.P. MyGreen: An IoT-Enabled Smart Greenhouse for Sustainable Agriculture. *IEEE Consum. Electron. Mag.* **2021**, *10*, 57–62. [CrossRef]
88. Vu, V.A.; Cong Trinh, D.; Truvant, T.C.; Dang Bui, T. Design of automatic irrigation system for greenhouse based on LoRa technology. In Proceedings of the 2018 International Conference on Advanced Technologies for Communications (ATC), Ho Chi Minh City, Vietnam, 18–20 October 2018; Bao, V.N.Q., Duy, T.T., Eds.; IEEE Computer Society: New York, NY, USA, 2018; pp. 72–77.
89. Verdouw, C.; Tekinerdogan, B.; Beulens, A.; Wolfert, S. Digital Twins in Smart Farming. *Agric. Syst.* **2021**, *189*, 103046. [CrossRef]
90. Muñoz, M.; Gil, J.D.; Roca, L.; Rodríguez, F.; Berenguel, M. An Iot Architecture for Water Resource Management in Agroindustrial Environments: A Case Study in Almería (Spain). *Sensors* **2020**, *20*, 596. [CrossRef] [PubMed]
91. Sengupta, A.; Debnath, B.; Das, A.; De, D. FarmFox: A Quad-Sensor Based IoT Box for Precision Agriculture. *IEEE Consum. Electron. Mag.* **2021**, *10*, 63–68. [CrossRef]
92. Valecce, G.; Strazzella, S.; Radesca, A.; Grieco, L.A. Solarfertigation: Internet of things architecture for smart agriculture. In Proceedings of the 2019 IEEE International Conference on Communications Workshops (ICC Workshops), Shanghai, China, 20–24 May 2019; Institute of Electrical and Electronics Engineers Inc.: New York, NY, USA, 2019.
93. Shafi, U.; Mumtaz, R.; García-Nieto, J.; Hassan, S.A.; Zaidi, S.A.R.; Iqbal, N. Precision Agriculture Techniques and Practices: From Considerations to Applications. *Sensors* **2019**, *19*, 3796. [CrossRef] [PubMed]
94. Heideker, A.; Ottolini, D.; Zyrianoff, I.; Neto, A.T.; Salmon Cinotti, T.; Kamienski, C. IoT-based measurement for smart agriculture. In Proceedings of the 2020 IEEE International Workshop on Metrology for Agriculture and Forestry (MetroAgriFor), Trento, Italy, 4–6 November 2020; Institute of Electrical and Electronics Engineers Inc.: New York, NY, USA, 2020; pp. 68–72.
95. Wu, F.; Ma, J. Evolution Dynamics of Agricultural Internet of Things Technology Promotion and Adoption in China. *Discrete Dyn. Nat. Soc.* **2020**, *2020*, 1854193. [CrossRef]
96. Dutta, P.; Chakraborti, S.; Chaudhuri, K.M.; Mondal, S. Physiological responses and resilience of plants to climate change. In *New Frontiers in Stress Management for Durable Agriculture*; Springer: Singapore, 2020; pp. 3–20. ISBN 9789811513220.
97. El Ansari, L.; Chenoune, R.; Yigezu, Y.A.; Gary, C.; Belhouchette, H. Trade-Offs between Sustainability Indicators in Response to the Production Choices of Different Farm Household Types in Drylands. *Agronomy* **2020**, *10*, 998. [CrossRef]
98. Githiru, M.; Mutwiwa, U.; Kasaine, S.; Schulte, B. A Spanner in the Works: Human-Elephant Conflict Complicates the Food-Water-Energy Nexus in Drylands of Africa. *Front. Environ. Sci.* **2017**, *5*, 69. [CrossRef]
99. Ennouri, K.; Kallel, A.; Albano, R. Remote Sensing: An Advanced Technique for Crop Condition Assessment. *Math. Probl. Eng.* **2019**, *2019*, 9404565. [CrossRef]
100. Chhogyel, N.; Kumar, L.; Bajgai, Y. Consequences of Climate Change Impacts and Incidences of Extreme Weather Events in Relation to Crop Production in Bhutan. *Sustainability* **2020**, *12*, 4319. [CrossRef]
101. Rout, S.K.; Gulathi, J.M.L.; Parasar, B. Smart Agriculture System: Review. *Int. J. Mod. Agric.* **2020**, *9*, 674–680.

102. Kumari, S.; Patil, Y. Achieving climate smart agriculture with a sustainable use of water: A conceptual framework for sustaining the use of water for agriculture in the era of climate change. In *Climate Change and Environmental Concerns: Breakthroughs in Research and Practice*; IGI Global: Hershey, PA, USA, 2018; pp. 111–133. ISBN 9781522554882.
103. Aryal, J.P.; Sapkota, T.B.; Rahut, D.B.; Jat, M.L. Agricultural Sustainability under Emerging Climatic Variability: The Role of Climate-Smart Agriculture and Relevant Policies in India. *Int. J. Innov. Sustain. Dev.* **2020**, *14*, 219–245. [CrossRef]
104. Ait Issad, H.; Aoudjit, R.; Rodrigues, J.J.P.C. A Comprehensive Review of Data Mining Techniques in Smart Agriculture. *Eng. Agric. Environ. Food* **2019**, *12*, 511–525. [CrossRef]
105. Morimoto, E. What is cyber-physical system driven agriculture?—Redesign of big data for outstanding farmer management. In Proceedings of the ASABE Annual International Meeting, Detroit, MI, USA, 29 July–1 August 2018; American Society of Agricultural and Biological Engineers: St. Joseph, MI, USA, 2018.
106. Anderson, J.R. Concepts of food sustainability. In *Encyclopedia of Food Security and Sustainability*; Elsevier: New York, NY, USA, 2018; pp. 1–8. ISBN 9780128126882.
107. Fiore, M.; Monasterolo, I.; Jones, A.; Contò, F. Understanding Limits to Data Informative Power for Sustainable Food Policies in Transition and Post-Transition Countries. *Int. J. Glob. Small Bus.* **2015**, *7*, 300–317. [CrossRef]
108. Msangi, J.P. *Food Security among Small-Scale Agricultural Producers in Southern Africa*; Springer International Publishing: Cham, Germany, 2014; Volume 9783319094953, p. 173. ISBN 9783319094953.
109. Makate, C. Effective Scaling of Climate Smart Agriculture Innovations in African Smallholder Agriculture: A Review of Approaches, Policy and Institutional Strategy Needs. *Environ. Sci. Policy* **2019**, *96*, 37–51. [CrossRef]
110. Were, K.; Gelaw, A.M.; Singh, B.R. Smart strategies for enhanced agricultural resilience and food security under a changing climate in sub-saharan Africa. In *Climate Change and Multi-Dimensional Sustainability in African Agriculture: Climate Change and Sustainability in Agriculture*; Springer International Publishing: Cham, Germany, 2016; pp. 431–453. ISBN 9783319412382.
111. Manik, B.K. Revisit to Policy Formulation for Climate-Smart Agriculture in India. *Int. J. Innov. Technol. Explor. Eng.* **2019**, *8*, 144–151.
112. Saroar, M.M.; Filho, W.L. *Adaptation Through Climate Smart Agriculture: Status and Determinants in Coastal Bangladesh*; Climate Change Management; Springer: Berlin/Heidelberg, Germany, 2016; pp. 157–178.
113. Delian, E.; Chira, A.; Badea, M.L.; Chira, L. Sustainable Agriculture Systems to Mitigate Climate Change Effects: A Brief Overview. *Sci. Pap. Ser. Manag. Econ. Eng. Agric. Rural Dev.* **2019**, *19*, 127–134.
114. Ajayi, A.E.; Horn, R. Transformation of Ex-Arable Land to Permanent Grassland Promotes Pore Rigidity and Mechanical Soil Resilience. *Ecol. Eng.* **2016**, *94*, 592–598. [CrossRef]
115. Cochran, F.V.; Brunsell, N.A.; Suyker, A.E. A Thermodynamic Approach for Assessing Agroecosystem Sustainability. *Ecol. Indic.* **2016**, *67*, 204–214. [CrossRef]
116. Berhanu, Y.; Angassa, A.; Aune, J.B. A System Analysis to Assess the Effect of Low-Cost Agricultural Technologies on Productivity, Income and GHG Emissions in Mixed Farming Systems in Southern Ethiopia. *Agric. Syst.* **2021**, *187*, 102988. [CrossRef]
117. Dabkienė, V.; Baležentis, T.; Štreimikienė, D. Calculation of the Carbon Footprint for Family Farms Using the Farm Accountancy Data Network: A Case from Lithuania. *J. Clean. Prod.* **2020**, *262*, 121509. [CrossRef]
118. Fischer, H.W.; Reddy, N.L.N.; Rao, M.L.S. Can More Drought Resistant Crops Promote More Climate Secure Agriculture? Prospects and Challenges of Millet Cultivation in Ananthapur, Andhra Pradesh. *World Dev. Perspect.* **2016**, *2*, 5–10. [CrossRef]
119. Fungo, B.; Buyinza, J.; Sekatuba, J.; Nansereko, S.; Ongodia, G.; Kwaga, P.; Mudondo, S.; Eryau, K.; Akelem, R.; Musinguzi, P.; et al. Forage Biomass and Soil Aggregate Carbon under Fodder Banks with Contrasting Management Regimes. *Agrofor. Syst.* **2020**, *94*, 1023–1035. [CrossRef]
120. Hao, P.-F.; Qiu, C.-W.; Ding, G.; Vincze, E.; Zhang, G.; Zhang, Y.; Wu, F. Agriculture Organic Wastes Fermentation CO2 Enrichment in Greenhouse and the Fermentation Residues Improve Growth, Yield and Fruit Quality in Tomato. *J. Clean. Prod.* **2020**, *275*, 123885. [CrossRef]
121. Huang, Y.-P.; Wang, T.-H.; Basanta, H. Using Fuzzy Mask R-CNN Model to Automatically Identify Tomato Ripeness. *IEEE Access* **2020**, *8*, 207672–207682. [CrossRef]
122. Jat, H.S.; Datta, A.; Choudhary, M.; Sharma, P.C.; Yadav, A.K.; Choudhary, V.; Gathala, M.K.; Jat, M.L.; McDonald, A. Climate Smart Agriculture Practices Improve Soil Organic Carbon Pools, Biological Properties and Crop Productivity in Cereal-Based Systems of North-West India. *CATENA* **2019**, *181*, 104059. [CrossRef]
123. Kimaro, A.A.; Mpanda, M.; Rioux, J.; Aynekulu, E.; Shaba, S.; Thiong'o, M.; Mutuo, P.; Abwanda, S.; Shepherd, K.; Neufeldt, H.; et al. Is Conservation Agriculture "climate-Smart" for Maize Farmers in the Highlands of Tanzania? *Nutr. Cycl. Agroecosyst.* **2016**, *105*, 217–228. [CrossRef]
124. Rahman, M.M.; Aravindakshan, S.; Hoque, M.A.; Rahman, M.A.; Gulandaz, M.A.; Rahman, J.; Islam, M.T. Conservation Tillage (CT) for Climate-Smart Sustainable Intensification: Assessing the Impact of CT on Soil Organic Carbon Accumulation, Greenhouse Gas Emission and Water Footprint of Wheat Cultivation in Bangladesh. *Environ. Sustain. Ind.* **2021**, *10*, 100106. [CrossRef]
125. Lang, C.; Mission, E.G.; Ahmad Fuaad, A.A.-H.; Shaalan, M. Nanoparticle Tools to Improve and Advance Precision Practices in the Agrifoods Sector towards Sustainability—A Review. *J. Clean. Prod.* **2021**, *293*, 126063. [CrossRef]
126. Pramanik, P.; Ray, P.; Maity, A.; Das, S.; Ramakrishnan, S.; Dixit, P. Nanotechnology for improved carbon management in soil. In *Carbon Management in Tropical and Sub-Tropical Terrestrial Systems*; Springer: Singapore, 2019; pp. 403–415. ISBN 9789811396281.

127. Makate, C.; Wang, R.; Makate, M.; Mango, N. Impact of Drought Tolerant Maize Adoption on Maize Productivity, Sales and Consumption in Rural Zimbabwe. *Agrekon* **2017**, *56*, 67–81. [CrossRef]
128. Mazhar, R.; Ghafoor, A.; Xuehao, B.; Wei, Z. Fostering Sustainable Agriculture: Do Institutional Factors Impact the Adoption of Multiple Climate-Smart Agricultural Practices among New Entry Organic Farmers in Pakistan? *J. Clean. Prod.* **2021**, *283*, 124620. [CrossRef]
129. Musa, F.B.; Kamoto, J.F.M.; Jumbe, C.B.L.; Zulu, L.C. Adoption and the Role of Fertilizer Trees and Shrubs as a Climate Smart Agriculture Practice: The Case of Salima District in Malawi. *Environments* **2018**, *5*, 122. [CrossRef]
130. Mylona, P.; Sakellariou, M.; Giannakopoulos, C.; Psiloglou, B.; Kitsara, G. Terrace landscapes as green infrastructures for a climate-smart agriculture to mitigate climate change impacts. Presented at 9th International Conference on Information and Communication Technologies in Agriculture, Food & Environment (HAICTA 2020), Thessaloniki, Greece, 24–27 September 2020; pp. 236–243. Available online: http://ceur-ws.org/Vol-2761/HAICTA_2020_paper35.pdf (accessed on 15 May 2021).
131. Nair, P.K.R.; Kumar, B.M.; Naresh Kumar, S. Climate change, carbon sequestration, and coconut-based ecosystems. In *The Coconut Palm (Cocos nucifera L.) Research and Development Perspectives*; Springer: Singapore, 2019; pp. 779–799. ISBN 9789811327544.
132. Nair, P.K.R.; Viswanath, S.; Lubina, P.A. Cinderella Agroforestry Systems. *Agrofor. Syst.* **2017**, *91*, 901–917. [CrossRef]
133. Naughton, S.; Kavanagh, S.; Lynch, M.; Rowan, N.J. Synchronizing Use of Sophisticated Wet-Laboratory and in-Field Handheld Technologies for Real-Time Monitoring of Key Microalgae, Bacteria and Physicochemical Parameters Influencing Efficacy of Water Quality in a Freshwater Aquaculture Recirculation System: A Case Study from the Republic of Ireland. *Aquaculture* **2020**, *526*, 735377. [CrossRef]
134. Nyamadzawo, G.; Wuta, M.; Nyamangara, J.; Nyamugafata, P.; Chirinda, N. Optimizing Dambo (Seasonal Wetland) Cultivation for Climate Change Adaptation and Sustainable Crop Production in the Smallholder Farming Areas of Zimbabwe. *Int. J. Agric. Sustain.* **2015**, *13*, 23–39. [CrossRef]
135. Ronga, D.; Parisi, M.; Barbieri, L.; Lancellotti, I.; Andreola, F.; Bignami, C. Valorization of Spent Coffee Grounds, Biochar and Other Residues to Produce Lightweight Clay Ceramic Aggregates Suitable for Nursery Grapevine Production. *Horticulturae* **2020**, *6*, 58. [CrossRef]
136. Singh, R.; Singh, G.S. Traditional Agriculture: A Climate-Smart Approach for Sustainable Food Production. *Energy Ecol. Environ.* **2017**, *2*, 296–316. [CrossRef]
137. Singh, V.K.; Rathore, S.S.; Singh, R.K.; Upadhyay, P.K.; Shekhawat, K. Integrated Farming System Approach for Enhanced Farm Productivity, Climate Resilience and Doubling Farmers' Income. *Indian J. Agric. Sci.* **2020**, *90*, 1378–1388.
138. Snyder, C.S. Enhanced Nitrogen Fertiliser Technologies Support the "4R" Concept to Optimise Crop Production and Minimise Environmental Losses. *Soil Res.* **2017**, *55*, 463–472. [CrossRef]
139. Sosa-Hernández, M.A.; Leifheit, E.F.; Ingraffia, R.; Rillig, M.C. Subsoil Arbuscular Mycorrhizal Fungi for Sustainability and Climate-Smart Agriculture: A Solution Right under Our Feet? *Front. Microbiol.* **2019**, *10*, 744. [CrossRef]
140. Steinke, J.; Mgimiloko, M.G.; Graef, F.; Hammond, J.; van Wijk, M.T.; van Etten, J. Prioritizing Options for Multi-Objective Agricultural Development through the Positive Deviance Approach. *PLoS ONE* **2019**, *14*, e0212926. [CrossRef] [PubMed]
141. Subbarao, G.V.; Arango, J.; Masahiro, K.; Hooper, A.M.; Yoshihashi, T.; Ando, Y.; Nakahara, K.; Deshpande, S.; Ortiz-Monasterio, I.; Ishitani, M.; et al. Genetic Mitigation Strategies to Tackle Agricultural GHG Emissions: The Case for Biological Nitrification Inhibition Technology. *Plant Sci.* **2017**, *262*, 165–168. [CrossRef]
142. Tolga, A.C.; Basar, M. Hydroponic System Evaluation in Urban Farming via Fuzzy EDAS and TODIM Methods. *J. Intell. Fuzzy Syst.* **2020**, *39*, 6325–6337. [CrossRef]
143. Tolga, A.C.; Gamsiz, B.; Basar, M. *Evaluation of Hydroponic System in Vertical Farming via Fuzzy EDAS Method*; International Conference on Intelligent and Fuzzy Systems, INFUS 2019; Springer: Berlin/Heidelberg, Germany, 2020; Volume 1029, pp. 745–752.
144. Venkatramanan, V.; Shah, S.; Rai, A.K.; Prasad, R. Nexus Between Crop Residue Burning, Bioeconomy and Sustainable Development Goals Over North-Western India. *Front. Energy Res.* **2021**, *8*, 105. [CrossRef]
145. Verburg, R.; Rahn, E.; Verweij, P.; van Kuijk, M.; Ghazoul, J. An Innovation Perspective to Climate Change Adaptation in Coffee Systems. *Environ. Sci. Policy* **2019**, *97*, 16–24. [CrossRef]
146. Vanbergen, A.J.; Aizen, M.A.; Cordeau, S.; Garibaldi, L.A.; Garratt, M.P.D.; Kovács-Hostyánszki, A.; Lecuyer, L.; Ngo, H.T.; Potts, S.G.; Settele, J.; et al. *Transformation of Agricultural Landscapes in the Anthropocene: Nature's Contributions to People, Agriculture and Food Security*; Advances in Ecological Research; Academic Press Inc.: Amsterdam, The Netherlands, 2020; Volume 63, pp. 193–253. ISSN 00652504.
147. Thierfelder, C.; Chivenge, P.; Mupangwa, W.; Rosenstock, T.S.; Lamanna, C.; Eyre, J.X. How Climate-Smart Is Conservation Agriculture (CA)?—Its Potential to Deliver on Adaptation, Mitigation and Productivity on Smallholder Farms in Southern Africa. *Food Secur.* **2017**, *9*, 537–560. [CrossRef]
148. Clapp, J.; Newell, P.; Brent, Z.W. The Global Political Economy of Climate Change, Agriculture and Food Systems. *J. Peasant Stud.* **2018**, *45*, 80–88. [CrossRef]
149. Bai, H.; Tao, F. Sustainable Intensification Options to Improve Yield Potential and Eco-Efficiency for Rice-Wheat Rotation System in China. *Field Crops Res.* **2017**, *211*, 89–105. [CrossRef]
150. Baratella, V.; Trinchera, A. Organosilicone Surfactants as Innovative Irrigation Adjuvants: Can They Improve Water Use Efficiency and Nutrient Uptake in Crop Production? *Agric. Water Manag.* **2018**, *204*, 149–161. [CrossRef]

151. Bhatt, R.; Hossain, A.; Singh, P. Scientific interventions to improve land and water productivity for climate-smart agriculture in South Asia. In *Agronomic Crops: Volume 2: Management Practices*; Springer: Singapore, 2019; pp. 499–558. ISBN 9789813297838.
152. Bhatt, R.; Kaur, R.; Ghosh, A. Strategies to practice climate-smart agriculture to improve the livelihoods under the rice-wheat cropping system in South Asia. In *Sustainable Management of Soil and Environment*; Springer: Singapore, 2019; pp. 29–71. ISBN 9789811388323.
153. Kakraliya, S.K.; Singh, I.; Dadarwal, R.S.; Singh, L.K.; Jat, R.D.; Jat, H.S.; Jat, M.L. Impact of Climate-Smart Agricultural Practices on Growth and Crop Yields of Rice (Oryza Sativa)-Wheat (Triticum Aestivum) System in North-Western Indo-Gangetic Plains. *Indian J. Agric. Sci.* **2018**, *88*, 1543–1551.
154. Hossain, A.; Bhatt, R. Intervention of Climate Smart Technologies for Improving Water Productivity in an Enormous Water Use Rice-Wheat System of South-Asia. *Int. Lett. Nat. Sci.* **2019**, *75*, 27–35. [CrossRef]
155. Kiwia, A.; Kimani, D.; Harawa, R.; Jama, B.; Sileshi, G.W. Sustainable Intensification with Cereal-Legume Intercropping in Eastern and Southern Africa. *Sustainability* **2019**, *11*, 2891. [CrossRef] [PubMed]
156. Fuchs, L.E.; Orero, L.; Namoi, N.; Neufeldt, H. How to Effectively Enhance Sustainable Livelihoods in Smallholder Systems: A Comparative Study from Western Kenya. *Sustainability* **2019**, *11*, 1564. [CrossRef]
157. Cavanagh, C.J.; Vedeld, P.O.; Petursson, J.G.; Chemarum, A.K. Agency, Inequality, and Additionality: Contested Assemblages of Agricultural Carbon Finance in Western Kenya. *J. Peasant Stud.* **2020**. [CrossRef]
158. Chinseu, E.L.; Dougill, A.J.; Stringer, L.C. Strengthening Conservation Agriculture Innovation Systems in Sub-Saharan Africa: Lessons from a Stakeholder Analysis. *Int. J. Agric. Sustain.* **2021**. [CrossRef]
159. Gérardeaux, E.; Loison, R.; Palaï, O.; Sultan, B. Adaptation Strategies to Climate Change Using Cotton (Gossypium Hirsutum L.) Ideotypes in Rainfed Tropical Cropping Systems in Sub-Saharan Africa. A Modeling Approach. *Field Crops Res.* **2018**, *226*, 38–47. [CrossRef]
160. Machekano, H.; Mvumi, B.M.; Nyamukondiwa, C. Diamondback Moth, *Plutella Xylostella* (L.) in Southern Africa: Research Trends, Challenges and Insights on Sustainable Management Options. *Sustainability* **2017**, *9*, 91. [CrossRef]
161. Ofori, M.; El-Gayar, O. The state and future of smart agriculture: Insights from mining social media. In *Proceedings of the 2019 IEEE International Conference on Big Data (Big Data)*; Baru, C., Huan, J., Khan, L., Hu, X.T., Ak, R., Tian, Y., Barga, R., Zaniolo, C., et al., Eds.; Institute of Electrical and Electronics Engineers Inc.: New York, NY, USA, 2019; pp. 5152–5161.
162. Whitfield, S.; Benton, T.G.; Dallimer, M.; Firbank, L.G.; Poppy, G.M.; Sallu, S.M.; Stringer, L.C. Sustainability Spaces for Complex Agri-Food Systems. *Food Secur.* **2015**, *7*, 1291–1297. [CrossRef]
163. Whitfield, S.; Challinor, A.J.; Rees, R.M. Frontiers in Climate Smart Food Systems: Outlining the Research Space. *Front. Sustain. Food Syst.* **2018**, *2*, 2. [CrossRef]
164. Reidsma, P.; Bakker, M.M.; Kanellopoulos, A.; Alam, S.J.; Paas, W.; Kros, J.; de Vries, W. Sustainable Agricultural Development in a Rural Area in the Netherlands? Assessing Impacts of Climate and Socio-Economic Change at Farm and Landscape Level. *Agric. Syst.* **2015**, *141*, 160–173. [CrossRef]
165. Cesco, S.; Pii, Y.; Borruso, L.; Orzes, G.; Lugli, P.; Mazzetto, F.; Genova, G.; Signorini, M.; Brunetto, G.; Terzano, R.; et al. A Smart and Sustainable Future for Viticulture Is Rooted in Soil: How to Face Cu Toxicity. *Appl. Sci.* **2021**, *11*, 907. [CrossRef]
166. Imran, M.A.; Ali, A.; Ashfaq, M.; Hassan, S.; Culas, R.; Ma, C. Impact of Climate Smart Agriculture (CSA) through Sustainable Irrigation Management on Resource Use Efficiency: A Sustainable Production Alternative for Cotton. *Land Use Policy* **2019**, *88*, 104113. [CrossRef]
167. Moreno, A.; Bhattacharyya, A.; Jansen, L.; Arkeman, Y.; Hartanto, R.; Kleinke, M. Environmental engineering and sustainability for smart agriculture: The application of UAV-based remote sensing to detect biodiversity in oil palm plantations. In *Proceedings of the IOP Conference Series: Earth and Environmental Science*; Institute of Physics Publishing: Bristol, UK, 2019; Volume 335.
168. Nasser, F.; Maguire-Rajpaul, V.A.; Dumenu, W.K.; Wong, G.Y. Climate-Smart Cocoa in Ghana: How Ecological Modernisation Discourse Risks Side-Lining Cocoa Smallholders. *Front. Sustain. Food Syst.* **2020**, *4*, 73. [CrossRef]
169. Noponen, M.R.A.; Góngora, C.; Benavides, P.; Gaitán, A.; Hayward, J.; Marsh, C.; Stout, R.; Wille, C. Environmental Sustainability-Farming in the Anthropocene. In *The Craft and Science of Coffee*; Elsevier Inc.: New York, NY, USA, 2017; pp. 81–107. ISBN 9780128035580.
170. Mohanty, S.; Mohanta, B.; Nanda, P.; Sen, S.; Patnaik, S. *Smart Village Initiatives: An Overview*; Modeling and Optimization in Science and Technologies; Springer: Berlin/Heidelberg, Germany, 2020; Volume 17, pp. 3–24. ISSN 21967326.
171. Mukherjee, P.; Barik, R.K.; Pradhan, C. *A Comprehensive Proposal for Blockchain-Oriented Smart City*; Studies in Systems, Decision and Control; Springer Science and Business Media Deutschland GmbH: Berlin/Heidelberg, Germany, 2021; Volume 308, pp. 55–87.
172. Goparaju, L.; Ahmad, F. Analyzing the Risk Related to Climate Change Attributes and Their Impact, a Step towards Climate-Smart Village (CSV): A Geospatial Approach to Bring Geoponics Sustainability in India. *Spat. Inf. Res.* **2019**, *27*, 613–625. [CrossRef]
173. Amadu, F.O.; McNamara, P.E.; Miller, D.C. Understanding the Adoption of Climate-Smart Agriculture: A Farm-Level Typology with Empirical Evidence from Southern Malawi. *World Dev.* **2020**, *126*, 104692. [CrossRef]
174. Long, T.B.; Blok, V.; Coninx, I. The Diffusion of Climate-Smart Agricultural Innovations: Systems Level Factors That Inhibit Sustainable Entrepreneurial Action. *J. Clean. Prod.* **2019**, *232*, 993–1004. [CrossRef]
175. Greenland, S.; Levin, E.; Dalrymple, J.F.; O'Mahony, B. Sustainable Innovation Adoption Barriers: Water Sustainability, Food Production and Drip Irrigation in Australia. *Soc. Responsib. J.* **2019**, *15*, 727–741. [CrossRef]

176. Long, T.B.; Blok, V.; Coninx, I. Barriers to the Adoption and Diffusion of Technological Innovations for Climate-Smart Agriculture in Europe: Evidence from the Netherlands, France, Switzerland and Italy. *J. Clean. Prod.* **2016**, *112*, 9–21. [CrossRef]
177. Kibria, G.; Haroon, A.K.Y.; Nugegoda, D. An Innovative Model for Engagement of Rural Citizens/Community of Bangladesh with Climate Change. *J. Clim. Chang.* **2017**, *3*, 73–80. [CrossRef]
178. Osorio-Garcia, A.M.; Paz, L.; Howland, F.; Ortega, L.A.; Acosta-Alba, I.; Arenas, L.; Chirinda, N.; Martinez-Baron, D.; Bonilla Findji, O.; Loboguerrero, A.M.; et al. Can an Innovation Platform Support a Local Process of Climate-Smart Agriculture Implementation? A Case Study in Cauca, Colombia. *Agroecol. Sustain. Food Syst.* **2020**, *44*, 378–411. [CrossRef]
179. Vernooy, R.; Bouroncle, C.; Roque, V.S.; García, J.R. Sustainable Territories Adapted to the Climate: Insights from a New University Course Designed and Delivered in Guatemala. *Sustainability* **2020**, *12*, 4978. [CrossRef]
180. Castillo, K.; Cabrera-Rios, M.; Persans, M.W.; DeYoe, H.R. Engaging minority students in sustainable bioenergy and water quality through an education and research network. In Proceedings of the 2016 ASEE Annual Conference & Exposition, New Orleans, LA, USA, 26 June–29 June 2016; American Society for Engineering Education: Washington, DC, USA, 2016; Volume 2016.
181. Taneja, G.; Pal, B.D.; Joshi, P.K.; Aggarwal, P.K.; Tyagi, N.K. Farmers' preferences for climate-smart agriculture-an assessment in the indo-gangetic plain. In *Climate Smart Agriculture in South Asia: Technologies, Policies and Institutions*; Springer: Singapore, 2019; pp. 91–111. ISBN 9789811081712.
182. Ologeh, I.O.; Akarakiri, J.B.; Adesina, F.A. *Promoting Climate Smart Agriculture Through Space Technology in Nigeria*; Climate Change Management; Springer: Berlin/Heidelberg, Germany, 2016; pp. 99–112.
183. Drexler, K. Climate-Smart Adaptations and Government Extension Partnerships for Sustainable Milpa Farming Systems in Mayan Communities of Southern Belize. *Sustainability* **2021**, *13*, 3040. [CrossRef]
184. Dung, L.T. Factors Influencing Farmers' Adoption of Climate-Smart Agriculture in Rice Production in Vietnam's Mekong Delta. *Asian J. Agric. Dev.* **2020**, *17*, 109–124. [CrossRef]
185. Obasi, P.C.; Chikezie, C. *SMART Agriculture and Rural Farmers Adaptation Measures to Climate Change in Southeast Nigeria: Implications for Sustainable Food Security*; Climate Change Management; Springer: Berlin/Heidelberg, Germany, 2020; pp. 813–833.
186. Rampa, A.; Gadanakis, Y.; Rose, G. Land Reform in the Era of Global Warming—Can Land Reforms Help Agriculture Be Climate-Smart? *Land* **2020**, *9*, 471. [CrossRef]
187. Czekaj, M.; Adamsone-Fiskovica, A.; Tyran, E.; Kilis, E. Small Farms' Resilience Strategies to Face Economic, Social, and Environmental Disturbances in Selected Regions in Poland and Latvia. *Glob. Food Secur.* **2020**, *26*, 100416. [CrossRef]

agronomy

MDPI

Article

Extraction of Phenolic Compounds from Cherry Seeds: A Preliminary Study

Yuliya Dulyanska [1,*], Luísa Paula Cruz-Lopes [2], Bruno Esteves [2], José Vicente Ferreira [2], Idalina Domingos [2], Maria João Lima [1], Paula M. R. Correia [1], Manuela Ferreira [3], Anabela Fragata [4], Maria João Barroca [5,6], Aida Moreira da Silva [5,6] and Raquel P. F. Guiné [1,*]

1 CERNAS-IPV Research Centre and ESAV, Polytechnic Institute of Viseu, 3504-510 Viseu, Portugal; mjoaolima@esav.ipv.pt (M.J.L.); paulacorreia@esav.ipv.pt (P.M.R.C.)
2 CERNAS-IPV Research Centre and ESTGV, Polytechnic Institute of Viseu, 3504-510 Viseu, Portugal; lvalente@estgv.ipv.pt (L.P.C.-L.); bruno@estgv.ipv.pt (B.E.); jvf@estgv.ipv.pt (J.V.F.); ijd@estgv.ipv.pt (I.D.)
3 UICISA: E Research Centre and ESSV, Polytechnic Institute of Viseu, 3504-510 Viseu, Portugal; mmcferreira@gmail.com
4 CI&DEI Research Centre and ESTGV, Polytechnic Institute of Viseu, 3504-510 Viseu, Portugal; afragata@estgl.ipv.pt
5 Molecular Physical-Chemistry R&D Unit, Department of Chemistry, University of Coimbra, 3004-535 Coimbra, Portugal; mjbarroca@esac.pt (M.J.B.); aidams@esac.pt (A.M.d.S.)
6 Agrarian School, Polytechnic Institute of Coimbra, 3040-360 Coimbra, Portugal
* Correspondence: ydulyanska@esav.ipv.pt (Y.D.); raquelguine@esav.ipv.pt (R.P.F.G.)

Abstract: Agri-food waste has proved to be a valuable bioresource that can be used to obtain a variety of valuable materials, ingredients and chemicals. The optimum conditions for extracting bioactive compounds from sweet cherry seeds (SCS) with different solvents and temperatures were tested in this work. The choice criteria were based on the most efficient extracting capacity while looking for cleaner techniques with lower health or environmental impacts. Some extracting solvents (methanol, ethanol and water) were tested in different combinations and temperatures. The obtained extracts were evaluated for total phenolic compounds and some families of phenolics as well, using spectrophotometric methods. The results obtained showed that the highest extraction of total phenolic compounds was at 70 °C with 60:40 ratio water:ethanol (2.65 mg GAE/g), while maximum flavonoids were obtained at 80 °C and 50% ethanolic aqueous solution (7.26 mg QE/g). The highest value for ortho-diphenols was 21.47 mg GAE/g for 50 °C and water:ethanol 50:50 solution. The highest proanthocyanidins and flavonols were obtained for 50:50 solution at 70 °C (6.43 mg CE/g and 3.88 mg QE/g, respectively), while the same solution at 80 °C allowed obtaining maximum phenolic acids (1.68 mg CAE/g). The extraction of anthocyanins was found to vary significantly with concentration and temperature, being highest in the range 35–40 °C, when using an 80:20 water:ethanol solution. Hierarchical clustering showed three clusters, while factor analysis resulted in two factors and four groups of samples. In conclusion, it was found that extracts obtained from sweet cherry seeds have relevant bioactive compounds with applications in the food, pharmaceutical or cosmetic industries.

Keywords: extraction procedure; bioactive compounds; anthocyanins; phenolic compounds; cherry seeds

Citation: Dulyanska, Y.; Cruz-Lopes, L.P.; Esteves, B.; Ferreira, J.V.; Domingos, I.; Lima, M.J.; Correia, P.M.R.; Ferreira, M.; Fragata, A.; Barroca, M.J.; et al. Extraction of Phenolic Compounds from Cherry Seeds: A Preliminary Study. *Agronomy* **2022**, *12*, 1227. https://doi.org/10.3390/agronomy12051227

Academic Editor: Spyridon A. Petropoulos

Received: 14 March 2022
Accepted: 17 May 2022
Published: 20 May 2022

1. Introduction

Sweet cherry (*Prunus avium* L.) is cultivated widely [1], and consumers highly appreciate the fruits due to their excellent organoleptic characteristics [2]. Research has shown that sweet cherry has a high nutritional value and an important content of bioactive compounds [3]. The global sweet cherry production is around 4.0 million tons per year. The sweet cherry production share in 2021/2022 concerning the principal producing countries was Turkey (21.4%), followed by the European Union (16.6%), China (14.9%), Chile (9.9%) and the United States (9.8%) [4].

The extraction process is one of the fundamental steps for the recovery of phenolic compounds from the food industry and the related by-products or residues [5]. Different methodologies are available to perform this operation [6] but the choice will also depend on the kind of food product or residue under analysis [7]. The extraction of phenolic compounds can be performed by techniques such as infusion, decoction, maceration, percolation, digestion, Soxhlet extraction, aqueous alcoholic extraction by phytonics processes, supercritical fluid extraction, counter current extraction, ultrasound extraction, or by fermentation [8]. Undoubtedly, solid–liquid extraction has been the most extensively utilized to separate several phenolic compounds or families of compounds. However, the efficiency of the extraction process is highly variable and is dependent on the processing conditions, especially the type and concentration of solvent, the ratio of solvent-to-mass of the sample, the extraction time, and the temperature [9]. Naczk and Shahidi [10] showed that extraction times longer than 24 h could increase the oxidation process of phenolic compounds, therefore they should not be used in order to preserve the antioxidant properties of the phenolic compounds. Golpour et al. [6] evaluated the effect of different operating conditions on the extraction of phenolic compounds form strawberry, namely time, solvent concentration and volume/mass ratio. They found that total phenolic compounds were better extracted when using a volume/mass ratio of 12 mL/g, using a methanol aqueous solution with a concentration of 70% methanol and extraction time of 40 min. Guiné et al. [11] described the optimization of extraction of bioactive compounds from beetroot, and found that the ratio mass of sample/volume of solvent was a factor highly influencing the extraction efficiency. Ahmad et al. [12] studied the effect of solvent and temperature on the extraction of phenolic compounds from olive fruits, and reported that the type of solvent greatly influenced the extraction, while temperature did not. However, Beaufils et al. [13] reported the strong effect of temperature on the extraction of phenolic compounds, following the trends of other studies. Casagrande et al. [14] described an optimal temperature for the extraction of phenolics of 80 °C, and also Marete et al. [15] reported the best extraction temperatures in the range from 80 to 100 °C. However, some authors state that such high temperatures can have a negative effect on the compounds by degrading them and reducing their biological activity [16].

The quantification of phenolic compounds in cherry fruits has been performed by Kashyap et al. [17]. In their study, the authors used response surface methodology to optimize the extraction conditions, in particular solvent concentration (40 to 80%), solvent/mass ratio (10 to 30 mL/g), extraction time (90 to 240 s), and microwave power (300 to 600 W). The optimized extract resulted in a total phenolic content of 155.27 ± 2.76 mg GAE/g in the cherry pomace. Schmidt et al. [18] evaluated the phenolic profile of cherry fruits and found that the major phenolics were procyanidins (1380.64 to 1888.00 μg/g). Basanta et al. [19] investigated the preservation of the antioxidant effects of phenolic compounds in cherry surfaces and found that anthocyanins were more degraded by light, while flavonols remained more stable. The study of phenolic compounds from cherry seeds is scarce, but Oliveira et al. [20] studied some separation methods to isolate compounds with antioxidant activity from Brazilian cherry seeds. They found that temperature and time influenced the extraction yield of phenolic compounds.

Two main strategies are adequate to implement a circular economy in any industrial sector: reducing pollution levels and finding the most sustainable solution to manage the waste resulting from the processes of the agri-food industry. The large quantity of organic waste produced by the food industry, besides representing a significant loss of valuable materials, also poses serious management problems, both from economic and environmental points of view [21]. Waste recovery has been defined as the process of converting waste into more valuable products [22]. This constitutes a useful approach to addressing waste materials' management, thus increasing the competitiveness of agri-food industries, where a wide variety of products can be obtained using waste as a raw material [23]. Agri-food waste is a valuable bioresource that can originate a variety of convenient materials, ingredients and chemicals [24,25]. In addition, opportunities for the application of industrial symbiosis in the valorisation of wastes in the food manufacturing industry have already been identified [21]. Because food waste has great potential to provide beneficial social and environmental benefits, several countries are already promoting strategies for its valorisation [21]. In addition to the ecological benefits, there is also an economic gain for the companies, linked with the integration of recovered ingredients into human food supply chains [26].

The fresh-cut fruit industry discards large percentages of by-products, such as peels, seeds, or even flesh that is not used for specific quality reasons. However, these residues can present similar contents of bioactive compounds or even higher than the parts effectively used, such as the flesh. These bioactive compounds include, for example, phenolic compounds, carotenoids and vitamins [21]. Sweet cherry seeds result from processing sweet cherry for the production of sweets, juices and jams. Generally, seeds are considered as production waste, which has gained strong interest due to the environmental aspects related to waste disposal. Additionally, it is well documented that production waste, such as peels, seeds, and pomace, contains high-value bioactive compounds [27]. Hence, the present work focuses on using cherry seeds that are waste derived from food manufacturing industries. The extraction potential of various organic solvents and different extraction conditions was investigated to optimize the extraction parameters such as type of solvent and temperature for maximum yield of bioactive compounds extracted. The bioactive substances evaluated in the extracts were total phenolic compounds and different families of phenolics.

2. Materials and Methods

2.1. Samples

The waste management company Nutrofertil, located in Portugal, namely in the district of Viseu (Tondela), provided the Seeds of Sweet Cherry (SSC) for this study. Since the company uses cherries from different varieties for processing, the seeds were not from a single variety, but rather from mixed varieties. This is an industrial waste to be valued instead of being discarded. The samples used were pre-dried in the sun. After this, the samples were crushed in a mill (Retsch SMI—Retsch GmbH, Haan, Germany) and sieved for 30 min at 50 rpm (Retsch AS200 sieve—Retsch GmbH, Haan, Germany). A fraction of the particles with sizes over 0.425 mm (>35 mesh) were used for the tests. Then, the fractions obtained were again dried on a stove for 24 h at a temperature of 40 °C, for better preservation during storage until further usage.

2.2. Preparations of Extracts

The extracts were obtained from the contact between 5 g of sample with 50 mL of solvent for 40 min under various experimental conditions, as depicted in Figure 1. In a preliminary assay, extractions were performed at room temperature under magnetic stirring at medium speed for different extracting solutions. One more extraction was also performed with water using an ultrasonic bath. Each extract was filtered into a 50 mL volumetric flask using Whatman N$^{o.}$ 3 paper filter and allowed to stand in the dark at room temperature until analysis. Oxidation was prevented by leaving the flasks away from light in closed cabinets. The samples were left for a period of up to one hour, and the experiments were repeated three times. Based on the results of the preliminary assay, the experiments of the core work were made using a similar methodology, but different processing conditions, namely the extraction solvent, were now reduced to mixtures of water:ethanol but at various proportions (water:ethanol ratios—50:50; 60:40; 80:20; 100:0 %*v:v*), and the extraction was carried out at different temperatures of 35 °C, 50 °C, 70 °C and 80 °C, all under magnetic stirring for 40 min. Some extra experiments were made to further investigate results for the number of anthocyanins and total phenolic compounds, as depicted in Figure 1 under the additional assay.

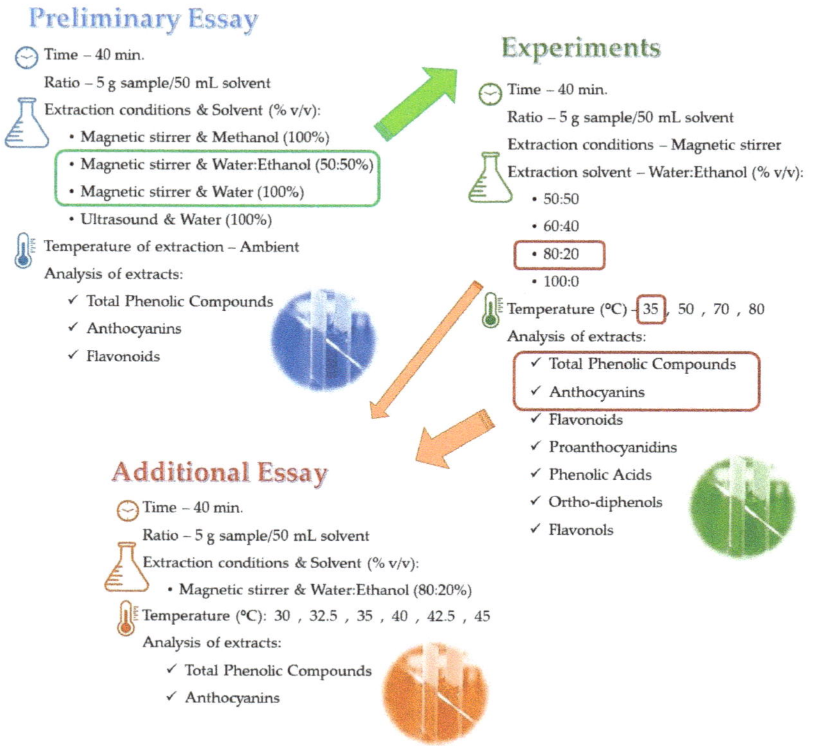

Figure 1. Schematic representation of the experimental conditions tested.

2.3. *Quantification of Total Phenols*

The Folin–Ciocalteu method is one of the relatively simple spectrophotometric assays used to determine the content of total phenolic compounds [6]. For measurements, 125 µL of samples (or standard solution) were mixed with 750 µL of distilled water and 125 µL of Folin–Ciocalteu reagent. After 6 min, 2 mL of sodium carbonate was added 5% (*v:v*) and the prepared solution was left for 60 min at room temperature in the dark for the reaction to occur. Then, the mixture was shaken, and the absorbance of the standards and samples was measured at $\lambda = 760$ nm. Autozero was performed with distilled water and the blank with 125 µL of distilled water instead of the samples. At the same time, the calibration curve was obtained with standard gallic acid solutions with different concentrations between 0 and 1 g/L. The results were calculated as mean value and standard deviation from three replicates of spectrophotometric measurements for each sample and expressed in milligrams of gallic acid equivalents per gram of sample (mg GAE/g).

2.4. *Quantification of Anthocyanins*

The extracts obtained (1 mL) were diluted to a final volume V with the solution of 1 mL of HCl (concentration of 37%), 70 mL of ethanol and 30 mL of H_2O, previously prepared in a 100 mL flask, with three replicates. The cells of 1 cm of optical path were filled, and the spectrophotometer was read at 540 nm. Quantitative determination of anthocyanins was carried out in slightly acidic media in the wavelength range of 490–550 nm [28]. This is possible because anthocyanins have a typical absorption band around this range. This band is far from the absorption bands of other phenols, which have spectral maxima in the UV range [29].

The samples were analysed in triplicate, and the anthocyanin content in the extract was determined using the following expression, taking into account the dilution factor (V): Total Anthocyanins (mg/mL) = Absorbance at 540 × 16.17 × V.

Final results were calculated as mean value and standard deviation from three repetitions and expressed as micrograms of malvidin-3-glucoside (Mv3G) equivalents per gram of sample of cherry seeds (μg Mv3G/g) [30].

2.5. Quantification of Proanthocyanidins

The quantitative determination of proanthocyanidins (condensed tannins) was performed using the vanillin assay, a method widely used in tests on vegetables and fruits. [31]. The vanillin/HCl assay measured proanthocyanidins content according to the procedure reported by Zam et al. [32]. Aliquots of 1 mL of samples were mixed with 2.5 mL of 1% vanillin-methanol solution and 2.5 mL of 8% HCl-methanol solution. Then, the reaction mixture was incubated in a water bath for 20 min at 30 °C. The absorbance was measured at 500 nm and blanked with methanol. Total proanthocyanidins contents were calculated as mean value and standard deviation from three repetitions and expressed as catechin equivalents in milligrams per gram (mg CE/g) using catechin (0.06–0.3 mg/mL) for the standard curve.

2.6. Quantification of Flavonoids

Aluminium chloride ($AlCl_3$) was used to quantify flavonoid content [33]. Total flavonoids were estimated according to the method described by Gonçalves et al. [34], after some adjustments. A calibration curve was performed with standard quercetin solutions with different concentrations (0.0; 0.1; 0.2 and 0.3 g/L) to quantify the flavonoid content. One millilitre of the sample was thoroughly mixed with 1 mL of 2% $AlCl_3{\cdot}6H_2O$ (2 g in 100 mL of methanol). The absorbance was measured at 370 nm after 10 min of rest. The total flavonoids content was calculated as mean value and standard deviation from three repetitions and expressed as quercetin equivalent (QE) in milligrams per gram of samples.

2.7. Quantification of Ortho-Diphenols

The ortho-diphenol content was determined colourimetrically by the complexation of ortho-diphenols with the molybdate ions [35].

The experimental procedure was adapted from Santos et al. [36]. In one tube, 4 mL of properly diluted samples and 1 mL of 5% (*v:v*) dehydrated sodium molybdate in ethanol/water (1:1) were added, stirred and remained at room temperature for 15 min. The samples were analysed by spectrophotometry at a wavelength of 370 nm. A blank was obtained by measuring a mixture of 4 mL of phenolic solution with 1 mL of ethanol/water (1:1). In parallel, a calibration curve was determined with standard solutions of gallic acid with different concentrations in the range 5–300 mg/L. The ortho-diphenol content of each sample was then determined by interpolating the values obtained colourimetrically in the calibration curve. The results were calculated as mean value and standard deviation from three repetitions and expressed as gallic acid equivalents per gram of samples (mg GAE/g).

2.8. Quantification of Phenolic Acids

Total phenolic acid determinations were carried out according to the Arnov method described by Gawlik-Dziki et al. [37]. One millilitre of the sample was mixed with 5 mL of distilled water, 1 mL 0.5 mol/L HCl, 1 mL of Arnov reagent (10 g sodium molybdate and 10 g sodium nitrite dissolved in 100 mL of distilled water) and 1 mL 1 mol/L NaOH and complete to 10 mL with distilled water. Total phenolic acids were determined by measuring the absorbance at 490 nm of the complex formed between phenolic acids and sodium molybdate—sodium nitrite as an equivalent of caftaric acid. The total phenolic acids content was calculated as mean value and standard deviation from three repetitions and expressed as caftaric acid equivalents (CAE) in milligrams per gram of extract.

2.9. Quantification of Flavonols

Series of reference quercetin solutions containing 0.05, 0.1, 0.15, 0.2, 0.3, 0.4, and 0.5 mg/mL of quercetin were prepared. Two millilitres of such reference were mixed with 2 mL of aluminium trichloride solution (20 g/L) in 95% (*v*/*v*) ethanol, and 6 mL of sodium acetate solution in ethanol (50 g/L) were added. The absorbance was read at 440 nm after 2.5 h at 20 °C. A calibration curve expressing the dependence of the absorbency on the concentration of quercetin was drawn. Seed extract samples were prepared under the same conditions using 2 mL of extract (10 g/L) in 95% ethanol instead of quercetin. All determinations were performed in triplicate. The percentage of flavanols was calculated by the formula:

$$Flavonols = \frac{C \times V}{m} \quad (1)$$

where C is the concentration of quercetin, determined from the calibration curve (mg QE/mL); V is the volume of plant extract (mL); m is the weight of pure plant extract (g) [38]. Final results were calculated as mean value and standard deviation from three repetitions and expressed as milligrams of quercetin equivalents per gram of sample of cherry seeds (mg QE/g).

2.10. Statistical Analysis

To test differences between mean values for multiple groups, the analysis of variance was undertaken (ANOVA), coupled with post hoc test Tukey to identify differences when significant at the level of 5%. Additionally, statistical techniques were used to classify the data and find grouping structures between them. A hierarchical cluster analysis was performed for values of total phenolic compounds and anthocyanins considering the data obtained from the three phases of the research, using squared Euclidean distance and the average linkage between groups method. Finally, principal component factor analysis with varimax rotation was undertaken. The suitability of the data for the application of the analysis was verified by the Kaiser–Meyer–Olkin measure of sample adequacy and the Bartletts' Test of sphericity, as well as on the values of the coefficients in the anti-image matrix. All statistical analyses were made with SPSS software (from IBM Inc., Armonk, NY, USA) version 26, and a significance level of 5% was considered ($p < 0.05$).

3. Results and Discussion

3.1. Preliminary Assay

Figure 2 presents the content of total phenolic compounds, anthocyanins and flavonoids in the extracts obtained in the preliminary assay, all performed at ambient temperature. The most efficient solvents for the recovery of polyphenols were the mixture water:ethanol and water extraction, using magnetic stirring (extracting approximately 1.3 mg GAE/g in both cases and without statistically significant differences between them). The least interesting extraction conditions were water using ultrasound stirring or methanol with magnetic stirring. Methanol was also inefficient in extracting flavonoids, which were maximum for 50% water:ethanol solutions (1.04 ± 0.06 mg QE/g). On the other hand, methanol removed the highest number of anthocyanins, more than 250 µg Mv3G/g. Taking the global results of this preliminary phase and considering the problems associated with methanol from an environmental point of view when compared with ethanol, as well as its highest negative health impact, the following experiments were performed with water:ethanol solutions at variable ratios and using different temperatures to maximize the contents of various phenolic compounds.

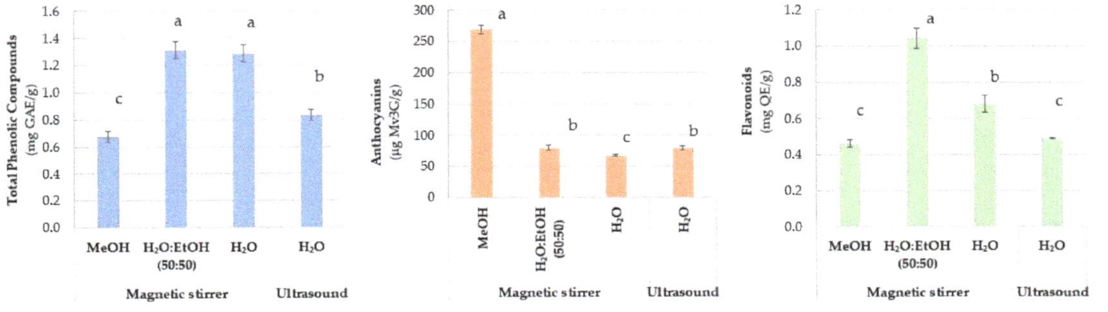

Figure 2. Content of total phenolic compounds, anthocyanins and flavonoids in the extracts obtained in the preliminary assay. Bars with the same letter are not statistically different (ANOVA with Tukey post hoc test, level of significance $p < 0.05$).

Because the samples used in the experiments consisted of waste possibly containing cherry seeds from different varieties, it is not possible to establish a direct association between the obtained results and a particular type of cherry seed, i.e., from a specific cultivar or variety. Nevertheless, it is possible that results from seeds of different cherry varieties could differ in terms of chemical composition or content of phenolic compounds. Results from other studies show that cherry kernel oil can differ in terms of composition [39] or texture [40] according to variety. Since the pits can be a waste possible of valorisation, they can be used both for the production of oil as well as extraction of bioactive compounds. Hence, differentiation of the chemical composition of pits, based on distinctive cultivars, could be of high relevance. On the other hand, considering that these wastes come from food processing industrial plants, most of the time they are not separated according to variety. So, additional research could be conducted on the basis of academic studies focusing on individual varieties. Górnaś et al. [41] evaluated the effect of the cultivar of sour cherry pits on oil yield and the content of individual fatty acids, tocochromanols, carotenoids, sterols and squalene. Similar studies could, in the future, focus on the evaluation of phenolic compounds according to the variety of cherry pits.

3.2. Core Experiments for Optimization of Extraction Conditions

Figure 3 shows the concentrations of total phenolic compounds according to the extraction conditions (temperature and solution ratio). Results show that the best ratios to recover total phenolic compounds at 35 °C were water:ethanol 50:50 (1.54 ± 0.14 mg GAE/g) and 60:40 (1.41 ± 0.01 mg GAE/g), with no significant differences between them. Higher content of water significantly decreased the total phenolic compounds in the extract. Similar results were obtained at 80 °C. On the other hand, for 50 °C and 70 °C the highest extraction was obtained with the 60:40 ratio, decreasing for higher water content. The highest amount was obtained at 70 °C and with a 60:40 ratio. Some increase in temperature allows extracting more phenolic compounds, either due to the increased permeability of cell walls and modifications in the structure or as a response to some degree of stress-induced in the biological material. However, higher temperatures do not favour the increase in total phenolics, probably due to the thermal degradation happening around 80 °C [36,42]. The results obtained for the total phenolic compounds removed by ethanol: water extraction are not that different from those reported by Afonso et al. [43] that presented amounts of total phenolics varying between 1.17 and 2.76 mg GAE/g for sweet cherry kernels extracted with 70% methanol at 70 °C.

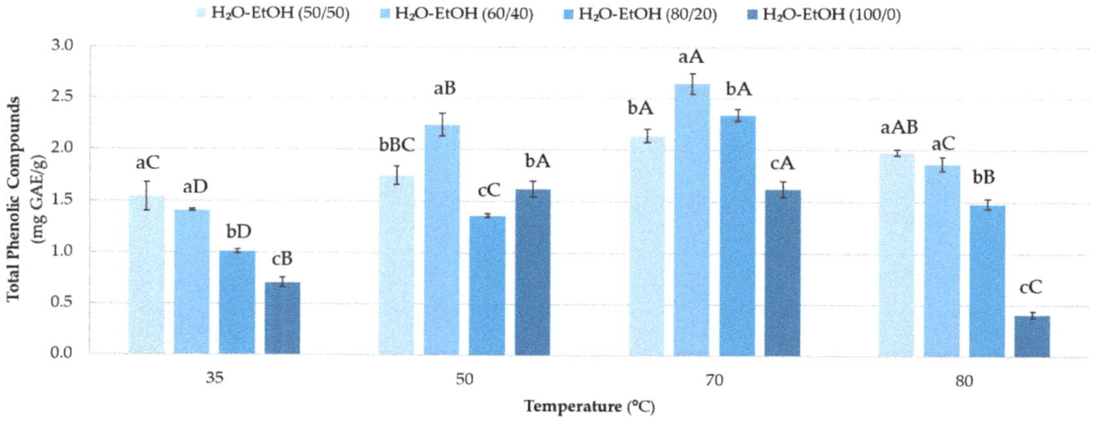

Figure 3. Total phenolic compounds in the different extracts according to the type of solvent and extraction temperature. For the same temperature, bars with the same lower letter are not statistically different (ANOVA with Tukey post hoc test, level of significance $p < 0.05$). For the same extracting solution, bars with the same capital letter are not statistically different (ANOVA with Tukey post hoc test, level of significance $p < 0.05$).

Table 1 presents the concentrations of the different groups of phenolic compounds according to the extraction conditions (temperature and solution ratio). It was found that a change in the concentration of water:ethanol solution significantly affected the final result obtained on extracted flavonoids from SCS at lower temperatures (35 and 50 °C). The largest amount of extractable components at 35 °C was obtained using a solution of water:ethanol with a concentration of 50:50 (4.74 ± 0.05 mg QE/g). With an increase in water concentration in the solution, the amount of the extractable compounds decreased very intensively to only 0.76 ± 0.22 mg QE/g when 100% water was used. Therefore, flavonoids were more easily removed by ethanol than water and the increase in the ethanol percentage allowed better recovery of these phenolics. These results were lower than those obtained by Engida et al. [44] with 40%, 60% and 80% ethanol solutions in the extraction of flavonoids from sarang semut (*Myrmecodia pendan*) (63.28 ± 1.75 mg QE/g) and by Domingués-Perles et al. [45] from red grape (*Vitis vinifera* L.) (37.34 ± 6.60 to 71.49 ± 2.33 mg CE/g). When extracting the compounds at a temperature of 50 °C, again, the amount of the extracted flavonoids decreased for a lower concentration of the ethanol:water solution used. The most significant number of flavonoids (4.45 ± 0.12 mg QE/g) was extracted for 50:50 water:ethanol ratio. Globally, it was found that increasing temperatures as well as increasing the percentage of ethanol yields higher concentrations of flavonoids, being maximum at 80 °C and 50:50 ratio (7.26 ± 0.05 mg QE/g). The solubilisation of compounds in a specific solvent is dependent on the polarity of the molecules of the solvent as well as those to be solubilized, which explains differences in the solubilisation capacity of water versus ethanol. According to Herrera-Pool et al. [46] there is a direct relation between the dielectric constant accounting for the polarity of the solvents and the extraction of phenolic compounds. The results obtained in the present work for the content of flavonoids are higher than those obtained by Afonso et al. [43] that presented amounts of total flavonoids varying between 0.23 ± 0.03 and 2.59 ± 0.44 mg CE/g for sweet cherry kernels extracted with 70% methanol at 70 °C but very similar with results obtained by Beghari et al. [47], who presented the total amount of flavonoids ranging from 1.45 to 7.30 mg CE/g for sweet cherry seeds extracted with a solution of 50:50 water:ethanol.

Table 1. Content of different families of phenolic compounds in the extracts according to the type of solvent and extraction temperature.

Compounds	Temperature (°C)	H_2O-EtOH (50/50)	H_2O-EtOH (60/40)	H_2O-EtOH (80/20)	H_2O (100%)
Flavonoids [1,2] (mg QE/g)	35	4.74 ± 0.05 aC	3.96 ± 0.14 bB	3.09 ± 0.18 cC	0.76 ± 0.22 dD
	50	4.45 ± 0.12 aD	3.63 ± 0.22 bB	1.94 ± 0.09 cD	1.56 ± 0.03 dC
	70	6.30 ± 0.12 aB	6.84 ± 0.37 aA	4.60 ± 0.13 bB	2.05 ± 0.09 cB
	80	7.26 ± 0.05 aA	6.99 ± 0.08 aA	5.25 ± 0.12 bA	2.49 ± 0.19 cA
Ortho-diphenols [1,2] (mg GAE/g)	35	19.97 ± 0.12 aA	19.91 ± 1.51 aA	17.57 ± 0.15 bA	3.55 ± 0.21 cC
	50	21.47 ± 0.40 aA	18.11 ± 0.09 bA	15.45 ± 1.35 cB	9.72 ± 0.40 dA
	70	20.92 ± 1.66 aA	18.11 ± 0.52 bA	12.19 ± 0.05 cC	9.47 ± 0.59 dA
	80	16.77 ± 0.94 aB	18.63 ± 1.38 aA	18.47 ± 0.39 aA	7.66 ± 0.15 bB
Proanthocyanidins [1,2] (mg CE/g)	35	6.19 ± 0.44 aA	4.91 ± 0.43 bBC	3.57 ± 0.13 cB	2.97 ± 0.14 cA
	50	4.85 ± 0.33 aB	4.48 ± 0.10 aC	3.37 ± 0.09 bB	2.07 ± 0.07 cB
	70	6.43 ± 0.38 aA	5.93 ± 0.22 aA	4.14 ± 0.20 bA	2.74 ± 0.14 cA
	80	6.01 ± 0.12 aA	5.56 ± 0.41 aAB	4.17 ± 0.28 bA	2.01± 0.11 cB
Flavonols [1,2] (mg QE/g)	35	2.06 ± 0.09 dB	2.50 ± 0.08 cC	2.87 ± 0.12 bA	3.32 ± 0.21 aA
	50	1.36 ± 0.04 dC	1.63 ± 0.06 cD	1.83 ± 0.07 bC	2.61 ± 0.07 aB
	70	3.88 ± 0.28 aA	3.06 ± 0.08 bA	1.84 ± 0.08 cC	1.37 ± 0.02 dD
	80	3.63 ± 0.01 aA	2.70 ± 0.04 bB	2.38 ± 0.09 cB	1.92 ± 0.07 dC
Phenolic acids [1,2] (mg CAE/g)	35	1.01 ± 0.07 aC	0.84 ± 0.05 bC	0.54 ± 0.04 cC	0.22 ± 0.01 dB
	50	1.01 ± 0.03 aC	0.88 ± 0.07 bC	0.61 ± 0.0 1cC	0.45 ± 0.03 dA
	70	1.48 ± 0.08 aB	1.29 ± 0.03 bA	0.86 ± 0.05 cA	0.47 ± 0.03 dA
	80	1.68 ±0.05 aA	1.06 ±0.01 bB	0.75 ± 0.02 cB	0.47 ± 0.03 dA
Anthocyanins [1,2] (µg Mv3G/g)	35	26.09 ± 0.37 bA	16.17 ± 0.65 dC	**66.08 ± 0.49 aA**	23.07 ± 1.35 cB
	50	26.63 ± 0.37 aA	25.54 ± 1.62 aA	23.71 ± 1.22 aB	26.52 ± 1.94 aA
	70	25.66 ± 1.95 aA	20.27 ± 0.99 bB	23.72 ± 0.67aB	12.29 ± 0.56 cC
	80	18.33 ± 1.14b B	16.38 ± 0.19 bcC	23.50 ± 1.95 aB	13.58 ± 0.86 cC

[1] Values in the same line with the same lower letter are not statistically different (ANOVA with Tukey post hoc test, level of significance $p < 0.05$). [2] Values in the same column, for each of the properties evaluated, with the same capital letter are not statistically different (ANOVA with Tukey post hoc test, level of significance $p < 0.05$).

The result for the extraction of ortho-diphenols at 35 °C showed similar values for the concentration of the water:ethanol solution 50:50 and 60:40 (around 20 mg GAE/g), but with an increase in the concentration of water in the solution; the amount of the extracted compounds decreased significantly in a similar trend as that observed previously for flavonoids. Similar results were presented by Domingués-Perles et al. [45] for red grape stems, where ortho-diphenols extraction was higher for higher concentrations of ethanol. At 50 °C and 70 °C, the concentrations of ortho-diphenols differed significantly for all used concentrations of the water:ethanol solution, decreasing with decreasing ethanol concentration. The highest values were obtained again for the 50:50 ratio solution (around 21 mg GAE/g). Unlike the trends observed for all other temperatures, for 80 °C there were no significant differences for the extracts obtained with ratios 50:50, 60:40 or 80:20 ratios, and only for the 100% water solution was the amount significantly lower (7.66 ± 0.15 mg GAE/g). Meanwhile, it should be noted that the sample with the largest amount of ortho-diphenols (21.47 ± 0.40 mg GAE/g) was obtained at 50 °C, and the most efficient solution was water:ethanol 50:50. This was higher than the results obtained by Afonso et al. [43] for seeds of sweet cherry extracted with a water:methanol solution, which varied from 0.09 ± 0.01 to 0.16 ± 0.02 mg CE/g. The differences in the values might be attributed to the standard used in each of the determinations: gallic acid in this case and catechin in the work by Afonso et al. [43].

The best extracting conditions to maximize the concentration of proanthocyanidins were using water:ethanol ratio of 50:50 at 70 °C (6.43 ± 0.38 mg CE/g), and also the 60:40 ratio (5.93 ± 0.22 mg CE/g) at the same temperature, with these conditions providing results statistically not significantly different. So, it was found that with the increase in water percentage, and regardless of temperature, there was a decrease in the amount of extracted proanthocyanidins, as previously observed for other families of phenolics. The range of values obtained in the present work is higher than the results obtained by Chaovanalikit et al. [48], who presented total amounts of proanthocyanidins ranging from 4.70 ± 0.81 to 22.5 ± 1.86 mg epicatechin equivalents/100 g for sweet cherry seeds extracted with 70% (*v:v*) acetone.

The results for flavonols extraction showed that the highest values were achieved with the 50:50 water:ethanol solution for higher temperatures (3.88 ± 0.28 and 3.63 ± 0.01 mg QE/g at 70 and 80 °C, respectively). However, for lower temperatures, the efficacy of extraction of flavonols increased with the ratio of water in the extracting solution, with relatively high amounts quantified at 35 and 50 °C when the solvent used was water at 100% (3.32 ± 0.21 and 2.61 ± 0.07 mg QE/g, respectively). The least efficient conditions for flavonols' extraction were 50 °C, with extracting solution water:ethanol 50:50 (1.36 ± 0.04 mg QE/g) and 70 °C with extracting solution water 100% (1.37 ± 0.02 mg QE/g). These results were in a higher range than those obtained by Chaovanalikit et al. [48], which presented total amounts of flavonols ranging from 1.59 ± 0.21 to 63.2 ± 4.17 mg rutin equivalents/100 g for sweet cherry seeds extracted with 70% (*v/v*) acetone.

The number of phenolic acids was lower than all the other phenolic compounds. The results show that there was, again, increased extraction for a higher percentage of ethanol in the solution. Extraction at 70 °C and 80 °C presented a significant yield compared to the extraction results at 35 °C and 50 °C. The highest amount was obtained at 80 °C in the water:ethanol solution 50:50 with a value of 1.68 ± 0.05 mg CAE/g, which was higher than the result presented by Chaovanalikit et al. [48] for total amounts of phenolic acids for sweet cherry seeds extracted with 70% (*v/v*) acetone (ranging from 5.58 ± 0.38 to 14.5 ± 1.28 mg chlorogenic acid/100 g).

Regarding the results for the extraction of total anthocyanins, it was observed that at 50 °C for all water:ethanol concentrations, the results were similar (in the range 23.71 ± 1.22 to 26.63 ± 0.37 μg Mv3G/g, with no significant differences), i.e., the proportion of ethanol in the solution did not influence the extraction capacity. However, for higher temperatures, 70 or 80 °C, the use of water (100%) becomes much less efficient in extracting the anthocyanins. At 80 °C the highest amount was obtained with a ratio of 80:20 (23.50 ± 1.95 μg Mv3G/g), and the lowest was obtained for 100% water (13.58 ± 0.86 μg Mv3G/g). The results, interestingly, showed that at 35 °C and using a solution with a ratio 80:20, maximum anthocyanins content was extracted (66.08 ± 0.49 μg Mv3G/g). On the other hand, the lowest amount was found in the assay performed with a ratio of 60:40 for the extracting solution (16.17 ± 0.65 μg Mv3G/g). The incredibly high value accounted for anthocyanins in the 80:20 extract compared with all other results led us to carry out a test to verify if there was a trend explaining this high value.

3.3. Additional Tests for Temperature Range 30–50 °C

To verify the results obtained previously for the anthocyanins, an additional test was carried out, for which the 80:20 ratio water:ethanol was chosen, and the extraction was performed at different temperatures around 35 °C (varying from 30 to 45 °C). These additional tests aimed to verify if the high value obtained for the anthocyanins with solution 80:20 and at a temperature of 35 °C (66.08 ± 0.49 μg Mv3G/g) was not considered an outlier, resulting from a possible experimental error, since it was considerably higher than for all other conditions (concentrations and temperatures), varying from a minimum of 12.29 ± 0.56 to a maximum of 26.63 ± 0.37 μg Mv3G/g.

Figure 4 shows the content of anthocyanins and also of total phenolic compounds in the extracts obtained in this additional test with the aforementioned conditions.

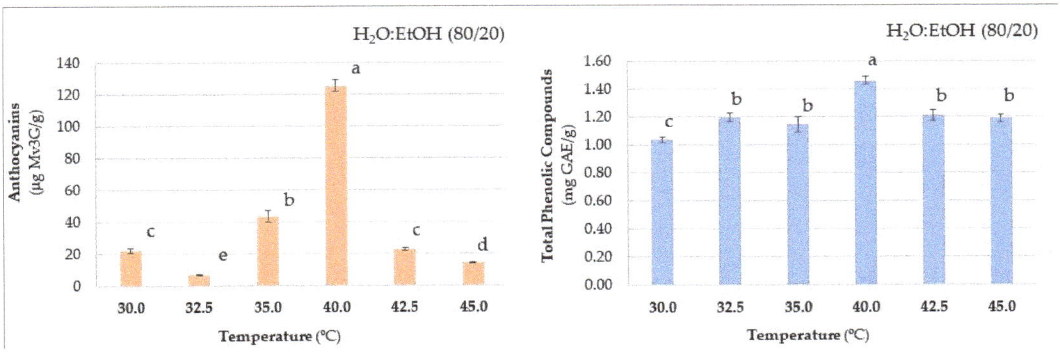

Figure 4. Content of anthocyanins and total phenolic compounds over a range of temperatures from 30 °C to 45 °C. Bars with the same letter are not statistically different (ANOVA with Tukey post hoc test, level of significance $p < 0.05$).

Regarding the contents of anthocyanins, a trend of increasing concentration with temperature until 40 °C was observed, followed by a decrease after that, with significant differences for practically all temperatures. As previously observed for the temperature of 35 °C, an even more considerable increase in the extracted amount was obtained at 40 °C (125.37 ± 3.90 µgMv3G/g), which seems to be the best temperature for the extraction of anthocyanins. In fact, for temperatures higher than 40 °C, the extraction efficiency decreased again, so that similar values were obtained at 30 °C and 42.5 °C (21.67 ± 1.48 and 22.64 ± 0.86 µgMv3G/g, respectively). The maximum extraction of anthocyanins was in a similar range when compared to the value of 10.4 mg of cyanodin-3-glucoside/100 g (equal to 104 µg/g) obtained by Chaovanalikit et al. [48] for cherry seeds of the cultivar Bing, but higher than the values for cherry seeds of cultivars Royal Ann, Rainier and Montmorency, determined by the same authors (0.1 to 0.8 mg/100 g corresponding to 1–8 µg/g).

Concerning the total phenolic compounds, it was found that the lowest value of 1.03 ± 0.02 mg GAE/g was obtained at 30 °C. However, the results increased with the increase in temperature up to 40 °C, as verified for the anthocyanins. For example, at 32.5 °C and 35 °C, the amounts were 1.20 ± 0.03 and 1.15 ± 0.06 mg GAE/g, respectively, without significant differences. It should be noted that 40 °C was the best temperature for extracting total phenolic compounds, with a value of 1.46 ± 0.03 mg GAE/g. However, it was observed that the values fall again for 42.5 °C and 45 °C, being 1.21 ± 0.04 and 1.19 ± 0.03 mg GAE/g, respectively. These are very similar to the values obtained at 32.5 °C and 35 °C. These results confirm that a temperature of 40 °C appears to be optimal for the recovery of total phenolic compounds and anthocyanins from the SCS.

3.4. Global Analysis of Extracts

Figure 5 shows the dendogram for hierarchical clustering obtained for total phenolic compounds and antocyanins using the average linkage between groups method and considering the data from all experiments, i.e., the preliminary assay, the core experiments and also the additional assay. The results confirm that measurements made at 40 °C, with magnetic stirring and using extracts obtained with aqueous solutions of ethanol (80:20 % *v*/*v*) constitute a separate cluster and that those obtained with similar conditions but for a slightly lower temperature of 35 °C constitute another isolated cluster. These results confirm the high contents found for these specific conditions, particularly in what concerns the anthocyanins.

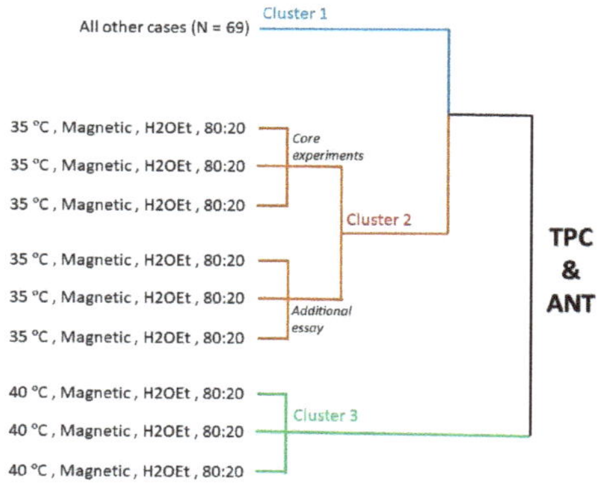

Figure 5. Dendogram for hierarchical clustering obtained for total phenolic compounds and antocyanins.

Considering all variables measured during phase two (the core experiments), factor analysis was also carried out. The Kaiser–Meyer–Olkin (KMO) measure of sample adequacy was reasonably high (KMO = 0.732), and the Bartletts' Test of sphericity was highly significant ($p < 0.0005$). The anti-image matrix showed practically all values higher than 0.5, the highest being for flavonoids (0.894) and lower for anthocyanins and flavonols. The results showed that two factors were extracted, which explained 75.65% of the variance (58.06% for factor 1 and 17.59% for factor 2). The rotated component matrix, obtained by principal component analysis and Varimax rotation in three iterations, indicated that anthocyanins and flavonols are associated with factor 2. In contrast, all other measured variables are associated with factor 1. The graph in Figure 6 clearly shows four groups. One group for the samples with water:thanol 80:20 (% v/v) for 35 °C (G1) corresponded to the extraction of the highest amounts of anthocyanins, thus corresponding to the highest values of factor 2. Another group accounts for the samples 100% water at 70 °C (G3), with particularly low values of factor 2, i.e., low contents of anthocyanins and flavonols. The other two groups are differentiated in terms of their values of factor 1, where group G2 refers to extracts with low contents of total phenolic compounds, flavonoids, proanthocyanidins, ortho-diphenols and phenolic acids. At the same time, G4 includes extracts with higher contents in those compounds.

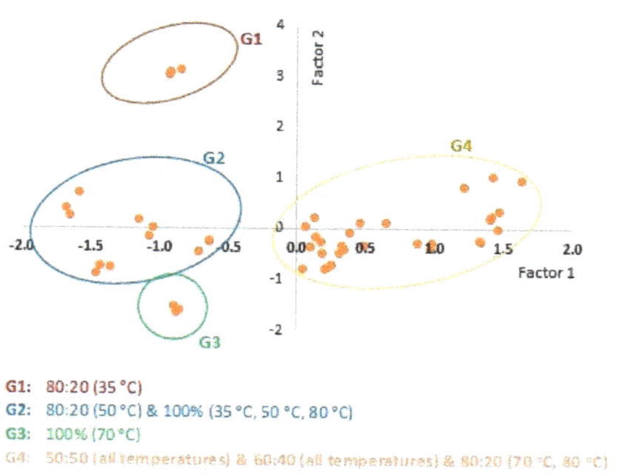

Figure 6. Grouping structure according to both factors extracted.

4. Conclusions

The main focus in the present work was the extraction of polyphenols from cherry seeds that are considered residues of food processing industries. The extraction potential of different conditions was investigated to optimize the extraction parameters, such as solvent and temperature, for the maximum yield of extracted bioactive compounds. The bioactive compounds evaluated were total phenolic compounds, anthocyanins, flavonoids, ortho-diphenols, proanthocyanidins, flavonols and phenolic acids. The results showed that the best extraction technique for phenolic compounds was an aqueous ethanol solution. The results indicated that temperatures that are too high, of 80 °C, are not beneficial and can degrade some families of phenolic compounds, reducing the extraction of total phenolic compounds, anthocyanins ortho-diphenols, proanthocyanidins and flavonols. On the contrary, these temperatures favour the recovery of flavonoids and phenolic acids. Concerning the composition of the extracting solution, most families of phenolics analysed were better extracted with water:ethanol solution with a ratio 50:50, except for total phenolic compounds (best for ratio 60:40) and anthocyanins (best for ratio 80:20). Given these results, and after discarding the use of methanol, which showed lower extracting capacity in a preliminary assay, ethanol is less pollutant, and the ratio of water can be 50% or even higher to obtain desirable results. Nonetheless, it was also concluded that using 100% water would not allow a proper extraction of the bioactive molecules studied. Nowadays, there are more attempts to use environmentally friendly extraction methods, using a more sustainable approach, resulting in the maximum recovery of bioactive compounds, using ecologically correct solvents and minimum processing costs. Extracts of bioactive compounds obtained from SCS can find other applications in the food, pharmaceutical or cosmetic industries, thus contributing to the eco-valorisation of agricultural or industrial residues, and the principles of circular economy.

Author Contributions: Conceptualization, R.P.F.G. and L.P.C.-L.; methodology, R.P.F.G. and Y.D.; software, R.P.F.G. and Y.D.; validation, R.P.F.G.; formal analysis, R.P.F.G. and Y.D.; investigation, R.P.F.G., L.P.C.-L. and Y.D.; resources, R.P.F.G., L.P.C.-L. and Y.D.; data curation, R.P.F.G.; writing—original draft preparation, R.P.F.G. and Y.D.; writing—review and editing, R.P.F.G., L.P.C.-L., B.E., J.V.F., I.D., M.J.L., P.M.R.C., M.F., A.F., M.J.B. and A.M.d.S.; supervision, R.P.F.G.; project administration, L.P.C.-L.; funding acquisition, R.P.F.G., L.P.C.-L., M.F. and A.F. All authors have read and agreed to the published version of the manuscript.

Funding: This work was funded by CERNAS Research Centre (Polytechnic Institute of Viseu, Portugal) in the ambit of the project VALCER (Ref. PROJ/IPV/ID&I/021). We received funding also form the FCT—Foundation for Science and Technology (Portugal) through projects Ref. UIDB/00681/2020, UIDB/05507/2020 and UIDB/00742/2020. The APC was funded by FCT through project Ref. UIDB/00681/2020, project Ref. UIDB/05507/2020, project Ref. UIDB/00742/2020.

Institutional Review Board Statement: Not applicable.

Informed Consent Statement: Not applicable.

Data Availability Statement: Data are available from the corresponding author upon request.

Acknowledgments: This work was supported by the FCT—Foundation for Science and Technology, I.P. Furthermore, we would like to thank the CERNAS, CI&DEI and UCISA:E Research Centres and the Polytechnic Institute of Viseu for their support. This research was developed in the ambit of project VALCER, with reference PROJ/IPV/ID&I/021.

Conflicts of Interest: The authors declare no conflict of interest.

References

1. Proietti, S.; Moscatello, S.; Villani, F.; Mecucci, F.; Walker, R.P.; Famiani, F.; Battistelli, A. Quality and Nutritional Compounds of *Prunus cerasus* L. Var. Austera Fruit Grown in Central Italy. *HortScience* **2019**, *54*, 1005–1012. [CrossRef]
2. Švarc-Gajić, J.; Cerdà, V.; Clavijo, S.; Suárez, R.; Mašković, P.; Cvetanović, A.; Delerue-Matos, C.; Carvalho, A.P.; Novakov, V. Bioactive Compounds of Sweet and Sour Cherry Stems Obtained by Subcritical Water Extraction. *J. Chem. Technol. Biotechnol.* **2018**, *93*, 1627–1635. [CrossRef]

3. Martínez-Esplá, A.; Zapata, P.J.; Valero, D.; García-Viguera, C.; Castillo, S.; Serrano, M. Preharvest Application of Oxalic Acid Increased Fruit Size, Bioactive Compounds, and Antioxidant Capacity in Sweet Cherry Cultivars (*Prunus avium* L.). *J. Agric. Food Chem.* **2014**, *62*, 3432–3437. [CrossRef] [PubMed]
4. USDA. *Fresh Peaches and Cherries: World Markets and Trade*; United States Department of Agriculture—Foreign Agricultural Service: Washington, DC, USA, 2021; pp. 3–5.
5. Lapornik, B.; Prošek, M.; Golc Wondra, A. Comparison of Extracts Prepared from Plant By-Products Using Different Solvents and Extraction Time. *J. Food Eng.* **2005**, *71*, 214–222. [CrossRef]
6. Golpour, I.; Ferrão, A.C.; Gonçalves, F.; Correia, P.M.R.; Blanco-Marigorta, A.M.; Guiné, R.P.F. Extraction of Phenolic Compounds with Antioxidant Activity from Strawberries: Modelling with Artificial Neural Networks (ANNs). *Foods* **2021**, *10*, 2228. [CrossRef]
7. Azmir, J.; Zaidul, I.S.M.; Rahman, M.M.; Sharif, K.M.; Mohamed, A.; Sahena, F.; Jahurul, M.H.A.; Ghafoor, K.; Norulaini, N.A.N.; Omar, A.K.M. Techniques for Extraction of Bioactive Compounds from Plant Materials: A Review. *J. Food Eng.* **2013**, *117*, 426–436. [CrossRef]
8. de Pinho Ferreira Guiné, R.; dos Reis Correia, P.M.; Ferrão, A.C.; Gonçalves, F.; Lerat, C.; El-Idrissi, T.; Rodrigo, E. Evaluation of Phenolic and Antioxidant Properties of Strawberry as a Function of Extraction Conditions. *Braz. J. Food Technol.* **2020**, *23*, e2019142. [CrossRef]
9. Ballesteros, L.F.; Teixeira, J.A.; Mussatto, S.I. Selection of the Solvent and Extraction Conditions for Maximum Recovery of Antioxidant Phenolic Compounds from Coffee Silverskin. *Food Bioprocess Technol.* **2014**, *7*, 1322–1332. [CrossRef]
10. Naczk, M.; Shahidi, F. Extraction and Analysis of Phenolics in Food. *J. Chromatogr. A* **2004**, *1054*, 95–111. [CrossRef]
11. Guiné, R.; Mendes, M.; Gonçalves, F. Optimization of Bioactive Compound's Extraction Conditions from Beetroot by Means of Artificial Neural Networks (ANN). *Agric. Eng. Int. CIGR E-J.* **2019**, *21*, 216–223.
12. Ahmad, R.; Ahmad, N.; Aljamea, A.; Abuthayn, S.; Aqeel, M. Evaluation of Solvent and Temperature Effect on Green Accelerated Solvent Extraction (ASE) and UHPLC Quantification of Phenolics in Fresh Olive Fruit (*Olea Europaea*). *Food Chem.* **2021**, *342*, 128248. [CrossRef] [PubMed]
13. Beaufils, N.; Boucher, J.; Peydecastaing, J.; Rigal, L.; Vilarem, G.; Villette, M.-J.; Candy, L.; Pontalier, P.-Y. The Effect of Time and Temperature on the Extraction of Xylose and Total Phenolic Compounds with Pressurized Hot Water from Hardwood Species Used for Pulp and Paper Production in the South of France. *Bioresour. Technol. Rep.* **2021**, *16*, 100832. [CrossRef]
14. Casagrande, M.; Zanela, J.; Wagner, A.; Busso, C.; Wouk, J.; Iurckevicz, G.; Montanher, P.F.; Yamashita, F.; Malfatti, C.R.M. Influence of Time, Temperature and Solvent on the Extraction of Bioactive Compounds of Baccharis Dracunculifolia: In Vitro Antioxidant Activity, Antimicrobial Potential, and Phenolic Compound Quantification. *Ind. Crops Prod.* **2018**, *125*, 207–219. [CrossRef]
15. Marete, E.N.; Jacquier, J.C.; O'Riordan, D. Effects of Extraction Temperature on the Phenolic and Parthenolide Contents, and Colour of Aqueous Feverfew (*Tanacetum parthenium*) Extracts. *Food Chem.* **2009**, *117*, 226–231. [CrossRef]
16. Yusoff, I.M.; Taher, Z.M.; Rahmat, Z.; Chua, L.S. A review of ultrasound-assisted extraction for plant bioactive compounds: Phenolics, flavonoids, thymols, saponins and proteins. *Food Res. Int.* **2022**, *157*, 111268. [CrossRef]
17. Kashyap, P.; Riar, C.S.; Jindal, N. Effect of Extraction Methods and Simulated in Vitro Gastrointestinal Digestion on Phenolic Compound Profile, Bio-Accessibility, and Antioxidant Activity of Meghalayan Cherry (*Prunus nepalensis*) Pomace Extracts. *LWT* **2022**, *153*, 112570. [CrossRef]
18. de Oliveira Schmidt, H.; Rockett, F.C.; Klen, A.V.B.; Schmidt, L.; Rodrigues, E.; Tischer, B.; Augusti, P.R.; de Oliveira, V.R.; da Silva, V.L.; Flôres, S.H.; et al. New Insights into the Phenolic Compounds and Antioxidant Capacity of Feijoa and Cherry Fruits Cultivated in Brazil. *Food Res. Int.* **2020**, *136*, 109564. [CrossRef]
19. Basanta, M.F.; Rojas, A.M.; Martinefski, M.R.; Tripodi, V.P.; De'Nobili, M.D.; Fissore, E.N. Cherry (*Prunus avium*) Phenolic Compounds for Antioxidant Preservation at Food Interfaces. *J. Food Eng.* **2018**, *239*, 15–25. [CrossRef]
20. Oliveira, A.L.; Destandau, E.; Fougère, L.; Lafosse, M. Isolation by Pressurised Fluid Extraction (PFE) and Identification Using CPC and HPLC/ESI/MS of Phenolic Compounds from Brazilian Cherry Seeds (*Eugenia uniflora* L.). *Food Chem.* **2014**, *145*, 522–529. [CrossRef]
21. Mirabella, N.; Castellani, V.; Sala, S. Current Options for the Valorization of Food Manufacturing Waste: A Review. *J. Clean. Prod.* **2014**, *65*, 28–41. [CrossRef]
22. Arancon, R.A.D.; Lin, C.S.K.; Chan, K.M.; Kwan, T.H.; Luque, R. Advances on Waste Valorization: New Horizons for a More Sustainable Society. *Energy Sci. Eng.* **2013**, *1*, 53–71. [CrossRef]
23. Venkata Mohan, S.; Nikhil, G.N.; Chiranjeevi, P.; Nagendranatha Reddy, C.; Rohit, M.V.; Kumar, A.N.; Sarkar, O. Waste Biorefinery Models towards Sustainable Circular Bioeconomy: Critical Review and Future Perspectives. *Bioresour. Technol.* **2016**, *215*, 2–12. [CrossRef] [PubMed]
24. Lin, C.S.K.; Pfaltzgraff, L.A.; Herrero-Davila, L.; Mubofu, E.B.; Abderrahim, S.; Clark, J.H.; Koutinas, A.A.; Kopsahelis, N.; Stamatelatou, K.; Dickson, F.; et al. Food Waste as a Valuable Resource for the Production of Chemicals, Materials and Fuels. Current Situation and Global Perspective. *Energy Environ. Sci.* **2013**, *6*, 426–464. [CrossRef]
25. Pfaltzgraff, L.A.; Bruyn, M.D.; Cooper, E.C.; Budarin, V.; Clark, J.H. Food Waste Biomass: A Resource for High-Value Chemicals. *Green Chem.* **2013**, *15*, 307–314. [CrossRef]
26. Garcia-Garcia, G.; Stone, J.; Rahimifard, S. Opportunities for Waste Valorisation in the Food Industry—A Case Study with Four UK Food Manufacturers. *J. Clean. Prod.* **2019**, *211*, 1339–1356. [CrossRef]

27. Tomsone, L.; Galoburda, R.; Kruma, Z.; Cinkmanis, I. Characterization of Dried Horseradish Leaves Pomace: Phenolic Compounds Profile and Antioxidant Capacity, Content of Organic Acids, Pigments and Volatile Compounds. *Eur. Food Res. Technol.* **2020**, *246*, 1647–1660. [CrossRef]
28. Khoddami, A.; Wilkes, M.A.; Roberts, T.H. Techniques for Analysis of Plant Phenolic Compounds. *Molecules* **2013**, *18*, 2328–2375. [CrossRef]
29. Dzomba, P.; Togarepi, E.; Mupa, M. Anthocyanin Content and Antioxidant Activities of Common Bean Species (*Phaseolus vulgaris* L.) Grown in Mashonaland Central, Zimbabwe. *AJAR* **2013**, *8*, 3330–3333. [CrossRef]
30. Matos, S.R. *Influência da Localização Geográfica de Produção nas Propriedades de Mirtilos "Vaccinium corymbosum L." cv Bluecrop*; Instituto Politécnico de Viseu: Viseu, Portugal, 2015.
31. Butler, L.G.; Price, M.L.; Brotherton, J.E. Vanillin Assay for Proanthocyanidins (Condensed Tannins): Modification of the Solvent for Estimation of the Degree of Polymerization. Available online: https://pubs.acs.org/doi/pdf/10.1021/jf00114a020 (accessed on 3 February 2022).
32. Zam, W.; Bashour, G.; Abdelwahed, W.; Khayata, W. Separation and Purification of Proanthocyanidins Extracted from Pomegranate's Peels (*Punica Granatum*). *Int. J. Pharm. Sci. Nanotechnol.* **2012**, *5*, 1808–1814. [CrossRef]
33. Costa, É.A.D. *Desenvolvimento de um Método de Análise de Catequina e Quercetina em Extratos de Espécies Vegetais e Fitoterápicos*; Universidade Federal de Uberlândia: Uberlândia, Brazil, 2018.
34. Gonçalves, F.; Gonçalves, J.C.; Ferrão, A.C.; Correia, P.; Guiné, R.P.F. Evaluation of Phenolic Compounds and Antioxidant Activity in Some Edible Flowers. *Open Agric.* **2020**, *5*, 857–870. [CrossRef]
35. Maestro Durán, R.; Borja Padilla, R.; Martín, A.; Fiestas Ros de Ursinos, J.A.; Alba Mendoza, J. Bíodegradación de los compuestos fenólicos presentes en el alpechín. *Biodegrad. Phenolic Compd. Present Oil-Mill Wastewater* **1991**, *42*, 271–276. [CrossRef]
36. Santos, S.C.R.V.L.; Guiné, R.P.F.; Barros, A. Effect of Drying Temperatures on the Phenolic Composition and Antioxidant Activity of Pears of Rocha Variety (*Pyrus communis* L.). *Food Meas.* **2014**, *8*, 105–112. [CrossRef]
37. Gawlik-Dziki, U. Dietary Spices as a Natural Effectors of Lipoxygenase, Xanthine Oxidase, Peroxidase and Antioxidant Agents. *LWT-Food Sci. Technol.* **2012**, *47*, 138–146. [CrossRef]
38. Ložienė, K.; Venskutonis, R.; Sipailiene, A. Radical Scavenging and Antibacterial Properties of the Extracts from Different *Thymus Pulegioides* L. Chemotypes. *Food Chem.* **2007**, *103*, 546–559. [CrossRef]
39. Aqil, Y.; Ouassor, I.; Belmaghraoui, W.; Hajjaji, S.E. Prunus Avium Kernel Oil Characterization: A Comparative Study of Four Varieties from Sefrou, Morocco. *OCL* **2020**, *27*, 24. [CrossRef]
40. Ropelewska, E. Classification of the Pits of Different Sour Cherry Cultivars Based on the Surface Textural Features. *J. Saudi Soc. Agric. Sci.* **2021**, *20*, 52–57. [CrossRef]
41. Górnaś, P.; Rudzińska, M.; Raczyk, M.; Mišina, I.; Soliven, A.; Segliņa, D. Composition of Bioactive Compounds in Kernel Oils Recovered from Sour Cherry (*Prunus cerasus* L.) by-Products: Impact of the Cultivar on Potential Applications. *Ind. Crops Prod.* **2016**, *82*, 44–50. [CrossRef]
42. Guiné, R.; Gonçalves, C.; Matos, S.; Gonçalves, F.; Costa, D.V.T.A.; Mendes, M. Modeling Through Artificial Neural Networks of the Phenolic Compounds and Antioxidant Activity of Blueberries. *Iran. J. Chem. Chem. Eng.* **2018**, *37*, 193–212.
43. Afonso, S.; Oliveira, I.V.; Meyer, A.S.; Aires, A.; Saavedra, M.J.; Gonçalves, B. Phenolic Profile and Bioactive Potential of Stems and Seed Kernels of Sweet Cherry Fruit. *Antioxidants* **2020**, *9*, 1295. [CrossRef]
44. Engida, A.M.; Kasim, N.S.; Tsigie, Y.A.; Ismadji, S.; Huynh, L.H.; Ju, Y.-H. Extraction, Identification and Quantitative HPLC Analysis of Flavonoids from Sarang Semut (*Myrmecodia pendan*). *Ind. Crops Prod.* **2013**, *41*, 392–396. [CrossRef]
45. Domínguez-Perles, R.; Teixeira, A.I.; Rosa, E.; Barros, A.I. Assessment of (Poly)Phenols in Grape (*Vitis vinifera* L.) Stems by Using Food/Pharma Industry Compatible Solvents and Response Surface Methodology. *Food Chem.* **2014**, *164*, 339–346. [CrossRef] [PubMed]
46. Herrera-Pool, E.; Ramos-Díaz, A.L.; Lizardi-Jiménez, M.A.; Pech-Cohuo, S.; Ayora-Talavera, T.; Cuevas-Bernardino, J.C.; García-Cruz, U.; Pacheco, N. Effect of Solvent Polarity on the Ultrasound Assisted Extraction and Antioxidant Activity of Phenolic Compounds from Habanero Pepper Leaves (*Capsicum chinense*) and Its Identification by UPLC-PDA-ESI-MS/MS. *Ultrason. Sonochem.* **2021**, *76*, 105658. [CrossRef] [PubMed]
47. Bagheri, S.; Alinejad, M.; Nejad, M.; Aliakbarian, B. Creating Incremental Revenue from Industrial Cherry Wastes. *Chem. Eng. Trans.* **2021**, *87*, 553–558. [CrossRef]
48. Chaovanalikit, A.; Wrolstad, R.E. Total Anthocyanins and Total Phenolics of Fresh and Processed Cherries and Their Antioxidant Properties. *J. Food Sci.* **2004**, *69*, FCT67–FCT72. [CrossRef]

agronomy

Article

Influence of Pre-Hydrolysis on the Chemical Composition of *Prunus avium* Cherry Seeds

Luísa Cruz-Lopes [1,*], Yuliya Dulyanska [1], Idalina Domingos [1], José Ferreira [1], Anabela Fragata [2], Raquel Guiné [1] and Bruno Esteves [1]

1 CERNAS Research Centre, Polytechnic Institute of Viseu, 3504-510 Viseu, Portugal; yuliyadulyanska@esav.ipv.pt (Y.D.); ijd@estgv.ipv.pt (I.D.); jvf@estgv.ipv.pt (J.F.); raquelguine@esav.ipv.pt (R.G.); bruno@estgv.ipv.pt (B.E.)
2 CI&DEI Research Centre, Polytechnic Institute of Viseu, 3504-510 Viseu, Portugal; afragata@estgl.ipv.pt
* Correspondence: lvalente@estgv.ipv.pt; Tel.: +351-232-480-500

Abstract: During the industrial processing of sweet cherry fruits, the seeds are considered agricultural waste and must be disposed of, typically through burning. In this context, it is intended to contribute to the scientific development of the ecovalorization of by-products and to provide new strategies for their transformation into value-added products obtained from sweet cherry seeds (SCS). This work aimed to establish the chemical characterization of SCS before and after several pre-hydrolysis steps in order to allow the solubilization of hemicelluloses that can later be used for the recovery of sugars. The higher percentage of cellulose and lignin remaining in the solid phase will allow its further processing for an integral valorization of the raw material. The temperature (160 and 170 °C) and time (0 and 180 min) of pre-hydrolysis were optimized to obtain the best liquefaction. The percentage of liquefied material was determined from the solid waste obtained at the time of filtration. The best liquefaction by the hydrolysis of SCS was obtained at 170 °C and 180 min, with a yield of 26.7%. The chemical analyses of SCS throughout hydrolysis showed the solubilization of hemicelluloses with increases in the time and temperature of the reactor. α-cellulose and lignin showed an increase both with temperature and time, increasing the material's potential for further processing in adhesives. FTIR analysis showed that there were significant changes in the spectra between the initial SCS, the solid residue, and the liquefied material. Pre-hydrolysis was proven to be an efficient process to improve the chemical composition of the material for further processing into adhesives or higher-mechanical-strength polyurethane foams.

Keywords: sweet cherry seeds; *Prunus avium* L.; chemical composition; pre-hydrolysis; ecovalorization; agro-industrial residues

Citation: Cruz-Lopes, L.; Dulyanska, Y.; Domingos, I.; Ferreira, J.; Fragata, A.; Guiné, R.; Esteves, B. Influence of Pre-Hydrolysis on the Chemical Composition of *Prunus avium* Cherry Seeds. *Agronomy* **2022**, *12*, 280. https://doi.org/10.3390/agronomy12020280

Academic Editor: Jose Maria Barrero

Received: 23 December 2021
Accepted: 18 January 2022
Published: 21 January 2022

1. Introduction

Sweet cherry, or *Prunus avium* L., is one of the most popular fruit of the *Rosaceae* family, *Prunoideae* subfamily, and *Prunus genus*, which have their origin in the Western Asian continent. Sweet cherry (*Prunus avium* L.) is the most cultivated species [1], and this popularity is based on its high appreciation by consumers due to its excellent quality [2]. The annual global sweet cherry production is around 2.2 million tons. The main sweet cherry producer in 2020/2021 concerning different countries was Turkey (23.4%), followed by the European Union (17.9%), United States (9.8%), China (11.5%), Ukraine, and Chile (7.7% and 7.3%, respectively) [3]. Portugal's climatic conditions have allowed the cherry tree to adapt very well, making its cultivation viable in various regions of the country. Studies conducted by the INE [4] (Portuguese Bureau of Statistics) showed that the Portuguese sweet cherry industry has an implementation area of 6387 ha, producing 9241 t of this fruit. Its production extends mainly through three regions: the North (area of 3177 ha and production of 6586 t), Centre (area of 3099 ha and production of 2510 t), and Alentejo (area of 64 ha and production of 57 t) [5]. There are several varieties of cherry in Portugal. The most important traditionally cultivated cherries are: De Saco da Cova da Beira, De Saco do Douro, Lisboeta, São Julião, Big Burlat, Maring, Napoleon Pé Comprido, and Big Windsor, and the first four cherries are of national origin [6]. Research shows that particular interest has focused not only on the nutritional value of this fruit, but also on its high health benefits due to its content of bioactive compounds, such as its antioxidant, anti-inflammatory, and anticancer properties [7–9].

When some fruits are processed by industry, wastes are inevitably produced. The seeds of the sweet cherry are not used for food and are generally considered agricultural waste by the processing industry, and must be disposed of, typically through burning. Thus, seed removal and disposal substantially raise production costs and contribute to pollution [10]. This has contributed to the increased research on this topic, addressing the possible uses of these residues to obtain high-value products in order to reduce environmental management costs [11]. Residues are also often used in composting processes [12]. Similarly to sour cherry, sweet cherry seeds are composed of two parts: the shell that corresponds to 75–80% in sour cherries and the kernel that represents the remainder (20–25%) [13]. According to these authors, the shell is used almost exclusively as fuel or considered waste. The kernel, however, is used for oil extraction. There have been some studies on sweet cherry seed oil, which, according to Bernardo-Gil et al. [14], contains more than 87% unsaturated fatty acids, such as oleic acid (43.7% by weight) and linoleic acid (41.8% by weight). New uses for sweet cherry seeds, mainly for their shell portion, are, therefore, of the utmost importance. There is a lack of knowledge on the chemical composition of sweet cherry seed shells and the overall composition of the seed. Results presented previously show that the main components of sweet cherry seeds are lignin, cellulose, and hemicelluloses [15–18].

Structural components are parts of cells; they are macromolecules of a polymeric nature, and this type of component includes cellulose, hemicelluloses, and lignin. Cellulose is a linear homopolymer formed by n repeated D-glucose units bound by β-1.4 glycolipid bonds ($n\ 10 \geq 000$). The glucose units are connected to each other by Van der Waals forces and hydrogen bonds [12].

Hemicelluloses are heterogeneous polysaccharides consisting of a mixture of pentoses (D-xylose and D-arabinose), hexoses (D-glucose, D-galactose, and D-mannose), and some hexuronic acids in small amounts, such as D-galacturonic acid, D-glucuronic acid, and its acid derivative 4-O-methylglucuronic acid [19]. In hemicellulose, the degree of polymerization ($50 \leq n \leq 300$) is lower than that in cellulose ($n \geq 10{,}000$), and hemicellulose is ramified and amorphous; therefore, it is less resistant to chemical degradation [20]. Unlike hemicelluloses, cellulose is linear and contains amorphous and crystalline zones [19].

Lignin is the second most abundant biopolymer in biomass after cellulose; it is a very complex and branched aromatic biopolymer [20]. The helical structure of lignin mainly results from the polymerization of three phenylpropanoid monomeric units: p-coumaryl alcohol, coniferyl alcohol, and sinapyl alcohol [20,21]. These phenylpropanoids give rise to the following phenolic substructures: p-hydroxyphenyl (H); guaiacyl (G) and syringyl (S), respectively [21]. Lignin has been used for the production of polyurethane foams and adhesives. Some authors state that the compression strength of PU foams increases with their lignin content [22,23]. Others state that, although the compressive properties of rigid polyurethane foams increase in the direction of foam rise with higher amounts of Kraft lignin, they decrease in the perpendicular direction [24]. Several authors reported higher hydrophobicity and flame retardancy with higher KL content [23,25,26], but also increased brittleness [23]. Adhesives from lignin have been studied for several years [27–32] and some authors have even stated that these adhesives can have a similar performance to phenol-formaldehyde resins [29]

There are mainly three different types of pre-hydrolysis: acid pre-hydrolysis, auto- pre-hydrolysis, and alkaline pre-hydrolysis. In accordance with Hamaguchi et al. [33], acid pre-hydrolysis, usually using sulfuric acid as a catalyst, leads to the production of oligomeric and monomeric sugars. Some authors, however, mention that acid pre-hydrolysis has corrosive effects on equipment and, at the same time, promotes extensive lignin condensation, and even a low yield due to some undesirable cellulose hydrolysis [34]. Auto hydrolysis is also a process under acidic conditions due to the acetic acid released by the cleavage of acetyl groups in hemicelluloses. Alkaline pre-hydrolysis is generally conducted with green or white liquor from the kraft process with strongly alkaline solutions under low temperatures [33,35,36]. Mechanical pre-treatment has also been used, generally to prepare samples for enzymatic or chemical hydrolysis with the objective to reduce the particle size and cellulose crystallinity so that enzymes or chemicals have better access [19,20,37,38]

Pre-hydrolysis with water (auto-hydrolysis) is the most used pre-treatment, in which biomass is subjected to compressed hot water, enabling the extensive solubilization of hemicelluloses [10]. The main reaction in this step is the depolymerization of hemicelluloses, leading to the formation of sugars and oligosaccharides [39,40]. The reaction is catalysed by hydronium ions resulting from water autoionization and in situ-generated organic acids. These acids act as a catalyst to hydrolyse the glycosidic bonds in hemicelluloses that are depolymerized into low-molecular-weight polysaccharides, oligosaccharides, monosaccharides, and other products, such as furfural and hydroxymethylfurfural. Due to the acid that is also released, some lignin and acetic acid can be found in the hydrolysate [26]. Additionally, pre-hydrolysis increases the surface area and decreases the crystallinity of cellulose, resulting in improved susceptibility towards future hydrolysis. Some authors state that pre-hydrolysis is a dynamic process, where, due to the acids released, the amount of hemicelluloses removed by acid hydrolysis in this process increases over time. However, at the same time, part of the obtained xylose is converted into furfural [41]. Therefore, the pre-hydrolysis time and temperature must be optimized to obtain the maximum amount of sugar without significant furfural conversion.

This work aims to establish the chemical characterization of sweet cherry seeds before and during the pre-hydrolysis of sweet cherry seeds, also aiming at obtaining an enriched lignin material that can be further processed into value-added products, such as adhesives.

2. Materials and Methods

2.1. Materials

The sweet cherry seeds (SCS) used in this study were wastes produced by the company Nutrofertil based in Portugal (Tondela), which is a residue management company. The samples were milled in a Retsch SMI mill (Retsch GmbH, Haan, Germany) and sieved using a Retsch AS200 (Retsch GmbH, Haan, Germany) for 20 min at 50 rpm. Four fractions, >40 mesh, 40–60 mesh, 60–80 mesh, and <80 mesh were obtained, but only the 40–60 mesh fraction was used for the tests.

2.2. Chemical Composition

The SCS were characterized for their ash content, extractives (in dichloromethane, ethanol, and hot water), α-cellulose, lignin, and hemicelluloses, Figure 1.

The 40–60 mesh fraction was dried at 105 °C for at least 24 h and afterwards used for the chemical analyses according to Tappi T 264 om-97 [42]. The average chemical composition of each sample was determined in triplicate. The extractives were determined by extraction with different solvents in sequential order of ascending polarity.

The ash content was determined by the calcination of the material at 525 °C according to the standard procedure Tappi T 211 om-93 [43]. The inorganic composition was determined by ICP after ash wet digestion in a Leco CHNS-932 Elemental Analyzer (St. Joseph, MI, USA). The extractive content consisted of the determination of dichloromethane, ethanol, and hot water extractives using Soxhlet extraction according to Tappi T 204 om-88 [44] as follows: approximately 10 g of the dried sample was used for Soxhlet extraction using 150 mL of dichloromethane, ethanol, and water as solvents. The extraction time was 6 h for dichloromethane and 16 h for ethanol and water, respecting a sequential extraction with increasing polarity. The extractive content was determined in relation to the dry material.

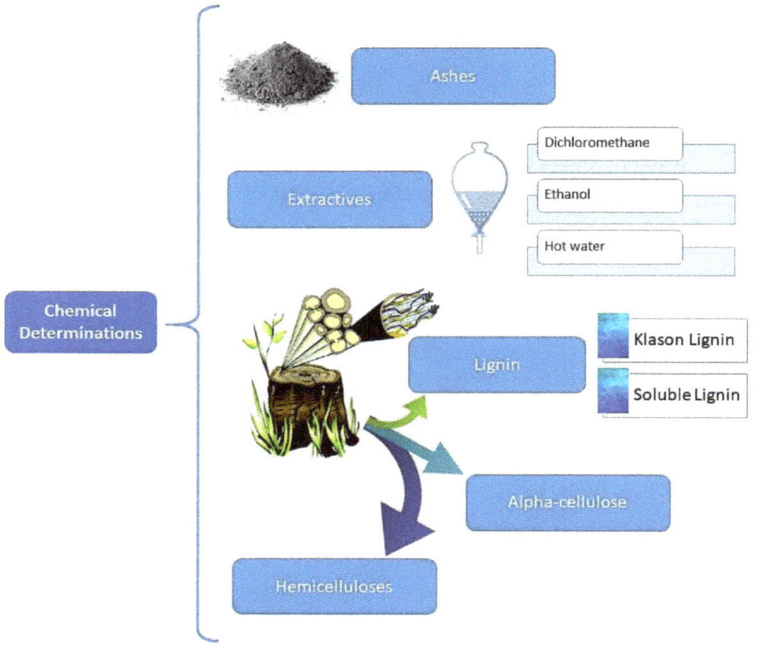

Figure 1. Scheme of the determination of the chemical composition of SCS.

The lignin content in SCS free of extractives was determined by the *Klason* method with 72% H_2SO_4 (according to Tappi T 204 om-88) [45]. The method consisted of two hydrolyses. The first hydrolysis was performed with 72% H_2SO_4 for 1 h, and the second with 3% H_2SO_4 in an autoclave at 120 °C for 1 h. The insoluble residue was obtained by filtering in a G2 glass crucible, and then dried to a constant weight. Then, the soluble lignin was analysed through spectrophotometry by measuring the absorption at 205 nm.

Holocellulose was determined by the acid chloride method in a water bath at 70 °C until the sample became white. The insoluble material was then filtered and dried [46]. The α-cellulose content was determined by hydrolysing the holocellulose with 2.5 mL of 17.5% NaOH in a thermal bath. The insoluble residue was filtered, washed with 8.3% NaOH and distilled water, finishing with 3.75 mL CH_3COOH 10% in a G2 glass crucible, and dried to a constant weight. The hemicellulose content was determined by the difference.

2.3. FTIR Analysis

The Fourier transform infrared spectroscopy (FTIR) technique was used to evaluate the functional groups present in the original sample of SCS and the sample that underwent pre-hydrolysis.

The initial dried SCS, the liquefied material, and the resulting solid residue were analysed by FTIR-ATR. The samples were previously crushed and dried in an oven at 100 °C for one week in order to ensure that water was completely removed.

FTIR-ATR spectra were taken using a Perkin Elmer UATR Spectrum Two (Waltham, MA, USA) with 72 scans/min and a resolution of 4.0 cm^{-1} over the 4000 to 400 cm^{-1} range. After performing the background, the sample was placed over the crystal. Solid samples were pressed against the crystal. An average of three spectra was used.

2.4. Pre-Hydrolysis

Pre-hydrolysis was carried out in a PARR LKT PED cylindrical glass reactor of 600 mL with a double coating. A 20 g sample (60 mesh) and 200 mL of distilled water were introduced into the reactor, and the automatic stirrer was set at 70 rpm. The temperature (160 and 170 °C, oil temperature on the shirt) and time (30, 60, 90, and 180 min) were optimized to obtain the best liquefaction percentage. After removal from the reactor, the mixture was filtered through a Buckner funnel with a paper filter, and the solid residue was separated from the liquefied fraction, oven dried, and weighed. The percentage of liquefied material was determined in accordance with Equation (1).

$$Liquefaction\ Yield\ (\%) = \frac{Initial\ dry\ mass\ (g) - Solid\ dry\ residue\ (g)}{Initial\ dry\ mass\ (g)} \times 100$$

To determine the chemical composition of the solid fraction, dried material was used, and the content of the following components was determined: Klason Lignin, holocellulose, α-cellulose, and hemicelluloses, using the 40–60 mesh fraction. The extractives were assumed to be mostly removed during pre-hydrolysis.

3. Results and Discussion

3.1. Chemical Composition

Table 1 presents the chemical composition of the SCS, where it can be observed that it was composed of 1.31% ashes, while total extractives represented around 4.69%, most of which were soluble in water (2.14%) and ethanol (2.04%), and only 0.49% in dichloromethane. The total lignin was 32.94%, most of which was insoluble lignin (31.67%) and the remaining 1.27% was soluble. α-Cellulose had a value of about 23.10% and 37.96% for hemicelluloses. The determination of the chemical composition will allow us to better understand the possible uses of SCS.

Table 1. Chemical composition of SCS.

Parameters		Content (% Dry Matter, *w/w*)
Ashes		1.31
	Na	0.10
	Mg	0.58
	P	0.95
	K	3.2
	Ca	0.93
	Fe	0.02
	Zn	<0.01
Extractives	Dichloromethane	0.49
	Ethanol	2.04
	Hot water	2.16
	Total	4.69
Lignin	Soluble	1.27
	Insoluble (*Klason*)	31.67
α-Cellulose		23.10
Hemicelluloses		37.96

Although the value obtained for the ashes (1.31%) is higher than the values obtained for SCS by Venegaz-Gomez et al. [47], Duman et al. [15], and Kamel et al. [48], with 0.24%, 1.16%, and 1.2%, respectively, the differences are not significant. The inorganic composition shows that SCS are more rich in potassium, with 3.2%, than all the other constituents, followed by calcium (Ca) and phosphorous (P), with 0.93 and 0.95%, respectively. There was also approximately 0.6% of magnesium (Mg) and a smaller amount of sodium (Na, 0.1%). Potassium was also the most representative element in grape stalks, followed by calcium, magnesium, zinc, and sodium [44]. Comparing the inorganic composition with those of black cherry seeds, almonds, and peanuts, all of these materials have higher percentages of potassium, followed by phosphorous; however, the third most representative mineral was magnesium, followed by calcium, with the exception of almonds, in which these inorganic compounds had the opposite order. Sodium and iron also had the lowest contents, similar to SCS [49].

Relative to hemicellulose in the present study, the value of 37.96% is higher compared with that in the studies by Gonzalez et al. [50], with 14.7%, and Duman et al. [15], with 28.59%. Nevertheless, these authors did not present the method used in their studies. The cellulose content was lower than the values obtained by Petrov et al. [16] and Gonzalez et al. [17], with 30% and 29.4%, respectively. The Klason lignin content is similar to the results obtained by Gonzalez et al. [50], González-Domínguez et al. [18] (30.7%), and Duman et al. [15] (29.08%), but lower than the results obtained by Petrov et al. [16] (40%).

Pre-hydrolysis with water (auto-hydrolysis) is an excellent method to remove hemicelluloses from lignocellulosic materials by a simple process without using chemical compounds. In order to optimize the liquefaction conditions of the pre-hydrolysis, tests were performed at different times (between 30 and 120 min) and temperatures of 160 °C and 170 °C. The results are presented in Figure 2, where it can be seen that the liquefaction yield increased with the temperature and time of the process. For example, using a temperature of 160 °C, the liquefied yield was around 6.3% for a 30 min liquefaction time, but it increased to 18.6% for 180 min. Similarly, at 170 °C, the liquefied yield for 180 min was about 26.6%, much higher than that obtained at 160 °C. A higher temperature might lead to better liquefaction percentages; however, some authors stated that 170°C was the optimum temperature to remove hemicelluloses without significant conversion of xylose to furfural [41]. The liquefaction percentages are higher than those obtained, for example, in the auto-hydrolysis of *Eucalyptus globulus* wood at 150 °C for 3 h (12.5%) [50] or *Eucalyptus urograndis* (10.8%) at 165 °C for 0.5 h [51]. Al-Dajani et al. obtained 19% liquefaction for aspen wood by auto-hydrolysis at 150 °C for 4.5 h [52].

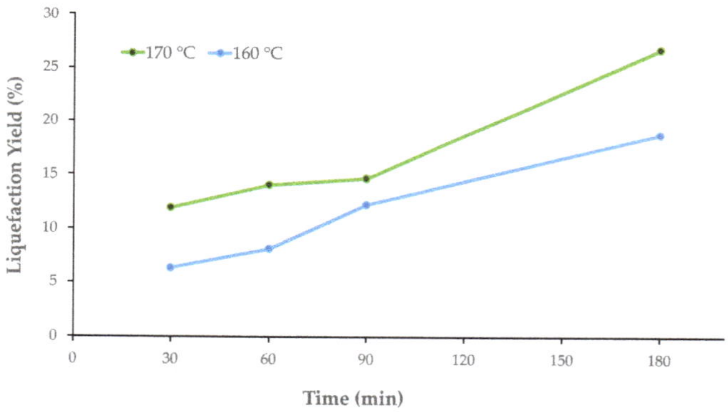

Figure 2. Pre-hydrolysis of SCS. Liquefaction yield with time.

Figure 3 presents the chemical composition variation over the several pre-hydrolysis steps at both temperatures of 160 °C and 170 °C.

For α-cellulose, the values obtained for the two temperatures are very different. There is a significant increase in α-cellulose from 23.1% in the original sample to 39.1% at 170 °C, but only a moderate increase at 160 °C to 27.4%. The increase in α-cellulose shows that this polymer is the most resistant to the pre-hydrolysis process, at least the crystalline part of the cellulose, since α-cellulose mostly represents this part. The reduction of hemicelluloses might even be higher since the solubilized cellulose counts toward the hemicellulose content.

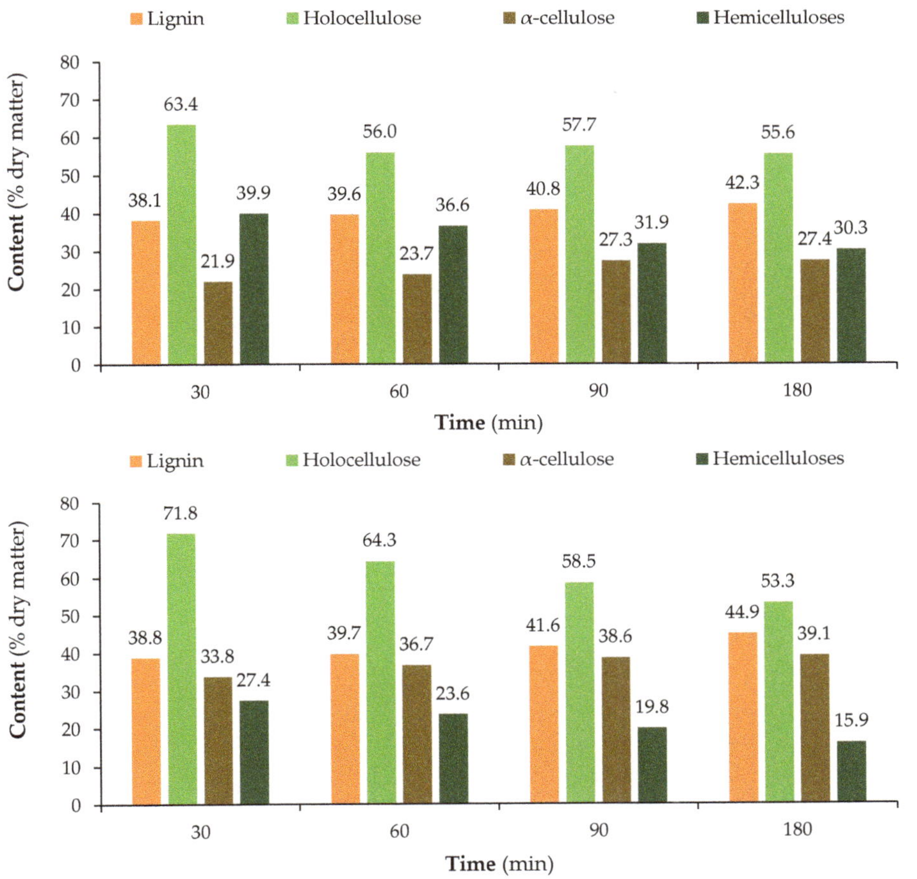

Figure 3. Variation in the chemical composition of the hydrolysed SCS with time at 160 °C (**top**) and 170 °C (**bottom**).

The influence of pre-hydrolysis temperature is better seen in Figure 4, where the differences between the initial material and solid residue after pre-hydrolysis are clearer, being expressed as a percentage of the initial content for each component. The hemicellulose content decreased to only 80% of the initial content at 160 °C, and even further to only 42% at 170 °C. On the other hand, the lignin and α-cellulose contents increased; particularly, the latter increased to 169% of the original value when treated at 170 °C. Holocellulose also decreased, and its reduction was lower than the relative increase in α-cellulose for both temperatures.

It was further observed that both the reaction time (Figure 3) and temperature (Figure 4) had a significant effect on the decrease in the hemicellulose content.

3.2. FTIR-ATR Analysis

FTIR-ATR of the SCS and of the liquefied and solid residues over the several pre-hydrolysis steps were performed. Figure 5 presents the spectra of the initial SCS and of the residue obtained after pre-hydrolysis at 160 °C and for a 30–180 min reaction time. The most characteristic peaks are marked for readability purposes.

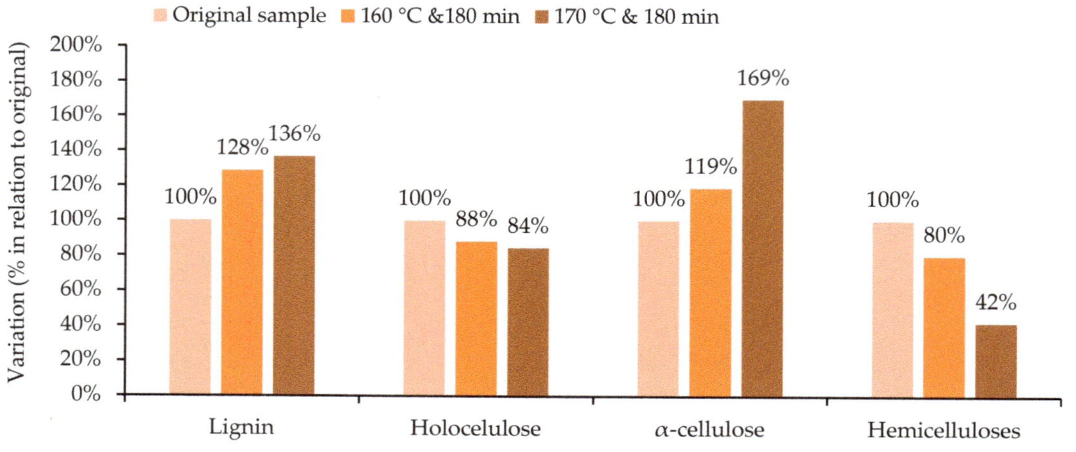

Figure 4. Variation in the chemical composition of the SCS hydrolysed in relation to the original sample, depending on the temperature at the end of the process (time of 180 min).

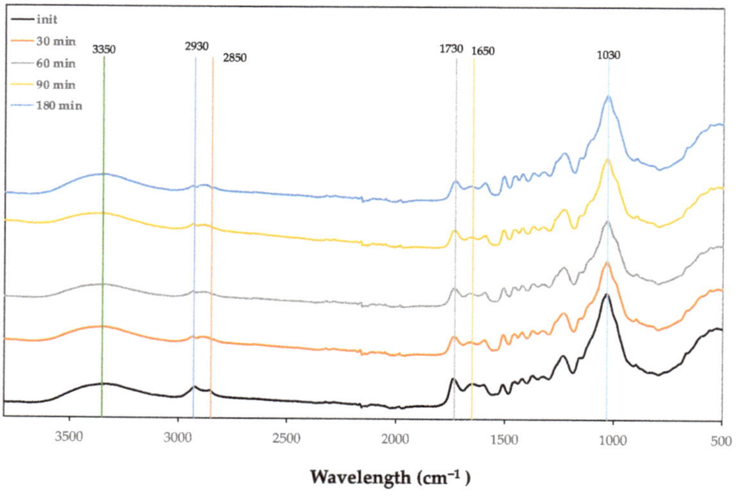

Figure 5. FTIR spectra of the original and solid residue from hydrolysed SCS at 160 °C.

Comparing the initial material with the residue, there seems to be a slight decrease and a narrowing of the OH band around 3350 cm^{-1}. The peaks at 2930 cm^{-1} and 2850 cm^{-1} have different behaviours. There is a decrease in the peak at 2930 cm^{-1} that becomes less distinguished from the peak at 2850 cm^{-1}. With the prolongation of the reaction time, this region divides into three different peaks, around 2930 cm^{-1}, 2885 cm^{-1}, and 2850 cm^{-1}. These bands are composed of the stretching asymmetric vibrations of -CH_3 (2970–2950 cm^{-1}) and -CH_2- (2935–2915 cm^{-1}) and stretching symmetric vibrations of -CH_3 (2880–2860 cm^{-1}) and -CH_2- (2865–2845 cm^{-1}) [53]. Usually, the asymmetric band is higher, which was true for the initial material but changed throughout the pre-hydrolysis treatment. This shows that there were some changes in the neighbourhoods of these bands. One of the possible reasons is the relative increase in the methoxyl groups in lignin due to the reduction of carbohydrates, since the CH_3 stretching vibrations of methoxyl have lower frequencies [54,55].

The bands at 1730 cm^{-1} and 1650 cm^{-1} were attributed to non-conjugated and conjugated C=O bonds. There seems to be a decrease specifically in the 1650 cm^{-1} peak. Lignin has been mentioned to have high absorption at 1600 cm^{-1} due to benzene ring stretching vibrations, but at this wavenumber there seemed to be no differences between the original material and the solid residue. No significant changes could be observed in the fingerprint region, although there seems to be a slight increase in the 1500 cm^{-1} peak. There is a decrease in the peak at 1360 cm^{-1} in relation to the peak at 1330 cm^{-1}.

There is also the appearance of a shoulder at around 1130 cm^{-1} that seems to be due to the slight narrowing of the 1030 cm^{-1} band. This might be due to the decreased absorption at around 1100 cm^{-1}, which was attributed to the C-O stretching vibrations in carbohydrates. At 170 °C, the changes in the spectrum are similar to those observed for 160 °C, with a slight decrease in the OH band and a decrease at 1650 cm^{-1}.

The absorptions of the liquefied material are much higher than those of the initial material, probably due to the air between the solid samples and ATR crystal resulting in a weaker absorbance signal for the initial material, and also due to the infrared radiation decrease with penetration depth [56].

The FTIR spectra of the original and liquified material from hydrolysed SCS at 170 °C are presented in Figure 6.

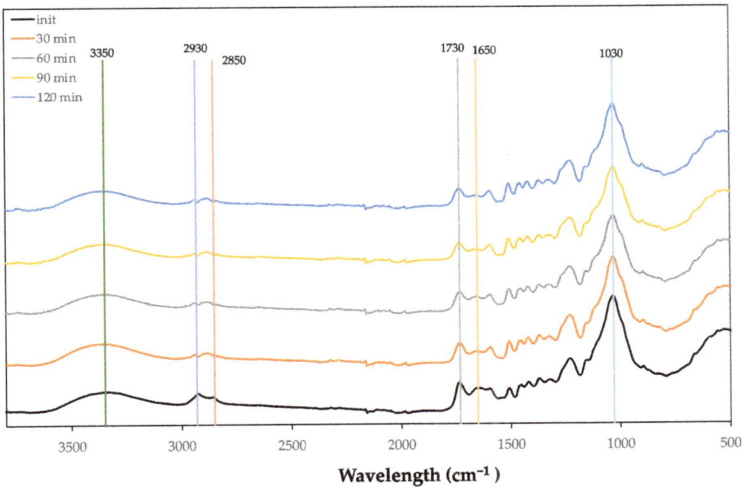

Figure 6. FTIR spectra of the original and liquified material from hydrolysed SCS at 170 °C.

The clearest change between the initial material and the hydrolysed material is the high increase in the OH band at around 3350 cm^{-1}, although there is no visible increase over the pre-hydrolysis time. This occurred at both 170 °C (Figure 6) and 160 °C (Figure 5). There is a shifting of the 2850 cm^{-1} peak to higher wavelengths at both temperatures, even starting at a 30 min reaction time. In relation to the 1730 cm^{-1} peak of non-conjugated C=O links, there seems to be an initial decrease for both temperatures, followed by an increase at 170 °C. This shows that several reactions are occurring at the same time during pre-hydrolysis. The C=O linkage has a strong absorption between 1750 and 1700 cm^{-1}, varying in accordance to the functional groups, with higher wavenumbers for aldehyde (1740–1720 cm^{-1}) followed by ketones (1725–1705 cm^{-1}) and carboxylic acids (1725–1700 cm^{-1}) [55]. Contrary to this, the peak at around 1600 cm^{-1} increased for the 30 min reaction, decreasing afterwards. This peak is mostly associated with benzene ring stretching vibrations, which could corroborate the wet analysis that shows that lignin increases in the residue, which means that there is a percentual decrease in the liquid. There was an increase followed by a decrease in the peak at around 1367 cm^{-1}. This decrease was not observed at 160 °C. Somewhat similar behaviour was observed for the peak at around 1230 cm^{-1}. A new shoulder appeared at 1060 cm^{-1} in the 1030 cm^{-1} band that was mainly attributed to the C-O stretching in holocellulose [57].

4. Conclusions

The chemical characterization of the sweet cherry seeds shows that this material has several components that allow the recovery of its residue. The chemical composition of samples that underwent pre-hydrolysis tests showed that, with an increase in time (from 30 to 180 min) and temperature (from 160 °C to 170 °C), there was an increase in the amounts of lignin and α-cellulose. In the initial sample, hemicelluloses represented the major component, while, after pre-hydrolysis, lignin and cellulose were the main polymers. The high lignin content shows that this residue has enormous potential to produce adhesives or rigid polyurethane foams with high mechanical resistance.

FTIR analysis showed that there were significant changes in the spectra from the initial material, solid residue, and liquefied material, the clearest being the increased OH band in the liquefied material, which was probably due to the solubilized hemicelluloses. The changes in the FTIR spectra of both the solid residue and the liquefied material throughout the pre-hydrolysis show that there were a lot of different reactions occurring at the same time, some increasing and others decreasing the same functional groups, making it difficult to track the changes over the pre-hydrolysis steps.

Potassium was the most representative inorganic compound, followed by calcium and phosphorous.

Author Contributions: Conceptualization, L.C.-L. and B.E.; methodology, L.C.-L., Y.D. and B.E.; formal analysis, L.C.-L. and Y.D.; investigation, L.C.-L., Y.D. and B.E.; resources, L.C.-L., Y.D., I.D., J.F., A.F., R.G. and B.E.; writing—original draft preparation, Y.D.; writing—review and editing, L.C.-L., R.G. and B.E.; supervision, L.C.-L. All authors have read and agreed to the published version of the manuscript.

Funding: This work is funded by National Funds through the FCT—Foundation for Science and Technology. I.P. within the scope of the project Ref UIDB/00681/2020 (CERNAS) and Ref UIDB/05507/2020 (CI&DEI).

Data Availability Statement: Not applicable.

Acknowledgments: We would like to thank the CERNAS and CI&DEI Research Centre and the Polytechnic Institute of Viseu for their support and the PROJ/IPV/ID&I/021, "Valorização de resíduos: potencial de aproveitamento do caroço de cereja (VALCER)" pela Bolsa de Investigação de Yulia.

Conflicts of Interest: The authors declare no conflict of interest.

References

1. Proietti, S.; Moscatello, S.; Villani, F.; Mecucci, F.; Walker, R.P.; Famiani, F.; Battistelli, A. Quality and Nutritional Compounds of *Prunus cerasus* L. Var. Austera Fruit Grown in Central Italy. *HortScience* **2019**, *54*, 1005–1012. [CrossRef]
2. Švarc-Gajić, J.; Cerdà, V.; Clavijo, S.; Suárez, R.; Mašković, P.; Cvetanović, A.; Delerue-Matos, C.; Carvalho, A.P.; Novakov, V. Bioactive compounds of sweet and sour cherry stems obtained by subcritical water extraction. *J. Chem. Technol. Biotechnol.* **2018**, *93*, 1627–1635. [CrossRef]
3. Publication | Stone Fruit: World Markets and Trade | ID: 0g354f20t | USDA Economics, Statistics and Market Information System. Available online: https://usda.library.cornell.edu/concern/publications/0g354f20t?locale=en (accessed on 29 August 2021).
4. Portal Do INE. Available online: https://www.ine.pt/xportal/xmain?xpid=INE&xpgid=ine_publicacoes&PUBLICACOESpub_boui=358629204&PUBLICACOESmodo=2 (accessed on 7 August 2021).
5. de Oliveira, T.S. O Marketing da Cereja da Cova da Beira. Ph.D. Thesis, University of Beira Interior, Covilhã, Portugal, 2015. Volume 73.
6. Produtos Tradicionais de Qualidade na Região Centro. Available online: https://ptqc.drapc.gov.pt/documentos/cereja_cova_beira.htm (accessed on 14 November 2021).
7. Delgado, J.; Terrón, M.P.; Garrido, M.; Barriga, C.; Paredes, S.D.; Espino, J.; Rodríguez, A.B. Systemic Inflammatory Load in Young and Old Ringdoves Is Modulated by Consumption of a Jerte Valley Cherry-Based Product. *J. Med. Food* **2012**, *15*, 707–712. [CrossRef] [PubMed]

8. McCune, L.M.; Kubota, C.; Stendell-Hollis, N.R.; Thomson, C.A. Cherries and Health: A Review. *Crit. Rev. Food Sci. Nutr.* **2010**, *51*, 1–12. [CrossRef] [PubMed]
9. Martínez-Esplá, A.; Zapata, P.J.; Valero, D.; García-Viguera, C.; Castillo, S.; Serrano, M. Preharvest Application of Oxalic Acid Increased Fruit Size, Bioactive Compounds, and Antioxidant Capacity in Sweet Cherry Cultivars (*Prunus avium* L.). *J. Agric. Food Chem.* **2014**, *62*, 3432–3437. [CrossRef]
10. Callahan, A.M.; Dardick, C.; Scorza, R. Characterization of 'Stoneless': A Naturally Occurring, Partially Stoneless Plum Cultivar. *J. Am. Soc. Hortic. Sci.* **2009**, *134*, 120–125. [CrossRef]
11. Aires, A.; Carvalho, R.; Saavedra, M.J. Valorization of solid wastes from chestnut industry processing: Extraction and optimization of polyphenols, tannins and ellagitannins and its potential for adhesives, cosmetic and pharmaceutical industry. *Waste Manag.* **2016**, *48*, 457–464. [CrossRef]
12. Ventorino, V.; Parillo, R.; Testa, A.; Viscardi, S.; Espresso, F.; Pepe, O. Chestnut green waste composting for sustainable forest management: Microbiota dynamics and impact on plant disease control. *J. Environ. Manag.* **2016**, *166*, 168–177. [CrossRef]
13. Yılmaz, F.M.; Görgüç, A.; Karaaslan, M.; Vardin, H.; Bilek, S.E.; Uygun, Ö.; Bircan, C. Sour Cherry By-products: Compositions, Functional Properties and Recovery Potentials—A Review. *Crit. Rev. Food Sci. Nutr.* **2019**, *59*, 3549–3563. [CrossRef]
14. Bernardo-Gil, G.; Oneto, C.; Antunes, P.; Rodrigues, M.F.; Empis, J.M. Extraction of lipids from cherry seed oil using supercritical carbon dioxide. *Eur. Food Res. Technol.* **2001**, *212*, 170–174. [CrossRef]
15. Duman, G.; Okutucu, C.; Ucar, S.; Stahl, R.; Yanik, J. The slow and fast pyrolysis of cherry seed. *Bioresour. Technol.* **2011**, *102*, 1869–1878. [CrossRef] [PubMed]
16. Petrov, N.; Budinova, T.; Razvigorova, M.; Minkova, V.; Vigouroux, R.; Björnbom, E. Preparation of activated carbons from cherry stones, apricot stones and grape seeds for removal of metal ions from water. In Proceedings of the 2nd Olle Indstorm Symposium on Renewable Energy-Bioenergy, Royal Institute of Technology, Stockholm, Sweden, 9–11 June 1999; pp. 46–550.
17. González, J.F.; Encinar, J.M.; Canito, J.L.; Sabio, E.; Chacón, M. Pyrolysis of cherry stones: Energy uses of the different fractions and kinetic study. *J. Anal. Appl. Pyrolysis* **2003**, *67*, 165–190. [CrossRef]
18. González-Domínguez, J.M.; Fernández-González, M.C.; Alexandre-Franco, M.; Gómez-Serrano, V. How does phosphoric acid interact with cherry stones? A discussion on overlooked aspects of chemical activation. *Wood Sci. Technol.* **2018**, *52*, 1645–1669. [CrossRef]
19. Kumari, D.; Singh, R. Pretreatment of lignocellulosic wastes for biofuel production: A critical review. *Renew. Sustain. Energy Rev.* **2018**, *90*, 877–891. [CrossRef]
20. Roy, R.; Rahman, M.S.; Raynie, D.E. Recent advances of greener pretreatment technologies of lignocellulose. *Curr. Res. Green Sustain. Chem.* **2020**, *3*, 100035. [CrossRef]
21. Laurichesse, S.; Avérous, L. Chemical modification of lignins: Towards biobased polymers. *Prog. Polym. Sci.* **2014**, *39*, 1266–1290. [CrossRef]
22. Wang, S.; Liu, W.; Yang, D.; Qiu, X. Highly resilient lignin-containing polyurethane foam. *Ind. Eng. Chem. Res.* **2018**, *58*, 496–504. [CrossRef]
23. Li, H.-Q.; Shao, Q.; Luo, H.; Xu, J. Polyurethane foams from alkaline lignin-based polyether polyol. *J. Appl. Polym. Sci.* **2016**, *133*. [CrossRef]
24. Haridevan, H.; Evans, D.A.; Ragauskas, A.J.; Martin, D.J.; Annamalai, P.K. Valorisation of technical lignin in rigid polyurethane foam: A critical evaluation on trends, guidelines and future perspectives. *Green Chem.* **2021**, *23*, 8725–8753. [CrossRef]
25. Flôres, C.C.; Rufino, T.d.C.; Oliveira, M.P. Effect of Kraft lignin and palm kernel oil as substitutes of petroleum-based polyols on the properties of viscoelastic polyurethane foams. *J. Polym. Res.* **2021**, *28*, 1–15. [CrossRef]
26. Cinelli, P.; Anguillesi, I.; Lazzeri, A. Green synthesis of flexible polyurethane foams from liquefied lignin. *Eur. Polym. J.* **2013**, *49*, 1174–1184. [CrossRef]
27. Ghaffar, S.H.; Fan, M. Lignin in straw and its applications as an adhesive. *Int. J. Adhes. Adhes.* **2014**, *48*, 92–101. [CrossRef]
28. Bertaud, F.; Tapin-Lingua, S.; Pizzi, A.; Navarrete, P.; Petit-Conil, M. Development of green adhesives for fibreboard manufacturing, using tannins and lignin from pulp mill residues. *Cell. Chem. Technol.* **2012**, *46*, 449–455.
29. Lee, S.J.; Kim, H.J.; Cho, E.J.; Song, Y.; Bae, H.-J. Isolation and characterization of lignin from the oak wood bioethanol production residue for adhesives. *Int. J. Biol. Macromol.* **2015**, *72*, 1056–1062. [CrossRef] [PubMed]
30. Olivares, M.; Guzmán, J.A.; Natho, A.; Saavedra, A. Kraft lignin utilization in adhesives. *Wood Sci. Technol.* **1988**, *22*, 157–165. [CrossRef]
31. Mansouri, N.-E.E.; Pizzi, A.; Salvado, J. Lignin-based polycondensation resins for wood adhesives. *J. Appl. Polym. Sci.* **2007**, *103*, 1690–1699. [CrossRef]
32. Pizzi, A. Recent developments in eco-efficient bio-based adhesives for wood bonding: Opportunities and issues. *J. Adhes. Sci. Technol.* **2006**, *20*, 829–846. [CrossRef]
33. Hamaguchi, M.; Kautto, J.; Vakkilainen, E. Effects of hemicellulose extraction on the kraft pulp mill operation and energy use: Review and case study with lignin removal. *Chem. Eng. Res. Des.* **2013**, *91*, 1284–1291. [CrossRef]
34. Liu, Z.; Fatehi, P.; Sadeghi, S.; Ni, Y. Application of hemicelluloses precipitated via ethanol treatment of pre-hydrolysis liquor in high-yield pulp. *Bioresour. Technol.* **2011**, *102*, 9613–9618. [CrossRef]
35. Al-Dajani, W.W.; Tschirner, U.W. Pre-extraction of hemicelluloses and subsequent kraft pulping Part I: Alkaline extraction. *Bioprod. Biosyst. Eng.* **2008**, *7*, 3–8.

36. Helmerius, J.; von Walter, J.V.; Rova, U.; Berglund, K.A.; Hodge, D.B. Impact of hemicellulose pre-extraction for bioconversion on birch Kraft pulp properties. *Bioresour. Technol.* **2010**, *101*, 5996–6005. [CrossRef] [PubMed]
37. Sidiras, D.K.; Koukios, E.G. Acid saccharification of ball-milled straw. *Biomass* **1989**, *19*, 289–306. [CrossRef]
38. Zhu, J.Y.; Pan, X.; Zalesny, R.S. Pretreatment of woody biomass for biofuel production: Energy efficiency, technologies, and recalcitrance. *Appl. Microbiol. Biotechnol.* **2010**, *87*, 847–857. [CrossRef] [PubMed]
39. Saeed, A.; Jahan, M.S.; Li, H.; Liu, Z.; Ni, Y.; van Heiningen, A. Mass balances of components dissolved in the pre-hydrolysis liquor of kraft-based dissolving pulp production process from Canadian hardwoods. *Biomass Bioenergy* **2012**, *39*, 14–19. [CrossRef]
40. Liu, C.; Wyman, C.E. Partial flow of compressed-hot water through corn stover to enhance hemicellulose sugar recovery and enzymatic digestibility of cellulose. *Bioresour. Technol.* **2005**, *96*, 1978–1985. [CrossRef] [PubMed]
41. Li, H.; Saeed, A.; Jahan, M.S.; Ni, Y.; van Heiningen, A. Hemicellulose removal from hardwood chips in the pre-hydrolysis step of the kraft-based dissolving pulp production process. *J. Wood Chem. Technol.* **2010**, *30*, 48–60. [CrossRef]
42. TAPPI. 264 om-97. In *Preparation of Wood in Chemical Analysis*; TAPPI: Peachtree Corners, GA, USA, 1997.
43. TAPPI. 211 om–93. In *Ash in Wood, Pulp, Paper and Paperboard: Combustion at 525 °C*; TAPPI: Peachtree Corners, GA, USA, 1993.
44. Prozil, S.O.; Evtuguin, D.V.; Lopes, L.P.C. Chemical composition of grape stalks of Vitis vinifera L. from red grape pomaces. *Ind. Crops Prod.* **2012**, *35*, 178–184. [CrossRef]
45. Neto, C.P.; Seca, A.; Fradinho, D.; Coimbra, M.A.; Domingues, F.; Evtuguin, D.; Silvestre, A.; Cavaleiro, J.A.S. Chemical composition and structural features of the macromolecular components of Hibiscus cannabinus grown in Portugal. *Ind. Crops Prod.* **1996**, *5*, 189–196. [CrossRef]
46. Browning, B.L. *Methods of Wood Chemistry Volume I & II*; John Wiley & Sons: Hoboken, NJ, USA, 1967.
47. Venegas-Gomez, A.; Gomez-Corzo, M.; Macías-García, A.; Carrasco-Amador, J.P. Charcoal obtained from cherry stones in different carbonization atmospheres. *J. Envirom. Chem. Eng.* **2020**, *8*, 103561. [CrossRef]
48. Kamel, B.S.; Kakuda, Y. Characterization of the seed oil and meal from apricot, cherry, nectarine, peach and plum. *J. Am. Oil Chem. Soc.* **1992**, *69*, 492–494. [CrossRef]
49. García-Aguilar, L.; Rojas-Molina, A.; Ibarra-Alvarado, C.; Rojas-Molina, J.; Vázquez-Landaverde, P.; Luna-Vázquez, F.; Zavala-Sánchez, M. Nutritional Value and Volatile Compounds of Black Cherry (*Prunus serotina*) Seeds. *Molecules* **2015**, *20*, 3479–3495. [CrossRef] [PubMed]
50. Mendes, C.V.T.; Carvalho, M.; Baptista, C.; Rocha, J.M.S.; Soares, B.I.G.; Sousa, G.D.A. Valorisation of hardwood hemicelluloses in the kraft pulping process by using an integrated biorefinery concept. *Food Bioprod. Process.* **2009**, *87*, 197–207. [CrossRef]
51. Colodette, J.L.; Longue, D., Jr.; Pedrazzi, C.; Oliveira, R.C.; Gomide, J.L.; Gomes, F.J. Pulpability and bleachability of xylan-depleted eucalyptus wood chips. *Ind. Eng. Chem. Res.* **2011**, *50*, 1847–1852. [CrossRef]
52. Al-Dajani, W.W.; Tschirner, U.W.; Jensen, T. Pre-extraction of hemicelluloses and subsequent kraft pulping Part II: Acid-and autohydrolysis. *Bioprod. Biosyst. Eng.* **2009**, *8*, 30–37.
53. Esteves, B.; Marques, A.V.; Domingos, I.; Pereira, H. Chemical changes of heat treated pine and eucalypt wood monitored by FTIR. *Maderas Cienc. Tecnol.* **2013**, *15*, 245–258. [CrossRef]
54. Coates, J. Interpretation of Infrared Spectra, a Practical Approach. In *Encyclopedia of Analytical Chemistry*; John Wiley & Sons: Hoboken, NY, USA, 2000.
55. Esteves, B.; Cruz-Lopes, L.; Ferreira, J.; Domingos, I.; Nunes, L.; Pereira, H. Optimizing Douglas-fir bark liquefaction in mixtures of glycerol and polyethylene glycol and KOH. *Holzforschung* **2018**, *72*, 25–30. [CrossRef]
56. Jelle, B.P.; Nilsen, T.-N.; Hovde, P.J.; Gustavsen, A. Accelerated climate aging of building materials and their characterization by Fourier transform infrared radiation analysis. *J. Build. Phys.* **2012**, *36*, 99–112. [CrossRef]
57. Lao, W.; Li, G.; Zhou, Q.; Qin, T. Quantitative analysis of biomass in three types of wood-plastic composites by FTIR spectroscopy. *BioResources* **2014**, *9*, 6073–6086. [CrossRef]

agronomy

MDPI

Article

Tissue-Specific Transcriptomes Outline Halophyte Adaptive Strategies in the Gray Mangrove (*Avicennia marina*)

David R. Nelson [1,2,*,†], Amphun Chaiboonchoe [3,†], Khaled M. Hazzouri [4], Basel Khraiwesh [1,‡], Amnah Alzahmi [1], Ashish Jaiswal [2], Guillermo Friis [1], John A. Burt [1], Khaled M. A. Amiri [4,5] and Kourosh Salehi-Ashtiani [1,2,*]

1 Center for Genomics and Systems Biology, New York University-Abu Dhabi, Abu Dhabi P.O. Box 129188, United Arab Emirates
2 Division of Science and Math, New York University-Abu Dhabi, Abu Dhabi P.O. Box 129188, United Arab Emirates
3 Siriraj Center of Research Excellence for Precision Medicine and Systems Pharmacology (SiCORE-PM&SP), Department of Pharmacology, Faculty of Medicine, Siriraj Hospital, Mahidol University, Bangkok 10400, Thailand
4 Khalifa Center for Genetic Engineering and Biotechnology, United Arab Emirates University, Al Ain P.O. Box 15551, United Arab Emirates
5 Biology Department, United Arab Emirates University, Al Ain P.O. Box 15551, United Arab Emirates
* Correspondence: drn2@nyu.edu (D.R.N.); ksa3@nyu.edu (K.S.-A.); Tel.: +971-2-628-5138 (D.R.N.); +971-2-628-4546 (K.S.-A.)
† These authors contributed equally to this work.
‡ Deceased.

Citation: Nelson, D.R.; Chaiboonchoe, A.; Hazzouri, K.M.; Khraiwesh, B.; Alzahmi, A.; Jaiswal, A.; Friis, G.; Burt, J.A.; Amiri, K.M.A.; Salehi-Ashtiani, K. Tissue-Specific Transcriptomes Outline Halophyte Adaptive Strategies in the Gray Mangrove (*Avicennia marina*). *Agronomy* **2022**, *12*, 2030. https://doi.org/10.3390/agronomy12092030

Academic Editors: António Dinis Ferreira, Raquel P. F. Guiné and António Moitinho Rodrigues

Received: 4 July 2022
Accepted: 23 August 2022
Published: 26 August 2022

Abstract: *Avicennia marina* forests fulfill essential blue carbon and ecosystem services, including halting coastal erosion and supporting fisheries. Genetic studies of *A. marina* tissues could yield insight into halophyte adaptive strategies, empowering saline agriculture research. We compare transcriptomes from *A. marina* pneumatophores, stems, leaves, flowers, seeds, and transcriptomes across four widely divergent environments in the Indo-Pacific (Red Sea, Arabian Gulf, Bay of Bengal, and Red River Delta) to decipher the shared and location-, tissue-, and condition-specific functions. On average, 4.8% of transcripts per tissue were uniquely expressed in that tissue, and 12.2% were shared in all five tissues. Flowers' transcript expression was the most distinct, with domain-centric gene ontology analysis showing high enrichment for stimulus-responsive processes, as well as genes implicated in flowering (hydroxygeraniol dehydrogenase, TPM = 3687) and floral scent biosynthesis (e.g., benzoyl_coenzyme_A, 2497.2 TPM). Pneumatophores highly expressed antioxidant genes, such as glutathione S-transferase (GST, TPM = 4759) and thioredoxin (TRX, TPM = 936.2), as well as proteins in the GO term 'Hydroquinone:oxygen oxidoreductase activity' (enrichment Z = 7.69, FDR-corr. $p = 0.000785$). Tissue-specific metabolic pathway reconstruction revealed unique processes in the five tissues; for example, seeds showed the most complete expression of lipid biosynthetic and degradation pathways. The leaf transcriptome had the lowest functional diversity among the expressed genes in any tissue, but highly expressed a catalase (TPM = 4181) and was enriched for the GO term 'transmembrane transporter activity' (GO:0015238; Z = 11.83; FDR-corr. $p = 1.58 \times 10^{-9}$), underscoring the genes for salt exporters. Metallothioneins (MTs) were the highest-expressed genes in all tissues from the cultivars of all locations; the dominant expression of these metal-binding and oxidative-stress control genes indicates they are essential for *A. marina* in its natural habitats. Our study yields insight into how *A. marina* tissue-specific gene expression supports halotolerance and other coastal adaptative strategies in this halophytic angiosperm.

Keywords: genomics; halophytes; salt tolerance; transcriptomics; gray mangrove

1. Introduction

Mangroves support coastal ecosystems despite salinity, oxygen, and temperature extremes [1–5]. They are comprised of distantly related tree species that have converged to dwell on saline shorelines [6,7]. Among mangroves, *Avicennia* species are particularly hardy in the face of environmental extremes. However, they have relatively shallow genetic diversity and are vulnerable to environmental perturbations from anthropogenic climate changes [8–11].

Avicennia forests provide blue carbon and ecosystem services for their surrounding populations [4]. Their ability to halt coastal erosion, support fisheries, and sequester atmospheric carbon make them outstanding contributors to global environmental sustainability. These services are enabled by the preservation and propagation of crucial species, such as *A. marina* [12]. Further genome-level studies will reinforce afforestation efforts. Establishing additional omics resources for *A. marina*, including genomics, transcriptomics, metabolomics, and subsequent comparative analyses, will inform critical decisions regarding cultivar choices, suitable afforestation locations, and possible transgenic species implementation.

The unique ecophysiology of *A. marina* has facilitated its survival in a wide variety of Indo-Pacific coastal biomes [8,10,13–22]. It has colonized coasts from Africa and Asia to Zealandia since the early Holocene [23]. Habitat environments range from hypo- to hypersaline, and annual average, low, and high temperatures in *A. marina*-colonized locations range from Africa to New Zealand. The species has adapted to survive in a wide range of intertidal biome types. Furthermore, unrelated mangrove lineages have undergone convergent evolution towards functionally similar adaptations [6,7]. Factors advantageous in a seafaring lifestyle include seed survival across ocean voyages, secondary growth through successive cambia formation [20,24], halotolerance [25–29], and multifunctional pneumatophores that cope with a vast array of abiotic stressors [15,30,31].

A. marina inhabits the coastline of the Red Sea [32], a region occasionally characterized by extreme heat and salinity [13,14,19,21,22,33]. Cultivars from this region are suitable models to study the resilience of these angiosperms in the face of intense abiotic stress. The mechanisms behind *A. marina*'s resilient biology, especially in areas such as the Red Sea, can inform translational agronomical research. Oxidative stress results from salt toxicity that prohibits saline agriculture of most commercial crops; increasing resistance will allow productive transgenic species to thrive using readily available saline irrigation. The implementation of seawater irrigation would promote food security in the face of anticipated shortages in freshwater supply [34–37], as well as expand saline agriculture into inarable regions [28,38–42].

The mechanisms used by *A. marina* leaves to maintain photosynthesis in the face of high salt levels is the subject of intensive research, as salt is toxic to photosystems [43–48] yet accumulates at high levels in *A. marina* leaves [26]. A recent study examined *Avicennia officinalis* leaves' response to salt stress and found 1404 genes expressed after salt treatment [30]. Salt stress can be mitigated by osmolyte accumulation [45,49–51] and active transport [42,52–57]. Membrane integral active transporters are essential for *A. marina* survival and responsible for the large salt crystals that commonly form on the surface of *A. marina* leaves [58]. Despite intensive research into *Avicennia*'s leaves and pneumatophores, research into its other tissues, including flowers, seeds, and developing shoots, is lacking. Insight into how these tissues support *Avicennia*'s seafaring lifestyle could be obtained by examining their transcriptomic expression profiles.

Here, we highlight the expression of genes involved in processes essential for these understudied tissues. We use *A. marina* transcriptomes from different tissues and geographies to find commonalities of intrinsic gene expression patterns responsible for its ability to survive in conditions that would be prohibitive for other angiosperms. This information will be useful in understanding what core genotypes can promote successful saline agronomy endeavors.

2. Materials and Methods

2.1. Sequence Sources

Our five-tissue transcriptome dataset consisted of tissue samples we previously isolated and used to annotate a high-quality *A. marina* reference [59]. We used the high-quality reference genome recently published by the NYUAD Marine Biology Laboratory [59] for chromosome-level read mapping (see Figure S1). Reference transcripts for expression analyses were from He et al. (2020) [7]. Finally, we used public RNAseq data from *A. marina* cultivars from the Red Sea (NCBI BioProject: PRJNA591919), Vietnam (NCBI Bioproject: PRJDB6605, Department of Biotechnology, Graduate School of Engineering, Osaka University, Japan), India (NCBI Bioproject: PRJNA283781, M.S. Swaminathan Research Foundation, Chennai, India), and China (NCBI Bioproject: PRJNA350568, Lingnan Normal University, Zhanjiang, China) to further relate our tissue-specific transcriptomes to natural gene expression patterns.

2.2. Sequence Processing and Alignment

Adapters and filter low-quality bases (<Q20) were removed with the CLC Genomics Workbench (v11.0; Qiagen, Hilden, DE) trim tool. Filtered reads were used for mapping to the reference transcriptome [7] to determine the differentially expressed genes (DEGs, see Table S1) using the CLC Genomics Workbench (v11.0; Qiagen, Hilden, DE). Mapping of RNAseq reads to the reference genome [59] was done in HISAT2 [60], similarly to [61], to observe the global expression patterns (see also Dataset S1). Mapping to the reference transcripts was done in the CLC Genomics Workbench (v11.0; Qiagen, Hilden, DE) in the same manner as [62]. Briefly, genes were extracted from the reference genome [59] (using annotations of the type gene). Other annotations on the gene sequences were preserved (e.g., coding sequences, introns, exons, and untranslated regions). Then, the annotated transcripts (using annotations of type mRNA) were extracted. If there were several annotated splice variants, they were all extracted. The mRNA annotation type was used for deciphering exon–exon boundaries. Reads were mapped against all the transcripts plus the entire gene, so all the exon–exon junctions were joined in the extracted transcript. Mapped reads were categorized and assigned to the genes (elaborated later in this section), and expression values for each gene and transcript were calculated. Then, putative exons were identified, following guidelines [63–65]. The default parameters for alignment were used, including no more than two mismatches per aligned read, ten maximum hits for a read, a 90% minimum length fraction mapped, and a minimum similarity fraction of 0.8.

Differential expression was calculated for each gene using transcripts-per-kilobase million (TPM) for accurate comparison between samples ($TPM = (R * 10^6)/(T * L)$, where T is the sum of all length-normalized transcript counts, R is the number of reads mapped to a transcript, and L is the length of the transcript) [66–69]. Transcripts with a TPM value of ten or higher were classified as actively expressed unless stated otherwise, and those with TPM values of less than ten were deemed unexpressed. This filtering strategy for a classification-based feature selection resolved 400–600 'uniquely' expressed genes in each tissue. A caveat of this study was the lack of triplicate samples for each condition; thus, qualitative comparative analyses comprised methods used here. Expression values for TPM, FPKM, RPKM, and raw values are in Table S1 (Red Sea, five tissues) and Table S2 (samples from various geographies).

2.3. Uniform Manifold Approximation Projection

The DEG TPMs in each tissue informed the Uniform manifold approximation projection (UMAP, [70]) in TensorFlow (tensorflow.org; https://projector.tensorflow.org/, accessed on 12 March 2021) to cluster different annotated genes by their tissue-specific expression profiles. Although UMAP is primarily used to decipher single-cell RNAseq expression data, its adept embedding clustering algorithm lends itself to a wide variety of applications. Briefly, UMAP clusters data points from variables in multiple dimensions using fuzzy simplices to preserve local relationship information while approximating their embedding. For the projection shown in Figure S1, the '3D' visualization mode was selected, and the 15 nearest neighbors for each data point were used to calculate the approximations. Graphs were manually optimized for visual clarity. Data points represented individual transcripts and were colored according to the tissue showing the highest expression in TPM.

2.4. Hierarchical Clustering

Hierarchical clustering was done using Pearson's correlation coefficient or 1-cosine similarity (Figure S3) distance metrics and by using an average agglomeration method [71] in Morpheus (https://software.broadinstitute.org/morpheus, accessed on 12 March 2021).

2.5. Intersection Analyses

Interactive Venn [72] and UpsetR (Figure S2) [73] were used to discern the shared and uniquely expressed genes and KEGG orthologs. Uniquely expressed genes >10 TPM are listed for each tissue in Table S1.

2.6. Domain-Centric GO Enrichment Analysis

The tissue-specific enrichment analysis was carried out using Domain-centric Gene Ontology (dcGO, [74]). Preliminary HMMsearch (hmmer.org) was performed on translated reference transcripts with the command 'hmmsearch –noali -E 0.000000001 –cpu 28 –domtblout $OUT $IN.aa.fa'. The software dcGO statistically infers associations between GO terms and combinations of PFAM domains. Unique transcripts were used as inputs for each tissue; the discovered GO terms shown are non-unique. An additional dcGO analysis with unique GO terms in each tissue from uniquely expressed genes (doubly-unique) is shown in Figure S4. A false discovery rate (FDR) threshold of 0.001 was used for detection of enrichment.

2.7. GO Network

The baseline GO network was constructed from the full GO list downloaded from http://geneontology.org/docs/download-ontology/ (accessed on 5 September 2020). The 'core ontology' file (go.obo http://purl.obolibrary.org/obo/go.obo, accessed on 3 July 2022; Core ontology (OBO Format)) was used to create the network infrastructure in Cytoscape (cytoscape.org, accessed on 3 July 2022). Expression data were mapped to the nodes on the network corresponding to the GO terms.

2.8. Metabolic Map Reconstruction

Metabolic maps were reconstructed using the Interactive Pathways Explorer (iPath) (a web-based tool for the visualization, analysis, and customization of various pathway maps; https://pathways.embl.de/, accessed on 23 November 2021). Briefly, KEGG ortholog (KO) designation lists corresponding to genes expressed at >10 TPM were uploaded for each tissue (see also Table S3).

2.9. Manhattan and Violin Plots

Sequence reads from public RNAseq data from A. marina cultivars in Vietnam (NCBI Bioproject: PRJDB6605, Department of Biotechnology, Graduate School of Engineering, Osaka University, Japan), India (NCBI Bioproject: PRJNA283781, M.S. Swaminathan Research Foundation, Chennai, India), and China (NCBI Bioproject: PRJNA350568, Lingnan Normal University, Zhanjiang, China) were mapped to reference transcripts [7], to obtain the TPMs, and plotted in Plotly (https://plotly.com/, accessed on 23 November 2021, px.scatter, y=violin), to show the highest-expressed transcripts and overall expression distributions2.10. Graphics

Figure 1B was provided by NASA. The International Space Station Program supports the laboratory as part of the ISS National Lab to help astronauts take pictures of Earth that will be of great value to scientists and the public, and to make those images freely available on the Internet. Astronaut photograph ISS052-E-15927 was acquired on 14 July

2017, with a Nikon D4 digital camera using an 1150 mm lens and is provided by the ISS Crew Earth Observations Facility and the Earth Science and Remote Sensing Unit, Johnson Space Center. The image was taken by a member of the Expedition 52 crew. The image has been cropped and enhanced to improve contrast, and lens artifacts have been removed. Caption by Andi Hollier, Hx5, JETS Contract at NASA-JSC.

Figure 1. Overview of the RNAseq data source tissues and geographies. (**A**) Locations (stars) of the *A. marina* RNAseq samples. Map and location visualized in Plotly. (**B**) Jeddah Port, showing the industrialization near the Red Sea mangrove sampling site. Image courtesy of NASA. (**C**) *A. marina* flowers. (**D**) *A. marina* seeds. (**E**) *A. marina* leaves. (**F**) *A. marina* stems. (**G**) *A. marina* pneumatophores. Mangrove tissue photos reproduced with permission from Dr. Alexey Sergeev.

Automatic generation of images was carried out in Plotly (Plotly express; https://plotly.com/python/plotly-express/, accessed on 23 November 2021) and R (https://www.r-project.org, accessed on 23 November 2021), ggplot2 (https://ggplot2.tidyverse.org, accessed on 23 November 2021), and processed with Adobe Illustrator (v25.2.3) and Adobe Photoshop CC (v20.0.6) for clarity and annotation. Circos (http://circos.ca/, accessed on 4 April 2020 [75]) was used to plot the TPM expression values of the RNAseq from various tissues mapped to the reference chromosomes for Figure S1 [59] with HISAT2 [60]. Hierarchical clustering was done in Morpheus (https://software.broadinstitute.org/morpheus, accessed on 19 May 2021). The GO network was created in Cytoscape (cytoscape.org, accessed on 3 July 2022).

3. Results and Discussion

We sampled sequencing reads obtained from *A. marina* in various locations in the Indo-Pacific region and collated them into a database. These reads were the basis for transcriptome comparisons between tissue types and geographies (Figure 1). Our characterization of the expressed functions was facilitated by the reference genome and coding sequence annotations [7]. In addition, we used the high-quality reference genome recently published by the NYUAD Marine Biology Laboratory [59] for mapping reads to chromosomes (Figure S1; Dataset S1). Finally, we used public RNAseq data from *A. marina* cultivars in Vietnam (NCBI Bioproject: PRJDB6605, Department of Biotechnology, Graduate School of Engineering, Osaka University, Japan), India (NCBI Bioproject: PRJNA283781, M.S. Swaminathan Research Foundation, Chennai, India), and China (NCBI Bioproject: PRJNA350568, Lingnan Normal University, Zhanjiang, China) to further relate our tissue-specific transcriptomes to natural gene expression patterns in *A. marina*. Together, these datasets provide a basis to study organ-specific transcriptomes in *A. marina*.

3.1. *A. marina* Coding Potential and Transcriptome

A. marina can survive extreme conditions because of the genes transcribed by its relatively compact, unchanging genome. We examined de novo-predicted genes, de novo-assembled transcriptomes, and evaluated expression profiles from two published reference sequence sources. Compared to other angiosperms, all *A. marina* transcriptomes published so far show high expression of a collection of genes implicated in heat and drought tolerance. Thus, their transcriptomic profiles contain the blueprints for angiosperm survival in harsh conditions.

Here, we analyze five separate transcriptome profiles from pneumatophore, leaf, flower, seed, and shoot tissue samples of *A. marina*. We compare these profiles with each other as well as transcriptomes from various tissues and locations throughout the Indo-Pacific, including India, Vietnam, and China. Investigation into these transcriptomic profiles yielded insight into the functional roles of each tissue and how they support *A. marina*'s seafaring lifestyle.

One of the most remarkable findings from our studies was that genes for metallothionein showed the highest expression regardless of tissue type or cultivar location (see Table S1). This finding provides strong evidence that metallothioneins are key components for *A. marina* antioxidant systems regardless of location, climate, or environment. Other significant findings include arrays of desiccation-related genes, especially in seawater-immersed tissues, and flower-specific genes involved in pollinator attraction processes. Below, we detail the features of the sampled transcriptomes. Tissue-specific transcriptomes are described in the order of their expression track rings in Figure S1.

3.2. *A. marina* Tissue-Specific Transcript Expression

Expressed genes were determined by having a coverage depth of >10 transcripts per kilobase million (TPM) unless stated otherwise. Downstream analyses on defined gene sets were uniform manifold approximation projection (UMAP [70]; Figure S1), hierarchical bi-clustering (Figure 2), Hidden Markov Model (HMM) alignment to known protein families (PFAMs [76,77]), and gene ontology (GO) enrichment analyses by tissue and location. The five tissues shared major chromosomal regions of high expression, with notable exceptions (Figure S1). These regions were in the outer arms; centromeric regions displayed canonical low transcript expression (Figure S1; Dataset S1).

Hierarchical bi-clustering of DEGs highlighted tissue-specific gene clusters (Figure 2). Uniquely-expressed (>10 TPM) clusters of DEGs (Table S1) for each tissue are described in the following sections, with implications for *A. marina*'s abiotic stress resistance. For insight into the relationship of different *A. marina* gene expressions at a snapshot of the Red Sea cultivar, Uniform Manifold Approximation Projection (UMAP; [70]) was applied (Figure S1) to the TPM profiles from five tissues (Table S1). Root-specific genes clustered near the seed and leaf genes; flowers and stems clustered tightly but were isolated from other tissues. Leaf-dominant genes were centrally projected, implying a median expression level for highly differentially expressed genes between the flower-stem and pneumatophore-seed groups. Roots and seeds are the only two *A. marina* tissues with prolonged seawater exposure; thus, similar gene expression networks may be required by these tissues to survive in these conditions.

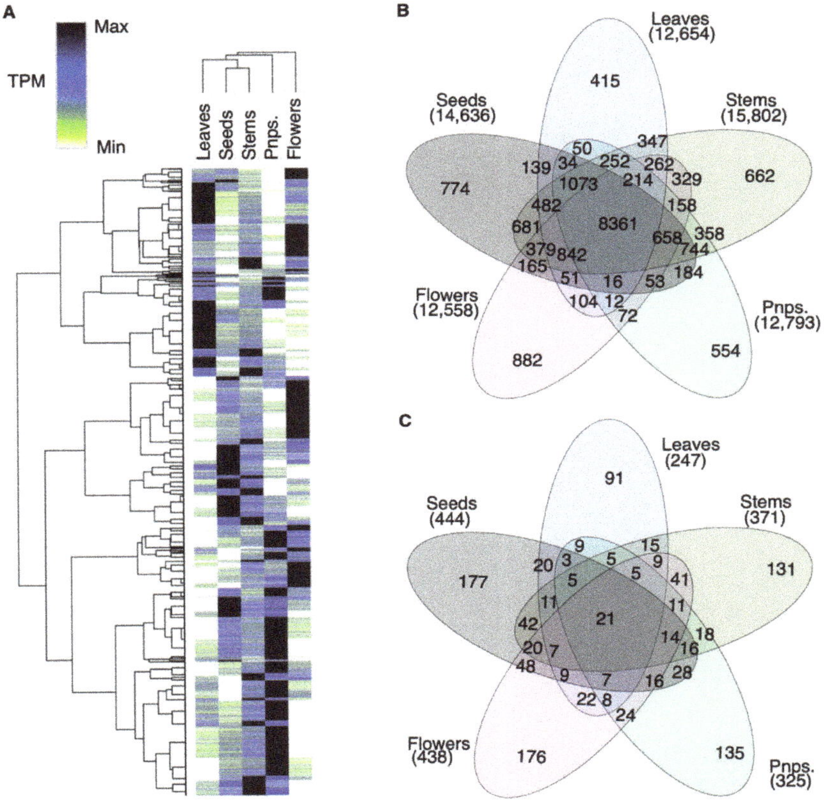

Figure 2. Segregation of Red Sea *A. marina* gene clusters based on tissue-specific expression patterns. Pnps. = Pneumatophores. (**A**) Hierarchical bi-clustering of transcripts-per-million (TPM) normalized read counts from the five mangrove tissues using Pearson's correlation (see Table S1). The y-axis indicates transcripts. Darker clusters in each tissue correspond with GO enrichment networks. (**B**) Venn diagram showing reference transcripts >10 TPM from five tissues. Leaves had the smallest unique set, while stems expressed the most transcripts >10 TPM. Figure S2 shows additional intersections not shown here. (**C**) Unique and shared KEGG Orthology (KO) in the unique gene sets from (**B**). Approximately 60% of KOs from the uniquely expressed genes in each tissue are shared with one or more other tissues. Nearly one-third of (**C**) KOs in any given tissue are unique to that tissue, while the ratio of unique (**B**) transcripts is 1/15th the expressed transcript count. This discrepancy suggests a high redundancy in function coded for by *A. marina* transcripts.

We used unique KEGG KO terms (see References [6,7]) as input for each tissue, with red lines representing unique pathways for flowers, leaves, pneumatophores, seeds, and stems (see Figure 3). The reference transcripts were used as queries in HMMsearch to detect PFAMs (E < 1×10^{-9}). These PFAMs were used to discover domain-centric gene ontology (dcGO; [74]) functional enrichment using either one or two exclusion filters. The first filtering step removed transcripts expressed in more than one tissue; uniquely expressed transcripts were used for dcGO [74] enrichment (Figure 3). The second filter removed any GO terms shared amongst tissues.

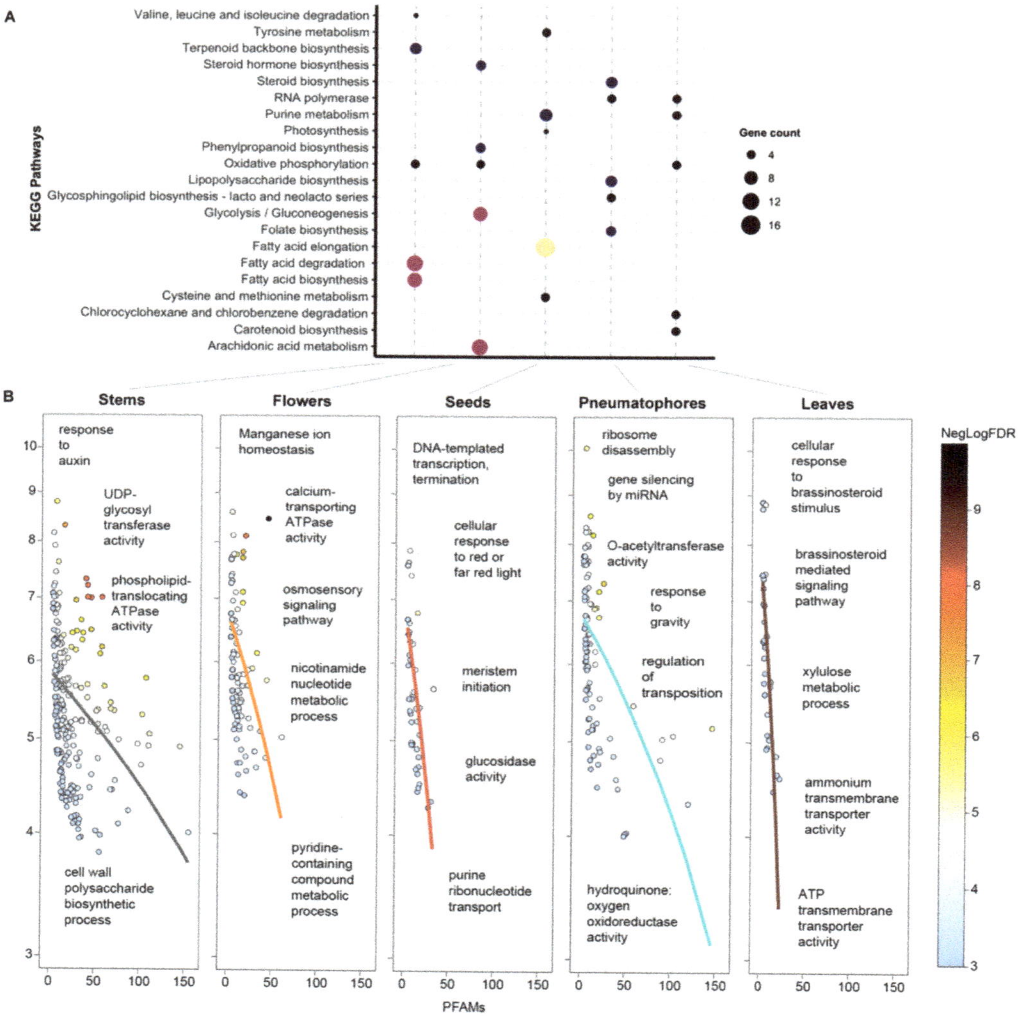

Figure 3. Tissue-specific KEGG pathways and GO term enrichment in *A. marina*. (**A**) Top five KEGG pathways with regard to uniquely expressed transcripts for each tissue. The y-axis represents the pathway name. The size of the bubble indicates the number of transcripts in the KEGG pathway. The color represents the gene count, with yellow representing a higher value and black representing a lower value. Transcripts expressed at >10 TPM (see Table S1) in each tissue were used to retrieve the KEGG orthologs. (**B**) Domain-centric gene ontology (dcGO; [74]) enrichment in uniquely expressed transcripts in the Red Sea cultivar.

3.3. Flower-Specific Transcript Expression

A. marina flowers accumulate relatively large amounts of sweet nectar and are pollinated by insects. Pollination by long-distance-traveling insects is crucial for species stability. Honeybees are the dominant pollinator of temperate *A. marina*, and pollinator-mediated competition has been reported [78]. Recent studies have hinted at a hidden world of mangrove insect interactions, with up to 160 distinct species per sampled site [79]. Insect-flower interactions can be interrogated at the molecular level in *A. marina* by examining the genes involved in nectar formation and chemical signaling. We examined the expressed genes involved in these processes.

Flower-specific genes yielded 5398 PFAMs, which contributed to enrichment in 577 GO terms (FDR corr. $p = 0.001$). Gene ontology analyses of flower-specific genes showed enrichment for stimulus-responsive processes; for example, the term 'response to hormone' (GO:0009725) was highly enriched in flowers (Z = 14.81; FDR-corr. $p = 3.85 \times 10^{-25}$; see also Table S3).

A large portion of flower-specific transcripts coded for proteins with known roles in flowering timing, such as cytochromes [44,80–85]. Putative cytochromes and ferredoxins were uniquely expressed in the flower transcriptome (Table S1), corresponding to the proposed roles of these enzymes in season-responsive flower developmental timing [86–89]. We note that multi-copper-containing glycoproteins (e.g., laccases; EC 1.10.3.2; i.e., *p*-diphenol: dioxygen oxidoreductase) were uniquely expressed in flowers. Mutations in *Arabidopsis* homologs of these genes have shortened the time to flowering [90]. An 8-hydroxygeraniol dehydrogenase (HGD; AM_02435) was also highly expressed in flowers (TPM = 3687). Gene ontology analysis with flower-specific transcripts revealed enrichment in processes related to the aforementioned redox reactions (e.g., 'cation homeostasis' (GO:0055080; Z = 8.83; FDR-corr. $p = 7.36 \times 10^{-11}$) and 'response to metal ion' (GO:0010038; Z = 8.96; FDR-corr. $p = 3.72 \times 10^{-10}$) (see also Table S3).

The flower RNAseq evidence is consistent with a pollinator-dependent flowering strategy in *A. marina*. Genes in pathways for nectar formation were uniquely expressed in flowers. Polygalacturonases (AM_00332 and AM_0031), three pectate lyases (AM_01234, AM_03511, AM_07809), and a pectinesterase (AM_01717) were uniquely expressed in flowers. These genes are implicated in nectar formation to attract pollinators, including honeybees [78]. Cheminformatic research has established molecular links between phytochemical signaling interventions and the regulation of the multi-enzyme arachidonic acid (AA) metabolic network [91]. The cysteine-rich peptide family gibberellic acid (GA)-stimulated *Arabidopsis* (GASA)/GA-stimulated transcript (GAST) plays critical roles in plant growth, development, and responses to abiotic and biotic stresses [92–94]. The GA-stimulated transcript (GAST1) was found in *Arabidopsis* roots and shoots, whereas our unique transcripts (AM_24369) were found in the flowers.

Pollination is a crucial yet understudied aspect of *A. marina*'s dispersal. Volatile organic carbon compounds produced by flowers attract and regulate pollinator visitations [95]. Insight into the genetic mechanisms governing *A. marina*'s pollination preferences would assist conservation and afforestation endeavors. Here, we show the uniquely expressed genes in flowers with putative roles in floral scent biosynthesis and nectar formation. Benzoyl_coenzyme_A (benzyl_alcohol_benzoyl_transferase, AM_23483, TPM = 2497.2) was one of the highest-expressed flower genes. This enzyme is responsible for the floral scent in a wide variety of ornamental plants (REFs). Thus, its role in *Avicennia* is likely to attract pollinators. 8-Hydroxygeraniol dehydrogenase (HGD; AM_02435) was highly expressed in flowers (TPM = 3687). In plants, monoterpene 8-hydroxygeraniol is an intermediate between geraniol and secologanin, among other floral compounds [96]. Secologanin is a branchpoint monoterpene in terpene indole alkaloid biosynthesis; activity through this pathway produces bioactive alkaloid secondary metabolites that sources the cornucopia of volatile organic carbons to attract pollinators [97–102].

A. marina has a short flowering period; the high expression of HGD is consistent with a burst of floral volatile organic carbon (VOC) production, which could signal migratory insects for pollination. Pollination by migratory insects is especially prevalent in desert regions, where resources are scarce or are only seasonally available. The sampling site and flanking areas (i.e., MENA countries and territories) occupy the largest deserts on earth; migratory pollinators in these areas likely provide significant benefit for *A. marina* propagation and, perhaps, evolution of the species. Finally, an MIP-family (NIP) aquaporin (AM_01631) was uniquely expressed in flowers (TPM = 4.3). Although its expression was lower than the 10 TPM cutoff used in this study, this aquaporin may have a flower-specific role, with implications for flower stability (i.e., wilt prevention) in high-heat conditions.

3.4. Seed-Specific Transcript Expression

Avicennia species have less capacity for local dispersion by clonal propagation compared to other mangrove species (e.g., *Rhizophora* [103]) but propagate to distant shores through ocean currents [104] with resilient, cryptoviviparic seeds. This shift in energy expenditure investment from clonal propagation to seed potency has given them an evolutionary edge to live in otherwise inhabitable shorelines. Seeds expressed 14,636 genes with TPM >10; 774 of these were unique. Seed-expressed genes had putative roles in development, including coordinating the timing for senescence and ripening. Unique seed-expressed genes included an egg cell-secreted protein (AM_33231). Orthologs of this gene are known to mediate gamete interactions during fertilization in *Arabidopsis* [105,106]. We found an ethylene_receptor_isoform_1 unique transcription factor (AM_06004) in seeds. Ethylene is a simple two-carbon gas that is required for a variety of plant processes, such as seed germination, seedling growth, leaf and petal abscission, organ senescence, and pathogen responses. Genes for Jumonji (AM_23304) and Myb (AM_16379), among other (AM_18056, AM_08631) transcription factors, were uniquely expressed (>10 TPM) in seeds. These genes play roles in chromatin remodeling and temporal- and temperature-dependent transcription orchestration [107–109], which are processes implicated in seed senescence and ripening timing [110].

A type 2 metallothionein (MT-2; AM_26466) was expressed at relatively low levels (96 TPM) in seeds; however, its expression was high in the other four tissues from the Red Sea cultivar: leaves (35,065 TPM), stems (30,600 TPM), pneumatophores (18,035 TPM), and flowers (9652 TPM; also see Table S1). An MT-3 protein had similarly high expression in all tissues except flowers, consistent with tissue-specific suppression [111]. Other genes with potential roles in ripening include the uniquely expressed isoform of an ethylene receptor (AM_06004; TPM = 6237 in seeds) and a metalloendopeptidase (AM_05873) in seeds. This transcript was expressed at 194 TPM in seeds but less than 10 TPM in other tissues. This digestive enzyme may promote ripening in *A. marina* seeds. Leaf and seed expression profiles clustered together based on Pearson correlation scores (Figure 2A); transcript expression similarity between these two tissues is seen in other plant species [112]. A G3P dehydrogenase was highly expressed in seeds; this was one of the rare examples of a primary metabolic gene with an expression like leaves (TPM = 2614 in seeds, TPM = 3082 in leaves; expressed as AM_02817 in both tissues). Fatty acid (FA) biosynthesis and degradation pathways showed higher activity in seeds and pneumatophores.

3.5. Leaf-Specific Transcript Expression

Leaves had less overall diversity of transcript expression; however, photosynthesis-associated genes were highly expressed in leaves. For example, rubisco subunits were highly expressed (TPM 15,000–20,000; see Table S1). The expression of transcripts involved in photosynthetic processes was an order of magnitude higher in leaf tissue, which can be attributed to their abundant chloroplasts [113]. Transcript expression was suppressed on several leaf chromosomes (e.g., right arms of Chr 21 and Chr 12), and leaves had the lowest number of expressed CDSs with annotated PFAMs. Overall, 12,645 genes were expressed at TPM >10 in leaves, and 415 of these were unique to leaves (Table S1). The leaf transcript with the highest expression was AM_26466 (35,065 TPM in leaves), annotated as an MT-2 [18]. Leaves had relatively high expression of central metabolism genes. The AM_26466 transcript was also expressed at high levels in the other four tissue and likely plays a critical role in *A. marina* survival. Glyceraldehyde-3-phosphate dehydrogenase was expressed at 1911 TPM in leaves but at less than 400 TPM in the other tissues, and fructose-bisphosphate aldolase was expressed at 5404 TPM in leaves but at less than 1000 TPM in other tissues (Table S1). These results are consistent with high metabolic activity in leaves, exemplifying the capacity for energy generation in these chloroplast-rich tissues. Reconstructed GO networks also display strong enrichment for 'plastid' and related processes in leaves (Figure 4).

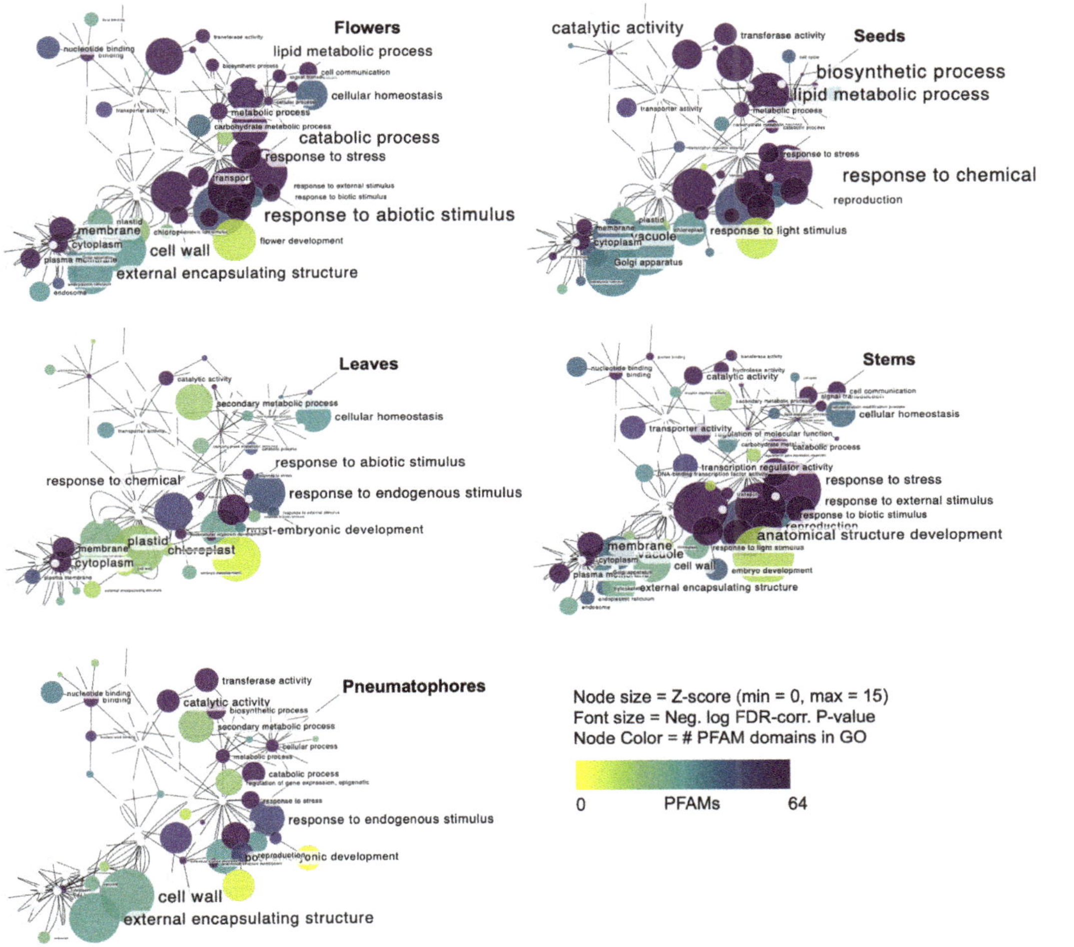

Figure 4. Domain-centric gene ontology (dcGO, [74]) enrichment in uniquely expressed transcripts in the Red Sea cultivar visualized in Cytoscape using the base GO core ontology association network. Panels show GO enrichment networks from uniquely expressed genes in each of five *A. marina* tissues. Hypergeometric distributions were the null hypotheses; false discovery rate (FDR, [114])-corrected *p* values (colorscale) are from Fischer's exact tests and Benjamini–Hochberg correction for multiple hypotheses against the hypergeometric null background. Z-scores indicate enrichment and are calculated by subtracting matches and dividing by the standard deviation with all UniProt entries following a true-path rule. A least-squares regression trendline is overlaid on each plot. Leaves showed a sharp bias in Z-score compared to the PFAM counts, indicating specialization. Z-score indicates enrichment or distance of the overlapping PFAM group with the constituents of the GO term. All *p*-values shown are < 0.001. Protein family domain counts in each GO term are shown in the color scale. The node size corresponds to the Z-score from dcGO [74] enrichment analysis.

Active salt efflux relies on ATP and other intracellular energy stores. *Avicennia* modulates salt influx and efflux through leaf pores; rhythmic fluctuations in salt secretion rates correspond to environmental salinity [29]. Candidate salt gland genes were expressed in concordance with [27]. The leaf-specific expression of ATPases may assist salt efflux in this tissue. The ATPc subunit was uniquely expressed in leaves at >10 TPM. This subunit has a rapid turnover rate in vivo [115,116], explaining its high expression in leaves. The GO term 'transmembrane transporter activity' (GO:0015238) was highly enriched in leaves ($Z = 11.83$; FDR-corr. $p = 1.58 \times 10^{-9}$, 13 PFAMs in this GO term), highlighting a possible role for the constituent genes in salt efflux.

Dehydrins have characteristically high expression in salt glands [27]; we observed the CDSs AM_1543 (Dehydrin, TPM = 67.2) and AM_31570 (Dehydrin_DHN1, TPM = 1504) expressed in leaves. Stems (TPM = 2701) and seeds (TPM = 2099) also expressed this dehydrin isoform at high levels, indicating that salt glands may be present and active in these non-leaf *A. marina* tissues.

Although the Red Sea samples were collected in the winter, we observed the high expression of genes involved in photorespiration in leaves. For example, a photorespiration enzyme, glycolate oxidase (GLO; AM_06008), was highly expressed in leaves (TPM = 1603). Although GLO isoforms have divergent roles [48], the overexpression of at least one GLO isoform has been reported to confer improved photosynthetic capacity under high heat and light conditions [117].

We searched for the expression of genes involved in cuticle formation. This type of secondary growth may be essential for *A. marina*'s desiccation tolerance, especially in periods of high temperatures, such as in the summers in the Middle East and Africa. A variety of wax and lipid biosynthetic genes are responsible for cuticle formation, although this process has not been well-studied in *A. marina*. Leaves expressed 39 lipid metabolism CDSs at TPM >10, many with putative roles in cuticle synthesis (see Figure S3). However, the only wax synthase we observed, 'wax synthase isoform 3′ (AM_21438), was highly expressed in flowers (TPM = 79.3) rather than in the leaves (TPM = 6.1). Hierarchal clustering of extracted lipid metabolic genes revealed two clusters of genes uniquely and highly expressed in leaves (Figure S3). Genes in the leaf-specific clusters included a non-specific lipid-transfer protein-like protein (AM_27366), plastid lipid-associated proteins (AM_10546, AM_04326), a lipid-binding protein (AM_04326), a StAR-related lipid transfer protein (AM_18353), calcium lipid-binding proteins (AM_28011, AM_13767), a galactolipid galactosyltransferase (AM_07797), and a constitutive plastid-lipid associated protein (AM_04709; see also Figure S3).

A type-2 MT, together with a collection of photosynthetic genes, was expressed far more than other leaf-expressed genes (Table S1). Type-2 MTs respond to oxidative stress (Mir, 2004), bind metals [118], and mitigate oxidative stress in *Avicennia* [119] (Babaei-Bondarti and Shahpiri, 2020); however, MTs are less studied than their mammalian counterparts [120] (Freisinger, 2008). Transgenic *E. coli* expressing a recombinant type-2 MT (GST-OsMTI-2b) accumulated more Pb^{2+}, Ni^{2+}, Cd^{2+}, Zn^{2+}, and Cu^{2+} than the controls and had a higher tolerance to these metals [121]. Metallothioneins mitigate metal-induced oxidative stress [118,120,122–124], but recent studies have shown that MT isoforms can mitigate salt-induced oxidative stress as well. For example, a salt-responsive, date palm metallothionein 2A (PdMT2A) conferred salt tolerance in transgenic experiments on *Saccharomyces cerevisiae* and *Arabidopsis thaliana* [125].

Over the last decade, several strategies for improving photosynthesis and crop yield have been extensively discussed, including leaf morphology and light interception. Leaf morphology and physiology are inextricably linked to light capture efficiency and temperature regulation. The leaf-specific transcription is related to alpha-form rubisco activase (AM 33275). Increased rubisco activase levels may thus improve photosynthetic efficiency by increasing the amount of rubisco activated for CO_2 fixation.

The antioxidant capacity of *A. marina* leaves must be sufficient to defend cellular integrity, especially of cellular membranes, in the face of seawater immersion from tidal fluxes. Visibly dehydrated salts from tidal fluctuations are regularly observed on *A. marina* leaves, as well as large salt crystals from cellular efflux from salt glands. A transcript coding for a catalase was highly expressed in leaves (AM_25336; TPM = 4181). Similar catalases facilitate radical detoxification (e.g., H_2O_2 removal [46,126–129]). A ferric-chelate reductase expressed at high levels (AM_22465; leaf TPM = 2578) may also be involved in maintaining redox homeostasis in leaves.

3.6. Stem-Specific Transcript Expression

Stems had the highest number of genes expressed at >10 TPM (n = 15,802 transcripts). Gene ontologies with high counts of contributing PFAMs were seen, indicating that *A. marina* stems retain a generalist expression profile. The expression profile in stem tissue represented an intermediate state between leaf and flower expression profiles (also see Figure S1). The correlation of expression patterns in these three tissues may be due to their relatively dry lifestyle; in contrast, pneumatophores and seeds endure long seawater immersion periods. Metabolic pathways related to the mitochondrial beta-oxidation pathway were prominently featured among the unique KEGG terms. This process is a major source of energy from the degradation of fatty acids, which are abundant in stems and seeds.

Many highly expressed, stem-specific transcripts had putative antioxidant functions (Table S1). Although metallothioneins (MTs) were among the most highly expressed genes in all five tissues, stems had the widest variety of MT types expressed. The highest-expressed stem transcript, an MT-2 (AM_26466; TPM = 30,600), had more than twice as high the expression of the next-ranked gene (Uncharacterized protein; AM_04606; TPM = 4708). A BlastP search with the 262 amino acid protein from AM_04606 yielded a match with a homolog in *Olea europaea var. sylvestris* (E-value = $9e \times 10^{-77}$, 47.04% identity) that was annotated as an acid phosphatase. An Interproscan (ebi.ac.uk/interpro) search revealed that AM_04606 was a haloacid dehydrogenase (HAD). The HAD superfamily includes a variety of detoxification-related proteins [130]. These proteins respond to heavy metal (e.g., cadmium [131]) and salinity and alkaline [37] stress in other angiosperms, and their high expression here indicates that the HAD salinity and alkaline stress response mechanisms are conserved and amplified in *A. marina*. The implementation of HAD in transgenic, phosphate-deficient wild-type rice (*Oryza sativa*) [132] led to phosphate-accumulating rice able to overcome nutrient shortages.

Dehydrins are essential for plant response and adaptation to abiotic stresses. Dehydrins are highly hydrophilic, thermostable proteins that accumulate in vegetative plant tissues during drought or salt stress, or in seeds during maturation and drying [133]. A hydrophilic, glycine-rich dehydrin (AM_31570; DHN1; TPM = 2700) was highly expressed in stem tissue. These short (~100 amino acid) proteins are induced in response to dehydration, elevated salt, and low temperature [134]. Our data suggest that they preferentially occupy hardy stem tissue in *A. marina*. Although two lignin-forming anionic peroxidases were unique to stems (AM_07165, AM_07166), lignin glucosyltransferases (AM_31811, AM_31812) were uniquely expressed in pneumatophores. The high expression of HAD, MT, and other detoxification and antioxidant transcripts in stems (e.g., HSP70 (AM_04996; TPM = 1107), gamma-thionin/defensin (AM_11821; TPM = 1204), and thioredoxin (AM_24260; TPM = 1870.9)) outline the genetic mechanisms for abiotic stress resistance in *A. marina*.

3.7. Pneumatophore-Specific Transcript Expression

The pneumatophores of *A. marina* develop in oxygen-poor soils and filter out salt while absorbing water, essential nutrients, and oxygen [135]. They grow as pencil-width tubes to penetrate dense, low-oxygen mud (Figure 1). We found 554 genes uniquely expressed (>10 TPM) in *A. marina* pneumatophores compared to the other four tissues from the Red Sea cultivar, BioProject: PRJNA591919. Transcripts with putative roles in lignin biosynthesis outline a route to hardy, successful pneumatophore formation necessary for roots in low-oxygen soils. The pneumatophore transcriptome had the unique expression of several genes with putative roles in halogen degradation, indicating a possible mechanism to degrade otherwise harmful environmental xenobiotics.

A. marina must cope with salt stress while absorbing water and nutrients from the surroundings. A key mechanism for absorbing water while repelling excess salt is ultrafiltration, a process carried out by aquaporins and similar integral membrane transporters [42,136–140]. Six transcripts coding for aquaporins (NIP1.1) were expressed in pneumatophores: AM_29290 and AM_02471 were expressed at >10 TPM; AM_29291, AM_24990, AM_32021, and AM_01631 were expressed at sub-threshold levels. These aquaporins likely contribute to *A. marina*'s uptake of water from hypersaline waters, although their full ecophysiology is yet to be elucidated. Various transporter proteins, including H^+ -ATPases, ATP-binding cassette transporters, and aquaporins, were found in proteomic studies of *A. officinalis* membranes [141]. NIP1.1 aquaporin expression was found to contribute to hydrogen peroxide sensitivity in *Arabidopsis* [142]; thus, *A. marina* may tightly regulate the expression of its NIP1.1 isoforms to maintain osmotic balance. The low expression of four NIP1.1 genes in *A. marina* may reflect this delicate balance between water uptake and membrane lipid peroxidation mitigation. Aquaporins are essential for water filtration in *A. marina* [29,42,57,136–143]; our data suggest that low levels of expression are sufficient in natural conditions.

The *Avicennia*–seawater interface is dominated by thick mucilages, comprising the pneumatophore extra cellular matrix (ECM). Avicennia ECMs support the maintenance of osmotic balance between the seawater and inner tissues. A xyloglucan endo-transglycosylase (AM_08706) was uniquely expressed in pneumatophores (TPM = 134.97). These enzymes remodel cell wall hemicellulose molecules and likely help to generate salt-resistant ECMs for Avicennia.

The chemistry of mangrove root environments is highly reducing and rich in hydrogen sulfides and other toxic molecules [31,144]. Furthermore, they are highly anaerobic compared with the root environments of most other angiosperms that are waterlogged in their natural environments. For example, poplar has a robust short-term waterlogging response, invoking hydrological conductance detours in a shift to fermentative metabolism [145]. *A. marina* root environments are hostile to most angiosperms and necessitate robust molecular defenses by plant roots. Here, we investigated the pneumatophore-specific expression for responsible genes.

The *A. marina* rhizosphere is an assemblage of microorganisms exchanging nutrient and waste molecules and providing physical, stabilizing interactions for the *A. marina* pneumatophore system [14]. The *A. marina* rhizosphere is scantly studied, although microbiome interactions, and potential fauna interactions, have critical roles in *A. marina* pneumatophores and their greater mangrove ecosystems. The assimilation of other nutrients, especially organic molecules that may require processing for a bioactive form, is likely facilitated by halotolerant pneumatophore microbiome community members [3,14,15]. Uniquely expressed transcripts in pneumatophores included genes predicted with roles in *A. marina*'s rhizosphere interactions (as detailed in Reference [3], e.g., using purine (nitrogen) transporters and cytokines). An aspartate-semialdehyde dehydrogenase (ASD, AM_32125, EC 1.2.1.11) was uniquely expressed in pneumatophores (TPM = 11.7). The ASD enzyme is not well-conserved in higher plants [146–151]; but, its presence indicates potential rhizosphere feeding strategies. The ASD reaction represents a critical branch point in nitrogenous amino acid synthesis with possible involvement in nitrogenous compound transformation in relationship to surrounding microbial producers. It forms an early branch point in the biosynthetic pathway to lysine, methionine, leucine, and isoleucine from aspartate. The ASD also synthesizes microbial cell wall constituents, including diaminopimelate. Neighboring microbes may feed *A. marina* nitrogenous compounds with a payback of more complex molecular building blocks for specialized cell walls as in other plants [152–155]. Additionally, the biosynthesis of steroids and steroid precursor molecules was also prominently highlighted in our root-specific KEGG analyses, and these may form the building blocks for microbe–host communication in *A. marina*'s rhizosphere.

Protein-L-isoaspartate O-methyltransferase (AM_26147; EC:2.1.1.77) was highly expressed in pneumatophores (TPM = 110.9); PRIM proteins in *Arabidopsis thaliana* respond to salt and other abiotic stress [156]. *Arabidopsis* PRIMs promote seed longevity and germination vigor; however, a similar role in *A. marina* was not supported by our transcriptomic evidence. MTs are crucial components of the cellular antioxidant network [18,157]. Transgenic experiments using *A. marina* MTs in *E. coli* have considerably boosted its antioxidant capacity [119]. Our data suggest that the expression of these proteins should be considered the primary defense mechanism of *A. marina* against oxidative stress in each of the five tissues we sampled and in leaf tissues from cultivars in the Indian and Pacific oceans.

Although MTs were shared, high-expression outliers in most samples, a variety of other putative antioxidant genes were also highly expressed. A glutathione S-transferase (GST) transcript had relatively high expression (TPM = 4759) in pneumatophores. We previously reported the relatively high copy number of GSTs in coastal photosynthetic microbes [158]; the duplication and increased expression of GSTs may also facilitate coastal habitation. Protein families in the GO term 'Hydroquinone:oxygen oxidoreductase activity' were uniquely enriched in pneumatophores (Z-score = 7.69, FDR-corr. $p = 0.00079$). Quinone cofactors are crucial components of redox cycling; many quinone derivatives can form a redox cycle in conjunction with glutathione to generate oxygen [159]. Salt accelerates oxidative stress, and the salinity in the Red Sea can reach 40.5–41.8 ppt [13]. Strong redox buffer cycles, such as the quinone regeneration processed detected in our enrichment analysis, are necessary for oxidative stability at the tide-delimited border between anoxia and hyperoxia in *A. marina* pneumatophores.

3.8. Comparison of A. marina Transcriptomes across Geographies and Climates

Transcripts with the highest expression were common in *A. marina* samples from different geographic locations, solidifying the role of these critical proteins as supportive of *A. marina* growth and survival in diverse habitats (Figure 5). For example, AM_26466, a metallothionein (Huang and Wang, 2010), was highly expressed in nearly every sample. Interproscan searches with this small (72 amino acids) protein matched IPR000347, which has a class II metallothionein protein family domain (PF01439) and the GO molecular function 'metal ion binding' (GO:0046872). *A. marina* leaf samples from the Red Sea, India, and Vietnam showed higher expression of AM_26466 than heavy-metal-treated lab cultivars, indicating that the high expression of this gene is inherent to *A. marina*'s natural transcriptome. Another short protein (AM_31570, 203 amino acids) was predicted to function as a water-responsive dehydrin. The only protein observed to be highly expressed specifically in treated (either PEG or heavy metals) *A. marina* leaves was AM_20119 (GO terms: 'glycolytic process' (GO:0006096), 'fructose-bisphosphate aldolase activity' (GO:0004332), and 'catalytic activity' (GO:0003824)). The only high-molecular-weight (HMW) protein (1662 amino acids) with considerable expression (920.5 TPM) was annotated as a multifunctional redox/glutamate synthase, expressed exclusively in Red Sea leaves. A larger protein (AM_33275) was highly expressed in Red Sea *A. marina* leaves as well as heavy-metal and control leaves from Shanghai. The AM_33275 protein was predicted to have ATPase activity (IPR003959, PF00004) and membrane localization (SignalP-TM).

Transcripts coding for MTs were highly expressed in *A. marina* leaves from distant geographies and habitats. These proteins regulate heavy metal homeostasis in the human immune system [123] and act to chelate toxic metals in plants and their microbial community members [118,120,122,124,160]. Several studies have highlighted the potential for bioaccumulation of pollutants, especially heavy metals, in the Red Sea [161–164]. The toxicity of heavy metals near the Jeddah coast has been documented in fish [161–164]. Heavy metals cause damaging oxidative stress to cells through redox interactions to form free radicals [165–167]. Various biomolecules, including MTs [118,120,168], function to protect cells from damaging free radicals. The ubiquitous high expression of MTs in *A. marina* transcriptomes from various sources highlights its importance for survival across a broad range of abiotic stressors.

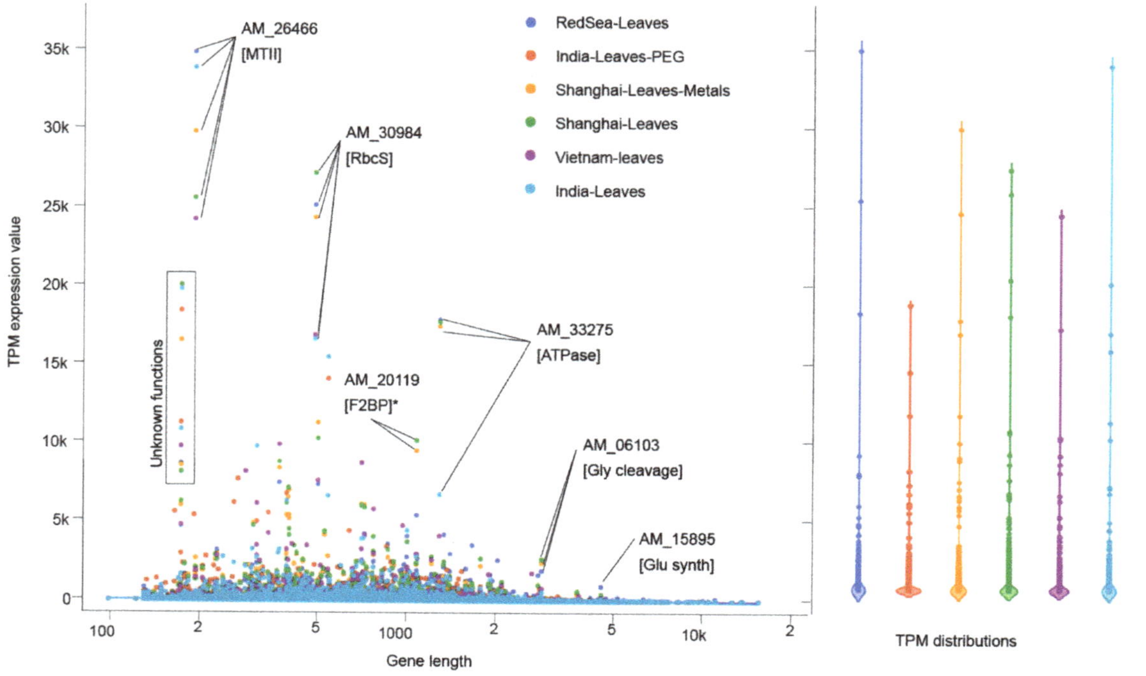

Figure 5. Comparison of the *A. marina* expression profiles in distant Indo-Pacific samples. Expression values for *A. marina* leaves from the Red Sea and India, Shanghai, and Vietnam coasts, as well as their distributions, are shown along the y-axis. High expression outliers are noted with annotation names and functions (MTII = type-2 metallothionein; RbcS = rubisco small subunit; F2BP = fructose 2,6-bisphospatase; Gly cleavage = glycine cleavage; Glu synth = glutamine synthesis). Metallothioneins and rubisco subunits had the highest expression in most samples, indicating that an antioxidant defense system fueled by high carbon input is characteristic of *A. marina* independent of geography or climate.

We sought to discover the genetic bases for the unique phenotypes of *A. marina*'s organs. Overall, the RNAseq data yields insight into how tissue-specific expression forms the basis for the stress resistance and broad propagation of *A. marina*. Genes such as dehydrins and metallothioneins form the backbone of *Avicennia*'s environmental resilience, and our data shows that their expression is high across a variety of tissues and geographical locations. Our analyses showed quite divergent expression profiles in the different tissues, and this reveals that many collections of expressed genes are missed when sampling single tissues for organismal analysis. Sampling multiple tissues will need to become a more prevalent practice in mangrove studies, to establish a comprehensive overview of these model halophytes.

4. Conclusions

We investigated the transcriptomic profiles of *A. marina* tissues and from different ecotypes to understand the genetic basis for its ability to survive a seafaring lifestyle. The expression of genes such as MTs, dehydrins, salt efflux pumps, cytochromes, HADs, and lignin glucosyltransferases were prominent features in the tissues implicated in sustaining growth in the face of seawater immersion. The foremost expression of MTs in all samples suggests their primary role in halotolerance. Although many of the *A. marina* genes we described have biotechnology potential for saline agriculture, optimizing the use of transgenic MTs should provide a strong backbone for non-halotolerant plants to tolerate seawater irrigation. In conclusion, we demonstrated the distinct transcripts and functions that can be used for future research into the use of *A. marina*, to determine their potential application in improving plant stress tolerance. Our analyses of tissue-specific analyses of *A. marina* transcriptomes will constitute a valuable resource for researching specific processes, functional descriptions, and pathways.

Supplementary Materials: The following supporting information can be downloaded at: https://www.mdpi.com/article/10.3390/agronomy12092030/s1, Figure S1: Circos and UMAP plots of tissue-specific transcriptomes. Figure S2: UpsetR [73] plot highlighting shared genes expressed at >10 TPM in the five tissues from the Red Sea cultivar. Figure S3: Expression patterns of lipid metabolic genes in the five Red Sea tissues. Table S1: Differentially-expressed genes mapped to the reference CDSs. Table S2: Expression values (TPM) for the mapping of external datasets (.fastqs) downloaded from NCBI. Table S3: dcGO enrichment from PFAMs in tissue-specific, uniquely-expressed transcripts (≥10 TPM). And Supplementary datasets for "Tissue-specific transcriptomes outline halophyte adaptive strategies in the gray mangrove, Avicennia marina" are available at https://zenodo.org/record/6793097#.YsJ7xexBxew (accessed on 3 July 2022).

Author Contributions: Conceptualization, K.S.-A., B.K. and D.R.N.; methodology, A.C., D.R.N., A.C. and K.S.-A.; software, A.C. and D.R.N.; formal analysis, D.R.N. and K.M.H.; investigation, D.R.N. and K.M.H.; resources, K.S.-A., J.A.B. and K.M.A.A.; data curation, A.J. and D.R.N.; writing—original draft preparation, A.A. and D.R.N.; writing—review and editing, J.A.B., G.F., D.R.N. and K.S.-A.; visualization, D.R.N.; supervision, K.S.-A., K.M.A.A. and J.A.B.; project administration, K.S.-A., J.A.B.; D.R.N. and A.C. contributed equally to this paper. All authors have read and agreed to the published version of the manuscript.

Funding: This research was supported by KAUST Laboratory Baseline Funds and by New York University Abu Dhabi (NYUAD) Faculty Research fund AD060 and NYUAD Research Institute grant 73 71210 CGSB9. JB was supported by the NYUAD Water Research Center grant CG007, while GF was supported by the NYUAD-CGSB grant CGSB5.

Data Availability Statement: Supplementary Dataset S1 are available at Zenodo (10.5281/zenodo.6793097). The underlying RNAseq datasets used for the analyses in this study are publicly available. We used public RNAseq data from *A. marina* specimens from the Red Sea (NCBI BioProject: PRJNA591919), Vietnam (NCBI Bioproject: PRJDB6605, Department of Biotechnology, Graduate School of Engineering, Osaka University, Japan), India (NCBI Bioproject: PRJNA283781, M.S. Swaminathan Research Foundation, Chennai, India), and China (NCBI Bioproject: PRJNA350568, Lingnan Normal University, Zhanjiang, China).

Acknowledgments: We thank the New York University Abu Dhabi High-Performance Computing Facility for computational resources. We thank Alexey Sergeev for permission to use the *Avicennia* photographs and NASA for the Jeddah satellite image. This manuscript is dedicated to the memory of Basel Khraiwesh.

Conflicts of Interest: The authors declare no conflict of interest.

References

1. Cavanaugh, K.C.; Dangremond, E.M.; Doughty, C.L.; Williams, A.P.; Parker, J.D.; Hayes, M.A.; Rodriguez, W.; Feller, I.C. Climate-driven regime shifts in a mangrove-salt marsh ecotone over the past 250 years. *Proc. Natl. Acad. Sci. USA* **2019**, *116*, 21602–21608. [CrossRef]
2. Charles, S.P.; Kominoski, J.S.; Armitage, A.R.; Guo, H.; Weaver, C.A.; Pennings, S.C. Quantifying how changing mangrove cover affects ecosystem carbon storage in coastal wetlands. *Ecology* **2019**, *101*, e02916. [CrossRef] [PubMed]
3. Inoue, T.; Shimono, A.; Akaji, Y.; Baba, S.; Takenaka, A.; Tuck Chan, H. Mangrove-diazotroph relationships at the root, tree and forest scales: Diazotrophic communities create high soil nitrogenase activities in Rhizophora stylosa rhizospheres. *Ann. Bot.* **2019**, *125*, 131–144. [CrossRef] [PubMed]
4. Sasmito, S.D.; Taillardat, P.; Clendenning, J.N.; Cameron, C.; Friess, D.A.; Murdiyarso, D.; Hutley, L.B. Effect of land-use and land-cover change on mangrove blue carbon: A systematic review. *Glob. Change Biol.* **2019**, *25*, 4291–4302. [CrossRef] [PubMed]
5. Theuerkauff, D.; Rivera-Ingraham, G.A.; Lambert, S.; Mercky, Y.; Lejeune, M.; Lignot, J.H.; Sucre, E. Wastewater bioremediation by mangrove ecosystems impacts crab ecophysiology: In-situ caging experiment. *Aquat. Toxicol.* **2019**, *218*, 105358. [CrossRef] [PubMed]

6. He, Z.; Li, X.; Yang, M.; Wang, X.; Zhong, C.; Duke, N.C.; Wu, C.-I.; Shi, S. Speciation with gene flow via cycles of isolation and migration: Insights from multiple mangrove taxa. *Natl. Sci. Rev.* **2019**, *6*, 275–288. [CrossRef] [PubMed]
7. He, Z.; Xu, S.; Zhang, Z.; Guo, W.; Lyu, H.; Zhong, C.; Boufford, D.E.; Duke, N.C.; Shi, S.; The International Mangrove Consortium. Convergent adaptation of the genomes of woody plants at the land–sea interface. *Natl. Sci. Rev.* **2020**, *7*, 978–993. [CrossRef]
8. Arnaud-Haond, S.; Teixeira, S.; Massa, S.I.; Billot, C.; Saenger, P.; Coupland, G.; Duarte, C.M.; Serrão, E.A. Genetic structure at range edge: Low diversity and high inbreeding in Southeast Asian mangrove (*Avicennia marina*) populations. *Mol. Ecol.* **2006**, *15*, 3515–3525. [CrossRef]
9. Dodd, R.S.; Afzal-Rafii, Z.; Kashani, N.; Budrick, J. Land barriers and open oceans: Effects on gene diversity and population structure in *Avicennia germinans* L. (Avicenniaceae). *Mol. Ecol.* **2002**, *11*, 1327–1338. [CrossRef]
10. Hazarika, D.; Thangaraj, M.; Sahu, S.K.; Kathiresan, K. Genetic diversity in three populations of *Avicennia marina* along the eastcoast of India by RAPD markers. *J. Environ. Biol.* **2013**, *34*, 663–666.
11. Maguire, T.L.; Peakall, R.; Saenger, P. Comparative analysis of genetic diversity in the mangrove species *Avicennia marina* (Forsk.) Vierh. (Avicenniaceae) detected by AFLPs and SSRs. *Theor. Appl. Genet.* **2002**, *104*, 388–398. [CrossRef]
12. Friis, G.; Burt, J.A. Evolution of mangrove research in an extreme environment: Historical trends and future opportunities in Arabia. *Ocean Coast. Manag.* **2020**, *195*, 105288. [CrossRef]
13. Almahasheer, H.; Duarte, C.M.; Irigoien, X. Phenology and Growth dynamics of *Avicennia marina* in the Central Red Sea. *Sci. Rep.* **2016**, *6*, 37785. [CrossRef]
14. Alzubaidy, H.; Essack, M.; Malas, T.B.; Bokhari, A.; Motwalli, O.; Kamanu, F.K.; Jamhor, S.A.; Mokhtar, N.A.; Antunes, A.; Simoes, M.F.; et al. Rhizosphere microbiome metagenomics of gray mangroves (*Avicennia marina*) in the Red Sea. *Gene* **2016**, *576*, 626–636. [CrossRef]
15. Chen, C.; Han, S.; Zhu, Z.; Fu, G.; Wang, R.; Zhang, Q.; Ye, Y.; Ren, Y.; Yan, C.; Xu, L.; et al. Idiomarina mangrovi sp. nov., isolated from rhizosphere soil of a mangrove *Avicennia marina* forest. *Int. J. Syst. Evol. Microbiol.* **2019**, *69*, 1662–1668. [CrossRef]
16. Clarke, P.J.; Allaway, W.G. The regeneration niche of the grey mangrove (*Avicennia marina*): Effects of salinity, light and sediment factors on establishment, growth and survival in the field. *Oecologia* **1993**, *93*, 548–556. [CrossRef]
17. Ganesan, G.; Sankararamasubramanian, H.; Narayanan, J.M.; Sivaprakash, K.; Parida, A. Transcript level characterization of a cDNA encoding stress regulated NAC transcription factor in the mangrove plant *Avicennia marina*. *Plant Physiol. Biochem.* **2008**, *46*, 928–934. [CrossRef]
18. Huang, G.Y.; Wang, Y.S. Expression and characterization analysis of type 2 metallothionein from grey mangrove species (*Avicennia marina*) in response to metal stress. *Aquat. Toxicol.* **2010**, *99*, 86–92. [CrossRef]
19. Khraiwesh, B.; Pugalenthi, G.; Fedoroff, N.V. Identification and analysis of red sea mangrove (*Avicennia marina*) microRNAs by high-throughput sequencing and their association with stress responses. *PloS ONE* **2013**, *8*, e60774. [CrossRef]
20. Schmitz, N.; Robert, E.M.; Verheyden, A.; Kairo, J.G.; Beeckman, H.; Koedam, N. A patchy growth via successive and simultaneous cambia: Key to success of the most widespread mangrove species *Avicennia marina*? *Ann. Bot.* **2008**, *101*, 49–58. [CrossRef]
21. Simoes, M.F.; Antunes, A.; Ottoni, C.A.; Amini, M.S.; Alam, I.; Alzubaidy, H.; Mokhtar, N.A.; Archer, J.A.; Bajic, V.B. Soil and Rhizosphere Associated Fungi in Gray Mangroves (*Avicennia marina*) from the Red Sea–A Metagenomic Approach. *Genom. Proteom. Bioinform.* **2015**, *13*, 310–320. [CrossRef] [PubMed]
22. Usman, A.R.; Alkredaa, R.S.; Al-Wabel, M.I. Heavy metal contamination in sediments and mangroves from the coast of Red Sea: *Avicennia marina* as potential metal bioaccumulator. *Ecotoxicol. Environ. Saf.* **2013**, *97*, 263–270. [CrossRef] [PubMed]
23. Mildenhall, D.C.; Brown, L.J. An early Holocene occurrence of the mangrove *Avicennia marina* in Poverty Bay, North Island, New Zealand: Its climatic and geological implications. *N. Z. J. Bot.* **1987**, *25*, 281–294. [CrossRef]
24. Schmitz, N.; Verheyden, A.; Kairo, J.G.; Beeckman, H.; Koedam, N. Successive cambia development in *Avicennia marina* (Forssk.) Vierh. is not climatically driven in the seasonal climate at Gazi Bay, Kenya. *Dendrochronologia* **2007**, *25*, 87–96. [CrossRef]
25. Cheng, B.; Li, H. Impact of climate change and human activities on economic values produced by ecosystem service functions of rivers in water shortage area of Northwest China. *Environ. Sci. Pollut. Res. Int.* **2020**, *27*, 26570–26578. [CrossRef]
26. Cheng, H.; Inyang, A.; Li, C.D.; Fei, J.; Zhou, Y.W.; Wang, Y.S. Salt tolerance and exclusion in the mangrove plant *Avicennia marina* in relation to root apoplastic barriers. *Ecotoxicology* **2020**, *29*, 676–683. [CrossRef]
27. Jyothi-Prakash, P.A.; Mohanty, B.; Wijaya, E.; Lim, T.-M.; Lin, Q.; Loh, C.-S.; Kumar, P.P. Identification of salt gland-associated genes and characterization of a dehydrin from the salt secretor mangrove Avicennia officinalis. *BMC Plant Biol.* **2014**, *14*, 291. [CrossRef]
28. Song, J.; Wang, B. Using euhalophytes to understand salt tolerance and to develop saline agriculture: Suaeda salsa as a promising model. *Ann. Bot.* **2015**, *115*, 541–553. [CrossRef]
29. Tan, W.-K.; Lin, Q.; Lim, T.-M.; Kumar, P.; Loh, C.-S. Dynamic secretion changes in the salt glands of the mangrove tree species Avicennia officinalis in response to a changing saline environment. *Plant Cell Environ.* **2013**, *36*, 1410–1422. [CrossRef]
30. Krishnamurthy, P.; Mohanty, B.; Wijaya, E.; Lee, D.Y.; Lim, T.M.; Lin, Q.; Xu, J.; Loh, C.S.; Kumar, P.P. Transcriptomics analysis of salt stress tolerance in the roots of the mangrove Avicennia officinalis. *Sci. Rep.* **2017**, *7*, 10031. [CrossRef]
31. Kryger, L.; Lee, S.K. Effects of Mangrove Soil Ageing on the Accumulation of Hydrogen Sulphide in Roots of *Avicennia* spp. *Biogeochemistry* **1996**, *35*, 367–375. [CrossRef]
32. Arshad, M.; Eid, E.M.; Hasan, M. Mangrove health along the hyper-arid southern Red Sea coast of Saudi Arabia. *Environ. Monit. Assess.* **2020**, *192*, 189. [CrossRef]

33. Masoud, M.S.; Abdel-Halim, A.M.; El Ashmawy, A.A. Seasonal variation of nutrient salts and heavy metals in mangrove (*Avicennia marina*) environment, Red Sea, Egypt. *Environ. Monit. Assess.* **2019**, *191*, 425. [CrossRef]
34. Gao, X.; Liu, Y.; Sun, B. Water shortage risk assessment considering large-scale regional transfers: A copula-based uncertainty case study in Lunan, China. *Environ. Sci. Pollut. Res. Int.* **2018**, *25*, 23328–23341. [CrossRef]
35. Kummu, M.; Guillaume, J.H.; de Moel, H.; Eisner, S.; Florke, M.; Porkka, M.; Siebert, S.; Veldkamp, T.I.; Ward, P.J. The world's road to water scarcity: Shortage and stress in the 20th century and pathways towards sustainability. *Sci. Rep.* **2016**, *6*, 38495. [CrossRef]
36. Nabi, G.; Ali, M.; Khan, S.; Kumar, S. The crisis of water shortage and pollution in Pakistan: Risk to public health, biodiversity, and ecosystem. *Environ. Sci. Pollut. Res. Int.* **2019**, *26*, 10443–10445. [CrossRef]
37. Xu, Z.; Cai, X.; Yin, X.; Su, M.; Wu, Y.; Yang, Z. Is water shortage risk decreased at the expense of deteriorating water quality in a large water supply reservoir? *Water Res.* **2019**, *165*, 114984. [CrossRef]
38. Hayat, K.; Zhou, Y.; Menhas, S.; Bundschuh, J.; Hayat, S.; Ullah, A.; Wang, J.; Chen, X.; Zhang, D.; Zhou, P. Pennisetum giganteum: An emerging salt accumulating/tolerant non-conventional crop for sustainable saline agriculture and simultaneous phytoremediation. *Environ. Pollut.* **2020**, *265*, 114876. [CrossRef]
39. Illera-Vives, M.; Lopez-Mosquera, M.E.; Salas-Sanjuan Mdel, C.; Lopez-Fabal, A. Leaching techniques for saline wastes composts used as growing media in organic agriculture: Assessment and modelling. *Environ. Sci. Pollut. Res. Int.* **2015**, *22*, 6854–6863. [CrossRef]
40. Jin, Y.; Weining, S.; Nevo, E. A MAPK gene from Dead Sea fungus confers stress tolerance to lithium salt and freezing-thawing: Prospects for saline agriculture. *Proc. Natl. Acad. Sci. USA* **2005**, *102*, 18992–18997. [CrossRef]
41. Qadir, M.; Oster, J.D. Crop and irrigation management strategies for saline-sodic soils and waters aimed at environmentally sustainable agriculture. *Sci. Total Environ.* **2004**, *323*, 1–19. [CrossRef]
42. Qin, S.; Liu, Y.; Han, Y.; Xu, G.; Wan, S.; Cui, F.; Li, G. Aquaporins and their function in root water transport under salt stress conditions in *Eutrema salsugineum*. *Plant Sci.* **2019**, *287*, 110199. [CrossRef]
43. Guo, R.Z.; Yan, H.Y.; Li, X.X.; Zou, X.X.; Zhang, X.J.; Yu, X.N.; Ci, D.W.; Wang, Y.F.; Si, T. Green leaf volatile (Z)-3-hexeny-1-yl acetate reduces salt stress in peanut by affecting photosynthesis and cellular redox homeostasis. *Physiol. Plant* **2020**, *10*, 785. [CrossRef]
44. Lekklar, C.; Suriya-Arunroj, D.; Pongpanich, M.; Comai, L.; Kositsup, B.; Chadchawan, S.; Buaboocha, T. Comparative genomic analysis of rice with contrasting photosynthesis and grain production under salt stress. *Genes* **2019**, *10*, 562. [CrossRef]
45. Nounjan, N.; Chansongkrow, P.; Charoensawan, V.; Siangliw, J.L.; Toojinda, T.; Chadchawan, S.; Theerakulpisut, P. High Performance of Photosynthesis and Osmotic Adjustment Are Associated With Salt Tolerance Ability in Rice Carrying Drought Tolerance QTL: Physiological and Co-expression Network Analysis. *Front. Plant Sci.* **2018**, *9*, 1135. [CrossRef]
46. Wang, Z.; Yang, H.; Ramesh, A.; Roberts, L.J., 2nd; Zhou, L.; Lin, X.; Zhao, Y.; Guo, Z. Overexpression of Cu/Zn-superoxide dismutase and/or catalase accelerates benzo(a)pyrene detoxification by upregulation of the aryl hydrocarbon receptor in mouse endothelial cells. *Free Radic. Biol. Med.* **2009**, *47*, 1221–1229. [CrossRef]
47. Wasai, S.; Kanno, N.; Matsuura, K.; Haruta, S. Increase of Salt Tolerance in Carbon-Starved Cells of Rhodopseudomonas palustris Depending on Photosynthesis or Respiration. *Microorganisms* **2018**, *6*, 4. [CrossRef]
48. Zhang, Z.; Li, X.; Cui, L.; Meng, S.; Ye, N.; Peng, X. Catalytic and functional aspects of different isozymes of glycolate oxidase in rice. *BMC Plant Biol.* **2017**, *17*, 135. [CrossRef]
49. Chang, B.; Yang, L.; Cong, W.; Zu, Y.; Tang, Z. The improved resistance to high salinity induced by trehalose is associated with ionic regulation and osmotic adjustment in Catharanthus roseus. *Plant Physiol. Biochem.* **2014**, *77*, 140–148. [CrossRef]
50. Frank, W.; Ratnadewi, D.; Reski, R. Physcomitrella patens is highly tolerant against drought, salt and osmotic stress. *Planta* **2005**, *220*, 384–394. [CrossRef]
51. Pires, R.M.O.; Avila, M.A.B.; Leite, D.G.; Santos, H.O.; Souza, G.A.; Von Pinho, E.V.R. Physiological and enzymatic alterations in sesame seeds submitted to different osmotic potentials. *Genet. Mol. Res.* **2017**, *16*, gmr16039425. [CrossRef] [PubMed]
52. Empitu, M.A.; Kadariswantiningsih, I.N. Modelling salt transport disorders of human kidney in zebrafish: The grain of salt. *J. Physiol.* **2019**, *597*, 5529–5530. [CrossRef] [PubMed]
53. Grosell, M.; Heuer, R.M.; Wu, N.C.; Cramp, R.L.; Wang, Y.; Mager, E.M.; Dwyer, R.G.; Franklin, C.E. Salt-water acclimation of the estuarine crocodile *Crocodylus porosus* involves enhanced ion transport properties of the urodaeum and rectum. *J. Exp. Biol.* **2020**, *223*, jeb210732. [CrossRef] [PubMed]
54. Stolting, G.; Fahlke, C. Chloride channels in renal salt and water transport. *Acta Physiol.* **2017**, *219*, 11–13. [CrossRef]
55. Wu, H.; Li, Z. The Importance of Cl^- Exclusion and Vacuolar Cl^- Sequestration: Revisiting the Role of Cl^- Transport in Plant Salt Tolerance. *Front. Plant Sci.* **2019**, *10*, 1418. [CrossRef]
56. Zhang, Y.; Fang, J.; Wu, X.; Dong, L. Na^+/K^+ Balance and Transport Regulatory Mechanisms in Weedy and Cultivated Rice (*Oryza sativa* L.) Under Salt Stress. *BMC Plant Biol.* **2018**, *18*, 375. [CrossRef]
57. Zhou, Y.; Tao, J.; Ahammed, G.J.; Li, J.; Yang, Y. Genome-wide identification and expression analysis of aquaporin gene family related to abiotic stress in watermelon. *Genome* **2019**, *62*, 643–656. [CrossRef]
58. Severi, E.; Hosie, A.H.; Hawkhead, J.A.; Thomas, G.H. Characterization of a novel sialic acid transporter of the sodium solute symporter (SSS) family and in vivo comparison with known bacterial sialic acid transporters. *FEMS Microbiol. Lett.* **2010**, *304*, 47–54. [CrossRef]

59. Friis, G.; Vizueta, J.; Smith, E.G.; Nelson, D.R.; Khraiwesh, B.; Qudeimat, E.; Salehi-Ashtiani, K.; Ortega, A.; Marshell, A.; Duarte, C.M.; et al. A high-quality genome assembly and annotation of the gray mangrove, *Avicennia marina*. *G3 Genes Genomes Genet.* **2021**, *11*, jkaa025. [CrossRef]
60. Kim, D.; Paggi, J.M.; Park, C.; Bennett, C.; Salzberg, S.L. Graph-based genome alignment and genotyping with HISAT2 and HISAT-genotype. *Nat. Biotechnol.* **2019**, *37*, 907–915. [CrossRef]
61. Pertea, M.; Kim, D.; Pertea, G.M.; Leek, J.T.; Salzberg, S.L. Transcript-level expression analysis of RNA-seq experiments with HISAT, StringTie and Ballgown. *Nat. Protoc.* **2016**, *11*, 1650–1667. [CrossRef]
62. Mortazavi, A.; Williams, B.A.; McCue, K.; Schaeffer, L.; Wold, B. Mapping and quantifying mammalian transcriptomes by RNA-Seq. *Nat. Meth.* **2008**, *5*, 621–628. [CrossRef]
63. Conesa, A.; Madrigal, P.; Tarazona, S.; Gomez-Cabrero, D.; Cervera, A.; McPherson, A.; Szczesniak, M.W.; Gaffney, D.J.; Elo, L.L.; Zhang, X.; et al. A survey of best practices for RNA-seq data analysis. *Genome Biol.* **2016**, *17*, 13. [CrossRef]
64. Stark, R.; Grzelak, M.; Hadfield, J. RNA sequencing: The teenage years. *Nat. Rev. Genet.* **2019**, *20*, 631–656. [CrossRef]
65. Wang, Z.; Gerstein, M.; Snyder, M. RNA-Seq: A revolutionary tool for transcriptomics. *Nat. Rev. Genet.* **2009**, *10*, 57–63. [CrossRef]
66. Lee, S.; Seo, C.H.; Lim, B.; Yang, J.O.; Oh, J.; Kim, M.; Lee, S.; Lee, B.; Kang, C.; Lee, S. Accurate quantification of transcriptome from RNA-Seq data by effective length normalization. *Nucleic Acids Res.* **2011**, *39*, e9. [CrossRef]
67. Li, P.; Piao, Y.; Shon, H.S.; Ryu, K.H. Comparing the normalization methods for the differential analysis of Illumina high-throughput RNA-Seq data. *BMC Bioinform.* **2015**, *16*, 347. [CrossRef]
68. Wu, P.Y.; Phan, J.H.; Zhou, F.; Wang, M.D. Evaluation of Normalization Methods for RNA-Seq Gene Expression Estimation. In Proceedings of the 2011 IEEE International Conference on Bioinformatics and Biomedicine Workshops (BIBMW), Atlanta, GA, USA, 12–15 November 2011; pp. 50–57. [CrossRef]
69. Zyprych-Walczak, J.; Szabelska, A.; Handschuh, L.; Gorczak, K.; Klamecka, K.; Figlerowicz, M.; Siatkowski, I. The Impact of Normalization Methods on RNA-Seq Data Analysis. *Biomed. Res. Int.* **2015**, *2015*, 621690. [CrossRef]
70. McInnes, L.; Healy, J.; Melville, J. UMAP: Uniform Manifold Approximation and Projection for Dimension Reduction. *arXiv* **2018**, arXiv:1802.03426.
71. Warnes, G.R.; Bolker, B.; Bonebakker, L.; Gentleman, R.; Huber, W.; Liaw, A.; Lumley, T.; Maechler, M.; Magnusson, A.; Moeller, S. gplots: Various R Programming Tools for Plotting Data. R Package Version 2009, 2. Available online: https://scholar.google.com/citations?view_op=view_citation&hl=en&user=o7GsfJwAAAAJ&citation_for_view=o7GsfJwAAAAJ:u-x6o8ySG0sC (accessed on 3 July 2022).
72. Heberle, H.; Meirelles, G.V.; da Silva, F.R.; Telles, G.P.; Minghim, R. InteractiVenn: A web-based tool for the analysis of sets through Venn diagrams. *BMC Bioinform.* **2015**, *16*, 169. [CrossRef]
73. Conway, J.R.; Lex, A.; Gehlenborg, N. UpSetR: An R package for the visualization of intersecting sets and their properties. *Bioinformatics* **2017**, *33*, 2938–2940. [CrossRef]
74. Fang, H.; Gough, J. A domain-centric solution to functional genomics via dcGO Predictor. *BMC Bioinform.* **2013**, *14* (Suppl. S3), S9. [CrossRef]
75. Krzywinski, M.I.; Schein, J.E.; Birol, I.; Connors, J.; Gascoyne, R.; Horsman, D.; Jones, S.J.; Marra, M.A. Circos: An information aesthetic for comparative genomics. *Genome Res.* **2009**, *19*, 1639–1645. [CrossRef]
76. Sonnhammer, E.L.; Eddy, S.R.; Durbin, R. Pfam: A comprehensive database of protein domain families based on seed alignments. *Proteins* **1997**, *28*, 405–420. [CrossRef]
77. Mistry, J.; Chuguransky, S.; Williams, L.; Qureshi, M.; Salazar, G.A.; Sonnhammer, E.L.L.; Tosatto, S.C.E.; Paladin, L.; Raj, S.; Richardson, L.J.; et al. Pfam: The protein families database in 2021. *Nucleic Acids Res.* **2021**, *49*, D412–D419. [CrossRef]
78. Landry, C.L. Pollinator-mediated competition between two co-flowering Neotropical mangrove species, Avicennia germinans (Avicenniaceae) and Laguncularia racemosa (Combretaceae). *Ann. Bot.* **2013**, *111*, 207–214. [CrossRef]
79. Liu, L.; Guo, Z.; Zhong, C.; Shi, S. DNA barcoding reveals insect diversity in the mangrove ecosystems of Hainan Island, China. *Genome* **2018**, *61*, 797–806. [CrossRef]
80. De Vetten, N.; ter Horst, J.; van Schaik, H.P.; de Boer, A.; Mol, J.; Koes, R. A cytochrome b5 is required for full activity of flavonoid 3', 5'-hydroxylase, a cytochrome P450 involved in the formation of blue flower colors. *Proc. Natl. Acad. Sci. USA* **1999**, *96*, 778–783. [CrossRef]
81. Holton, T.A.; Brugliera, F.; Lester, D.R.; Tanaka, Y.; Hyland, C.D.; Menting, J.G.; Lu, C.Y.; Farcy, E.; Stevenson, T.W.; Cornish, E.C. Cloning and expression of cytochrome P450 genes controlling flower colour. *Nature* **1993**, *366*, 276–279. [CrossRef]
82. Liu, Z.; Boachon, B.; Lugan, R.; Tavares, R.; Erhardt, M.; Mutterer, J.; Demais, V.; Pateyron, S.; Brunaud, V.; Ohnishi, T.; et al. A Conserved Cytochrome P450 Evolved in Seed Plants Regulates Flower Maturation. *Mol. Plant.* **2015**, *8*, 1751–1765. [CrossRef]
83. Martsinkovskaya, A.I.; Poghosyan, Z.P.; Haralampidis, K.; Murphy, D.J.; Hatzopoulos, P. Temporal and spatial gene expression of cytochrome B5 during flower and fruit development in olives. *Plant Mol. Biol.* **1999**, *40*, 79–90. [CrossRef] [PubMed]
84. Petkova-Andonova, M.; Imaishi, H.; Ohkawa, H. CYP92B1, A cytochrome P450, expressed in petunia flower buds, that catalyzes monooxidation of long-chain fatty acids. *Biosci. Biotechnol. Biochem.* **2002**, *66*, 1819–1828. [CrossRef] [PubMed]
85. Shahak, Y.; Posner, H.B.; Avron, M. Evidence for a Block between Plastoquinone and Cytochrome f in a Photosynthetic Mutant of Lemna with Abnormal Flowering Behavior. *Plant Physiol.* **1976**, *57*, 577–579. [CrossRef] [PubMed]

86. Ito, A.; Saito, T.; Nishijima, T.; Moriguchi, T. Effect of extending the photoperiod with low-intensity red or far-red light on the timing of shoot elongation and flower-bud formation of 1-year-old Japanese pear (*Pyrus pyrifolia*). *Tree Physiol.* **2014**, *34*, 534–546. [CrossRef]
87. Koornneef, M. Plant development: Timing when to flower. *Curr. Biol.* **1997**, *7*, R651–652. [CrossRef]
88. Lord, E.M.; Crone, W.; Hill, J.P. Timing of events during flower organogenesis: *Arabidopsis* as a model system. *Curr. Top Dev. Biol.* **1994**, *29*, 325–356. [CrossRef] [PubMed]
89. Wagner, D. Flower morphogenesis: Timing is key. *Dev. Cell.* **2009**, *16*, 621–622. [CrossRef] [PubMed]
90. Cai, X.; Davis, E.J.; Ballif, J.; Liang, M.; Bushman, E.; Haroldsen, V.; Torabinejad, J.; Wu, Y. Mutant identification and characterization of the laccase gene family in Arabidopsis. *J. Exp. Bot.* **2006**, *57*, 2563–2569. [CrossRef] [PubMed]
91. Salekeen, R.; Mou, S.N.; Islam, M.E.; Ahmed, A.; Billah, M.M.; Rahman, S.M.M.; Islam, K.M.D. Predicting multi-enzyme inhibition in the arachidonic acid metabolic network by Heritiera fomes extracts. *J. Biomol. Struct. Dyn.* **2022**, *40*, 4259–4272. [CrossRef]
92. Aubert, D.; Chevillard, M.; Dorne, A.M.; Arlaud, G.; Herzog, M. Expression patterns of GASA genes in *Arabidopsis thaliana*: The GASA4 gene is up-regulated by gibberellins in meristematic regions. *Plant Mol. Biol.* **1998**, *36*, 871–883. [CrossRef]
93. Roxrud, I.; Lid, S.E.; Fletcher, J.C.; Schmidt, E.D.; Opsahl-Sorteberg, H.G. GASA4, one of the 14-member Arabidopsis GASA family of small polypeptides, regulates flowering and seed development. *Plant Cell Physiol.* **2007**, *48*, 471–483. [CrossRef]
94. Rubinovich, L.; Weiss, D. The Arabidopsis cysteine-rich protein GASA4 promotes GA responses and exhibits redox activity in bacteria and in planta. *Plant J. Cell Mol. Biol.* **2010**, *64*, 1018–1027. [CrossRef]
95. Carvalho, A.T.; Dotterl, S.; Schlindwein, C. An aromatic volatile attracts oligolectic bee pollinators in an interdependent bee-plant relationship. *J. Chem. Ecol.* **2014**, *40*, 1126–1134. [CrossRef]
96. Ilc, T.; Parage, C.; Boachon, B.; Navrot, N.; Werck-Reichhart, D. Monoterpenol Oxidative Metabolism: Role in Plant Adaptation and Potential Applications. *Front. Plant Sci.* **2016**, *7*, 509. [CrossRef]
97. Awadasseid, A.; Li, W.; Liu, Z.; Qiao, C.; Pang, J.; Zhang, G.; Luo, Y. Characterization of Camptotheca acuminata 10-Hydroxygeraniol oxidoreductase and iridoid synthase and their application in biological preparation of nepetalactol in Escherichia coli featuring NADP$^+$-NADPH cofactors recycling. *Int. J. Biol. Macromol.* **2020**, *162*, 1076–1085. [CrossRef]
98. Carbonezi, C.A.; Martins, D.; Young, M.C.; Lopes, M.N.; Furlan, M.; Rodriguez Filho, E.; Bolzani Vda, S. Iridoid and seco-iridoid glucosides from Chioccoca alba (Rubiaceae). *Phytochemistry* **1999**, *51*, 781–785. [CrossRef]
99. Hofer, R.; Dong, L.; Andre, F.; Ginglinger, J.F.; Lugan, R.; Gavira, C.; Grec, S.; Lang, G.; Memelink, J.; Van der Krol, S.; et al. Geraniol hydroxylase and hydroxygeraniol oxidase activities of the CYP76 family of cytochrome P450 enzymes and potential for engineering the early steps of the (seco)iridoid pathway. *Metab. Eng.* **2013**, *20*, 221–232. [CrossRef]
100. Miettinen, K.; Dong, L.; Navrot, N.; Schneider, T.; Burlat, V.; Pollier, J.; Woittiez, L.; van der Krol, S.; Lugan, R.; Ilc, T.; et al. The seco-iridoid pathway from Catharanthus roseus. *Nat. Commun.* **2014**, *5*, 3606. [CrossRef] [PubMed]
101. Valletta, A.; Trainotti, L.; Santamaria, A.R.; Pasqua, G. Cell-specific expression of tryptophan decarboxylase and 10-hydroxygeraniol oxidoreductase, key genes involved in camptothecin biosynthesis in Camptotheca acuminata Decne (Nyssaceae). *BMC Plant Biol.* **2010**, *10*, 69. [CrossRef]
102. Willems, M. Quantitative Determination of Seco-Iridoid Glucosides from the Fruits of Ligustrum vulgare by HPLC. *Planta Med.* **1988**, *54*, 66–68. [CrossRef]
103. Do, B.T.N.; Koedam, N.; Triest, L. *Avicennia marina* maintains genetic structure whereas Rhizophora stylosa connects mangroves in a flooded, former inner sea (Vietnam). *Estuar. Coast. Shelf Sci.* **2019**, *222*, 195–204. [CrossRef]
104. Geng, Q.; Wang, Z.; Tao, J.; Kimura, M.K.; Liu, H.; Hogetsu, T.; Lian, C. Ocean Currents Drove Genetic Structure of Seven Dominant Mangrove Species Along the Coastlines of Southern China. *Front. Genet.* **2021**, *12*, 615911. [CrossRef] [PubMed]
105. Rademacher, S.; Sprunck, S. Downregulation of egg cell-secreted EC1 is accompanied with delayed gamete fusion and polytubey. *Plant Signal. Behav.* **2013**, *8*, e27377. [CrossRef] [PubMed]
106. Sprunck, S.; Rademacher, S.; Vogler, F.; Gheyselinck, J.; Grossniklaus, U.; Dresselhaus, T. Egg cell-secreted EC1 triggers sperm cell activation during double fertilization. *Science* **2012**, *338*, 1093–1097. [CrossRef] [PubMed]
107. Deveson, I.W.; Holleley, C.E.; Blackburn, J.; Marshall Graves, J.A.; Mattick, J.S.; Waters, P.D.; Georges, A. Differential intron retention in Jumonji chromatin modifier genes is implicated in reptile temperature-dependent sex determination. *Sci. Adv.* **2017**, *3*, e1700731. [CrossRef] [PubMed]
108. Gan, E.S.; Xu, Y.; Wong, J.Y.; Goh, J.G.; Sun, B.; Wee, W.Y.; Huang, J.; Ito, T. Jumonji demethylases moderate precocious flowering at elevated temperature via regulation of FLC in Arabidopsis. *Nat. Commun.* **2014**, *5*, 5098. [CrossRef]
109. Zheng, S.; Hu, H.; Ren, H.; Yang, Z.; Qiu, Q.; Qi, W.; Liu, X.; Chen, X.; Cui, X.; Li, S.; et al. The *Arabidopsis* H3K27me3 demethylase JUMONJI 13 is a temperature and photoperiod dependent flowering repressor. *Nat Commun* **2019**, *10*, 1303. [CrossRef]
110. Leon, K.E.; Aird, K.M. Jumonji C Demethylases in Cellular Senescence. *Genes* **2019**, *10*, 33. [CrossRef]
111. Bentz, A.B.; Thomas, G.W.C.; Rusch, D.B.; Rosvall, K.A. Tissue-specific expression profiles and positive selection analysis in the tree swallow (Tachycineta bicolor) using a de novo transcriptome assembly. *Sci. Rep.* **2019**, *9*, 15849. [CrossRef]
112. Klepikova, A.V.; Penin, A.A. Gene Expression Maps in Plants: Current State and Prospects. *Plants* **2019**, *8*, 309. [CrossRef]
113. Kubínová, Z.; Janáček, J.; Lhotáková, Z.; Kubínová, L.; Albrechtová, J. Unbiased estimation of chloroplast number in mesophyll cells: Advantage of a genuine three-dimensional approach. *J. Exp. Bot.* **2014**, *65*, 609–620. [CrossRef]
114. Benjamini, Y.; Hochberg, Y. Controlling the False Discovery Rate: A Practical and Powerful Approach to Multiple Testing. *J. R. Stat. Soc. Ser. B Methodol.* **1995**, *57*, 289–300. [CrossRef]

115. Lunde, C.; Drew, D.P.; Jacobs, A.K.; Tester, M. Exclusion of Na+ via sodium ATPase (PpENA1) ensures normal growth of Physcomitrella patens under moderate salt stress. *Plant Physiol.* **2007**, *144*, 1786–1796. [CrossRef]
116. Qudeimat, E.; Frank, W. Ca^{2+} signatures: The role of Ca^{2+}-ATPases. *Plant Signal. Behav.* **2009**, *4*, 350–352. [CrossRef]
117. Cui, L.-L.; Lu, Y.-S.; Li, Y.; Yang, C.; Peng, X.-X. Overexpression of Glycolate Oxidase Confers Improved Photosynthesis under High Light and High Temperature in Rice. *Front. Plant Sci.* **2016**, *7*, 1165. [CrossRef]
118. Fernandez, L.R.; Vandenbussche, G.; Roosens, N.; Govaerts, C.; Goormaghtigh, E.; Verbruggen, N. Metal binding properties and structure of a type III metallothionein from the metal hyperaccumulator plant Noccaea caerulescens. *Biochim. Biophys. Acta* **2012**, *1824*, 1016–1023. [CrossRef]
119. Babaei-Bondarti, Z.; Shahpiri, A. A metallothionein type 2 from *Avicennia marina* binds to iron and mediates hydrogen peroxide balance by activation of enzyme catalase. *Phytochemistry* **2020**, *176*, 112396. [CrossRef]
120. Freisinger, E. Plant MTs-long neglected members of the metallothionein superfamily. *Dalton. Trans.* **2008**, 6663–6675. [CrossRef]
121. Pirzadeh, S.; Shahpiri, A. Functional characterization of a type 2 metallothionein isoform (OsMTI-2b) from rice. *Int. J. Biol. Macromol.* **2016**, *88*, 491–496. [CrossRef]
122. Cabral, A.C.S.; Jakovleska, J.; Deb, A.; Penner-Hahn, J.E.; Pecoraro, V.L.; Freisinger, E. Further insights into the metal ion binding abilities and the metalation pathway of a plant metallothionein from Musa acuminata. *J. Biol. Inorg. Chem.* **2018**, *23*, 91–107. [CrossRef]
123. Rice, J.M.; Zweifach, A.; Lynes, M.A. Metallothionein regulates intracellular zinc signaling during $CD4^+$ T cell activation. *BMC Immunol.* **2016**, *17*, 13. [CrossRef] [PubMed]
124. Singh, G.; Tripathi, S.; Shanker, K.; Sharma, A. Cadmium-induced conformational changes in type 2 metallothionein of medicinal plant Coptis japonica: Insights from molecular dynamics studies of apo, partially and fully metalated forms. *J. Biomol. Struct. Dyn.* **2019**, *37*, 1520–1533. [CrossRef] [PubMed]
125. Patankar, H.V.; Al-Harrasi, I.; Al Kharusi, L.; Jana, G.A.; Al-Yahyai, R.; Sunkar, R.; Yaish, M.W. Overexpression of a Metallothionein 2A Gene from Date Palm Confers Abiotic Stress Tolerance to Yeast and *Arabidopsis thaliana*. *Int. J. Mol. Sci.* **2019**, *20*, 2871. [CrossRef] [PubMed]
126. Hamelet, J.; Seltzer, V.; Petit, E.; Noll, C.; Andreau, K.; Delabar, J.M.; Janel, N. Cystathionine beta synthase deficiency induces catalase-mediated hydrogen peroxide detoxification in mice liver. *Biochim. Biophys. Acta* **2008**, *1782*, 482–488. [CrossRef] [PubMed]
127. Pheomphun, P.; Treesubsuntorn, C.; Jitareerat, P.; Thiravetyan, P. Contribution of Bacillus cereus ERBP in ozone detoxification by *Zamioculcas zamiifolia* plants: Effect of ascorbate peroxidase, catalase and total flavonoid contents for ozone detoxification. *Ecotoxicol. Environ. Saf.* **2019**, *171*, 805–812. [CrossRef] [PubMed]
128. Sahoo, R.; Bhattacharjee, A.; Majumdar, U.; Ray, S.S.; Dutta, T.; Ghosh, S. A novel role of catalase in detoxification of peroxynitrite in S. cerevisiae. *Biochem. Biophys. Res. Commun.* **2009**, *385*, 507–511. [CrossRef] [PubMed]
129. Stührwohldt, N.; Bühler, E.; Sauter, M.; Schaller, A. Phytosulfokine (PSK) precursor processing by subtilase SBT3.8 and PSK signaling improve drought stress tolerance in Arabidopsis. *J. Exp. Bot.* **2021**, *72*, 3427–3440. [CrossRef]
130. Koonin, E.V.; Tatusov, R.L. Computer analysis of bacterial haloacid dehalogenases defines a large superfamily of hydrolases with diverse specificity. Application of an iterative approach to database search. *J. Mol. Biol.* **1994**, *244*, 125–132. [CrossRef]
131. Wang, Y.; Zheng, X.; He, X.; Lü, Q.; Qian, X.; Xiao, Q.; Lin, R. Effects of Pseudomonas TCd-1 on rice (*Oryza sativa*) cadmium uptake, rhizosphere soils enzyme activities and cadmium bioavailability under cadmium contamination. *Ecotoxicol. Environ. Saf.* **2021**, *218*, 112249. [CrossRef]
132. Pandey, B.K.; Mehra, P.; Verma, L.; Bhadouria, J.; Giri, J. OsHAD1, a Haloacid Dehalogenase-Like APase, Enhances Phosphate Accumulation. *Plant Physiol.* **2017**, *174*, 2316–2332. [CrossRef]
133. Hanin, M.; Brini, F.; Ebel, C.; Toda, Y.; Takeda, S.; Masmoudi, K. Plant dehydrins and stress tolerance: Versatile proteins for complex mechanisms. *Plant Signal. Behav.* **2011**, *6*, 1503–1509. [CrossRef]
134. Hinniger, C.; Caillet, V.; Michoux, F.; Ben Amor, M.; Tanksley, S.; Lin, C.; McCarthy, J. Isolation and characterization of cDNA encoding three dehydrins expressed during Coffea canephora (Robusta) grain development. *Ann. Bot.* **2006**, *97*, 755–765. [CrossRef]
135. Nath, B.; Birch, G.; Chaudhuri, P. Assessment of sediment quality in *Avicennia marina*-dominated embayments of Sydney Estuary: The potential use of pneumatophores (aerial roots) as a bio-indicator of trace metal contamination. *Sci. Total Environ.* **2014**, *472*, 1010–1022. [CrossRef]
136. Alishahi, M.; Kamali, R. Forced diffusion of water molecules through aquaporin-5 biomembrane; a molecular dynamics study. *Biophys. Physicobiol.* **2018**, *15*, 255–262. [CrossRef]
137. Merlaen, B.; De Keyser, E.; Ding, L.; Leroux, O.; Chaumont, F.; Van Labeke, M.C. Physiological responses and aquaporin expression upon drought and osmotic stress in a conservative vs prodigal *Fragaria x ananassa* cultivar. *Plant Physiol. Biochem.* **2019**, *145*, 95–106. [CrossRef]
138. Sahebi, S.; Fadaie, N.; Mirshekar, L.; Kamarehie, B.; Mohammadi, T. Assessing biomimetic aquaporin membrane for forward osmosis desalination process: A dataset. *Data Brief* **2019**, *26*, 104482. [CrossRef]
139. Wei, X.; Jin, X.; Ndayambaza, B.; Min, X.; Zhang, Z.; Wang, Y.; Liu, W. Transcriptome-wide characterization and functional identification of the aquaporin gene family during drought stress in common Vetch. *DNA Cell Biol.* **2019**, *38*, 374–384. [CrossRef]
140. Zou, Z.; Yang, J. Genome-wide comparison reveals divergence of cassava and rubber aquaporin family genes after the recent whole-genome duplication. *BMC Genom.* **2019**, *20*, 380. [CrossRef]

141. Krishnamurthy, P.; Tan, X.F.; Lim, T.K.; Lim, T.-M.; Kumar, P.P.; Loh, C.-S.; Lin, Q. Proteomic analysis of plasma membrane and tonoplast from the leaves of mangrove plant Avicennia officinalis. *Proteomics* **2014**, *14*, 2545–2557. [CrossRef]
142. Sadhukhan, A.; Kobayashi, Y.; Nakano, Y.; Iuchi, S.; Kobayashi, M.; Sahoo, L.; Koyama, H. Genome-wide Association Study Reveals that the Aquaporin NIP1;1 Contributes to Variation in Hydrogen Peroxide Sensitivity in *Arabidopsis thaliana*. *Mol. Plant* **2017**, *10*, 1082–1094. [CrossRef]
143. Yusupov, M.; Razzokov, J.; Cordeiro, R.M.; Bogaerts, A. Transport of reactive oxygen and nitrogen species across aquaporin: A molecular level picture. *Oxid. Med. Cell Longev.* **2019**, *2019*, 2930504. [CrossRef] [PubMed]
144. Nickerson, N.H.; Thibodeau, F.R. Association between Pore Water Sulfide Concentrations and the Distribution of Mangroves. *Biogeochemistry* **1985**, *1*, 183–192. [CrossRef]
145. Kreuzwieser, J.; Rennenberg, H. Molecular and physiological responses of trees to waterlogging stress: Responses of tree to waterlogging. *Plant Cell Environ.* **2014**, *37*, 2245–2259. [CrossRef] [PubMed]
146. Amala, M.; Rajamanikandan, S.; Prabhu, D.; Surekha, K.; Jeyakanthan, J. Identification of anti-filarial leads against aspartate semialdehyde dehydrogenase of Wolbachia endosymbiont of Brugia malayi: Combined molecular docking and molecular dynamics approaches. *J. Biomol. Struct. Dyn.* **2019**, *37*, 394–410. [CrossRef]
147. Cohen, G.N. Aspartate-semialdehyde dehydrogenase from Escherichia coli. *Methods Enzymol.* **1985**, *113*, 600–602. [CrossRef]
148. Cox, R.J.; Gibson, J.S.; Hadfield, A.T. Design, synthesis and analysis of inhibitors of bacterial aspartate semialdehyde dehydrogenase. *Chembiochem* **2005**, *6*, 2255–2260. [CrossRef]
149. Dahal, G.P.; Viola, R.E. Structural insights into inhibitor binding to a fungal ortholog of aspartate semialdehyde dehydrogenase. *Biochem. Biophys. Res. Commun.* **2018**, *503*, 2848–2854. [CrossRef]
150. Evitt, A.S.; Cox, R.J. Synthesis and evaluation of conformationally restricted inhibitors of aspartate semialdehyde dehydrogenase. *Mol. Biosyst.* **2011**, *7*, 1564–1575. [CrossRef]
151. Gayler, K.R.; Morgan, W.R. An NADP-dependent Glutamate Dehydrogenase in Chloroplasts from the Marine Green Alga *Caulerpa simpliciuscula*. *Plant Physiol.* **1976**, *58*, 283–287. [CrossRef]
152. Born, T.L.; Blanchard, J.S. Structure/function studies on enzymes in the diaminopimelate pathway of bacterial cell wall biosynthesis. *Curr. Opin. Chem. Biol.* **1999**, *3*, 607–613. [CrossRef]
153. Mengin-Lecreulx, D.; van Heijenoort, J.; Park, J.T. Identification of the mpl gene encoding UDP-N-acetylmuramate: L-alanyl-gamma-D-glutamyl-meso-diaminopimelate ligase in *Escherichia coli* and its role in recycling of cell wall peptidoglycan. *J. Bacteriol.* **1996**, *178*, 5347–5352. [CrossRef]
154. Park, J.T.; Raychaudhuri, D.; Li, H.; Normark, S.; Mengin-Lecreulx, D. MppA, a periplasmic binding protein essential for import of the bacterial cell wall peptide L-alanyl-gamma-D-glutamyl-meso-diaminopimelate. *J. Bacteriol.* **1998**, *180*, 1215–1223. [CrossRef]
155. Wehrmann, A.; Phillipp, B.; Sahm, H.; Eggeling, L. Different modes of diaminopimelate synthesis and their role in cell wall integrity: A study with *Corynebacterium glutamicum*. *J. Bacteriol.* **1998**, *180*, 3159–3165. [CrossRef]
156. Villa, S.T.; Xu, Q.; Downie, A.B.; Clarke, S.G. Arabidopsis protein repair l-isoaspartyl methyltransferases: Predominant activities at lethal temperatures. *Physiol. Plant.* **2006**, *128*, 581–592. [CrossRef]
157. Gonzalez-Mendoza, D.; Moreno, A.Q.; Zapata-Perez, O. Coordinated responses of phytochelatin synthase and metallothionein genes in black mangrove, Avicennia germinans, exposed to cadmium and copper. *Aquat. Toxicol.* **2007**, *83*, 306–314. [CrossRef]
158. Nelson, D.R.; Chaiboonchoe, A.; Fu, W.; Hazzouri, K.M.; Huang, Z.; Jaiswal, A.; Daakour, S.; Mystikou, A.; Arnoux, M.; Sultana, M.; et al. Potential for Heightened Sulfur-Metabolic Capacity in Coastal Subtropical Microalgae. *iScience* **2019**, *11*, 450–465. [CrossRef]
159. Cotelle, N.; Moreau, S.; Cotelle, P.; Catteau, J.P.; Bernier, J.L.; Henichart, J.P. Generation of free radicals by simple prenylated hydroquinone derivatives, natural antitumor agents from the marine urochordate Aplidium californicum. *Chem. Res. Toxicol.* **1991**, *4*, 300–305. [CrossRef]
160. Naik, M.M.; Pandey, A.; Dubey, S.K. Pseudomonas aeruginosa strain WI-1 from Mandovi estuary possesses metallothionein to alleviate lead toxicity and promotes plant growth. *Ecotoxicol. Environ. Saf.* **2012**, *79*, 129–133. [CrossRef]
161. Attia, O.E.; Ghrefat, H. Assessing heavy metal pollution in the recent bottom sediments of Mabahiss Bay, North Hurghada, Red Sea, Egypt. *Environ. Monit. Assess.* **2013**, *185*, 9925–9934. [CrossRef]
162. Idris, A.M.; Said, T.O.; Omran, A.A.; Fawy, K.F. Combining multivariate analysis and human risk indices for assessing heavy metal contents in muscle tissues of commercially fish from Southern Red Sea, Saudi Arabia. *Environ. Sci. Pollut. Res. Int.* **2015**, *22*, 17012–17021. [CrossRef]
163. Mohamed, S.A.; Elshal, M.F.; Kumosani, T.A.; Mal, A.O.; Ahmed, Y.M.; Almulaiky, Y.Q.; Asseri, A.H.; Zamzami, M.A. Heavy Metal Accumulation is Associated with Molecular and Pathological Perturbations in Liver of Variola louti from the Jeddah Coast of Red Sea. *Int. J. Environ. Res. Public Health* **2016**, *13*, 342. [CrossRef]
164. Saleh, Y.S.; Marie, M.A. Assessment of metal contamination in water, sediment, and tissues of Arius thalassinus fish from the Red Sea coast of Yemen and the potential human risk assessment. *Environ. Sci. Pollut. Res. Int.* **2015**, *22*, 5481–5490. [CrossRef]
165. Euceda, N.; Jahnke, J.; Espinal, A.; Louis, M.F.; Bashkin, E.; Roccanova, P.; Espaillat, A.; Fuentes, G.V.; Nieto, F.; Gao, R. Thioguanine restoration through type I photosensitization-superoxide oxidation-glutathione reduction cycles. *Physical. Chem. Chem. Phys.* **2021**, *23*, 5069–5073. [CrossRef]

166. Simunkova, M.; Barbierikova, Z.; Jomova, K.; Hudecova, L.; Lauro, P.; Alwasel, S.H.; Alhazza, I.; Rhodes, C.J.; Valko, M. Antioxidant vs. Prooxidant Properties of the Flavonoid, Kaempferol, in the Presence of Cu(II) Ions: A ROS-Scavenging Activity, Fenton Reaction and DNA Damage Study. *Int. J. Mol. Sci.* **2021**, *22*, 1619. [CrossRef]
167. Tsang, T.; Davis, C.I.; Brady, D.C. Copper biology. *Curr. Biol.* **2021**, *31*, R421–R427. [CrossRef]
168. Mir, G. A plant type 2 metallothionein (MT) from cork tissue responds to oxidative stress. *J. Exp. Bot.* **2004**, *55*, 2483–2493. [CrossRef]

agronomy

MDPI

Review

The PGPR Mechanisms of Salt Stress Adaptation and Plant Growth Promotion

Rakiba Shultana [1,2,*], Ali Tan Kee Zuan [2,*], Umme Aminun Naher [3], A. K. M. Mominul Islam [4,*], Md. Masud Rana [1], Md. Harun Rashid [4], Israt Jahan Irin [5], Shams Shaila Islam [6], Adiba Afrin Rim [3] and Ahmed Khairul Hasan [4]

1 Agronomy Division, Bangladesh Rice Research Institute, Gazipur 1701, Bangladesh
2 Department of Land Management, Faculty of Agriculture, Universiti Putra Malaysia, Serdang 43400, Selangor, Malaysia
3 Soil Science Division, Bangladesh Rice Research Institute, Gazipur 1701, Bangladesh
4 Department of Agronomy, Bangladesh Agricultural University, Mymensingh 2202, Bangladesh
5 Department of Agronomy, Khulna Agricultural University, Khulna 9100, Bangladesh
6 Department of Agronomy, Hajee Mohammad Danesh Science and Technology University, Dinajpur 5200, Bangladesh
* Correspondence: rakibabrri@gmail.com (R.S.); tkz@upm.edu.my (A.T.K.Z.); akmmominulislam@bau.edu.bd (A.K.M.M.I.); Tel.: +880-1717-852-931 (R.S.)

Abstract: Worldwide crop productivity hampers severely due to the adverse effects of salinity. Global warming causes a rapid escalation of the salt-affected area, and new agricultural land is affected through saltwater intrusion. The ever-growing human population impulses to utilize the saline area for crop cultivation to ensure food security. Salinity resistance crops could be a promising substitute but with minor success because inappropriate tactics on saline soil management resulted in unsatisfactory yield. Salt-tolerant plant growth-promoting rhizobacteria (ST-PGPR) is considered an alternate way towards enhancing crop growth in saline ecosystems. It is reported that PGPR is enabled to produce exopolysaccharides which lead to biofilm formation and generate osmoprotectants and antioxidant enzymes that can significantly contribute to stimulating plant growth in the saline ecosystem. In addition, several plant growth-promoting characteristics of PGPR such as the acquisition of essential nutrients and upsurge hormone production could enhance plant growth simultaneously. In this review, we will explore the survival mechanisms of ST-PGPR and their influence on plant growth promotion in saline ecosystems.

Keywords: salinity; PGPR; mechanism; crop; growth

Citation: Shultana, R.; Zuan, A.T.K.; Naher, U.A.; Islam, A.K.M.M.; Rana, M.M.; Rashid, M.H.; Irin, I.J.; Islam, S.S.; Rim, A.A.; Hasan, A.K. The PGPR Mechanisms of Salt Stress Adaptation and Plant Growth Promotion. *Agronomy* **2022**, *12*, 2266. https://doi.org/10.3390/agronomy12102266

Academic Editors: Raquel P. F. Guiné, António Dinis Ferreira and António Moitinho Rodrigues

Received: 17 August 2022
Accepted: 18 September 2022
Published: 22 September 2022

Publisher's Note: MDPI stays neutral with regard to jurisdictional claims in published maps and institutional affiliations.

1. Introduction

Soil salinity is considered a major abiotic threat to agricultural production around the world [1]. Salinity is documented as a severe climatic menace affecting almost one billion hectares of land globally [2]. This cruel ecological anxiety causes an annual estimated economical loss in crop production is about USD 27.3 billion [3,4]. Furthermore, the risk of salinization at different latitudes is increasing due to the global warming scenarios and therefore a special attempt is required to obtain the maximum agricultural output from a saline ecosystem [5]. Annually 2500–5000 km^2 of crop production is lost due to salinity since it occupies more than 20% of the world-irrigated land [6]. Improper irrigation practices are expected to affect approximately 50% of the irrigated areas in the world following an annual expansion of up to 500,000 ha. These realities are an indicator of extreme global risk in achieving food security [7]. Lack of rainfall and increase in temperatures in most agricultural regions are the consequences of climate change which may lead to more arid and semi-arid zones [8,9]. To meet the rising food demand, the manipulation of saline areas for agricultural production is the way forward. Thus, coping with salinity is the ultimate target for rising food production [10]. Renovation of salt-affected lands for successful crop cultivation through effective management practices is the challenge that needs to be highlighted. Physical removals of salts from the soil surface or chemical application are expensive as well as have an adverse environmental impact and would be difficult to apply in huge areas for soil retrieval purposes. In this case, the manipulation of soil beneficial soil microorganisms in stress-prone areas is an important concern. Microbial inoculants could improve plant health in saline-affected soils by ameliorating salt stress, supporting plant growth, and controlling diseases [11–13]. Several studies have confirmed the positive effects of soil beneficial microbes that could increase plants' tolerance toward adverse salinity stresses [14,15]. Moreover, several studies are proving the hypothesis that PGPR facilitates plants' continuing crop production in stressed soil through exopolysaccharides production and biofilm formation, which facilitates bacterial aggregation and forms a protective cover to get rid of adverse climatic conditions [16]. Bacterial production of osmoprotectants, antioxidant enzymes, and volatile organic compounds can trigger bacterial survival under high osmotic conditions. Through the production of the ACC deaminase enzyme, these bacteria help slow down ethylene production and accelerate bacterial survival under saline conditions. Due to their unique mechanism to withstand under saline state, they consistently assist the plants to grow through the production of various traits related to plant growth, such as the production of growth hormone, fixation of atmospheric nitrogen, and solubilizing of inorganic phosphate. This updated information of review will be helpful outlines to explore the mechanisms of PGPR to alleviate salt stress in plants.

2. Soil Salinization

Soil salinity denotes the excess amount of soluble salt in the root zone of plants. Due to the elevated osmotic pressure, salinity affects plant growth by restricting the uptake of water and essential plant elements [17]. The accretion of available salts such as sodium (Na^+), calcium (Ca^{2+}), potassium (K^+), magnesium (Mg^{2+}), chloride (Cl^-), sulfate ($SO_4{}^{2-}$), carbonate ($CO_3{}^{2-}$), bicarbonate ($HCO_3{}^-$) is considered as soil salinization. Moreover, weathering of minerals is also the cause of salt deposition. In addition, anthropogenic factors such as irrigation of the crop with salt waters, poor cultural practices, and low precipitation are other causes of soil salinization. The frequent use of different inorganic fertilizers and amendments of soil with gypsum, composts, and manures also contributes to developing soil salinization [18].

3. Effect of Salinity on Plants

Worldwide salinity is considered the main abiotic stress, troubling the coastal agricultural system [9]. Soil salinity considerably affects the humid and sub-humid rice-cultivating zone where the rate of sea-level rise is projected to surge, thus having a dramatic effect on crop production, especially salt-sensitive rice genotypes, which could be lost 50% of yield [19–21]. The crop response to salinity depends on several factors (i) the climatic conditions (ii) stress intensity and (iii) the tolerance level of the genotype [22]. Salinity negatively affects rice stand establishment, panicles, tillers, spikelets, individual grain size, and crop maturity [23]. Other crops such as wheat, sorghum, and cowpea are mostly susceptible to salinity at vegetative and early reproductive stages [24]. The osmotic stress and ionic toxicity are the primary causes of secondary oxidative stress in plants under salinity stress [25,26].

The toxicity of salinity in plants often occurs through (i) osmotic imbalance (ii) toxicity of ions (iii) oxidative stress following disruption of the photosystem, and other physiological disparities [27]. The ions Na^+ and Cl^- are the causes of plant cell damage at both osmotic and ionic levels, which accumulate in the chloroplasts at a high concentration under salinity stress, consequently damaging thylakoid membranes [28]. In all rice genotypes, K^+ concentrations decreased with the rise of salinity concentration, thus hindering the photosynthetic rate by altering the ultra-structure of the organelles and various pigment concentrations, including connected enzymes and stomatal regulations [29].

4. Role of PGPR for Salt Stress Reduction

Plant growth-promoting rhizobacteria (PGPR), are a group of rhizospheric bacteria, first defined by Kloepper and Schroth [30], a zone where the plant roots are available and essential macro and micronutrients are extracted resulting from higher microbial activities [26]. PGPR takes part in (i) nutrient mobilization in soil (ii) production of plant growth regulators (iii) controlling phytopathogenic attack (iv) induced systemic resistance (v) improvement of soil structure and (vi) polluted soil remediation [31,32]. The application of these beneficial microbes to the soil–plant system is well studied and proven under stressed soil [33]. A salt-tolerant bacterial strain *Staphylococcus xylosus* ST-1 caused a 25% growth increment of seedling at 100 mM NaCl over control [34]. In another study, the osmotic balance of the cells was changed by *Bacillus mojavensis* VKAK1 through changing plant water relations [35]. *Azospirillum* AZ19 strain inoculation in wheat plants originated from saline or non-salinated conditions and showed increased grain yield [36]. Habib et al. [37] showed that isolates UPMR7 (*Bacillus* sp.), UPMR17 (*Citrobacter* sp.), and UPMR18 can resist high NaCl concentration (up to 6%), which helps their survivability in a saline environment. Some of the potential salt-tolerant bacteria associated with different crops were shown in Table 1.

Table 1. The salt-tolerant bacteria with its mechanisms of salt-stress reduction in different crops.

Name of the Bacteria	Plant Species	Major Mechanism	Reference
Bacillus megaterium A12	*Lycopersicon esculentum*	The upregulation of PIP aquaporin expression	[38]
Bacillus subtilis GB03	*Arabidopsis thaliana*	The upregulation of the sodium transporter HKT1	[39]
Pseudomonas syringae S5, *Pseudomonas fluorescens* S20, *Enterobacter aerogenes* S14,	*Zea mays*	ACC deaminase enzyme production	[40]
Pseudomonas fluorescens TDK1	*Arachis hypogaea*	ACC deaminase production	[41]
Enterobacter sp. EJ01	*Lycopersicum esculentum* *Arabidopsis thaliana*	The regulation of salt stress responsive genes such as DREB2b, RD29A, RD29B, and RAB18. The upregulation of proline biosynthetic genes (i.e., P5CS1 and P5CS2) and of genes related to priming processes (i.e., MPK3 and MPK6)	[42]
Pseudomonas syringae Mk1; *Pseudomonas fluorescens* Mk20 and *Pseudomonas fluorescens* Mk25	*Vigna radiata*	Auxin production, ACC deaminase production	[43]
Brachybacterium saurashtrense JG-06, *Brevibacterium casei* JG-08	*Arachis hypogaea*	Reduced oxidative stress through high proline and low MDA content in plants	[44]
Pseudomonas sp. PMDZnCd 2003	*Oryza sativa*	Indole-3-acetic acids (IAA) production, nitrogen fixation, and phosphate solubilization.	[45]
Alcaligenes sp. SB1.ACC2 and *Ochrobactrum* sp. SB2.ACC2	*Oryza sativa*	Production of ACC Deaminase enzyme production	[46]
Azospirillum sp.	*Brassica napus*	Regulation of antioxidant enzymes	[47]
Streptomyces sp. PGPA39	*Solanum lycopersicum*	Production of ACC Deaminase,	[48]
Serratia sp. SL-12	*Triticum aestivum*	Accumulation of osmolytes such as total soluble sugar and total protein content	[49]
Dietzia natronolimnaea STR1	*Triticum aestivum*	ABA-signaling cascade, as *TaABARE* and *TaOPR1* were upregulated	[50]

Table 1. *Cont.*

Name of the Bacteria	Plant Species	Major Mechanism	Reference
Azospirillum lipoferum FK1	*Cicer arietinum*	Modulating osmolytes, antioxidant machinery and stress-related genes expression.	[51]
Pseudomonas fluorescens PGU2-79, WBO-3, WKZ1-93 and WB1-7	*Triticum aestivum,*	ACC deaminase production	[52]
Pseudomonas fluorescens B10, B2-10, B2-11 and B4-6	*Hordeum vulgare*	ACC deaminase production	[53]
Pseudomonas PS01	*Arabidopsis thaliana*	Upregulation of LOX2	[54]
Aneurinbacillus aneurinilyticus ACC02 and *Paenibacillus* sp. ACC06	*Phaseolus vulgaris*	ACC deaminase activity	[55]
Burkholderia cenocepacia CR318	*Zea mays*	Phosphate and potassium solubilization and antimicrobial activity	[56]
Ochrobactrum sp. NBRISH6	*Zea mays*	Ion homeostasis	[57]
Bacillus sp. NBRI YN4.4	*Zea mays*	Improves photosynthetic pigments, soluble sugar content, enhances soil enzymes.	[58]
Aeromonas sp. SAL-17 and SAL-21	*Triticum aestivum*	Acyl homoserine lactone	[59]
Bacillus atrophaeus BR5, OR15, and RB13	*Arabidopsis thaliana, Triticum aestivum*	Increase proline, TSS, Antioxidant enzyme, decrease MDA	[60]
Bacillus paramycoides HB6J2, *Bacillus amyloliquefaciens* HB8P1 and *Bacillus pumilus* HB4N3	-	HCN production, phosphate solubilization, IAA and ammonia production	[61]
Azospirillum lipoferum SP2, *Bacillus coagulans* NCAIM B.01123, *Bacillus circulance* NCAIMB.02324, and *Bacillus subtilis* MF497446	*Triticum aestivum*	Reduced the uptake of Na+ resulted in an increment in superoxide dismutase (SOD), catalase (CAT), and peroxidase (POX) activities that lessened oxidative damage and improved the nutrient uptake (N, P, and K) of deficiently irrigated wheat plants under soil salinity.	[62]
Enterobacter cloacae PM23	*Zea mays*	Enhanced radical scavenging capacity, relative water content, soluble sugars, proteins, total phenolic, and flavonoid content	[63]

Soil bacteria are less tolerant to salinity than root-associated bacteria. In the rhizosphere, salinity stress is higher due to the higher uptake of water by the plant roots [64]. PGPR strains such as *P. chlororaphis* TSAU13, *P. extremorientalis* TSAU6, *P. extremorientalis* TSAU20, *P. putida* TSAU1, and *P. fluorescens* WCS356, can withstand up to 3% of NaCl [65]. The salinity stress on photosynthesis, essential nutrients, and antioxidant enzymes of basil plants was reduced through inoculation with *Pseudomonas* sp. and *Bacillus lentus*. Inoculation of *Azosprillium brasilense* NH, a halotolerant strain in wheat, enhanced germination and plant growth in salinated soil [66]. Abbaspoor et al. [67] also recorded that *P. fluorescens* 153 and *P. putida* 108 inoculations to wheat plants improved growth, grain yield, and 1000-grain weight. Stimulation of plant growth using salt-tolerant strains, *Exiguobacterium oxidotolerans* STR36 and *Bacillus pumilus* STR2 was noticed by Bharti et al. [68]. Vivekanandan et al. [69] inoculated five halo-tolerant bacterial strains on wheat seedlings at 80, 160, and 320 mM of NaCl, resulting in a considerable increase in biomass and root length compared with un-inoculated controls. *Hallobacillus* sp. S13 and *Bacillus halodenitrificans* PU62 inoculation in wheat seedlings showed more than a 90% increase in dry biomass compared with on-inoculated wheat plants at 320 mM of NaCl resulting in a remarkable decline in the toxic effects of NaCl.

5. Mechanisms of Plant Growth Promotion by PGPR under Saline Conditions

In saline conditions, plant growth could be hastened by PGPR by facilitating resource acquisition (nitrogen, phosphorus, potassium) and moderating plant hormone levels through producing ACC deaminase enzyme or indirectly by producing exopolysaccharides and biofilm, osmoprotectants, antioxidant enzymes, and volatile organic compounds (VOCs). All these properties help bacteria to survive and stimulate plant growth under saline conditions [70].

5.1. Nitrogen Fixation

In agricultural production, nitrogen is the major nutrient that has a remarkable effect on plant growth. The free-living and symbiotic bacteria in nature can fix atmospheric N_2 in salt stress conditions and contribute to plant growth. A nitrogen-fixing salt-tolerant bacterium, *Swaminathan halotolerant* PA51T, isolated from wild rice associated with the mangrove ecosystem [71] has the potential to fix atmospheric nitrogen. Five salt-tolerant strains of rhizobium (L-19, L-68, L-292, L-304, and L-335) isolated from saline soils were inoculated to lentil plants (*Lens culinaris*) under saline conditions. Among the isolated strains L-19 and L-304 produced higher nodulation, yield, and nitrogen fixation in lentils [72]. Silini-Cherif et al. [73] identified a nitrogen-fixing bacterium named *Pantoea agglomerans* Ima2 from the wheat rhizosphere, which can tolerate a salinity level of 100 to 400 mM, and its application increased IAA production, siderophore formation, and solubilization of phosphates. Kumar et al. [74] isolated *Mesorhizobium loti* MTCC2379 and MTCC2381 from acacia, a salt-tolerant strain showing efficient nitrogenase activity under salt stress conditions. The symbiosis of rhizobium–legume is the most essential system of nitrogen fixation. Some rhizobia could tolerate up to 1.8 M of NaCl concentration. With the morphological and metabolic changes along with structural modifications, these salt-tolerant rhizobia cope with and adapt to salt stress. Under salt stress conditions, some of the rhizobia can form a successful symbiosis with legumes [75].

5.2. Phosphate Solubilization

High salinity reduces the uptake of available phosphorus (P) by plant roots due to sorption processes to the soil colloid. P solubilizing bacteria even in stress conditions could solubilize fixed and applied P in soil [76]. Chookietwattana et al. [77] have found *Bacillus megaterium* A12 as the efficient halotolerant phosphate solubilizing bacteria. Son et al. [78] identified *Pantoea agglomerans* R-42, a phosphate solubilizing bacterium from a salt-stressed environment. The soybean (*Glycine max*) seeds inoculated with halo-tolerant phosphate solubilizing bacteria significantly increased germination percentage and germination index, especially within 30 and 90 mM NaCl concentrations. Hence, it was suggested that the salt-tolerant phosphate solubilizing bacteria might be useful to reclaim the salt stress toxicity in plants.

5.3. Plant Growth Regulators

The indole-3-acetic acid (IAA), commonly known as the auxins, are important hormones in plants regulated by PGPR that help to promote plant root development and alter root architecture [79,80]. Nakbanpote et al. [45] demonstrated that the production of IAA by *Pseudomonas* sp. PDMZnCd2003 was not affected by salinity stress at 4–16 dS m^{-1}. The auxin signaling plays a significant role in restructuring plant roots [81,82]. The halophyte strains of *Brevibacterium halotolerans* DSM8802, *Bacillus subtilis* h-g, *Brachybacterium saurashtrense* JG06, and *Pseudomonas* sp. JG010 can accelerate plant growth by producing indole acetic acid (IAA) [83–85]. Similarly, Shultana et al. [86] identified a promising strain, UPMRB9 (*Bacillus tequilensis*) based on the measurement of its IAA production showed a significant growth enhancement of three rice varieties in saline conditions.

5.4. ACC Deaminase Enzyme Production

The PGPR can produce ACC (1-aminocyclopropane-1-carboxylate) deaminase; thus, it can lower the ACC level in salt-stressed plants and reduce the quantity of ethylene synthesis in plants. Several studies reported that ACC deaminase-producing PGPR could help plants survive against salinity stress through the reduction of ethylene levels [87,88]. A number of PGPR genera, namely *Bacillus*, *Burkholderia*, *Azospirillum*, *Pseudomonas*, and *Rhizobium* are commonly known to synthesize ACC-deaminase enzyme [89–91]. Several reports showed that under axenic conditions, ACC deaminase-producing bacteria trigger plant growth [92,93]. The salt-tolerant and ACC deaminase-producing bacterium augment root development through the increased surface area for better water and nutrient accumulation [94]. Salt-tolerant bacteria associated with ACC deaminase production are shown in Table 2.

Table 2. ACC deaminase-producing salt-tolerant bacteria.

Name of the Bacteria	Plant Species	Reference
Pseudomonas syringae S5, *Pseudomonas fluorescens* S20 *Enterobacter aerogenes* S14	*Zea mays*	[40]
Raoultella planticola Rs-2	*Gossypium hirsutum*	[95]
Pseudomonas fluorescens EU647703.1	*Brassica napus*	[96]
Enterobacter cloaceae AJS-15	*Aerva javanica*	[97]
Bacillus mojavensi K78	*Triticum aestivum*	[98]
Pseudomonas migulae 8R6 and *Pseudomonas* sp. UW4	*Camelina sativa*	[99]
Bacillus megaterium NMp082	*Medicago* spp.; *Arabidopsis thaliana*	[100]
Bacillus cereus KP027636.1, *Serratia odorifera* NR037110.1, *Lelliottia amnigena* KM114915.1, *Arthrobacter arilaitensis* CP012750.1, *Pseudomonas putida* GQ2008822.1	*Triticum aestivum*	[101]
Enterobacter sp. PR 14	-	[102]
B. safensis HB-5	*Cicer arietinum*	[103]
Enterobacter cloacae ZNP-4	*Triticum aestivum*	[104]
Enterobacter ludwigii B30	*Cynodon dactylon*	[105]

5.5. Exo-Polysaccharide Production

Bacterial exo-polysaccharide production is recognized as a strategy for the existence under saline conditions reported by several researchers [106–108] where at high salt levels bacteria can retain a mini assembly to hold water level around the cells. Exopolysaccharides (EPSs) help to enable bacterial survival from inhospitable conditions [88] through chelating sodium ion (Figure 1) and reduce its availability for plants [109]. Bacterial polysaccharides are considered as a diverse range of macromolecules which includes peptidoglycan, lipopolysaccharides, capsules, and exopolysaccharides which are water-soluble acids, participate in the host–pathogen interaction and also the components of the structural cell wall (e.g., peptidoglycan) and facilitate the bacterium to survive in unfavorable environments [110,111]. These compounds were recognized as biologically active substances that promote the growth of bacteria and other plant species and also help their adhesion to surfaces and prevent desiccation [110]. The bacterial cells could discharge extra-cellular polysaccharides (EPS) into the atmosphere. EPS is environmentally important since it affects the microbial diversity and carbon cycle [112].

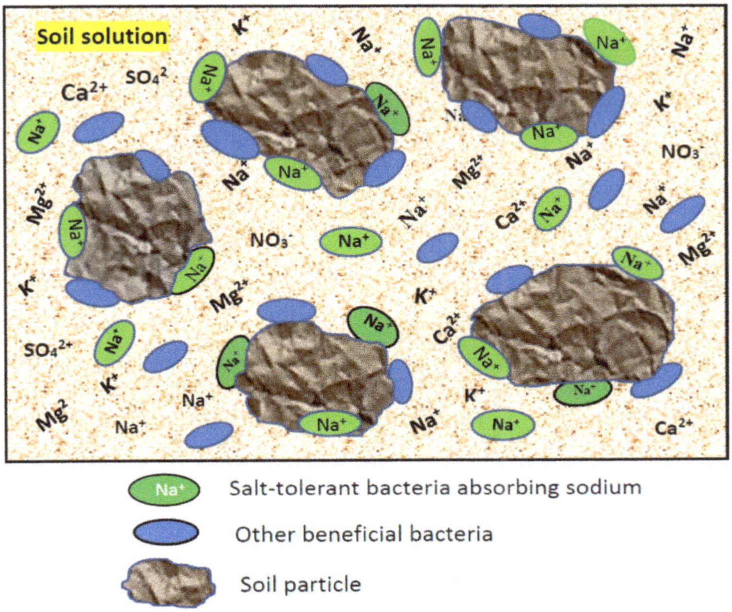

Figure 1. Sodium absorption by the EPS producing salt-tolerant bacteria under saline soil.

Exo-polysaccharides (EPS) influence the formation of rhizosheath around the plant roots [113]. The micro-organisms that live in the proximity of plant roots can synthesize or release EPS in soil. The EPS-synthesizing rhizobacteria take part in the aggregation of soil and rhizosheath (biofilm) formation around the roots of the plants [113,114].

EPS functioned as a blockade within cells and the neighboring environment and thus plays a shielding role against dehydration, UV radiations, and salinity [115]. EPS enhances the retention of water and dispersion of carbon in the bacterial community. Recent findings showed that salinity tolerance of *Suaeda fruticose* markedly increased by the inoculation of *Glutamicibacter* sp. MK847981 and *Pseudomonas* sp. MK087034 through sinking the concentration of Na^+ and increasing K^+, consequently increasing the ratio of K^+/Na^+ [116]. The content of Na^+ in soybean was reduced because of the application of EPS-releasing bacterial strains in salinized soil. The progressive increase in mineral contents along with the reduction in Na^+ and Cl^- concentrations in maize were noticed through bacterial inoculation in saline soil. Vivas et al. [117] informed that *Bacillus* sp. inoculated lettuce plants showed higher N, P, and K concentrations under stress conditions which were increased by 5, 70, and 50%, respectively, compared with control.

Under the saline condition, uptake of Na^+ restricts by wheat roots through EPS-producing bacteria since EPS can alter the microenvironment and protects bacteria from desiccation [118]. In addition, EPS-producing rhizobium strain inoculation to plant roots improved soil properties [119]. Exopolysaccharide linkages help to bind microorganisms together growing in the free planktonic state. Microbial EPS are rich in monosaccharides such as glucose, fructose, mannose, xylose, etc., that serve as a signal for root colonization [120]. In addition, the major functional groups in bacterial EPS such as hydroxyl, carboxyl, phosphate, sulfhydryl, and amino groups are the prime factors chelating Na^+ under saline soil conditions, thereby reducing the exposure of plants to the salt ions [121].

5.6. Exopolysaccharides and Biofilm Formation

Bacterial biofilm formation is closely linked with EPS production, which essentially contributes to bacterial colonization around plant roots [122]. The PGPR in soil participates in removing contaminants and toxicants from soil and water [34,123]. EPS-driven biofilm protects bacteria embedded with the EPS layer from uncongenial conditions such as the presence of salinity, antibiotics, and radiations [124]. Several studies showed that microbial biofilms attached to roots significantly enhance soil fertility [120]. The salt-tolerant PGPR is enabled to synthesize biofilm containing extracellular polysaccharides with high water holding capacity [125]. Apart from increasing the effective root colonization, the bacteria also have the competitive advantage of osmo-tolerance under salt stress. Previously, a study proved that root colonization and plant growth-promoting activities of PGPR did not interfere with salinity [126]. The production of exopolysaccharides, biofilm formation and accumulation of intracellular osmolytes govern the osmo-tolerance of PGPR. Ashraf et al. [127] found that the inoculation of EPS producing bacterial strains to the roots of wheat plants in salt-affected soils provides a "blanket salt-tolerant cover". Bacterial species that enable the production of exopolysaccharide and biofilm are shown in Table 3.

Table 3. The exopolysaccharide and biofilm producing salt-tolerant bacteria.

Name of the Bacteria	Plant Species	Reference
Halomonas variabilis (HT1) and *Planococcus rifietoensis* (RT4)	*Cicer arietinum*	[106]
Pseudomonas fluorescens, *Bacillus amyloliquefaciens* and *Bacillus polymyxa*	*Triticum aestivum*	[128]
Bacillus amyloliquefaciens MAS4, *Bacillus insolitus* MAS10 and MAS26, *Pseudomonas syringae* MAS129, *Microbacterium* sp MAS133.	*Triticum aestivum*	[127]
Shewanella putrefaciens (isolates No.603)	-	[129]
Bacillus iodinum RS16, and *Bacillus aryabhattai* RS341	*Capsicum annuum*	[130]
Bacillus sp. SKU5, *Burkholderia cepacian* (SKU6), *Microbacterium* sp. (SKU9), *Enterobacter* sp. (SKU9), and *Paenibacillus macerans* (SKU10)	*Triticum aestivum*	[131–133]
Pseudomonas aeruginosa (Pa2), *Proteus penneri* (Pp1), and *Alcaligenes faecalis* (AF3)	*Zea mays*	[134]
Bacillus tequilensis UPMRB9, *Bacillus aryabhattai* UPMRE6	*Oryza sativa*	[121]
Brevibacterium sediminis S4-57	*Oryza sativa*	[135]

6. Salinity Tolerance of *Bacillus* sp.

The gram-positive bacteria *Bacillus* is widely familiar with rhizobacteria. Some important member of a genus under *Bacillus* includes *B. licheniformis* HSW-16, *B. amyloliquefaciens* SN13, *B. megaterium* A12, *B. subtilis* SU47, and *B. pumilus* HB4N3 are reported for plant growth, and stress management [61,131,133,136,137]. The PGPR, *Bacillus subtilis* 93,151 inoculated transgenic *Arabidopsis thaliana* showed enhanced proline synthesis with proBA genes that can upsurge the plant's salinity tolerance [138]. Root hydraulic conductivity of maize plants was increased by the inoculation of *Bacillus megaterium* compared to the uninoculated plants under 2.59 dSm^{-1} of salinity. Wheat seed treated with *B. aquimaris* SU8 strains increased higher shoot biomass, and NPK accumulation through the higher synthesis of total soluble sugars, reducing sugars, and Na reduction in leaves under 5.2 dSm^{-1} of salinity in field conditions [139]. Inoculation of *B. subtilis* BERA71 to chickpea plants improved the upregulation of antioxidant systems through the reduction of ROS and increased nutrient absorption [140,141]. Improved systemic acquired resistance (SAR) in wheat by the inoculation of strain *B. licheniformis* HSW-16 exhibited enhanced ammonium assimilation, nitrogen fixation, and phosphate and potassium uptake under saline conditions [137].

7. Osmoprotectants

Osmoprotectants, commonly known as a compatible solute, traveled from producers to consumers. The osmotic adjustment of bacterial cells largely depends on various kinds of osmoprotectants required for bacterial cells for osmotic adjustment and thus cells can be protected against high temperature, oxygen radicals, and desiccation [142]. Proline, glycine betaine (GB), proline betaine, and choline, a precursor of glycine betaine, stimulates bacterial growth and nitrogen fixation when added to media of elevated osmotic strength and proline overproduction also enhances osmo-tolerance [143,144]. Among the compatible solutes, glycine betaine plays a protective function under saline condition [115]. Glycine betaine is electrically neutral and dipolar at physiological pH. The essential role of GB in salinity stress is the stabilization of RuBisCO, protection of photosynthetic apparatus, foraging of reactive oxygen species (ROS), and osmotic adjustment [144]. It is widely accepted that GB at low concentrations protect nucleic acids, lipids and proteins and also performed as pools of nitrogen and carbon sources [145]. Only a few microorganisms secrete GB that can be transported actively and accumulate osmoprotectant [146].

The bacterial membrane is penetrable to water but creates an active blockade for various solutes in the medium and metabolites in the cytoplasm. To cope with osmotic stresses, the cells gather organic solutes under hyperosmotic conditions and releases under hypoosmotic conditions. The amino acids (e.g., proline and glutamate), the amino acid derivatives (peptides and N-acetylated amino acids), sugars (e.g., trehalose and sucrose), amines (e.g., carnitine, glycine betaine), tetrahydropyrimidines and K^+ [147] comprises compatible solutes. These compatible solutes originate by de novo synthesis (synthesis of complex molecules from simple molecules) or shifted with the major cellular system without interference. In rhizobial cells, the accretion of poly-b-hydroxyl butyrate usually acts as a defensive measure during elevated salinity stress [75]. Paul and Nair [126] observed the de novo synthesis of osmolyte by PGPR strain, *Pseudomonas fluorescens* MSP-393 such as alanine, serine, glycine, glutamic acid, threonine, and aspartic acid in their cytosol. The correct folding of polypeptides supported by compatible solutes under denaturing conditions both in vivo and in vitro consequently stabilizes proteins [148].

8. Induced Antioxidative Activity

In saline conditions, the antioxidant activities could be altered through the generation of ROS as a form of the hydroxyl radical (OH^-), superoxide radical (O^{2-}), and hydrogen peroxide (H_2O_2). ROS damages plant cells' DNA, proteins, and lipids. [149]. The strains *B. subtilis* BERA71 produces different antioxidant enzymes such as SOD, POX, and CAT as well as non-enzymatic antioxidants such as tocopherol, ascorbate peroxidase (APX), ascorbate, and glutathione which take parts in scavenging cycle [140]. An improvement of salinity tolerance in potato plants (*Solanum tuberosum*) is due to inoculation of *Bacillus pumilus* DH-11 and *B. firmus* str. 40, ACC deaminase producer and phosphate solubilizers, respectively [150]. Higher antioxidant enzymes in bacteria inoculated canola plants were also reported by Neshat et al. [151] who determined the higher production of SOD, POD, and CAT with the inoculation of *Enterobacter* sp. S16-3 and *Pseudomonas* sp. C16-20 under salt stress conditions. This is because of the accelerated photosynthetic rate, higher accumulation of proline, improved expression of mRNA, and the activities of antioxidant enzymes. Likewise, Kim et al. [42] testified an IAA and ACC deaminase producer *Enterobacter* sp. EJ01 strain inoculation shows an increase in dry weight and plant height of tomato and augmentation of ROS detoxifying enzyme in aerial plant tissue under salt stress.

9. Volatile Organic Compounds (VOCs)

A complex blend of volatiles could be released by PGPR [152,153]. Volatiles are organic compounds at room temperature that contains a high vapor pressure. The VOCs have odors or scents, derivatives of various nitrogen and sulfur-containing compounds such as phenylpropanoids, terpenoids, and fatty acids [154]. The PGPR-generated VOCs, change physical and chemical properties in plants and consequently enhance plant salinity tolerance [155]. The PGPR strain *Bacillus subtilis* GB03 mediated VOCs confers salt tolerance and plant growth promotion in *Arabidopsis thaliana* through recirculation and reduction of Na^+ levels in the entire plant under saline conditions through buildup HKT1, a high-affinity potassium transporter that facilitates Na^+ transportation, expression upregulated in shoots and downregulated in roots [39].

10. Molecular Mechanisms and Gene Expression of PGPR in Response to Salinity Stress

A higher concentration of NaCl stimulates bacteria towards showing an expression of a specific gene, which is denoted as a set of proteins produced in higher amounts in response to stress [126,156]. In the bioinformatics era, proteomics is considered a suitable tactic for disclosing the vibrant expressions of whole cells proteins and their interactions. Large numbers of specific proteins have been reported, which shows an increase in their level of expression. To identify and elucidate the genes responding to relative physiological actions, differentially displayed proteins could be used as nutrient transport, metabolism, and responses to stress, chemotaxis, motility, sporulation, and biosynthesis of teichuronic acid [157]. Diby et al. [158] confirmed that many genes are responsive to salt stress in a PGPR strain, *Pseudomonas pseudoalcaligenes* MSP-538. Under salt shock conditions, peptide mass fingerprinting analysis of *P. fluorescens* MSP-393 exposed various stress-related proteins [159]. A bacterium, Bacillus *subtilis* JH642, responsive to salt stress expressed the induction of upregulated 123 genes and downregulated 101 genes by the transcriptional profiling at 1.2 M NaCl [160]. Under salinity stress, *Escherichia coli* MC4100 has been shown to produce multiple up-regulated genes involved in the process of cellular metabolism, amino acid biosynthesis, and transportation [161]. A salt-responsive protein K^+ uptake kup/trkD was highly expressed in response to salt stress [162]. Previously, the involvement of non-coding RNA Yfr1 for salt sensing was explored [163]. Paul [164] has recently reviewed the mechanisms of salt tress adaptions in rhizobacteria.

A stress-related PGPR responsive protein named a chaperone is known to bind particularly denatured proteins and prevent degradation [165]. In another study, it was declared that in eubacteria and eukaryotic organelles, numerous enzyme-folding functions were regulated by chaperonin 60 [166]. Again, Holland et al. [167] reported that the seedlings of *N. tabacum* showed resistance against prolonged darkness, salt, and cold due to the buildup of chaperonin 60.

In rice plants, the differential expression of thioredoxin proteins was observed with *P. fluorescens* KH-1. The tolerance of methionine sulfoxide and H_2O_2 was noticed in *Saccharomycess* train EMY63 because of the expression of *Arabidopsis thioredoxin* [168]. The protein 10i showed high homology to the enzyme glutamine synthetase, which is required for osmolyte distribution and played a significant role in glutamate synthesis, a prominent osmolyte in bacteria [169].

The induced proteins 26i and 42i were found to be associated with membrane proteins, and it could also be corroborated with the high root-colonization potential of the strain even in salinated soils [106]. Protein 41i is a survival protein (SurE), essential for the survival of osmotic stress (2.5 M NaCl) in bacteria [170]. Kandasamy et al. [171] assumed an essential role of the GSTs gene for its overexpression which might be involved in the ISR for protecting cells from oxidative damage.

11. Knowledge Gaps and Future Prospects

Many unrevealed areas exist on the performance of these beneficial microbes in stressed soil and also concerning their interactions with the host plant. In-depth studies are required to know the role of abiotic factors in changing the activity of rhizobacteria and managing plant–microbe interactions, concerning their compliance to stress environments. There are a few recommendations for future work:

i. Identification of genetic and environmental factors responsible for higher bacterial EPS synthesis under salt stress conditions.

ii. Identification of stress-responsive proteins involved in signaling, gene expression, and metabolism during plant–microbe interaction under salt stress conditions.
iii. The mutual sharing of osmoprotectants and antioxidant enzymes of PGPR and plants for maximum plant–microbe interactions under salt stress conditions.
iv. Evaluation of crop performance inoculated with salt-tolerant PGPR in actual saline ecosystems is a prerequisite to observing the consistent field performance of the potential salt-tolerant PGPR.

12. Conclusions

Worldwide, there is a rising demand for the cultivation of crops in saline-affected areas by taking into account compatible, ecologically sound, and environmentally friendly tactics. The development of stress-tolerant crops is a desirable option but considered a long-drawn and expensive process, whereas soil manipulation, using microbial strains to alleviate plant stress recognized as a low-cost and environment-friendly option that could be achieved in a shorter time frame. The PGPR mechanisms of osmo-tolerance offer multiple advantages to plants cultivated in salinized soils. The salt-affected areas are expected to utilize for increasing crop productivity through a proper understanding of PGPR mechanisms on salt tolerance. This review has shown and suggested the function of salt-resistant plant growth-promoting microorganisms as an environmentally friendly and more economical to improve crop production in saline-affected areas. In the future, extensive research needs to be emphasized in this area, particularly on the field performance of potential microorganisms as a source of bio-fertilizers in stressed soil.

Author Contributions: Conceptualization: R.S.; formal analysis: R.S., A.T.K.Z. and U.A.N.; investigation: R.S. and A.T.K.Z.; resources: R.S., A.T.K.Z. and A.K.M.M.I.; supervision, project administration and funding acquisition: R.S., A.T.K.Z. and A.K.M.M.I.; writing—original draft preparation: R.S., M.M.R., I.J.I., S.S.I., A.A.R. and U.A.N.; writing—review and editing: A.K.M.M.I., M.H.R. and A.K.H.; visualization: R.S. and A.K.M.M.I. All authors have read and agreed to the published version of the manuscript.

Funding: Fundamental Research Grant Scheme (FRGS) (FRGS/1/2020/STG01/UPM/02/6) by the Ministry of Higher Education, Malaysia and Putra Grant (GP-IPS/2022/9709700) by the Universiti Putra Malaysia.

Institutional Review Board Statement: Not applicable.

Informed Consent Statement: Not applicable.

Data Availability Statement: The data presented in this study are available on request from the corresponding authors. The data are not publicly available as they represent the authors' own field material.

Acknowledgments: The authors sincerely acknowledge the Universiti Putra Malaysia (UPM) and OWSD (Organization for Women in Science for the Developing World) and Swedish International Development Cooperation Agency (SIDA) for the fellowship award.

Conflicts of Interest: The authors declare that they have no conflict of interest. The funders had no role in the design of the study; in the collection, analyses, or interpretation of data; in the writing of the manuscript, or in the decision to publish the results.

References

1. Ranjbar, F.; Jalali, M. The combination of geostatistics and geochemical simulation for the site-specific management of soil salinity and sodicity. *Comput. Electron. Agric.* **2016**, *121*, 301–312. [CrossRef]
2. Fageria, N.K.; Stone, L.F.; Santos, A.B.D. Breeding for salinity tolerance. *Plant Breed. Abiotic Stress Toler.* **2012**, *9783642305535*, 103–122. [CrossRef]
3. Munns, R.; Tester, M. Mechanisms of salinity tolerance. *Annu. Rev. Plant Biol.* **2008**, *59*, 651–681. [CrossRef] [PubMed]
4. Qadir, M.; Quillérou, E.; Nnagia, V.; Nurtaza, G.; Singh, M.; Thomas, R.J.; Drechsel, P.; Nable, A.D.; Noble, A.D. Economics of salt-induced land degradation and restoration. *Nat. Resour. Forum.* **2014**, *38*, 282–295. [CrossRef]
5. Turral, H.; Burke, J.; Faurès, J.M. *Climate Change, Water and Food Security*; Food and Agriculture Organization of the United Nations (FAO): Rome, Italy, 2011.

6. Nellemann, C.; MacDevette, M. (Eds.) *The Environmental Food Crisis: The Environment's Role in Averting Future Food Crises: A UNEP Rapid Response Assessment*; UNEP/Earthprint: Nairobi, Kenya, 2009.
7. Ondrasek, G.; Rengel, Z.; Romic, D.; Poljak, M.; Romic, M. Accumulation of non/essential elements in radish plants grown in salt-affected and cadmium-contaminated environment. *Cereal Res. Commun.* **2009**, *37*, 9–12.
8. Othman, Y.; Al-Karaki, G.; Al-Tawaha, A.R.; Al-Horani, A. Variation in germination and ion uptake in barley genotypes under salinity conditions. *World J. Agric. Sci.* **2006**, *2*, 11–15.
9. Ghosh, B.; Md, N.A.; Gantait, S. Response of rice under salinity stress: A review update. *Rice Res.* **2016**, *4*, 1–8. [CrossRef]
10. Shanker, A.; Venkateswarlu, B. (Eds.) *Abiotic Stress in Plants: Mechanisms and Adaptations*; InTech Publisher: JanezaTridne Rijeka, Croatia, 2011; p. 428.
11. Lugtenberg, B.J.; Malfanova, N.; Kamilova, F.; Berg, G. Plant growth promotion by microbes. *Mol. Microb. Ecol. Rhizosphere* **2013**, *2*, 561–573.
12. Lugtenberg, B.; Malfanova, N.; Kamilova, F.; Berg, G. Microbial control of plant root diseases. In *Molecular Microbial Ecology of the Rhizosphere*; De Bruijn, F.J., Ed.; Wiley-Blackwell: Hoboken, NJ, USA, 2013; pp. 575–586.
13. Pliego, C.; Kamilova, F.; Lugtenberg, B. Plant growth-promoting bacteria: Fundamentals and exploitation. In *Bacteria in Agrobiology: Crop Ecosys*; Springer: Berlin/Heidelberg, Germany, 2011; pp. 295–343.
14. Dodd, I.C.; Perez-Alfocea, F. Microbial alleviation of crop salinity. *Plant Mol. Bio.* **2012**, *63*, 3415–3428.
15. Berg, G.; Alavi, M.; Schmidt, C.S.; Zachow, C.; Egamberdieva, D.; Kamilova, F.; Lugtenberg, B. Biocontrol and Osmoprotection for Plants under Salinated Conditions. In *Molecular Microbial Ecology of the Rhizosphere*; De Bruijn, F.J., Ed.; Wiley-Blackwell: Hoboken, NJ, USA, 2013; pp. 561–573.
16. Qurashi, A.W.; Sabri, A.N. Osmoadaptation and plant growth promotion by salt tolerant bacteria under salt stress. *Afr. J. Microbiol. Res.* **2011**, *5*, 3546–3554.
17. Tester, M.; Davenport, R. Na+ tolerance and Na+ transport in higher plants. *Ann. Bot.* **2003**, *91*, 503–527. [CrossRef] [PubMed]
18. Kotuby-Amacher, J.; Koenig, K.; Kitchen, B. Salinity and plant tolerance. *AG-SO* **2000**, *3*, 1.
19. Selamat, A.; Ismail, M.R. Growth and production of rice for the increased Malaysian population as affected by global warming trends: Forecast for 2057. *Trans. Malays. Soc. Plant Physiol.* **2008**, *17*, 20–34.
20. Ismail, A.M.; Thapa, B.; Egdane, J. Salinity tolerance in rice: Physiological bases and implications on management strategies for better crop establishment. In *Improving Productivity and Livelihood for Fragile Environments*; IRRI Technical Bull. 13; International Rice Research Institute: Los Banos, Philippines, 2009; pp. 8–13.
21. Hakim, M.A.; Juraimi, A.S.; Ismail, M.R.; Hanafi, M.M.; Selamat, A. A survey on weed diversity in coastal rice fields of Sebarang Perak in peninsular Malaysia. *J. Anim. Plant Sci.* **2013**, *23*, 534–542.
22. Kanawapee, N.; Sanitchon, J.; Srihaban, P.; Theerakulpisut, P. Physiological changes during development of rice (*Oryza sativa* L.) varieties differing in salt tolerance under saline field condition. *Plant Soil.* **2013**, *370*, 89–101. [CrossRef]
23. Rad, H.E.; Aref, F.; Rezaei, M. Evaluation of salinity stress affects rice growth and yield components in Northern Iran. *Am. J. Sci. Res.* **2012**, *54*, 40–51.
24. Patil, A.D. Alleviating Salt Stress in Crop Plants through Salt Tolerant Microbes. *Int. J. Sci. Res.* **2015**, *4*, 1297–1302.
25. Ding, M.; Hou, P.; Shen, X.; Wang, M.; Deng, S.; Sun, J. Salt-induced expression of genes related to Na^+/K^+ and ROS homeostasis in leaves of salt resistant and salt sensitive poplar species. *Plant Mol. Biol.* **2010**, *73*, 251–269. [CrossRef]
26. Ahmad, F.; Ahmad, I.; Khan, M.S. Screening of free-living rhizospheric bacteria for their multiple plant growth promoting activities. *Braz. J. Microbiol.* **2008**, *163*, 173–181. [CrossRef]
27. Ali, Y.; Aslam, Z.; Ashraf, M.Y.; Tahir, G.R. Effect of salinity on chlorophyll concentration, leaf area, yield and yield components of rice genotypes grown under saline environment. *Int. J. Environ. Sci. Technol.* **2004**, *1*, 221–225. [CrossRef]
28. Omoto, E.; Taniguchi, M.; Miyake, H. Effects of salinity stress on the structure of bundle sheath and mesophyll chloroplasts in NAD-malic enzyme and PCK type C4 plants. *Plant Prod. Sci.* **2010**, *13*, 169–176. [CrossRef]
29. Ashraf, M.; Harris, P.J.C. Photosynthesis under stressful environments: An overview. *Photosynthetica* **2013**, *51*, 163–190. [CrossRef]
30. Kloepper, J.W. Plant growth-promoting rhizobacteria on radishes. In Proceedings of the 4th Internet Conference on Plant Pathogenic Bacter, Station de Pathologie Vegetale et. Phytobacteriologie, INRA, Angers, France; 1978; Volume 2, pp. 879–882.
31. Hayat, R.; Ali, S.; Amara, U.; Khalid, R.; Ahmed, I. Soil beneficial bacteria and their role in plant growth promotion: A review. *Ann. Microbiol.* **2010**, *60*, 579–598. [CrossRef]
32. Son, J.S.; Sumayo, M.; Hwang, Y.J.; Kim, B.S.; Ghim, S.Y. Screening of plant growth-promoting rhizobacteria as elicitor of systemic resistance against gray leaf spot disease in pepper. *Appl. Soil Ecol.* **2014**, *73*, 1–8. [CrossRef]
33. Bhattacharyya, P.N.; Jha, D.K. Plant growth-promoting rhizobacteria (PGPR): Emergence in agriculture. *World J. Microbiol. Biotechnol.* **2012**, *28*, 1327–1350. [CrossRef] [PubMed]
34. Afrasayab, S.; Faisal, M.; Hasnain, S. Comparative study of wild and transformed salt tolerant bacterial strains on *Triticum aestivum* grown under salt stress. *Braz. J. Microbiol.* **2010**, *41*, 946–955. [CrossRef]
35. Kapadia, C.; Patel, N.; Rana, A.; Vaidya, H.; Alfarraj, S.; Ansari, M.J.; Gafur, A.; Poczai, P.; Sayyed, R.Z. Evaluation of Plant Growth-Promoting and Salinity Ameliorating Potential of Halophilic Bacteria Isolated From Saline Soil. *Front. Plant Sci.* **2022**, *13*, 946217. [CrossRef]

36. García, J.E.; Maroniche, G.; Creus, C.; Suárez-Rodríguez, R.; Ramirez-Trujillo, J.A.; Groppa, M.D. In vitro PGPR properties and osmotic tolerance of different Azospirillum native strains and their effects on growth of maize under drought stress. *Microbiol. Res.* **2017**, *202*, 21–29. [CrossRef]
37. Habib, S.H.; Kausar, H.; Saud, H.M.; Ismail, M.R.; Othman, R. Molecular characterization of stress tolerant plant growth promoting rhizobacteria (PGPR) for growth enhancement of rice. *Int. J. Agric. Biol.* **2016**, *18*, 184–191. [CrossRef]
38. Akram, W.; Aslam, H.; Ahmad, S.R.; Anjum, T.; Yasin, N.A.; Khan, W.U.; Ahmad, A.; Guo, J.; Wu, T.; Luo, W.; et al. Bacillus megaterium strain A12 ameliorates salinity stress in tomato plants through multiple mechanisms. *J. Plant Interact.* **2019**, *14*, 506–518. [CrossRef]
39. Zhang, H.; Kim, M.S.; Sun, Y.; Dowd, S.E.; Shi, H.; Paré, P.W. Soil bacteria confer plant salt tolerance by tissue-specific regulation of the sodium transporter HKT1. *Mol. Plant Microbe Interact.* **2008**, *21*, 737–744. [CrossRef] [PubMed]
40. Nadeem, S.M.; Zahir, Z.A.; Naveed, M.; Arshad, M. Preliminary investigations on inducing salt tolerance in maize through inoculation with rhizobacteria containing ACC deaminase activity. *Can. J. Microbiol.* **2007**, *53*, 1141–1149. [CrossRef] [PubMed]
41. Saravanakumar, D.; Samiyappan, R. ACC deaminase from Pseudomonas fluorescens mediated saline resistance in groundnut (Arachis hypogea) plants. *J. Appl. Microbiol.* **2007**, *102*, 1283–1292. [CrossRef] [PubMed]
42. Kim, K.; Jang, Y.J.; Lee, S.M.; Oh, B.T.; Chae, J.C.; Lee, K.J. Alleviation of salt stress by *Enterobacter* sp. EJ01 in tomato and Arabidopsis is accompanied by upregulation of conserved salinity responsive factors in plants. *Mol. Cells* **2014**, *37*, 109–117.
43. Ahmad, M.; Zahir, Z.A.; Nazli, F.; Akram, F.; Arshad, M.; Khalid, M. Effectiveness of halo-tolerant, auxin producing Pseudomonas and Rhizobium strains to improve osmotic stress tolerance in mung bean (*Vigna radiata* L.). *Braz. J. Microbiol.* **2013**, *44*, 1341–1348. [CrossRef]
44. Shukla, P.S.; Agarwal, P.K.; Jha, B. Improved salinity tolerance of *Arachis hypogaea* (L.) by the interaction of halotolerant plant-growth-promoting rhizobacteria. *J. Plant Growth Regul.* **2012**, *31*, 195–206. [CrossRef]
45. Nakbanpote, W.; Panitlurtumpai, N.; Sangdee, A.; Sakulpone, N.; Sirisom, P.; Pimthong, A. Salt-tolerant and plant growth-promoting bacteria isolated from Zn/Cd contaminated soil: Identification and effect on rice under saline conditions. *J. Plant Interact.* **2014**, *9*, 379–387. [CrossRef]
46. Bal, H.B.; Nayak, S.; Dasand, T.; Adhya, K. Isolation of ACC deaminase producing PGPR from rice rhizosphere and evaluating their plant growth promoting activity under salt stress. *Plant Soil* **2013**, *366*, 93–105. [CrossRef]
47. Noorieh, B.; Arzanesh, H.; Mahlegha, G.; Maryam, S. The effect of plant growth promoting rhizobacteria on growth parameters, antioxidant enzymes and microelements of canola under salt stress. *J. Appl. Environ. Biol. Sci.* **2013**, *3*, 17–27.
48. Palaniyandi, S.A.; Damodharan, K.; Yang, S.H.; Suh, J.W. *Streptomyces* sp. strain PGPA39 alleviates salt stress and promotes growth of 'Micro Tom' tomato plants. *J. Appl. Microbiol.* **2014**, *117*, 766–773. [CrossRef]
49. Singh, R.P.; Jha, P.N. Alleviation of salinity-induced damage on wheat plant by an ACC deaminase-producing halophilic bacterium *Serratia* sp. SL- 12 isolated from a salt lake. *Symbiosis* **2016**, *69*, 101–111. [CrossRef]
50. Bharti, N.; Pandey, S.S.; Barnawal, D.; Patel, V.K.; Kalra, A. Plant growth promoting rhizobacteria Dietzia natronolimnaea modulates the expression of stress responsive genes providing protection of wheat from salinity stress. *Sci. Rep.* **2016**, *6*, 34768. [CrossRef] [PubMed]
51. El-Esawi, M.A.; Alaraidh, I.A.; Alsahli, A.A.; Alamri, S.A.; Ali, H.M.; Alayafi, A.A. Bacillus firmus (SW5) augments salt tolerance in soybean (*Glycine max* L.) by modulating root system architecture, antioxidant defense systems and stress-responsive genes expression. *Plant Physiol. Biochem.* **2018**, *132*, 375–384. [CrossRef] [PubMed]
52. Safari, D.; Jamali, F.; Nooryazdan, H.R.; Bayat, F. Evaluation of ACC deaminase producing Pseudomonas fluorescens strains for their effects on seed germination and early growth of wheat under salt stress. *Aust. J. Crop Sci.* **2018**, *12*, 413–421. [CrossRef]
53. Azadikhah, M.; Jamali, F.; Nooryazdan, H.R.; Bayat, F. Growth promotion and yield enhancement of barley cultivars using ACC deaminase producing Pseudomonas fluorescens strains under salt stress. *Spanish J. Agric. Res.* **2019**, *17*, e0801. [CrossRef]
54. Chu, T.N.; Tran, B.T.H.; Van Bui, L.; Hoang, M.T.T. Plant growth-promoting rhizobacterium Pseudomonas PS01 induces salt tolerance in *Arabidopsis thaliana*. *BMC Res. Notes* **2019**, *12*, 11. [CrossRef]
55. Gupta, S.; Pandey, S. ACC Deaminase Producing Bacteria with Multifarious Plant Growth Promoting Traits Alleviates Salinity Stress in French Bean (Phaseolus vulgaris) Plants. *Front. Microbiol.* **2019**, *10*, 1506. [CrossRef]
56. You, M.; Fang, S.; MacDonald, J.; Xu, J.; Yuan, Z.C. Isolation and characterization of Burkholderia cenocepacia CR318, a phosphate solubilizing bacterium promoting corn growth. *Microbiol. Res.* **2020**, *233*, 126395. [CrossRef]
57. Mishra, S.K.; Khan, M.H.; Misra, S.; Dixit, V.K.; Gupta, S.; Tiwari, S.; Chandra Gupta, S.; Chauhan, P.S. Drought tolerant Ochrobactrum sp. inoculation performs multiple roles in maintaining the homeostasis in *Zea mays* L. subjected to deficit water stress. *Plant Physiol. Biochem.* **2020**, *150*, 1–14. [CrossRef]
58. Dixit, V.K.; Misra, S.; Mishra, S.K.; Tewari, S.K.; Joshi, N.; Chauhan, P.S. Characterization of plant growth-promoting alkalotolerant Alcaligenes and Bacillus strains for mitigating the alkaline stress in Zea mays. *Antonie Van Leeuwenhoek* **2020**, *113*, 889–905. [CrossRef]
59. Nawaz, M.S.; Arshad, A.; Rajput, L.; Fatima, K.; Ullah, S.; Ahmad, M.; Imran, A. Growth-Stimulatory Effect of Quorum Sensing Signal Molecule N-Acyl-Homoserine Lactone-Producing Multi-Trait *Aeromonas* spp. on Wheat Genotypes Under Salt Stress. *Front. Microbiol.* **2020**, *11*, 553621. [CrossRef] [PubMed]
60. Kerbab, S.; Silini, A.; Bouket, A.C.; Cherif-Silini, H.; Eshelli, M.; Rabhi, N.E.H.; Belbahri, L. Mitigation of NaCl Stress in Wheat by Rhizosphere Engineering Using Salt Habitat Adapted PGPR Halotolerant Bacteria. *Appl. Sci.* **2021**, *11*, 1034. [CrossRef]

61. Sharma, A.; Dev, K.; Sourirajan, A.; Choudhary, M. Isolation and characterization of salt-tolerant bacteria with plant growth-promoting activities from saline agricultural fields of Haryana, India. *J. Genet. Eng. Biotechnol.* **2021**, *19*, 1–10. [CrossRef] [PubMed]
62. Omara, A.E.D.; Hafez, E.M.; Osman, H.S.; Rashwan, E.; El-Said, M.A.A.; Alharbi, K.; Abd El-Moneim, D.; Gowayed, S.M. Collaborative Impact of Compost and Beneficial Rhizobacteria on Soil Properties, Physiological Attributes, and Productivity of Wheat Subjected to Deficit Irrigation in Salt Affected Soil. *Plants* **2022**, *11*, 877. [CrossRef] [PubMed]
63. Ali, B.; Wang, X.; Saleem, M.H.; Sumaira; Hafeez, A.; Afridi, M.S.; Khan, S.; Zaib-Un-nisa; Ullah, I.; Amaral, A.T.D., Jr.; et al. PGPR-Mediated Salt Tolerance in Maize by Modulating Plant Physiology, Antioxidant Defense, Compatible Solutes Accumulation and Bio-Surfactant Producing Genes. *Plants* **2022**, *11*, 345. [CrossRef] [PubMed]
64. Tripathi, A.K.; Mishra, B.M.; Tripathi, P. Salinity stress responses in the plant growth promoting rhizobacteria, *Azospirillum* spp. *J. Biosci.* **1998**, *23*, 463–471. [CrossRef]
65. Egamberdiyeva, D.; Kucharova, Z. Selection for root colonizing bacteria stimulating wheat growth in saline soils. *Biol. Fertil. Soils* **2009**, *45*, 563–571. [CrossRef]
66. Nabti, E.; Sahnoune, M.; Ghoul, M.; Fischer, D.; Hofmann, A.; Rothballer, M.; Hartmann, A. Restoration of growth of durum wheat (*Triticum durum* var. waha) under saline conditions due to inoculation with the rhizosphere bacterium *Azospirillum brasilense* NH and extracts of the marine alga Ulva lactuca. *J. Plant Growth Regul.* **2010**, *29*, 6–22. [CrossRef]
67. Abbaspoor, A.; Zabihi, H.R.; Movafegh, S.; Asl, M.A. The efficiency of plant growth promoting rhizobacteria (PGPR) on yield and yield components of two varieties of wheat in salinity condition. *Am. Eurasian J. Sustain. Agric.* **2009**, *3*, 824–828.
68. Bharti, N.; Yadav, D.; Barnawal, D.; Maji, D.; Kalra, A. Exiguobacterium oxidotolerans, a halotolerant plant growth promoting rhizobacteria, improves yield and content of secondary metabolites in *Bacopa monnieri* (L.) pennell under primary and secondary salt stress. *World J. Microbiol. Biotechnol.* **2013**, *29*, 379–387. [CrossRef]
69. Vivekanandan, M.; Karthik, R.; Leela, A. Improvement of crop productivity in saline soils through application of saline-tolerant rhizosphere bacteria–current perspective. *Int. J. Adv. Res.* **2015**, *3*, 1273–1283.
70. Glick, B.R. Plant growth-promoting bacteria: Mechanisms and applications. *Scientifica* **2012**, *2012*, 963401. [CrossRef]
71. Loganathan, P.; Nair, S. Swaminathania salitolerants gen. nov., sp. nov., a salt-tolerant, nitrogen-fixing and phosphate-solubilizing bacterium from wild rice (Proteresia corctataTateoka). *Int. J. Syst. Evol. Microbiol.* **2004**, *54*, 1185–1190. [CrossRef]
72. Islam, M.Z.; Sattar, M.A.; Ashrafuzzaman, M.; Zulkerami, B.; Shamsuddoha, A.T.M. Evaluating some Salinity Tolerant Rhizobacterial Strains to Lentil Production under Salinity Stress. *Int. J. Agric. Biol.* **2013**, *15*, 499–504.
73. Silini-Cherif, H.; Silini, A.; Ghoul, M.; Yadav, S. Isolation and characterization of plant growth promoting traits of a rhizobacteria: *Pantoea agglomerans* lma2. *Pak. J. Biol. Sci.* **2012**, *15*, 267–276. [CrossRef]
74. Kumar, H.; Arora, N.K.; Kumar, V.; Maheshwari, D.K. Isolation, characterization and selection of salt tolerant rhizobia nodulating Acacia catechu and A. nilotica. *Symbiosis* **1999**, *26*, 279–288.
75. Arora, N.K.; Singhal, V.; Maheshwari, D.K. Salinity-induced accumulation of poly-b- hydroxyl butyrate in rhizobia indicating its role in cell protection. *World J. Microbiol. Biotechnol.* **2006**, *22*, 603–606. [CrossRef]
76. Banerjee, G.; Scott-Craig, J.S.; Walton, J.D. Improving enzymes for biomass conversion: A basic research perspective. *BioEnergy Res.* **2010**, *3*, 82–92. [CrossRef]
77. Chookietwattana, K.; Maneewan, K. Screening of efficient halotolerant phosphate solubilizing bacterium and its effect on promoting plant growth under saline conditions. *World Appl. Sci. J.* **2012**, *16*, 1110–1117.
78. Son, H.J.; Park, G.T.; Cha, M.S.; Heo, M.S. Solubilization of insoluble inorganic phosphates by a novel salt-and pH-tolerant *Pantoea agglomerans* R-42 isolated from soybean rhizosphere. *Bioresour. Technol.* **2006**, *97*, 204–210. [CrossRef]
79. Mantelin, S.; Touraine, B. Plant growth-promoting bacteria and nitrate availability: Impacts on root development and nitrate uptake. *J. Exp. Bot.* **2004**, *55*, 27–34. [CrossRef] [PubMed]
80. Kloepper, J.W.; Gutierrez-Estrada, A.; McInroy, J.A. Photoperiod regulates elicitation of growth promotion but not induced resistance by plant growth-promoting rhizobacteria. *Can. J. Microbiol.* **2007**, *53*, 159–167. [CrossRef]
81. Iglesias, M.J.; Terrile, M.C.; Casalongué, C.A. Auxin and salicylic acid signalings counteract the regulation of adaptive responses to stress. *Plant Signal. Behav.* **2011**, *6*, 452–454. [CrossRef]
82. Wang, Y.; Li, K.; Li, X. Auxin redistribution modulates plastic development of root system architecture under salt stress in Arabidopsis thaliana. *J. Plant Physiol.* **2009**, *166*, 1637–1645. [CrossRef] [PubMed]
83. Jha, B.; Gontia, I.; Hartmann, A. The roots of the halophyte Salicornia brachiata are a source of new halotolerant diazotrophic bacteria with plant growth-promoting potential. *Plant Soil* **2012**, *356*, 265–277. [CrossRef]
84. Piccoli, P.; Travaglia, C.; Cohen, A.; Sosa, L.; Cornejo, P.; Masuelli, R.; Bottini, R. An endophytic bacterium isolated from roots of the halophyte Prosopis strombulifera produces ABA, IAA, gibberellins A1 and A3 and jasmonic acid in chemically-defined culture medium. *Plant Growth Regul.* **2011**, *64*, 207–210. [CrossRef]
85. Sgroy, V.; Cassán, F.; Masciarelli, O.; Del Papa, M.F.; Lagares, A.; Luna, V. Isolation and characterization of endophytic plant growth-promoting (PGPB) or stress homeostasis-regulating (PSHB) bacteria associated to the halophyte *Prosopis strombulifera*. *Appl. Microbiol. Biotechnol.* **2009**, *85*, 371–381. [CrossRef]
86. Shultana, R.; Othman, R.; Zuan, A.T.K.; Yusop, M.R. Growth and nutrients uptake of rice at early seedling stage as inoculated with *Bacillus* spp. *Plant Arch.* **2019**, *19*, 1995–2001.

87. Zahir, Z.A.; Ghani, U.; Naveed, M.; Nadeem, S.M.; Asghar, H.N. Comparative effectiveness of Pseudomonas and *Serratia* sp. containing ACC-deaminase for improving growth and yield of wheat (*Triticum aestivum* L.) under salt-stressed conditions. *Arch. Microbiol.* **2009**, *191*, 415–424. [CrossRef]
88. Ali, S.; Charles, T.C.; Glick, B.R. Amelioration of damages caused by high salinity stress by plant growth-promoting bacterial endophytes. *Plant Physiol. Biochem.* **2014**, *80*, 160–167. [CrossRef]
89. Shaharoona, B.; Arshad, M.; Zahir, Z.A.; Khalid, A. Performance of *Pseudomonas* spp. containing ACC-deaminase for improving growth and yield of maize (*Zea mays* L.) in the presence of nitrogenous fertilizer. *Soil Biol. Biochem.* **2006**, *38*, 2971–2975. [CrossRef]
90. Sharma, P.; Khanna, V.; Kumari, P. Efficacy of aminocyclopropane-1-carboxylic acid (ACC)-deaminase-producing rhizobacteria in ameliorating water stress in chickpea under axenic conditions. *Afric. J. Microbiol. Res.* **2013**, *7*, 5749–5757.
91. Nascimento, F.X.; Rossi, M.J.; Soares, C.R.; McConkey, B.J.; Glick, B.R. New insights into 1-aminocyclopropane-1-carboxylate (ACC) deaminase phylogeny, evolution and ecological significance. *PLoS ONE* **2014**, *9*, e99168. [CrossRef] [PubMed]
92. Arif, M.S.; Akhtar, M.J.; Asghar, H.N.; Ahmad, R. Isolation and screening of rhizobacteria containing ACC-Daminase for growth promotion of sunflower seedlings under axenic conditions. *Soil Environ.* **2010**, *29*, 199–205.
93. Shahzad, S.M.; Khalid, A.; Arshad, M. Screening rhizobacteria containing ACC-deaminase for growth promotion of chickpea seedlings under axenic conditions. *Soil Environ.* **2010**, *29*, 38–46.
94. Siddikee, M.A.; Glick, B.R.; Chauhan, P.S.; JongYim, W.; Sa, T. Enhancement of growth and salt tolerance of red pepper seedlings (*Capsicum annuum* L.) by regulating stress ethylene synthesis with halotolerant bacteria containing 1-aminocyclopropane-1-carboxylic acid deaminase activity. *Plant Physiol. Biochem.* **2011**, *49*, 427–434. [CrossRef]
95. Wu, Z.; Yue, H.; Lu, J.; Li, C. Characterization of rhizobacterial strain Rs-2 with ACC deaminase activity and its performance in promoting cotton growth under salinity stress. *World J. Microbiol. Biotechnol.* **2012**, *28*, 2383–2393. [CrossRef]
96. Akhgar, A.R.; Arzanlou, M.; Bakker, P.A.H.M.; Hamidpour, M. Characterization of 1-aminocycloprane-1-carboxylate (ACC) deaminase containing *Pseudomonas* spp. in the rhizosphere of salt-stressed canola. *Pedosphere* **2014**, *24*, 461–468. [CrossRef]
97. Singh, R.P.; Jha, P.N. Plant growth promoting potential of ACC deaminase rhizospheric bacteria isolated from Aerva javanica: A plant adapted to saline environments. *Int. J. Curr. Microbiol. Appl. Sci.* **2015**, *4*, 142–152.
98. Pourbabaee, A.A.; Bahmani, E.; Alikhani, H.A.; Emami, S. Promotion of wheat growth under salt stress by halotolerant bacteria containing ACC deaminase. *J. Agric. Sci. Technol.* **2016**, *18*, 855–864.
99. Heydarian, Z.; Gruber, M.; Glick, B.R.; Hegedus, D.D. Gene expression patterns in roots of camelina sativa with enhanced salinity tolerance arising from inoculation of soil with plant growth promoting bacteria producing 1-aminocyclopropane-1-carboxylate deaminase or expression the corresponding acds gene. *Front. Microbiol.* **2018**, *9*, 1297. [CrossRef] [PubMed]
100. Chinnaswamy, A.; Coba de la Peña, T.; Stoll, A.; de la Peña Rojo, D.; Bravo, J.; Rincón, A.; Lucas, M.M.; Pueyo, J.J. A nodule endophytic Bacillus megaterium strain isolated from Medicago polymorpha enhances growth, promotes nodulation by Ensifer medicae and alleviates salt stress in alfalfa plants. *Ann. Appl. Biol.* **2018**, *172*, 295–308. [CrossRef]
101. Ateş; Kivanç, M. Isolation of ACC deaminase producing rhizobacteria from wheat rhizosphere and determinating of plant growth activities under salt stress conditions. *Appl. Ecol. Environ. Res.* **2020**, *18*, 5997–6008. [CrossRef]
102. Sagar, A.; Sayyed, R.Z.; Ramteke, P.W.; Sharma, S.; Marraiki, N.; Elgorban, A.M.; Syed, A. ACC deaminase and antioxidant enzymes producing halophilic Enterobacter sp. PR14 promotes the growth of rice and millets under salinity stress. *Physiol. Mol. Biol. Plants* **2020**, *26*, 1847–1854. [CrossRef] [PubMed]
103. Nagaraju, Y.; Mahadevaswamy; Naik, N.M.; Gowdar, S.B.; Narayanarao, K.; Satyanarayanarao, K. ACC Deaminase-positive halophilic bacterial isolates with multiple plant growth-promoting traits improve the growth and yield of chickpea (*Cicer arietinum* L.) Under Salinity Stress. *Front. Agron.* **2021**, *3*, 85. [CrossRef]
104. Singh, R.P.; Pandey, D.M.; Jha, P.N.; Ma, Y. ACC deaminase producing rhizobacterium Enterobacter cloacae ZNP-4 enhance abiotic stress tolerance in wheat plant. *PLoS ONE* **2022**, *17*, e0267127. [CrossRef]
105. Wei, H.; He, W.; Li, Z.; Ge, L.; Zhang, J.; Liu, T. Salt-tolerant endophytic bacterium Enterobacter ludwigii B30 enhance bermudagrass growth under salt stress by modulating plant physiology and changing rhizosphere and root bacterial community. *Front. Plant Sci.* **2022**, *13*, 2592. [CrossRef]
106. Qurashi, A.W.; Sabri, A.N. Bacterial exopolysaccharide and biofilm formation stimulate chickpea growth and soil aggregation under salt stress. *Braz. J. Microbiol.* **2012**, *43*, 1183–1191. [CrossRef]
107. Zhang, S.H.; Zhang, X.H.; Lv, L.; Wang, Q.; Jiang, Q. Formation of aerobic granular sludge under adverse conditions: Low DO and high ammonia. *J. Environ. Biol.* **2013**, *34*, 409.
108. Wang, W.; Liu, W.; Wang, L.; Yang, T.; Li, R. Characteristics and distribution research on extracellular polymer substance extracted from sewage sludge. *J. Environ. Biol.* **2016**, *37*, 305.
109. Vurukonda, S.S.K.P.; Vardharajula, S.; Shrivastava, M.; SkZ, A. Enhancement of drought stress tolerance in crops by plant growth promoting rhizobacteria. *Microbiol. Res.* **2016**, *184*, 13–24. [CrossRef] [PubMed]
110. Haggag, W.M. Isolation of bioactive antibiotic peptides from *Bacillus brevis* and *Bacillus polymyxa* against Botrytis grey mould in strawberry. *Arch. Phytopathol. Plant Prot.* **2008**, *41*, 477–491. [CrossRef]
111. Sandhya, V.S.K.Z.; Ali, S.Z.; Grover, M.; Reddy, G.; Venkateswarlu, B. Effect of plant growth promoting *Pseudomonas* spp. on compatible solutes, antioxidant status and plant growth of maize under drought stress. *Plant Growth Regul.* **2010**, *62*, 21–30. [CrossRef]

112. Giroldo, D.; Ortolano, P.I.; Vieira, A.A. Bacteria–algae association in batch cultures of phytoplankton from a tropical reservoir: The significance of algal carbohydrates. *Freshw. Biol.* **2007**, *52*, 1281–1289. [CrossRef]
113. Czarnes, S.; Hallett, P.D.; Bengough, A.G.; Young, I.M. Root- and microbial-derived mucilages affect soil structure and water transport. *Eur. J. Soil Sci.* **2000**, *51*, 435–443. [CrossRef]
114. Vanhaverbeke, C.; Heyraud, A.; Mazeau, K. Conformational analysis of the exopolysaccharide from Burkholderiacaribensis strain MWAP71: Impact on the interaction with soils. *Biopolymers* **2003**, *69*, 480–497. [CrossRef]
115. Chen, T.; Murata, N. Glycinebetaine: An effective protectant against abiotic stress in plants. *Trends Plant Sci.* **2008**, *13*, 499–505. [CrossRef]
116. Hidri, R.; Mahmoud, O.M.B.; Zorrig, W.; Mahmoudi, H.; Smaoui, A.; Abdelly, C.; Azcon, R.; Debez, A. Plant Growth-Promoting Rhizobacteria Alleviate High Salinity Impact on the Halophyte Suaeda fruticosa by Modulating Antioxidant Defense and Soil Biological Activity. *Front. Plant Sci.* **2022**, *13*, 1623. [CrossRef]
117. Vivas, A.; Marulanda, A.; Ruiz-Lozano, J.M.; Barea, J.M.; Azcón, R. Influence of a *Bacillus* sp. on physiological activities of two arbuscular mycorrhizal fungi and on plant responses to PEG-induced drought stress. *Mycorrhiza* **2003**, *13*, 249–256. [CrossRef]
118. Han, H.S.; Lee, K.D. Physiological responses of soybean-inoculation of *Bradyrhizobium japonicum* with PGPR in saline soil conditions. *Res. J. Agric. Biol. Sci.* **2005**, *1*, 216–221.
119. Roberson, E.B.; Firestone, M.K. Relationship between desiccation and exopolysaccharide production in a soil *Pseudomonas* sp. *Appl. Environ. Microbiol.* **1992**, *58*, 1284–1291. [CrossRef] [PubMed]
120. Ross, I.L.; Alami, Y.; Harvey, P.R.; Achouak, W.; Ryder, M.H. Genetic diversity and biological control activity of novel species of closely related pseudomonads isolated from wheat field soils in South Australia. *Appl. Environ. Microbiol.* **2000**, *66*, 1609–1616. [CrossRef]
121. Shultana, R.; Kee Zuan, A.T.; Yusop, M.R.; Saud, H.M. Characterization of salt-tolerant plant growth-promoting rhizobacteria and the effect on growth and yield of saline-affected rice. *PLoS ONE* **2020**, *15*, e0238537.
122. Chen, Y.; Yan, F.; Chai, Y.; Liu, H.; Kolter, R.; Losick, R.; Guo, J.H. Biocontrol of tomato wilt disease by Bacillus subtilis isolates from natural environments depends on conserved genes mediating biofilm formation. *Environ. Microbiol.* **2013**, *15*, 848–864. [CrossRef] [PubMed]
123. Nawaz, K.; Ashraf, M. Exogenous application of glycine betaine modulates activities of antioxidants in maize plants subjected to salt stress. *J. Agron. Crop Sci.* **2010**, *196*, 28–37. [CrossRef]
124. Wijman, J.G.; de Leeuw, P.P.; Moezelaar, R.; Zwietering, M.H.; Abee, T. Air-liquid interface biofilms of Bacillus cereus: Formation, sporulation, and dispersion. *Appl. Environ. Microbiol.* **2007**, *73*, 1481–1488. [CrossRef]
125. Xiang, W.; Guo, J.; Feng, W.; Huang, M.; Chen, H.; Zhao, J.; Zhang, Z.; Sun, Q. Community of extremely halophilic bacteria in historic Dagong Brine Well in southwestern China. *World J. Microbiol. Biotechnol.* **2008**, *24*, 2297–2305. [CrossRef]
126. Paul, D.; Nair, S. Stress adaptations in a plant growth promoting rhizobacterium (PGPR) with increasing salinity in the coastal agricultural soils. *J. Basic Microbiol.* **2008**, *48*, 378–384. [CrossRef]
127. Ashraf, M.; Hasnain, S.; Berge, O. Effect of exo-polysaccharides producing bacterial inoculation on growth of roots of wheat (*Triticum aestivum*) plants grown in a salt-affected soil. *Int. J. Environ. Sci. Technol.* **2006**, *3*, 43–51. [CrossRef]
128. Wafaa, M.W.; Haggag, M.M.; Hussein, H.M.; Mehanna, M.M.; El-Moneim, H.M.D. Bacteria polysaccharides elicit resistance of wheat against some biotic and abiotic stress. *Int. J. Pharm. Sci. Res.* **2014**, *50*, 292–298.
129. Trafny, E.A.; Lewandowski, R.; Zawistowska-Marciniak, I.; Stępińska, M. Use of MTT assay for determination of the biofilm formation capacity of microorganisms in metalworking fluids. *World J. Microbiol. Biotechnol.* **2013**, *29*, 1635–1643. [CrossRef] [PubMed]
130. Hong, B.H.; Joe, M.M.; Selvakumar, G.; Kim, K.Y.; Choi, J.H.; Sa, T.M. Influence of salinity variations on exocellular polysaccharide production, biofilm formation and flocculation in halotolerant bacteria. *J. Environ. Biol.* **2017**, *38*, 657. [CrossRef]
131. Upadhyay, S.K.; Singh, J.S.; Singh, D.P. Exopolysaccharide-producing plant growth-promoting rhizobacteria under salinity condition. *Pedosphere* **2011**, *21*, 214–222. [CrossRef]
132. Ruppel, S.; Franken, P.; Witzel, K. Properties of the halophyte microbiome and their implications for plant salt tolerance. *Funct. Plant Biol.* **2013**, *40*, 940–951. [CrossRef]
133. Upadhyay, S.K.; Singh, J.S.; Saxena, A.K.; Singh, D.P. Impact of PGPR inoculation on growth and antioxidant status of wheat under saline conditions. *Plant Biol.* **2012**, *14*, 605–611. [CrossRef] [PubMed]
134. Naseem, H.; Bano, A. Role of plant growth-promoting rhizobacteria and their exopolysaccharide in drought tolerance of maize. *J. Plant Interact.* **2014**, *9*, 689–701. [CrossRef]
135. Mahmud-Ur-Rahman; Naser, I.B.; Mahmud, N.U.; Sarker, A.; Hoque, M.N.; Islam, T. A highly salt-tolerant bacterium brevibacterium sediminis promotes the growth of rice (*Oryza sativa* L.) seedlings. *Stresses* **2022**, *2*, 275–289. [CrossRef]
136. Nautiyal, C.S.; Bhadauria, S.; Kumar, P.; Lal, H.; Mondal, R.; Verma, D. Stress induced phosphate solubilization in bacteria isolated from alkaline soils. *FEMS Microbiol. Lett.* **2000**, *182*, 291–296. [CrossRef]
137. Singh, R.P.; Jha, P.N. A halotolerant bacterium Bacillus licheniformis HSW-16 augments induced systemic tolerance to salt stress in wheat plant (*Triticum aestivum*). *Front. Plant Sci.* **2016**, *7*, 1890. [CrossRef]
138. Chen, M.; Wei, H.; Cao, J.; Liu, R.; Wang, Y.; Zheng, C. Expression of *Bacillus subtilis* proBA genes and reduction of feedback inhibition of proline synthesis increases proline production and confers osmotolerance in transgenic Arabidopsis. *BMB Rep.* **2007**, *40*, 396–403. [CrossRef]

139. Upadhyay, S.K.; Singh, D.P. Effect of salt-tolerant plant growth-promoting rhizobacteria on wheat plants and soil health in a saline environment. *Plant. Biol.* **2015**, *17*, 288–293. [CrossRef] [PubMed]
140. Abd_Allah, E.F.; Alqarawi, A.A.; Hashem, A.; Radhakrishnan, R.; Al-Huqail, A.A.; Olyan, F.; Al-Otibi, N.; Malik, A.; Alharbi, I.; Egamberdieva, D. Endophytic bacterium Bacillus subtilis (BERA 71) improves salt tolerance in chickpea plants by regulating the plant defense mechanisms. *J. Plant Interact.* **2018**, *13*, 37–44. [CrossRef]
141. Ahmad, P.; Abdel, L.A.A.; Abd_Allah, E.F.; Hashem, A.; Sarwat, M.; Anjum, N.A. Calcium and potassium supplementation enhanced growth, osmolyte, secondary metabolite production, and enzymatic antioxidant machinery in cadmium-exposed chickpea (*Cicer arietinum* L.). *Front. Plant Sci.* **2016**, *7*, 513. [CrossRef] [PubMed]
142. Fernandez-Aunión, C.; Ben-Hamouda, T.F.; Iglesias-Guerra, M.; Argandona Reina-Bueno, M.; Nieto, J.J.; Aouani, M.E.; Vargas, C. Biosynthesis of compatible solutes in rhizobial strains isolated from *Phaseolus vulgaris* nodules in Tunisian fields. *BMC Microbiol.* **2010**, *10*, 192. [CrossRef]
143. Le Rudulier, D.; Strøm, A.R.; Dandekar, A.M.; Smith, L.T.; Valentine, R.C. Molecular biology of osmoregulation. *Science* **1984**, *224*, 1064–1068. [CrossRef]
144. Wani, S.H.; Singh, N.B.; Haribhushanand, A.; Mir, J.I. Compatible solute engineering in plants for abiotic stress tolerance—Role of glycine betaine. *Curr. Genom.* **2013**, *14*, 157–165. [CrossRef] [PubMed]
145. Umezawa, T.; Fujita, M.; Fujita, Y.; Yamaguchi-Shinozaki, K.; Shinozaki, K. Engineering drought tolerance in plants: Discovering and tailoring genes to unlock the future. *Curr. Opin. Biotech.* **2006**, *17*, 113–122. [CrossRef] [PubMed]
146. Csonka, L.N.; Hanson, A.D. Prokaryotic osmoregulation: Genetics and physiology. *Annu. Rev. Microbiol.* **1991**, *45*, 569–606. [CrossRef]
147. Galinski, E.A.; Trüper, H.G. Microbial behaviour in salt-stressed ecosystems. *FEMS Microbiol. Rev.* **1994**, *15*, 95–108. [CrossRef]
148. Street, T.O.; Bolen, D.W.; Rose, G.D. A molecular mechanism for osmolyte-induced protein stability. *Proc. Natl. Acad. Sci. USA* **2006**, *103*, 13997–14002. [CrossRef]
149. Del Rio, L.A.; Corpas, J.F.; Sandalio, L.; Palma, J.; Barroso, D.J. Plant peroxisomes, reactive oxygen metabolism and nitric oxide. *IUBMB Life* **2003**, *55*, 71–81. [CrossRef]
150. Gururani, M.A.; Upadhyaya, C.P.; Baskar, V.; Venkatesh, J.; Nookaraju, A.; Park, S.W. Plant growth-promoting rhizobacteria enhance abiotic stress tolerance in Solanum tuberosum through inducing changes in the expression of ROS-scavenging enzymes and improved photosynthetic performance. *J. Plant Growth Regul.* **2013**, *32*, 245–258. [CrossRef]
151. Neshat, M.; Abbasi, A.; Hosseinzadeh, A.; Sarikhani, M.R.; Dadashi Chavan, D.; Rasoulnia, A. Plant growth promoting bacteria (PGPR) induce antioxidant tolerance against salinity stress through biochemical and physiological mechanisms. *Physiol. Mol. Biol. Plants* **2022**, *28*, 347–361. [CrossRef] [PubMed]
152. Groenhagen, U.; Baumgartner, R.; Bailly, A.; Gardiner, A.; Eberl, L.; Schulz, S.; Weisskopf, L. Production of bioactive volatiles by different *Burkholderia ambifaria* strains. *J. Chem. Ecol.* **2013**, *39*, 892–906. [CrossRef]
153. Garbeva, P.; Hordijk, C.; Gerards, S.; de Boer, W. Volatile-mediated interactions between phylogenetically different soil bacteria. *Front. Microbiol.* **2014**, *5*, 289. [CrossRef] [PubMed]
154. Wenke, K.; Kai, M.; Piechulla, B. Below ground volatiles facilitate interactions between plant roots and soil organisms. *Planta* **2010**, *231*, 499–506. [CrossRef]
155. Farag, M.A.; Zhang, H.; Ryu, C.M. Dynamic chemical communication between plants and bacteria through airborne signals: Induced resistance by bacterial volatiles. *J. Chem. Ecol.* **2013**, *39*, 1007–1018. [CrossRef]
156. Völker, U.; Engelmann, S.; Maul, B.; Riethdorf, S.; Völker, A.; Schmid, R.; Hecker, M. Analysis of the induction of general stress proteins of Bacillus subtilis. *Microbiology* **1994**, *140*, 741–752. [CrossRef]
157. Xie, S.; Wu, H.; Chen, L.; Zang, H.; Xie, Y.; Gao, X. Transcriptome profiling of Bacillus subtilis OKB105 in response to rice seedlings. *BMC Microbiol.* **2015**, *15*, 21. [CrossRef]
158. Diby, P.; Bharathkumar, S.; Sudha, N. Osmotolerance in biocontrol strain of pseudomonas pseudoalcaligenes MSP-538: A study using osmolyte, protein and gene expression profiling. *Ann. Microbiol.* **2005**, *55*, 243–247.
159. Paul, D.; Dineshkumar, N.; Nair, S. Proteomics of a plant growth-promoting rhizobacterium, Pseudomonas fluorescens MSP-393, subjected to salt shock. *World J. Microbiol. Biotechnol.* **2006**, *22*, 369–374. [CrossRef]
160. Steil, L.; Hoffmann, T.; Budde, I.; Völker, U.; Bremer, E. Genome-wide transcriptional profiling analysis of adaptation of *Bacillus subtilis* to high salinity. *J. Bacteriol.* **2003**, *185*, 6358–6370. [CrossRef]
161. Weber, A.; Jung, K. Profiling early osmostress-dependent gene expression in Escherichia coli using DNA macroarrays. *J. Bacteriol.* **2002**, *184*, 5502–5507. [CrossRef]
162. Shabala, S. Salinity and programmed cell death: Unravelling mechanisms for ion specific signalling. *J. Exp. Bot.* **2009**, *60*, 709–712. [CrossRef]
163. Georg, J.; Voss, B.; Scholz, I.; Mitschke, J.; Wilde, A.; Hess, W.R. Evidence for a major role of antisense RNAs in cyanobacterial gene regulation. *Mol. Sys. Biol.* **2009**, *5*, 305. [CrossRef]
164. Paul, D. Osmotic stress adaptations in rhizobacteria. *J. Basic Microbiol.* **2013**, *53*, 101–110. [CrossRef] [PubMed]
165. Rochester, D.E.; Winer, J.A.; Shah, D.M. The structure and expression of maize genes encoding the major heat shock protein hsp70. *EMBO* **1986**, *5*, 451–458. [CrossRef]
166. Lorimer, G.H. A quantitative assessment of the role of the chaperonin proteins in protein folding in vivo. *FASEB J.* **1996**, *10*, 5–9. [CrossRef]

167. Holland, N.; Belkind, A.; Holland, D.; Pick, U.; Edelman, M. Stress-responsive accumulation of plastid chaperonin 60 during seedling development. *Plant J.* **1998**, *13*, 311–316. [CrossRef]
168. Verdoucq, L.; Vignols, F.; Jacquot, J.P.; Chartier, Y.; Meyer, Y. In vivo characterization of a thioredoxin h target protein defines a new peroxiredoxin family. *J. Biol. Chem.* **1999**, *274*, 19714–19722. [CrossRef]
169. Patrice, M.; Robinson, A. The Enzymology of Osmoregulation in Archaea: Glutamine Synthetase and Beta-Glutamine Production. Ph.D. Thesis, Boston College, Chestnut Hill, MA, USA, 1999; p. 2677.
170. Visick, J.E.; Ichikawa, J.K.; Clarke, S. Mutations in the *Escherichia coli* surE gene increase isoaspartyl accumulation in a strain lacking the pcm repair methyltransferase but suppress stress-survival phenotypes. *FEMS Microbiol. Lett.* **1998**, *167*, 19–25. [CrossRef] [PubMed]
171. Kandasamy, S.; Loganathan, K.; Muthuraj, R.; Duraisamy, S.; Seetharaman, S.; Thiruvengadam, R.; Ramasamy, S. Understanding the molecular basis of plant growth promotional effect of Pseudomonas fluorescens on rice through protein profiling. *Proteome Sci.* **2009**, *7*, 47. [CrossRef] [PubMed]

agronomy

MDPI

Article

Characterization of Beekeepers and Their Activities in Seven European Countries

Raquel P. F. Guiné [1,2,*], Sónia Mesquita [1,*], Jorge Oliveira [1,2], Catarina Coelho [1,3], Daniela Teixeira Costa [1,2], Paula Correia [1,2], Helena Esteves Correia [1,2], Bjørn Dahle [4], Melissa Oddie [4], Risto Raimets [5], Reet Karise [5], Luis Tourino [6], Salvatore Basile [7], Emilio Buonomo [7], Ivan Stefanic [8] and Cristina A. Costa [1,2]

1 Agrarian School of Viseu, Polytechnic Institute of Viseu, 3500-606 Viseu, Portugal; joliveira@esav.ipv.pt (J.O.); ccoelho@esav.ipv.pt (C.C.); daniela@esav.ipv.pt (D.T.C.); paulacorreia@esav.ipv.pt (P.C.); hecorreia@esav.ipv.pt (H.E.C.); amarocosta@esav.ipv.pt (C.A.C.)
2 CERNAS Research Centre, Polytechnic Institute of Viseu, 3504-510 Viseu, Portugal
3 CECAV, Animal and Veterinary Research Center, University of Trás-os-Montes and Alto Douro, 5000-801 Vila Real, Portugal
4 Norwegian Beekeepers Association, 2040 Kløfta, Norway; bjorn.dahle@norbi.no (B.D.); melissa.oddie@nordi.no (M.O.)
5 Institute of Agricultural and Environmental Sciences, Estonian University of Life Sciences, 51014 Tartu, Estonia; ristorai@gmail.com (R.R.); reet.karise@emu.ee (R.K.)
6 Eosa Estrategia y Organización SA, 36202 Vigo, Spain; ltourino@eosa.com
7 Bio-Distretto Cilento, 84052 Ceraso, Italy; presidente@ecoregions.eu (S.B.); emiliobuonomo@gmail.com (E.B.)
8 Tera Tehnopolis, 31000 Osijek, Croatia; istefanic@fazos.hr
* Correspondence: raquelguine@esav.ipv.pt (R.P.F.G.); soniasmmsa@gmail.com (S.M.)

Citation: Guiné, R.P.F.; Mesquita, S.; Oliveira, J.; Coelho, C.; Costa, D.T.; Correia, P.; Correia, H.E.; Dahle, B.; Oddie, M.; Raimets, R.; et al. Characterization of Beekeepers and Their Activities in Seven European Countries. *Agronomy* **2021**, *11*, 2398. https://doi.org/10.3390/agronomy11122398

Academic Editor: Christos Athanassiou

Received: 13 October 2021
Accepted: 22 November 2021
Published: 25 November 2021

Abstract: Beekeeping is an ancient activity that is gaining interest among practitioners and society in general. It is as an activity with positive impacts in the environmental, social and economic spheres, with the potential to reconnect these dimensions and contribute to sustainable development. Thus, it is important to determine the profiles of beekeepers across the world, and to understand the main social, economic or ecological drivers that shape their activities. Hofstede cross-cultural dimensions were used to better explain differences between countries. A survey was undertaken of beekeepers in different countries (Croatia, Estonia, Finland, Italy, Norway, Portugal, and Spain) in the native language of each nation. A total of 313 questionnaires, using an online platform or in paper form, was delivered face-to-face during training or dissemination events in 2019. Norway and Finland were the countries with the highest percentage of respondents with a university degree (>80%), while Spain (42%) and Croatia (48%) presented the lowest percentages. Most participants were experienced beekeepers (59% had more than five years of experience) with more than 50 colonies. With the exception of Italy, beekeeping appears to be a hobby or an additional professional activity. The main beekeeping products for these beekeepers were honey, wax, colonies and propolis, with an average honey production per season of 24.5 kg/hive. Crossing socio-demographic characteristics and Hofstede cross-cultural dimensions showed a relation among countries with higher Power Distance Index (PDI; this value expresses how society accepts and expects a certain inequality of power) and lower annual income and educational level (Croatia, Portugal, and Spain). A strong correlation appeared between Masculinity Femininity Index (MFI; this value refers to gender effects in society, with feminine societies meaning that the dominant values are caring for others and quality of life, as opposed to masculine societies, which are driven by competition, achievement and success) and age, income and education, with Norway presenting the most feminine society, with more educated and older beekeepers. The Uncertainty Aversion Index (UAI; this value explains how members of a society feel when dealing with unknown situations) was strongly associated with education. The results showed that increasing beekeepers' income will contribute to balancing the distribution of power among members of society, and that this might be achieved by training, especially in Croatia, Portugal, Spain, and Italy.

Keywords: beekeeper; bee products; beekeeping sector; Hofstede dimensions

1. Introduction

Although several insect species play a significant role in pollination, honeybees are the most important managed pollinators, not only for agricultural crops but also for wild plants. Hence, managed honeybee colonies impact ecology and the economy to a great extent [1]. Beekeeping activities are ancient, dating back to 4500 BC, although nowadays, native beekeeping is complemented with professional, high knowledge input regarding colony management and the production of bee products [2,3].

Beekeeping is as an activity with positive impacts in the environmental, social and economic spheres, playing an extremely important role within family farming. It can also generate extra income or create job opportunities [4]. In many countries, beekeeping activities are transmitted from generation to generation. Uchiyama et al. [2] reported that beekeepers whose knowledge was transmitted from their ancestors tend to have more bee colonies, and also seem to better understand how ecological conditions are fundamental to sustainable beekeeping. Nevertheless, from the 1850s, owing to advances in apicultural science and new technologies, beekeeping became more efficient and profitable, rapidly shifting from family businesses to a commercial activity [5]. The importance of the traditional aspects associated with beekeeping is different among societies and countries [2,6,7], and, in many cases, reflects specific cultural factors [8,9].

Due to the actions focused on economic expansion, we have witnessed a growing imbalance in the social and environmental spheres. Therefore, humanity is faced with a problem that urgently needs to be solved: reconciling the economic dimension with the social and environmental dimensions [10]. The term Sustainability, according to the United Nations (1987), is defined as the ability "to satisfy the needs of the present without compromising the ability of future generations to satisfy their own needs". In this context, beekeeping satisfies all the necessary requirements for sustainability, having a direct relationship with plant biodiversity through pollination [11], as well as showing a capacity to increase the productivity of the primary sector, diversify and stimulate agricultural production, reduce unemployment and promote economic development.

About 35% of the plants used for human consumption depend on pollination, so the importance of preserving bees as pollinators is undeniable [12]. Three out of four crops that produce fruit or seeds for human consumption depend on pollinators, i.e., largely on bees. In view of the decline of pollinators, Food and Agriculture Organization of the United Nations (FAO) has developed initiatives to promote favorable practices in agricultural management, such as technical assistance to countries regarding the breeding of queen bees and sustainable solutions for the production and marketing of honey [13].

Thus, it is important to determine the profiles of beekeepers across the world, in terms of social, economic or ecological drivers, as well as the difficulties they experience, and to provide them with training on sustainable beekeeping practices to increase and encourage the practice of this activity that is so important for maintaining the balance of ecosystems and for sustainability.

According to Geert Hofstede, discrepancies in behaviors among countries can be attributed to cultural differences [14,15]. Hence, it is to be expected that different ways of interpreting, interacting, thinking or behaving might be related to cultural variances. These differences may also be present when someone interacts with people from other subcultures, as well as from different social classes, religions, gender or even from different regions within the same country. In an attempt to explain why people from other cultures seem to behave and think differently, Hofstede developed a theory based on studies he carried out in the 1980s involving more than 50 national cultures. His "Theory of Cultural Dimensions" offers a framework to examine how cultural values affect behaviors and give clues about the ways in which people in a certain cultural environment can act. According to his theory, six cultural dimensions exist: Power Distance Index (PDI); Individualism versus Collectivism Index (ICI); Masculinity *versus* Femininity Index (MFI); Uncertainty Avoidance Index (UAI); Term Orientation Index (TOI); and Indulgence Restraint Index (IRI) [14,15]. Hofstede provided scores for these variables in a number of countries which have been used in many fields. For example, they were used to examine cultural differences in the food sector [16,17], in tourism related activities [18,19] and in teaching and learning contexts [20,21].

Given the importance of beekeepers in the European Union, in particular nonprofessional and family beekeepers, it is relevant to understand how beekeeping activities develop in different socio-economic and cultural contexts, as a first step to sharing experiences across contexts in order to promote international cooperation, beekeeping innovation and the exchange of good practices. In this context, the aim of this research is to understand how socio-economic and cultural dimensions influence beekeepers' options and contribute to more sustainable practices. To this end, we first characterize beekeeping activities in seven European countries, and then investigate possible differences according to sociodemographic variables and variables characterizing beekeeping activities. Finally, we relate these variables with the cross-cultural dimensions defined by Hofstede.

2. Materials and Methods

The survey was sent to beekeepers in different countries as a part of the Beeb—beekeeping bridges project (2019-1-PT01-KA202-060782). The questionnaire was divided into different sections as follows: (I) Experience in beekeeping (comprising 10 questions); (II) Training needs (3 questions); (III) Experience in beekeeping training activities (3 questions); (IV) Use of distance learning technologies and tools (3 questions); (V) Distance learning tools (4 questions); and (VI) Sociodemographic characterization (6 questions). In this manuscript, parts (I) and (VI) of the questionnaire are addressed.

The questionnaire was translated into the languages of the different countries in which it was applied, i.e., Croatia, Estonia, Finland, Italy, Norway, Portugal, and Spain (Figure 1). The questionnaire, using an online platform or in paper, was delivered face-to-face during training or dissemination events organized by beekeeper associations or companies in each country.

For data treatment, the SPSS version 26 and Excel 2016 software were used. Basic statistical tools were utilized to describe the data, such as frequencies or mean values. To evaluate possible differences between the variables characterizing beekeeping activities among countries, crosstabs with chi-square tests were performed. The Cramer's V coefficient was considered when evaluating the strength of the relations between some of the variables under study. This coefficient varies from 0 to 1; for $V \approx 0.1$, the association was considered weak, while for $V \approx 0.3$ and $V \approx 0.5$ or over, the association was moderate and strong, respectively [22].

The obtained values for Hofstede's cross-cultural dimensions are shown in Table 1. Spearman correlations were determined in order to assess the relations between the six Hofstede cross-cultural dimensions and some other variables (sociodemographic variables and variables characterizing beekeeping activities). To this end, mean values of the tested variables were determined for each country (for ratio variables, i.e., age, income, time of activity, number of colonies, relevance of the activity) or percentage of positive responses (for categorical variables, i.e., being male, being a beekeeper, a technician, or a merchant, no one working on beekeeping project, family, friends, hired people).

The strength of the correlations was evaluated according to the following limits: if $\rho = 0$, there is no correlation; if $\rho \in [0.0, 0.2]$ the correlation is very weak; if $\rho \in [0.2, 0.4]$, the correlation is weak; if $\rho \in [0.4, 0.6]$, the correlation is moderate; if $\rho \in [0.6, 0.8]$, the correlation is strong; if $\rho \in [0.8, 1.0]$, the correlation is very strong; and if $\rho = 1$, the correlation is perfect [23,24]. The level of significance was 0.05 in all cases.

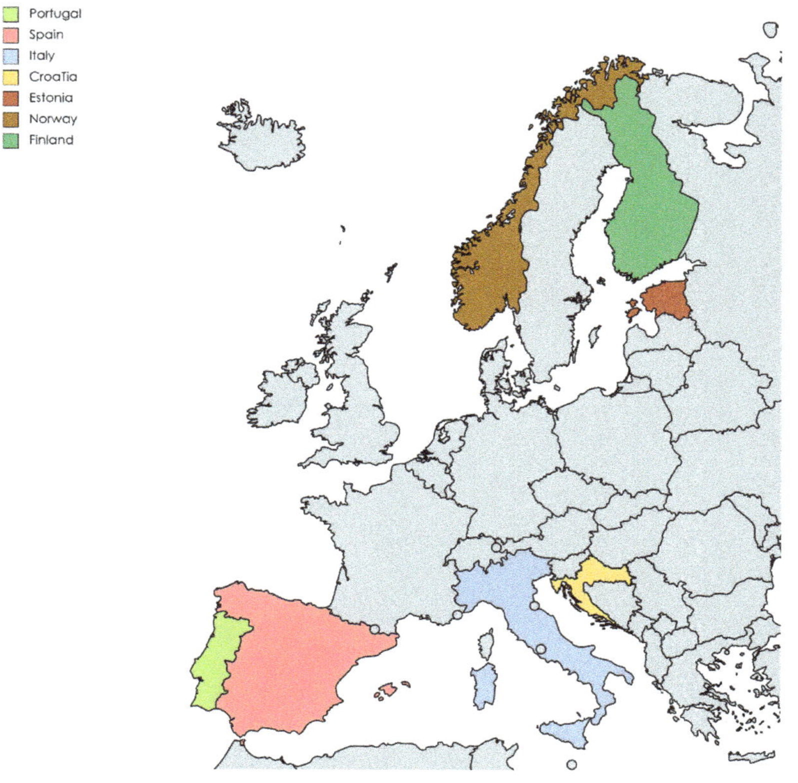

Figure 1. Geographical location of the countries included in the study.

Table 1. Scores of predictor variables (Hofstede's cross-cultural dimensions) for the seven countries included in the study [25].

Country	PDI	ICI	MFI	UAI	TOI	IRI
Croatia	73	33	40	80	58	33
Estonia	40	60	30	60	82	16
Finland	33	63	26	59	38	57
Italy	50	76	70	75	61	30
Norway	31	69	8	50	35	55
Portugal	63	27	31	99	28	33
Spain	57	51	42	86	48	44

Legend: Hofstede's cross-cultural dimensions: PDI—Power Distance Index; ICI—Individualism Collectivism Index; MFI—Masculinity—Femininity Index; UAI—Uncertainty Avoidance Index; TOI—Term Orientation Index; IRI—Indulgence Restraint Index. "Rule of the thumb"—If a score was under 48, the cultural score was relatively low on that scale, while if a score was over 52, the cultural scores were high on that scale. Scores between 48 and 52 were considered as intermediate. ☐ high; ☐ intermediate; ☐ low.

3. Results

3.1. Sample Characterization

Table 2 shows a sociodemographic characterization of the sample, which consisted of 313 participants from the seven countries included in the study. Norway was the country where most responses were collected (n = 74), followed by Croatia (n = 64), while those with the fewest responses were Finland and Italy (n = 15 and n = 16, respectively).

Table 2. Sociodemographic characterization of the sample (N = 313).

Variables	Categories	*n*	%
Country	Portugal	48	15.3
	Estonia	44	14.1
	Norway	74	23.6
	Spain	52	16.6
	Italy	16	5.1
	Croatia	64	20.4
	Finland	15	4.8
Age	Young (up to 30 y)	29	9.3
	Middle aged (31–59 y)	212	67.7
	Senior (age 60 or more)	57	18.2
	No answer	15	4.8
Sex	Female	73	23.3
	Male	231	73.8
	No answer	9	2.9
Education	Basic	10	3.2
	Secondary	110	35.1
	University	183	58.5
	No answer	10	3.2
Income	Less than 15,000 €/year	79	25.2
	Between 15,000 to 50,000 €/year	120	38.3
	More than 50,000 €/year	71	22.7
	No answer	43	13.7
Activity Type	Beekeeper	261	80.5
	Techician	39	12.0
	Merchant	24	7.4

Regarding age, participants were categorized as follows: young (up to 30 years old), middle-aged (between 31 and 59 years old) and senior (60 years or over). The distribution by age was 29 young respondents, 212 middle-aged respondents and 57 seniors (Table 2). From the 313 participants, 15 did not indicate their age. In terms of country distribution (Figure 2a), Italy was the country with highest percentage of young adults (31%), while Norway and Finland had the highest percentages of senior adults (37% and 29%, respectively).

As for gender, only nine participants did not specify their sex. The results indicated that the sample at study comprised mostly men, i.e., 231, with only 73 women (see Table 2). In Spain, Portugal and Finland, the differences between gender were more pronounced, with only 10%, 11% and 13% female participants, respectively.

Among the surveyed beekeepers, most had completed a university degree (n = 183) or secondary school (n = 110); a very low number, i.e., 10 participants, reported the lowest level of education (basic school only). Ten participants failed to respond to this question (Table 2).

Norway and Finland presented the highest percentage (>80%) of respondents that had completed a university education (Figure 2b), while Spain and Croatia presented the lowest percentage for this indicator (42% and 48%, respectively). Accordingly, Spain presented the higher percentage for participants who had completed only basic education, i.e., 12%.

Finally, for those who felt comfortable, one question addressed household annual income. For this question, a total of 270 answers were obtained, showing that for most participants (n = 120), annual income was between 15,000 and 50,000 €/year (Table 2). About 86% of participants from Norway had an annual income above 50,000 €, while few participants from Italy, Croatia, Spain or Portugal were within the highest income category (0%, 2%, 8% and 9%, respectively). Regarding the lowest annual income, i.e., under 15,000 €, Italy stood out with the highest percentage (63%), followed by Croatia (53%) (Figure 2c).

The participants were asked if they had beekeeping experience, and, if so, in which role(s), i.e., beekeeper, technician or merchant. The results revealed that participants were mostly beekeepers (n = 261), with few respondents identifying as technicians (n = 39) or merchants (n = 24).

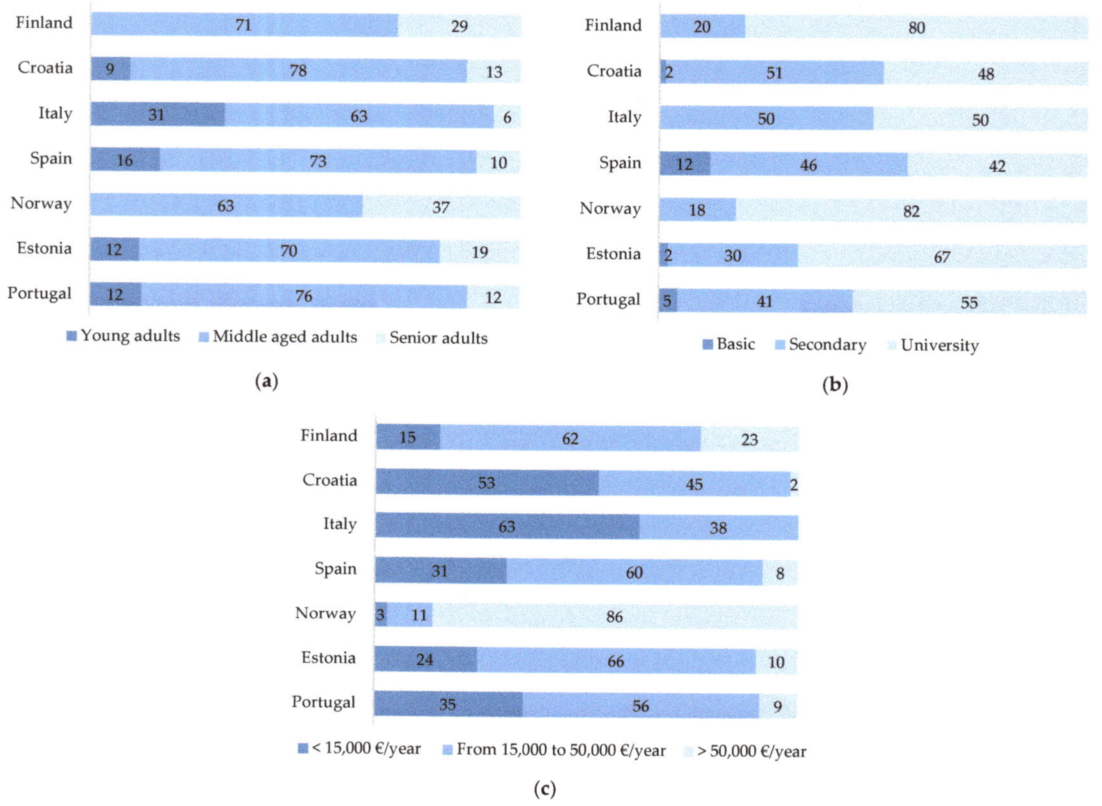

Figure 2. Relative distribution of age (**a**), education level (**b**), and income (**c**) by country (%).

3.2. Characterization of Beekeeping Activity

Most participants were experienced beekeepers, with only 9% reporting less than one year of activity, 32% with between 1 and 5 years of experience, 23% between 5 and 10 years of experience, and 36% reporting having worked in the field for more than 10 years. Twenty-six participants did not respond to this question. In Italy, the percentage of beekeepers with more than 10 years or experience was low (8%); Finland and Croatia reported the highest numbers of beekeepers with more than 10 years of experience (47% and 42%, respectively) (Figure 3).

The number of honeybee colonies per participant was highly variable, up to a maximum of 2400, distributed follows: up to 50 colonies (n = 178), between 51 and 150 (n = 56), between 151 and 300 (n = 21) and more than 301 (n = 17). From the 272 participants who indicated the number of colonies, the majority had a low number (65%). The result analyzed by country showed a similar trend, with participants being mostly beekeepers with stocks of up to 50 colonies (varying from 43% to 82%). This percentage was lowest for Finland. In contrast, Portugal had the highest percentage of beekeepers with 301 or more colonies (23%) (Figure 4).

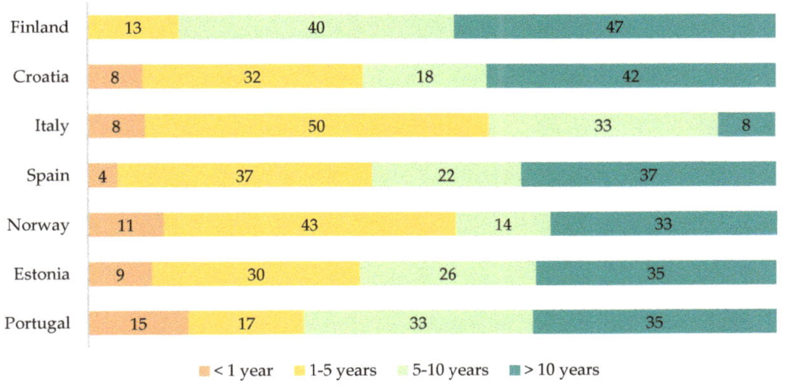

Figure 3. Relative distribution of time of activity by country (%).

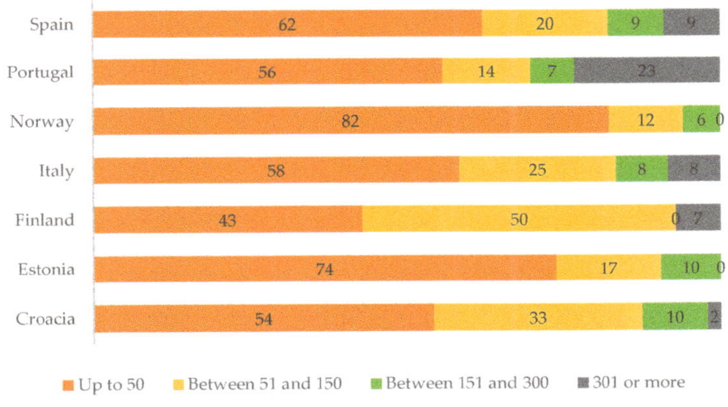

Figure 4. Relative distribution of number of honeybee colonies by country (%).

When asked about the relevance of the beekeeping activities, some said it was a hobby (120 participants), and as such, the products were mainly for self-consumption, family and gifts. Others said that it was a source of additional income for the family (118 participants). Finally, it was reported as a core business for a small number of participants, i.e., 43 participants, representing 15%. With the exception of Italy, where the majority (42%) cited beekeeping as their main source of income, in all countries, the prevailing trend was that beekeeping was a hobby or an additional source of income (Figure 5a).

In most cases (n = 217), all the work involved in the beekeeping activity was done by the individual, with help from unpaid collaborators. In fact, only a minority of the work was done by people hired specifically for a given job (only 2% of the beekeepers reported hiring workers). Family members or friends frequently participated, as reported by 49% and 9% of participants, respectively. A large number of participants revealed that they worked alone (40%). Figure 5b shows the relative importance of these stakeholders in beekeeping activities by country. The roles of the beekeepers and their families were shown to be pivotal in all countries, collectively representing a minimum of 79% in Spain and a maximum of 100% in Estonia. Italy was the country with the highest percentage of hired people, even though this still only represented 17%.

The participants were also asked to indicate the three most important products in their beekeeping activities; the results are presented in Table 3. The global results showed that the most important products were honey (285 positive answers), wax (103) and colonies (87).

Propolis ranked fourth, very close to colonies, being indicated by 82 participants as one of the three most important products in their beekeeping activities.

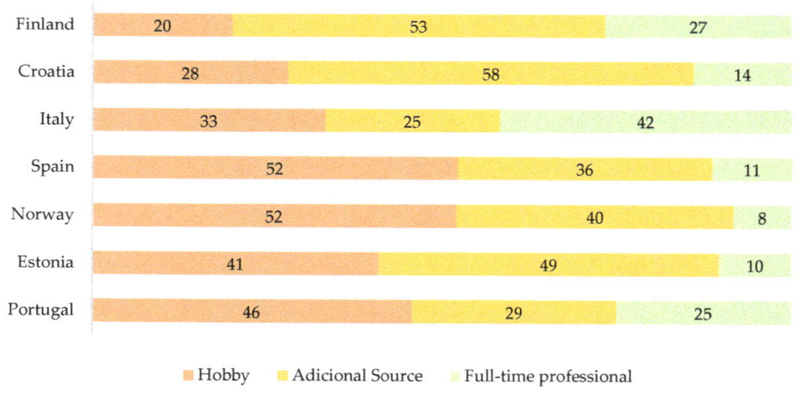

(a)

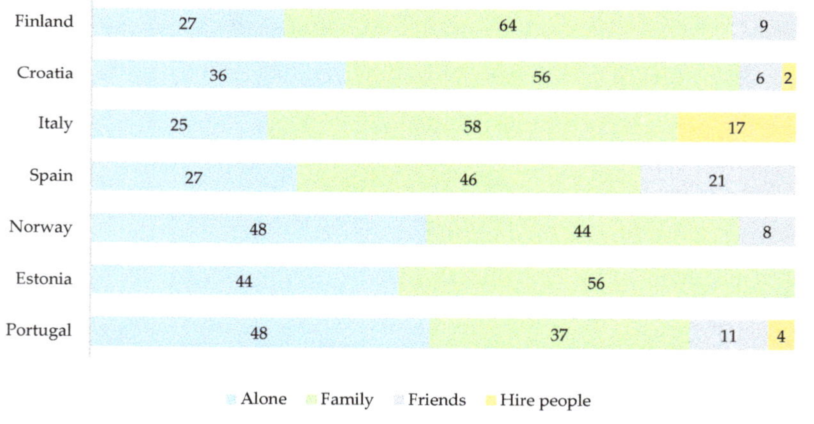

(b)

Figure 5. Relative distribution of the beekeeping activity relevance (**a**) and people who work in beekeeping by country (**b**) (%).

Table 3. Relevant products of beekeeping activities by country (Results presented as number of beekeepers).

	Portugal (N = 48)	Estonia (N = 44)	Norway (N = 74)	Spain (N = 52)	Italy (N = 16)	Croatia (N = 64)	Finland (N = 15)	Total
Honey	46	43	69	50	12	51	14	285
Wax	19	20	29	8	4	20	3	103
Colonies	15	10	24	15	2	17	4	87
Propolis	10	17	2	11	10	27	5	82
Polen	10	5	4	10	7	19	4	59
Queens	5	8	29	7	2	2	3	56
Polination services	8	1	8	4	0	3	3	27
Royal gelee	0	1	0	1	0	6	0	8
Apitoxin	0	0	0	2	0	1	0	3

From the participants in the survey, 92 indicated that they were migratory beekeepers while 194 were not. As for the mode of production, 74 declared themselves to be organic producers compared to 214 who were not. Among those who were organic, only 16 were certified.

The participants were asked about the average honey production per hive, considering the last season; the obtained responses were highly variable. The results for honey production per hive were classified as follows: 0 kg (n = 14); 1–10 kg (n = 44); 11–20 kg (n = 78); 21–50 kg (n = 95); 51–100 kg (n = 18); and greater than 100 kg (only one producer). In most cases, honey production between 21 and 50 kg per hive per season was reported. The results showed that the average honey production per hive by country was, in decreasing order, 38 kg in Estonia, 36 kg in Finland, 13.7 kg in Portugal, 32 kg in Croatia, 27 kg in Norway, 19 kg in Spain and 17 kg in Italy.

3.3. Country Differences in the Beekeeping Activity

Table 4 shows the relations between country for additional variables. Significant differences were observed for beekeeper age among countries (p < 0.0012), and the association was moderate (V = 0.456). Italy (36.25%) was the country with the highest number of young adults, followed by Spain (16.3%). In Croatia (78.1%), Portugal (75.6%) and Spain (73.5%), middle-aged adults represented the majority. Gender was significantly different among countries (p = 0.006), with a moderate association between variables (V = 0.244). The highest percentage of women beekeepers was reported in Italy (37.5%), followed by Estonia (35.7%). Income differences between countries were highly significant (p < 0.0000) and the association was strong (V = 0.573). Norwegian beekeepers had the highest income.

Table 4. Crosstabs and Chi-square tests among countries and some sociodemographic variables and variables associated with beekeeping activities. Percentages of positive answers for each variable.

Variables/Categories	Croacia	Estonia	Finland	Italy	Norway	Portugal	Spain
Age (p < 0.012, V = 0.456) (1)							
Young (up to 30 y)	9.4	11.6	0.0	31.25	0.0	12.2	16.3
Middle age (31–59 y)	78.1	69.8	71.4	62.5	63.4	75.6	73.5
Senior (60y or over)	12.5	18.6	28.6	6.25	36.6	12.2	10.2
Sex (p = 0.006, V = 0.244) (1)							
Female	25.0	35.7	13.3	37.5	32.9	11.4	10.0
Male	75.0	64.3	86.7	62.5	67.1	88.6	90.0
Education level (p = 0.0000, V = 0.263) (1)							
Basic	1.6	2.3	0.0	0.0	0.0	4.55	12.0
Secondary	50.8	30.2	20.0	50.0	18.1	40.9	46.0
University	47.6	67.5	80.0	50.0	81.9	54.55	42.0
Income (p < 0.0000, V = 0.573) (1)							
Less than 15000 €/year	58.2	24.4	15.4	62.5	3.1	35.3	31.3
15000 to 50,000 €/year	45.3	65.8	61.5	37.5	10.8	55.9	60.4
More than 50,000 €/year	1.9	9.8	23.1	0.0	86.1	8.8	8.3
Experience in beekeeping activities							
Beekeeper (p < 0.0000, V = 0.332) (1)	43.8	81.2	82.3	80.0	97.4	86.8	95.7
Beekeeping technician (p < 0.0000, V = 0.755) (1)	35.9	9.4	11.8	13.3	1.3	7.5	4.3
Beekeeping merchant (p < 0.0000, V = 0.738) (1)	20.3	9.4	5.9	6.7	1.3	5.7	0.0
Time of activity (p = 0.112, V = 0.172) (1)							
Less than 1 year	8.0	9.3	0.0	8.3	10.9	14.6	4.3
1 to 5 years	32.0	30.2	13.3	50.0	42.5	16.7	37.0
6 to 10 years	18.0	25.6	40.0	33.4	13.7	33.3	21.7
More than 10 years	42.0	34.9	46.7	8.3	32.9	35.4	37.0
Number of colonies (p < 0.000, V = 0.249) (1)							
Up to 50	54.2	73.8	42.9	58.4	82.3	55.8	62.2
Between 51 and 150	33.3	16.7	50.0	25.0	11.8	14.0	20.0
Between 151 and 300	10.4	9.5	0	8.3	5.9	7.0	8.9
301 or more	2.1	0	7.1	8.3	0	23.2	8.9

Table 4. *Cont.*

Variables/Categories	Croacia	Estonia	Finland	Italy	Norway	Portugal	Spain
Relevance of the beekeeping activity ($p = 0.006$, V = 0.221) [(1)]							
Hobby	28.0	37.2	20.0	33.3	52.1	45.8	52.2
Source of additional income	58.0	51.2	53.3	25.0	39.7	29.2	36.4
Main business	14.0	11.6	26.7	41.7	8.2	25.0	11.4
People who work in beekeeping							
No one ($p = 0.305$, V = 0.151) [(1)]	36.0	32.6	20.0	25.0	41.1	48.1	27.3
Family ($p = 0.05$, V = 0.197) [(1)]	56.0	55.8	46.7	58.3	45.2	37.0	47.7
Friends ($p = 0.013$, V = 0.227) [(1)]	6.0	2.3	13.3	0.0	12.3	11.1	25.0
Hired ($p = 0.001$, V = 0.264) [(1)]	2.0	9.3	20.0	16.7	1.4	3.7	0.0
Fraction of work done by you, your family or unpaid friend/neigbours ($p = 0.0000$, V = 0.264) [(1)]							
All (100%)	93.9	83.3	50.0	83.3	94.4	55.6	81.8
More than 50%	4.1	11.9	42.9	16.7	4.2	40.7	9.1
Between 25 and 50%	0.0	0.0	7.1	0.0	0.0	3.7	2.3
Less than 25%	2.0	4.8	0.0	0.0	1.4	0.0	6.8

[(1)] Chi-square test p-value (level of significance of 0.05) and Cramer's V coefficient.

Regarding experience as a beekeeper, technician or merchant, highly significant differences were found ($p < 0.0000$) in all cases, with a moderate association for beekeeper (V = 0.332) and strong associations for technician (V = 0.755) and merchant (V = 0.738). For the number colonies, differences between countries were significant and moderate ($p < 0.000$, V = 0.249). The relevance of beekeeping activity was very significantly different between countries ($p = 0.006$), and the association was moderate (V = 0.221). Regarding the types of people who work in beekeeping, significant differences were found between countries where beekeepers reported receiving help from friends ($p = 0.013$), with a moderate association (V = 0.227), and very significant differences were observed for beekeepers who hired personnel ($p = 0.001$), with a moderate association (V = 0.264). Regarding the portion of work done by the beekeeper, family or unpaid friends/neighbors and education level, highly significant differences were found ($p = 0.0000$) among counties with a moderate association (V = 0.264 and V = 0.263, respectively). However, for the variable "time of activity", no significant differences were found among countries ($p = 0.112$).

3.4. The Hofstede's Cross-Cultural Dimensions

In this work, the relation between some variables and the Hofstede cross-cultural dimensions was also investigated. The values of the six dimensions for the countries in this study are presented in Table 1, while Table 5 shows the associations between the Hofstede cultural dimensions and some other sociodemographic characteristics and beekeeping technical options.

As seen in Table 5, PDI was inversely significantly correlated with income and educational, and the association was very strong ($\rho = -0.786$ and $\rho = -0.865$). According to our Hofstede analysis, Norway (31%), Finland (33%) and Estonia (40%) were the countries with lowest PDI values (Table 1). These were also the countries where a higher number of beekeepers had more than 50,000 € of annual income (Norway—86.1%, Finland—23.1% and Estonia—9.8%) and higher educational level (Norway—81.9%, Finland—80.0% and Estonia—67.5%).

Regarding ICI [14], individualist countries were Italy (76%), Norway (69%) and Finland (63%), while Croatia (33%) and Portugal (27%) stood out as collectivist countries (Table 1). The results in Table 5 reveal that there were no significant correlations between ICI and the variables studied.

MFI was inversely significantly correlated with age ($\rho = -0.964$), income ($\rho = -0.893$) and education level ($\rho = -0.847$), with a very strong association (Table 5) [14]. Among the seven countries, only one revealed a high level of masculinity, i.e., Italy (70%) (Table 1). This country presented the lowest rate (6.25%) of senior beekeepers (60 years or over), and not a single beekeeper with more than 50,000 € of annual income. Regarding educational, 50% of Italian participants had secondary level and 50% superior level.

UAI was significant correlated with the number of colonies (ρ = 0.811), and was inversely significantly correlated with educational level (ρ = −0.847) [14]. The countries included in this research showed high UAI. Portugal was the country with the most participants reporting more than 301 colonies (more than 23%), but was also the country with the highest percentage (12.0%) of participants with basic education.

Table 5. Spearman correlations between the Hofstede scores of cultural dimensions and the variables that characterize beekeeping activities and the sociodemographic characteristics of beekeepers.

Variables/Categories	PDI	ICI	MFI	UAI	TOI	IRI
Sex	0.357	−0.679	0.000	0.536	−0.607	0.541
Age	−0.750	0.143	−0.964 **	−0.643	−0.464	0.541
Income	−0.786 *	0.214	−0.893 **	−0.679	−0.357	0.595
Education level	−0.865 *	0.541	−0.847 *	−0.847 *	−0.234	0.300
Experience in beekeeping activities						
Beekeeper	−0.464	0.036	−0.464	−0.143	−0.679	0.577
Beekeeping technician	0.429	0.000	0.429	0.107	0.536	−0.360
Beekeeping merchant	0.286	−0.036	0.143	−0.071	0.643	−0.577
Time of activity	0.286	−0.536	−0.143	0.179	−0.214	0.505
Number of colonies	0.631	−0.342	0.703	0.811 *	−0.180	−0.209
Relevance of the beekeeping activity	0.250	0.179	0.429	0.107	0.286	−0.198
People who work in the beekeeping						
No one	0.071	−0.536	−0.500	0.143	−0.500	−0.054
Family	−0.286	0.536	0.107	−0.429	0.607	−0.054
Friends	0.162	−0.487	−0.072	0.414	−0.739	0.655
Hired	0.571	−0.079	0.630	0.512	0.039	−0.477

Hofstede's cross-cultural dimensions: PDI—Power Distance Index; ICI—Individualism Collectivism Index; MFI—Masculinity—Femininity Index; UAI—Uncertainty Avoidance Index; TOI—Term Orientation Index; IRI—Indulgence Restraint Index. ** Correlation is significant at the 0.01 level; * Correlation is significant at the 0.05 level.

Regarding TOI and IRI, the results in Table 5 show that no significant correlations were found among the applied variables.

4. Discussion

This study presents recent sociodemographic information about beekeepers and beekeeping activities in seven European countries (Estonia, Croatia, Finland, Italy, Norway, Portugal and Spain). Among the seven countries, there were significant differences regarding age, sex, income, experience in beekeeping activities, fraction done by unpaid people and education level.

The participants in this study were mostly over 31 years of age, with a significant percentage being over 60 years old. Italy was the country with the youngest participants (31.25%), while Norway had the oldest ones (36.6%).

Information about European beekeepers is sparse. However, the Report on Prospects and Challenges for the EU Apiculture Sector [26], produced by the Committee on Agriculture and Rural Development, notes a particularly serious ageing problem in this sector, with only a small percentage of beekeepers in EU being aged under 50. In fact, as in our study, in the EPILOBEE project, only 4.35% of the beekeepers were less than 30 years old, and 23.1% were between 31 and 45 [27]. The ageing of the general population is a common problem in all European countries. Additionally, young people are not widely attracted to agriculture and agrarian activities like beekeeping. Therefore, it is urgent take measures to attract young people to this sector.

There is also very limited information about women beekeepers. Even with the numbers of new women beekeepers growing all over the world, they are always less represented than men [28–32]. In our study, women were also underrepresented, with Italy being the country with the highest percentage of women (37.5%) and Spain the country with the lowest (10%). Beekeeping is an activity mainly carried out by men, even though it has the potential to contribute to the empowerment of rural women, reinforcing their role in agriculture [33]. This is an activity that can be easily adjusted to the multifunctional role of rural women while, at the same time, providing an additional source of income.

The average annual income among beekeepers was lower than 50,000 € in all countries except Norway. More than 50% of the Italian and Croats reported earning less than 15,000 € per year, while more than 50% of Estonian, Finnish, Spanish and Portuguese participants reported earning between 15,000 € and 50,000 € per year. In Europe, beekeeping activities provide over 620,000 EU citizens with their main income or additional earnings [34].

Regarding educational level, most of the participants had higher or secondary education. In Finland and Norway, the level of education among beekeepers was highest. This was as expected, as the national levels of education are higher in these two nations (over 38%) (https://www.statista.com/statistics/1084737/eu-28-adults-with-tertiary-education-attainment/, accessed on 21 October 2021).

Most of the participants had more than five years of experience in beekeeping (over 60% in all countries) with the exception of Italy, where only 42% reported having more than five years of experience. This is possibly related to the fact that Italy was the country with the highest proportion of youngest participants (31.3%).

According to the EU [35], in 2019, the total number of beekeepers in Europe was over 600,000 with Italy being the country, among the countries included in this study, with the highest percentage (8.3%) of European beekeepers, followed by Spain (3.9%) which was also the country with the highest percentage of beehives (16.8%). More than 70% of respondents had up to 150 colonies. Portugal was the country in which the most participants had apiaries with more than 151 colonies (30.2%), but this result did not correspond to the real beekeeping scenario in the country, given that the percentage of beekeepers with more than 150 colonies was, in 2018, only 10.9% [36,37]. Portugal had also the highest average number of colonies per beekeeper (229), that is over the reported average number of colonies per beekeeper (67.9 in 2018) [35].

In all the countries, most of the beekeepers were still hobbyist or nonprofessional beekeepers, which is in line with the results obtained. In fact, the beekeeping activity is usually not the main source of income across European countries [38]. Norway was the country where more participants were only beekeepers. Once beekeeping is an activity that provides, on average, low annual income, it is natural beekeepers had another activity besides beekeeping, being mainly hobbyist or nonprofessional beekeepers.

About one third of the participants referred to work alone and only 8% hire other people to work with them. This is consistent with the percentage of professional beekeepers, both in our sample (14% when considering professional beekeepers with more than 150 colonies), and in Europe (4% of professional beekeepers) [39].

As mentioned before, Hofstede's cultural dimensions offers a framework for cross-cultural communication and analysis. It allows us to understand the effects of a society's culture on the values of its members, and how these values relate to behavior, using a structure derived from factor analysis, based on six dimensions previously described.

Power Distance Index expresses how the less powerful members of a society accept and expect a certain inequality of power, or in other words how the society handles inequalities among people [14]. In our sample, the Hofstede analysis shows that the countries with high Power Distance Index (PDI), the beekeepers had lower values of annual income and low educational level. Croatia, Portugal and Spain were the countries with high PDI while Norway, Finland and Estonia had low PDI values and higher values of annual income and higher education level. The members of societies with a high PDI tend not to question those who are at higher levels of power. In addition, they expect that more powerful members might serve as guides for their work. In cultures with a low PDI, the power is equally distributed among its members, and they move towards a higher status (through education, employment, income, etc.). It is therefore not surprising that the countries with high PDI values are the ones with the lowest level of education as well as the lower annual income.

In individualistic societies, where the Individualism Collectivism Index (ICI) is lower, members tend to make decisions in an independent way and care about themselves and their close family, while in collectivist societies (with higher ICI), group ties are strong and the family includes the entire family extension (uncles, aunts, cousins, etc.) [14]. In this case, we found no significant variables associated with this cross-cultural dimension. Nevertheless, it seems that women tend to be less individualist than man. Other studies have found a similar trend, with women being generally more collectivist than their males' counterparts, especially among small-holder businesses [40]. Even recognizing that through cooperation, the return from the beekeeping activity may increase, there are still numerous beekeepers that prefer to work alone and doesn't join any kind of cooperative or association among peers [41].

The Masculinity Femininity Index (MFI) refers to gender effects in the societies, with low scores (Feminine societies) on the dimension meaning that the dominant values are caring for others and quality of life as opposed to high scores (masculine societies) where the societies are driven by competition, achievement and success [14]. In societies with high masculinity, people are driven by competition and results, so people tend to be assertive and centered on material success. On the other hand, in societies with low masculinity or characteristics of femininity, people are focused on building good relationships and ensuring better quality of life for everyone. It is not so important to be the best, as long as everyone is happy. Among the groups of beekeepers at study, we found strong correlations between MFI, age, income and education. Norway was the most Feminine society, with older and more educated beekeepers and with higher values of annual income. On the opposite side, Italy presented a higher value of masculinity, and also younger beekeepers with lower income values. In fact, Norway has been a leading country implementing institutional solutions to attain equality between men and women, positioning the country as a world leader in gender equality [42–44]

The Uncertainty Aversion Index (UAI) was strongly related with education and professionalism (number of colonies). In fact, countries with high UAI (Portugal, Spain, Croatia, Italy, Estonia e Finland) had more colonies and low educational level. The Uncertainty Aversion Index explains how members of a society feel when dealing with unknown situations [14]. In cultures with strong UAI, people tend to avoid risks and unexpected situations, because they create great anxiety and stress, preferring to face predictable or controlled situations. Contrarily, members of a culture with a weak UAI tend to be more tolerant towards what they cannot control. Uncertainty is accepted as part of life and people are generally more relaxed and flexible when they must face unknown situations. Rossi and Sekhposyan [45] presented a macroeconomic forecast distribution of uncertainty among European countries, showing that Italy, Spain and Portugal were more avert to uncertainty specially during the sovereign debt crisis, while for the same period Netherlands was always more tolerant to uncertainty [45]. In fact, Norwegian beekeepers seems to be more resilient in face of uncertainty, which is probably related with their level of education (that is higher). Also, Norway is the country with a smaller number of colonies per beekeeper and with a higher number of nonprofessional beekeepers, thus resulting in beekeepers that are probably not so dependent on the revenue from the activity and more prepared to take risks.

Both for the Term Orientation Index (TOI) and Indulgence Restraint Index (IRI), we could not find significant correlations among the variables at study and the indexes. In fact, the TOI index refers to the links with the past while dealing with the challenges of the present and the future, and none of the variables at study shows a trend towards the past or the future of the beekeepers and its activities.

The Indulgence Restraint Index analyses the importance of happiness and control of life. Societies with high complacency (high IRI) allow people to freely satisfy their basic human needs and desires, especially those related to enjoying life and having fun. In societies with a high rate of repression (low IRI), people suppress their impulses through restrictive social norms. These societies have a high consideration for moral discipline and people tend to be more pessimistic [14].

5. Conclusions

The results of this research allow us to draw a few conclusions about the characteristic of beekeeping activities in seven European countries. First of all, there are significant differences among sociodemographic variables and experience in beekeeping between countries.

Most beekeepers who answered the questionnaire were over 31 years old, and were men, with Italy being the country with the highest percentage of young, female beekeepers. Norway was the only country with an average annual income over 50,000 €, while respondents from Italy and Croatia generally had incomes below 15,000 €. Most participants had secondary or higher education levels, with those from Norway and Finland having the highest level of education. About 90% of participants had up to 150 colonies and were mainly nonprofessionals or hobbyists; about one quarter were certified as organic ($n = 74$). Members of the family and friends play an important role in beekeeping activities, and just 2% of respondents reported having hired workers (most commonly in Italy). The main product of the hive is honey (75–98%), followed by wax (15–45%) and colonies (13–32%); these values varied for each country in the study. In most cases, the average honey production per hive was 21–50 kg, per season, with Estonia and Finland reporting the highest yields (38 kg and 36 kg, respectively) and Portugal the lowest (13.7 Kg).

Relating these values with the cross-cultural dimensions defined by Hofstede, we found no variables to be significantly associated with the Individualism Collectivism Index (ICI), Term Orientation Index (TOI) or Indulgence Restraint Index (IRI). However, in countries with a high Power Distance Index (PDI), beekeepers had lower annual income and educational level, e.g., in Croatia, Portugal and Spain, meaning that they are more available to accept and expect a certain inequality of power among society members. Furthermore, we found a strong correlation between the Masculinity Femininity Index (MFI), age, income and education, with Norway being the most Feminine society, i.e., with more educated and older beekeepers. The Uncertainty Aversion Index (UAI) was strongly related with education and number of colonies, with high values for six countries but not for Norway, meaning that Norwegian beekeepers seem to be more resilient to uncertainly, i.e., they are more inclined to take risks when faced with unexpected situations.

Education and training contribute to beekeepers' incomes, serving to balance the distribution of power among members of the society, and thus increasing their PDI. So, beekeeping training is important to reinforce the sector and its sustainability, especially in Croatia, Portugal, Spain, and Italy. Long-term oriented societies (i.e., with high TOI values) encourage people to invest in the future through education and economic development, so we might consider that support through education and incentives to innovate will empower beekeepers. Finally, this work contributes to our understanding of how beekeeping activities are developed, and the roles played by stakeholders in different countries in the European Union, considering different socio-economic and cultural contexts, as well as the promotion of innovation in beekeeping and the transfer of knowledge and good practices.

The present work had some limitations, namely, the number of respondents from each country was neither equal nor proportional due to difficulties in recruiting participants. Also, the distribution of sociodemographic groups was not the same among countries, and it was also not possible to recruit the same number of participants from different activities in the beekeeping sector. Still, even with these limitations, this work provides new insights into the characterization and understanding of the beekeeping sector in some countries, and will serve as a support for future research in this area.

Author Contributions: Conceptualization, R.P.F.G.; methodology, R.P.F.G. and J.O.; software, R.P.F.G.; validation, R.P.F.G.; formal analysis, R.P.F.G.; investigation, R.P.F.G., S.M., C.A.C., H.E.C., B.D., R.R., R.K., L.T., S.B., E.B., I.S., M.O., C.C. and J.O.; resources, C.A.C.; data curation, R.P.F.G.; writing—original draft preparation, D.T.C., P.C., J.O., C.C. and R.P.F.G.; writing—review and editing, R.P.F.G.; visualization, R.P.F.G.; supervision, R.P.F.G.; project administration, C.A.C.; funding acquisition, C.A.C. All authors have read and agreed to the published version of the manuscript.

Funding: This research was funded by project beeB—Foster for beekeeping bridges through innovative and participative training (ref. 2019-1-PT01-KA202-060782). The APC was funded by FCT—Foundation for Science and Technology, I.P., within the scope of the project Ref. UIDB/00681/2020.

Institutional Review Board Statement: This research was implemented taking care to ensure all ethical standards and followed the guidelines of the Declaration of Helsinki.

Informed Consent Statement: Informed consent was obtained from all subjects involved in the study.

Data Availability Statement: Data are available from the first author (R.P.F.G.) upon reasonable request.

Acknowledgments: This work was developed under project beeB—Foster for beekeeping bridges through innovative and participative training (ref. 2019-1-PT01-KA202-060782). The authors would also like to acknowledge support from the FCT—Foundation for Science and Technology, I.P., within the scope of the project Ref. UIDB/00681/2020, as well as the CERNAS Research Centre and the Polytechnic Institute of Viseu.

Conflicts of Interest: The authors declare no conflict of interest.

References

1. Brodschneider, R.; Brus, J.; Danihlík, J. Comparison of Apiculture and Winter Mortality of Honey Bee Colonies (Apis Mellifera) in Austria and Czechia. *Agric. Ecosyst. Environ.* **2019**, *274*, 24–32. [CrossRef]
2. Uchiyama, Y.; Matsuoka, H.; Kohsaka, R. Apiculture Knowledge Transmission in a Changing World: Can Family-Owned Knowledge Be Opened? *J. Ethn. Foods* **2017**, *4*, 262–267. [CrossRef]
3. Fels, D.I.; Blackler, A.; Cook, D.; Foth, M. Ergonomics in Apiculture: A Case Study Based on Inspecting Movable Frame Hives for Healthy Bee Activities. *Heliyon* **2019**, *5*, e01973. [CrossRef]
4. Popescu, C.R.G.; Popescu, G.N. *The Social, Economic, and Environmental Impact of Ecological Beekeeping in Romania in Agrifood Economics and Sustainable Development in Contemporary Society*; IGI Global: Hershey, PA, USA, 2019; pp. 75–96.
5. Andrews, E. 'The Main Objection to Numerous Small Bee Keepers': Biosecurity and the Professionalization of Beekeeping. *J. Hist. Geogr.* **2020**, *67*, 81–90. [CrossRef]
6. Paris, E.H.; Peraza Lope, C.; Masson, M.A.; Delgado Kú, P.C.; Escamilla Ojeda, B.C. The Organization of Stingless Beekeeping (Meliponiculture) at Mayapán, Yucatan, Mexico. *J. Anthropol. Archaeol.* **2018**, *52*, 1–22. [CrossRef]
7. Wagner, K.; Meilby, H.; Cross, P. Sticky Business—Why Do Beekeepers Keep Bees and What Makes Them Successful in Tanzania? *J. Rural Stud.* **2019**, *66*, 52–66. [CrossRef]
8. Devkota, K. Beekeeping: Sustainable Livelihoods and Agriculture Production in Nepal. In *Modern Beekeeping—Bases for Sustainable Production*; Ranz, R.E.R., Ed.; IntechOpen: London, UK, 2020; 11p, ISBN 978-1-83880-156-4.
9. Fedoriak, M.; Kulmanov, O.; Zhuk, A.; Shkrobanets, O.; Tymchuk, K.; Moskalyk, G.; Olendr, T.; Yamelynets, T.; Angelstam, P. Stakeholders' Views on Sustaining Honey Bee Health and Beekeeping: The Roles of Ecological and Social System Drivers. *Landsc. Ecol.* **2021**, *36*, 763–783. [CrossRef]
10. Lourenço, M.S.M.; de O Cabral, J.E. Apicultura e Sustentabilidade: Visão dos Apicultores de Sobral (CE). *Rev. Agro. Amb.* **2016**, *9*, 93. [CrossRef]
11. Khan, A.S.; de Matos, V.D.; Lima, P.V.P.S. Desempenho da apicultura no estado do Ceará: Competitividade, nível tecnológico e fatores condicionantes. *Rev. Econ. Sociol. Rural* **2009**, *47*, 651–676. [CrossRef]
12. FAO Journée Mondiale Des Abeilles: La Production Apicole et Les Bonnes Pratiques Adoptées Par Des Apiculteurs Pour Soutenir Leurs Moyens d'existence. Available online: http://www.fao.org/africa/news/detail-news/en/c/1278688/ (accessed on 24 May 2021).
13. FAO Organização Das Nações Unidas Para Agricultura e Alimentação: Dia Internacional Da Abelhas: Polinizadoras Essenciais Para o Futuro Dos Alimentos | FAO No Brasil | Food and Agriculture Organization of the United Nations. Available online: http://www.fao.org/brasil/noticias/detail-events/pt/c/1195001/ (accessed on 25 May 2021).
14. Hofstede, G.; Hofstede, G.J.; Minkov, M. *Cultures and Organizations—Software of the Mind*, 3rd ed.; McGraw-Hill: New York, NY, USA, 2010.
15. Hofstede, G. *Culture's Consequences—Comparing Values, Behaviors, Institutions and Organizations Across Nations*, 2nd ed.; SAGE Publishing—Academic Books: Thousand Oaks, CA, USA, 2001.

16. Djekic, I.; Kane, K.; Tomic, N.; Kalogianni, E.; Rocha, A.; Zamioudi, L.; Pacheco, R. Cross-Cultural Consumer Perceptions of Service Quality in Restaurants. *Nutr. Food Sci.* **2016**, *46*, 827–843. [CrossRef]
17. Nyarugwe, S.P.; Linnemann, A.R.; Ren, Y.; Bakker, E.-J.; Kussaga, J.B.; Watson, D.; Fogliano, V.; Luning, P.A. An Intercontinental Analysis of Food Safety Culture in View of Food Safety Governance and National Values. *Food Control* **2020**, *111*, 107075. [CrossRef]
18. Huang, S.S.; Crotts, J. Relationships between Hofstede's Cultural Dimensions and Tourist Satisfaction: A Cross-Country Cross-Sample Examination. *Tour. Manag.* **2019**, *72*, 232–241. [CrossRef]
19. Kang, D.S.; Mastin, T. How Cultural Difference Affects International Tourism Public Relations Websites: A Comparative Analysis Using Hofstede's Cultural Dimensions. *Public Relat. Rev.* **2008**, *34*, 54–56. [CrossRef]
20. Cronjé, J.C. Using Hofstede's Cultural Dimensions to Interpret Cross-Cultural Blended Teaching and Learning. *Comput. Educ.* **2011**, *56*, 596–603. [CrossRef]
21. Thowfeek, M.H.; Jaafar, A. Instructors' View about Implementation of E-Learning System: An Analysis Based on Hofstede's Cultural Dimensions. *Procedia Soc. Behav. Sci.* **2012**, *65*, 961–967. [CrossRef]
22. Witten, R.; Witte, J. *Statistics*, 9th ed.; Wiley: Hoboken, NJ, USA, 2009.
23. Maroco, J. *Análise Estatística Com o SPSS Statistics*, 5th ed.; ReportNumber: Sintra, Portugal, 2012.
24. Pestana, M.H.; Gageiro, J.N. *Análise de Dados Para Ciências Sociais—A Complementaridade Do SPSS*, 6th ed.; Edições Sílabo: Lisboa, Portugal, 2014.
25. Hofstede, G. Compare Countries—Hofstede Insights. Available online: https://www.hofstede-insights.com/product/compare-countries/ (accessed on 21 May 2021).
26. European Parliament. *Prospects and Challenges for the EU Apiculture Sector*; European Parliament: Brussels, Belgium, 2018. Available online: https://www.europarl.europa.eu/doceo/document/TA-8-2018-0057_EN.html/ (accessed on 21 October 2021).
27. Jacques, A.; Laurent, M.; Ribiere-Chabert, M.; Saussac, M.; Bougeard, S.; Hendrikx, P.; Chauzat, M.-P. Statistical Analysis on the EPILOBEE Dataset: Explanatory Variables Related to Honeybee Colony Mortality in EU during a 2 Year Survey. *EFSA Supporting Publ.* **2016**, *EN-883*, 228. [CrossRef]
28. Dassou, A.G.; Ogouchoro, D.; Vodouhe, F.G.; Dassou, H.G.; Dansi, A.; Tixier, P. Ethnoapicultural Investigation to Improve Conservation Status of Threatened Melliferous Agroforestry Species with High Medicinal and Food Values in Benin. *Agroforest Syst.* **2020**, *94*, 539–553. [CrossRef]
29. Olana, T.; Demrew, Z. The Role of Women in Beekeeping Activities and the Contribution of Bee-Wax and Honey Production for Livelihood Improvement. *Livest. Res. Rural Dev.* **2018**, *30*, 118.
30. Mburu, P.D.M.; Affognon, H.; Irungu, P.; Mburu, J.; Raina, S. Gender Roles and Constraints in Beekeeping: A Case from Kitui County, Kenya. *Bee World* **2017**, *94*, 54–59. [CrossRef]
31. Horn, T. *Beeconomy: What Women and Bees Can Teach Us About Local Trade and the Global Market*; University Press of Kentucky: Lexington, KY, USA, 2012; ISBN 978-0-8131-3435-2.
32. Gross, B. Women in Beekeeping: Impacts of a Beekeeper Educational Program. Master's Thesis, University of Nebraska-Lincoln, St, Lincoln, NE, USA, 2020.
33. Pocol, C.B.; McDonough, M. Women, Apiculture and Development: Evaluating the Impact of a Beekeeping Project on Rural Women's Livelihoods. *Bull. Univ. Agric. Sci. Vet. Med. Cluj-Napoca. Hortic.* **2015**, *72*, 487–492. [CrossRef]
34. Erdős, N. *Report on Prospects and Challenges for the EU Apiculture Sector*; Committee on Agriculture and Rural Development: Brussels, Belgium, 2018; p. 32.
35. European Commission. National Apiculture Programmes 2020–2022. EU Beekeeping Sector. Available online: https://ec.europa.eu/info/sites/default/files/food-farming-fisheries/animals_and_animal_products/documents/honey-apiculture-programmes-overview-2020-2022.pdf (accessed on 10 November 2021).
36. GPP. *Programa Apícola Nacional 2017–2019 (Portugal)*; Gabinete de Planeamento, Políticas e Administração Geral: Lisboa, Portugal, 2016; p. 99.
37. Portugal: National Apiculture Programme 2020–22. Available online: https://ec.europa.eu/info/system/files/food-farming-fisheries/animals_and_animal_products/documents/nap-pt_pt.pdf (accessed on 10 November 2021).
38. Chauzat, M.-P.; Laura, C.; Roy, L.; FRANCO, S.; Hendrikx, P.; Ribière-Chabert, M. Demographics of the European Apicultural Industry. *PLoS ONE* **2013**, *8*, e79018. [CrossRef]
39. Rossi, S. *The EU's Beekeeping Sector—At a Glance 2017*; European Parliamentary Research Service: Brussels, Belgium, 2017. Available online: https://www.europarl.europa.eu/RegData/etudes/ATAG/2017/608786/EPRS_ATA(2017)608786_EN.pdf (accessed on 10 November 2021).
40. Zeffane, R. Gender, Individualism–Collectivism and Individuals' Propensity to Trust: A Comparative Exploratory Study. *J. Manag. Organ.* **2020**, *26*, 445–459. [CrossRef]
41. Frick, M.; Grgić, Z.; Franić, R.; Štefanić, I.; Kezić, N. Cooperative Business Potential for Beekeepers in Croatia. *J. Apic. Res.* **2006**, *45*, 223–229. [CrossRef]
42. Tomczak, D.A. Gender Equality Policies and Their Outcomes in Norway. *Zarz. Publiczne* **2016**, *4*, 379–391. [CrossRef]
43. Hutt, R. These 10 Countries Are Closest to Achieving Gender Equality. Available online: https://www.weforum.org/agenda/2019/12/gender-gap-equality-women-parity-countries/ (accessed on 28 September 2021).

44. Statistics Norway Norway—World Leader in Gender Equality. Available online: https://www.ssb.no/en/befolkning/artikler-og-publikasjoner/norway-world-leader-in-gender-equality (accessed on 28 September 2021).
45. Rossi, B.; Sekhposyan, T. Macroeconomic Uncertainty Indices for the Euro Area and Its Individual Member Countries. *Empir. Econ.* **2017**, *53*, 41–62. [CrossRef]

MDPI
St. Alban-Anlage 66
4052 Basel
Switzerland
www.mdpi.com

Agronomy Editorial Office
E-mail: agronomy@mdpi.com
www.mdpi.com/journal/agronomy

www.ingramcontent.com/pod-product-compliance
Lightning Source LLC
Chambersburg PA
CBHW061825210926
51850CB00006B/805
* 9 7 8 3 0 3 6 5 9 6 0 2 0 *